Tunnel Visions

Tunnel Visions

The Rise and Fall of the Superconducting Super Collider

MICHAEL RIORDAN,
LILLIAN HODDESON,
AND ADRIENNE W. KOLB

The University of Chicago Press

CHICAGO AND LONDON

The University of Chicago Press, Chicago 60637
The University of Chicago Press, Ltd., London

Published 2015
Paperback edition 2018
Printed in the United States of America

27 26 25 24 23 22 21 20 19 18   1 2 3 4 5

ISBN-13: 978-0-226-29479-7 (cloth)
ISBN-13: 978-0-226-59890-1 (paper)
ISBN-13: 978-0-226-30583-7 (e-book)
DOI: https://doi.org/10.7208/chicago/9780226305837.001.0001

Library of Congress Cataloging-in-Publication Data

Riordan, Michael, 1946– author.
Tunnel visions : the rise and fall of the superconducting super collider / Michael Riordan, Lillian Hoddeson, and Adrienne W. Kolb.
pages cm
Includes bibliographical references and index.
ISBN 987-0-226-29479-7 (cloth : alk. paper)—ISBN 978-0-226-30583-7 (e-book)
1. Supercolliders—United States. 2. Physics—United States. I. Hoddeson, Lillian, author. II. Kolb, Adrienne W., author. III. Title
QC787.S83.R56 2015
539.7'360973—dc23
2015011972

∞ This paper meets the requirements of ANSI/NISO Z39.48-1992 (Permanence of Paper).

To our spouses,
Donna, Peter, and Rocky:
their support and encouragement
throughout this effort was invaluable

Contents

Preface

In October 1993 the US Congress terminated the Superconducting Super Collider—at the time the largest basic-science project ever attempted, with a total cost then estimated to exceed $10 billion. It was a stunning blow, a terrible loss for the nation's high-energy physics community, which until that moment had perched for decades at the pinnacle of American science. Since that fateful vote, this once-dominant scientific community has been in steady decline. With the 2010 startup of research on the CERN Large Hadron Collider and the 2011 shutdown of the Fermilab Tevatron, world leadership in high-energy physics crossed the Atlantic and returned to Europe.[1] The 2012 discovery of the Higgs boson at CERN only underscored this epochal transition.[2]

For more than three decades, US physicists had dominated the search for the fundamental particles at the heart of matter. Crucial to their hegemony was the ability to build ever-larger accelerators capable of producing controlled, high-energy electron and proton beams for use in experiments that examined matter at ever-smaller distances. In the aftermath of World War II and during the early decades of the Cold War, they managed to secure the federal support needed to construct these costly machines by virtue of the association of their research with national security.[3] This connection weakened during the 1970s, however, and European scientists began to pull even in high-energy physics, building equally powerful machines and beating their US counterparts to important discoveries.

Stung by these losses, US high-energy physicists hitched their hopes for recovering this leadership to the Reagan administration's determined efforts to reassert American competitiveness and fight the Cold War. But doing so meant building a new particle collider more than ten times larger and costlier than any machine this scientific community had previously attempted.[4] Committing to such a huge and expensive project required physicists to make uncomfortable compromises and to form new, unfamiliar alliances with the US Congress, the Department of Energy (DOE), Texas politicians and business leaders, and firms from the US military-industrial complex. Access to the billions of taxpayer dollars necessary to build the SSC came with strings attached, however, and with an unprecedented level of public scrutiny that few, if any, high-energy physicists could have anticipated.

The combination of this attention, continuing SSC cost overruns, and the widespread perceptions of project mismanagement by DOE and the physicists involved led to its demise nearly five years after it began. Its termination was a watershed event—a turning point not only in the history of physics but also for that of science in general. This death raises questions of whether Big Science has become so big and so expensive that maintaining public commitment during a major facility's lengthy construction phase can be taken for granted. Another question is whether academic scientists and their government overseers can effectively manage such an enormous undertaking. For science historians, the case of the Superconducting Super Collider therefore offers important lessons about the conditions required to build and sustain such a large scientific laboratory. Its rise and fall also serves as a cautionary tale about the viability of a research community that came to depend as much as did US high-energy physics upon a single facility of such unprecedented scale.

Major historical studies of research laboratories have thus far focused mainly on their successes. Examples of this genre in high-energy and nuclear physics include Heilbron and Seidel's *Lawrence and His Laboratory*;[5] Crease's *Making Physics*;[6] Hoddeson, Kolb, and Westfall's *Fermilab*;[7] and Hermann, Krige, Mersits, and Pestre's *History of CERN*.[8] Highlighting success is the dominant recipe followed in historical writings about laboratories in other fields—for example, histories of Bell Laboratories that focused on the transistor[9] and of Los Alamos that concentrated on the atomic bomb.[10] These histories examined the economic and physical resources, social relations, leadership, management philosophy, and other conditions that are apparently essential in establishing and running a successful research laboratory. But they cannot offer the degree of confidence that can be achieved from histories where the *absence* of a crucial condition or element is associated with failure.

Historical works on the cultures of creative laboratory spaces have also given valuable insights about aspects that foster creative research or invention.[11] One of these features is flexibility in the organization and spatial layout of the laboratory. Such creative spaces have a culture and physical environment that makes scientists eager to work there, thus enhancing their research productivity. And almost all these labs have a strong, respected, often charismatic leader who can effectively articulate and promote a clear, compelling mission while allowing individual scientists the freedom to pursue their research within a minimum of bureaucracy. But can these desirable features be maintained in projects as large as the SSC? The greater degree of regimentation and oversight occasioned by the need to spend billions of federal dollars responsibly in a gargantuan project like this may drive the best scientists to do their research elsewhere.

The demise of the SSC occurred against the political backdrop of changing scientific needs as the United States transitioned from a Cold War to a post–Cold War footing in the late 1980s and early 1990s. Government funding was growing steadily in fields closely aligned with economic advancement or human welfare. After the Reagan administration buildup of the 1980s, however, it began to level off or decline during the 1990s for defense research and in the basic physical sciences, which in the short run might contribute only to US national prestige and Nobel Prizes. How much this external political dynamic contributed to the SSC's fate is open to debate, but the impact was not small.[12] And the sheer size and cost of the Super Collider brought other political factors into play because of the financial pressure the huge project inevitably exerted upon other worthy scientific research.

By examining such a failed project, we can hope to discern the influences of both internal and external factors upon the evolution of a scientific laboratory. This approach resembles what research biologists do in trying to determine the conditions cells need to survive by withdrawing particular nutrients. There have been few studies of laboratory failures in the history of science, partly because of inadequate documentation.[13] Fortunately the history of the SSC has been well documented, and many of the important documents have been preserved in the Fermilab Archives. That record is supplemented by the many contemporaneous press accounts of this high-visibility project. In addition, we have recorded and transcribed over one hundred oral-history interviews with many of the key SSC participants, in both the high-energy physics community and the governmental agencies and bodies that interacted with the project, as well as many of its close observers.

It is difficult, perhaps impossible, to tell this history from the perspective of a single omniscient observer, although we attempt to present as objective an account as possible. There were diverse "communities of interest" (as we call them) deeply involved in the SSC: the high-energy physicists who eagerly designed the project; others who moved to Texas to begin constructing it; the DOE officials and bureaucrats who oversaw the project; engineers from the US military-industrial complex brought in to help manage it; and the members of Congress and their staffers who followed activities in Texas and Washington. In addition, there were physicists in other disciplines who opposed the project, reporters and science writers who covered it, and foreign physicists and their agency officials interested in becoming involved in SSC research. These groups did not share the same perspectives about the project. We try to recount their differing viewpoints without prematurely judging them.

This book is organized into two parts, corresponding to the emergence of the project and to its construction and demise. Part I includes chapters 1–3 on the SSC origins and conceptual design through late 1988, when Waxahachie, Texas, was chosen as the designated site. Chapter 1, covering the period up to mid-1983, is the work of all three authors.[14] Chapters 2 and 3 on collider design, early public perceptions of the SSC, and the national site-selection process are primarily the work of Hoddeson and Kolb.[15] The second part, including chapters 4–6 on the 1989–94 period, is largely the work of Riordan.[16] Chapter 4 covers the establishment of the SSC Laboratory in Texas;[17] Chapter 5 explores the interactions among the physicists, Washington officials, and foreign agents regarding attempts to internationalize the project; and chapter 6 recounts its demise through the 1993 termination and its aftermath. The final chapter, chapter 7, on reactions to and analyses of the SSC demise, including its impact on US high-energy physics, is the work of all three authors. In closing, we added an epilogue about the successful construction of the Large Hadron Collider at CERN and the 2012 discovery of the Higgs boson on this machine.

Two major historical accounts of the SSC have appeared thus far, one by science historian Daniel Kevles[18] and the other by physicist Stanley Wojcicki, a leading contributor to the initial, design phase of the project.[19] The first focused on how the project was embraced—or not embraced—in Washington, while the second is more of an "internalist" account that told the story of the SSC from the perspective of a highly knowledgeable insider. Both are worthy histories, but they do not offer the detailed insights or extensive analyses that are possible in a thoroughly researched, book-length history.

We believe that *Tunnel Visions* finally fills this gap in the growing body of literature about laboratory history. We hope it will be of great value to historians and other scholars of science, government officials and staffers, science-policy analysts and physicists—especially those who may contemplate attempting to build other multibillion-dollar "gigaprojects" like the SSC.

CHAPTER ONE

Origins of the Super Collider

In all failures, the beginning is certainly half the whole.
—GEORGE ELIOT, *Middlemarch*

During the late 1970s, US high-energy physicists could look back on three decades of unparalleled achievement. They had constructed a steady succession of particle accelerators with exponentially increasing energies and used them to probe the interior of the atomic nucleus, making one major discovery after another of the fundamental particles from which matter is made and about the forces binding them together. Combined with theoretical advances, in which US physicists also played a leading role, these discoveries culminated in the Standard Model of particle physics, the dominant paradigm to which almost all members of the discipline subscribed by 1980. This theory posits that ordinary matter is composed of basic building blocks called quarks and leptons—point-like particles, collectively called fermions, that carry half-integer spin. These constituents interact by exchanging other kinds of particles (such as the familiar photon) bearing integer spin called gauge bosons. The Standard Model succeeded in combining the seemingly dissimilar electromagnetic and weak forces into a single, unified force called the electroweak force. Unifications of fundamental forces are exceedingly rare and thus tremendously significant events in the history of physics, typically occurring about once every century.[1]

Largely because of their ability to construct ever more powerful accelerators and colliders, in which two particle beams clash to generate the highest collision energies attainable, US physicists had taken the lead in this research. In the 1970s, proton or electron machines at Brookhaven National Laboratory (BNL) and Cornell University in New York, the Fermi National Accelerator Laboratory (FNAL, also called Fermilab) in Illinois, and the Stanford Linear Accelerator Center (SLAC) in California supplied US physicists with a variety of high-energy beams. Europeans strove to keep pace, collaborating to construct competitive proton machines at the European Center for Nuclear Physics (CERN) near Geneva, Switzerland, and electron machines at the Deutsches Electronen Synchrotron (DESY) in Hamburg, West Germany. But the entrepreneurial spirit and risk taking that characterized American experimenters had generally won this competition, allowing them to make the lion's share of the important discoveries while their European counterparts were able only to confirm these breakthroughs. During the 1960s and 1970s, five quarks and two additional leptons turned up initially at laboratories west of the Atlantic.[2]

European physicists had evolved a more conservative tradition of building particle accelerators, colliders, and detectors, which were thoroughly engineered before construction began. While it meant that the equipment generally worked as designed from the outset, this approach also took longer to implement, giving more adventurous US physicists the inside track on important discoveries. On the other hand, it also meant that Europeans had the edge in constructing complicated particle detectors, such as the immense Gargamelle bubble chamber at CERN, which they used to obtain convincing evidence in 1973 for the weak neutral currents required by theories of electroweak unification.[3]

US HIGH-ENERGY PHYSICS AT A CROSSROADS

As the 1970s ended, however, Europe was pulling abreast of America in high-energy physics, often called particle physics. Due in part to US funding delays, DESY's collider PETRA began operations in 1979, more than a year before its SLAC equivalent PEP.[4] PETRA eventually allowed physicists to study the collisions of electrons with their antimatter counterparts, called positrons, at combined energies over 40 billion electron volts, or 40 GeV—about five times higher than previously attainable.[5] This advantage meant that European physicists (plus several groups of US physicists working at DESY) received credit that year for discovering the gluon—the gauge boson responsible for conveying the strong force that sequesters quarks together inside protons and neutrons.[6]

Moreover, the European particle-physics community, which had for years been steadily concentrating its efforts at CERN, was developing adventurous plans for the future as the 1980s began. CERN was adapting a large proton accelerator, the 300 GeV Super Proton Synchrotron, or SPS, to function also as a high-energy proton-antiproton collider able to produce the ultramassive W and Z particles predicted by the Standard Model and thought to be responsible for radioactivity.[7] And CERN was planning a far bigger machine, the Large Electron Positron (LEP) collider, which would occupy a 27 kilometer tunnel under the Swiss and French countryside and eventually allow electron-positron collisions to occur at energies up to 200 GeV.[8]

In contrast, US high-energy physics had been buffeted by shifting economic and political forces during the 1970s. In part because of the Arab oil embargo of 1973 and the attendant surge in energy costs, a major shakeup occurred in its principal federal funding agency. An outgrowth of the Manhattan Project, the Atomic Energy Commission, or AEC (which funded almost all US accelerator building through the mid-1970s), was dissolved in 1974 and its responsibilities segregated into the Energy Research and Development Administration (ERDA) and the Nuclear Regulatory Commission (NRC). ERDA in turn became part of the even larger Department of Energy (DOE) under the Carter administration in 1977. High-energy physics was subsequently just one part of a larger energy portfolio, which included billions of dollars for solar and renewable-energy projects during the late 1970s.[9] Funding for US high-energy physics was nearly flat (in constant dollars) during the latter half of the decade, while the costs of constructing its ever-larger particle accelerators, colliders, and detectors increased unabated.[10]

The Cold War rationale for building these expensive scientific facilities had declined during the 1970s, after the administration of Richard Nixon and the Soviet government of Leonid Brezhnev tacitly agreed to détente in their relationship, thus encouraging scientific exchanges and joint projects such as the 1975 docking of the Apollo and Soyuz space capsules.[11] High-energy physics had enjoyed a privileged status under the old AEC, whose General Advisory Council often made decisions in secret about proposed projects—which Congress then debated in closed sessions of the Joint Committee on Atomic Energy. But after the Energy Reorganization Act of 1974, the AEC ceased to exist. And congressional jurisdiction over energy projects was now assigned to separate House and Senate Appropriations Subcommittees on Energy and Water Development. Their members were much more interested in garnering lucrative projects for their districts and states than in helping to foster the research productivity of a relatively small (but still influential) group of physicists who worked mainly at national laboratories in California, Illinois, and New York.[12] During the mid-to-late 1970s, US high-energy physics lost its privileged status as the flagship of the AEC fleet and became just one among many petitioners for federal largesse.

In Europe, by contrast, high-energy physics continued to enjoy its special status well into the 1980s. Rather than becoming a marquee attraction of the Cold War competition between the US and USSR, it increasingly served a prominent role as the highest expression of postwar European integration. Ministers could point to CERN as a shining example of how competing European countries can successfully cooperate with one another on joint projects of scientific and cultural merit. And CERN had over the years evolved a governing structure, its Council, that effectively insulated its operations from the political vicissitudes of the participant nations. Combined with the widely shared European desire to foster its intellectual vitality and success, CERN's governance meant it could rely on steady funding of major projects, which remained firmly under control of the physicist-managers running the laboratory.[13]

The 1970s were also a decade of economic and industrial disruption in the United States. After the country abandoned the gold standard in 1971, the dollar plummeted against major foreign currencies, while the price of gold, oil, and other commodities soared.[14] After the Arab embargo of 1973–74, oil and gasoline prices more than tripled, triggering a deep recession in the United States. European and Japanese automakers took major market shares away from US firms, which proved unable to manufacture small, fuel-efficient vehicles economically. And Japanese firms such as Panasonic and Sony grabbed most of the US market for consumer electronics from their American competitors.[15] As the decade ended, the nation was deep in the throes of stagflation, with a flat or declining gross domestic product compounded by surging unemployment and double-digit inflation.

Amid such trying circumstances, it was difficult to maintain a healthy US high-energy physics program, given the ever-increasing price tags of the machines and equipment required to do this expensive research. Already constrained by the costs of the Vietnam War and of funding the Great Society programs of President Lyndon B. Johnson, the US budget for high-energy physics had peaked in 1970 during construction of the National Accelerator Laboratory (later renamed Fermilab) near Chicago.[16] After its $250 million particle accelerator—extending four miles in circumference across the Illinois plains and boosting protons to energies up to 200 GeV—was completed in 1972, funding for the discipline fell by almost 50 percent in real terms by mid-decade. And many other programs—at Brookhaven, SLAC and elsewhere—had been severely constricted during Fermilab's construction.[17]

Funding for US high-energy physics began to stabilize in the late 1970s, albeit haltingly. What might have seemed a modest budget increase when appropriated by Congress before the start of a fiscal year would often be eroded so much by inflation that it ended up being a *decrease* in terms of constant dollars by the end of that year. Thus the true budget for high-energy physics grew only marginally, if at all, in the latter half of the decade. New projects needed to keep the major national labs at the research frontier—such as SLAC's electron-positron collider PEP, Brookhaven's proposed Isabelle proton-proton collider, and a Fermilab project dubbed the Energy Doubler (later renamed the Energy Saver and then the Tevatron)—suffered from inevitable delays and postponements in this difficult funding climate.

And, despite the emergence of the Standard Model paradigm in the 1970s, there was still plenty of important research to be addressed. Besides the search for the all-important W and Z bosons, the carriers of the weak nuclear force in this theory, at least one predicted quark known as the top quark remained undiscovered. By 1980 nearly every high-energy physicist expected that these weighty particles would eventually turn up after particle colliders managed to attain the extremely high energies needed to create them (according to Einstein's $E = mc^2$). Producing and detecting sufficient numbers of these particles to prove their existence was a primary rationale for constructing ever larger and costlier equipment.

Another all-important target on high-energy physicists' wish lists was the so-called Higgs boson, named after the British physicist Peter Higgs, who conceived it in 1964 as the consequence of theories about how the W and Z bosons could acquire large masses.[18] In the following decade, the Higgs boson became the capstone of the emerging Standard Model, providing the consensus explanation of why quarks, leptons, and gauge bosons have the wide variety of masses observed. And it had to be unlike all of the other particles in that it could have no spin. In addition, particle theorists could not confidently predict what the Higgs mass might be—other than to say that it must come in below about 1,000 GeV, or 1 TeV. Otherwise, something was amiss with the Standard Model (see appendix 1 for more details).

If the Higgs boson were indeed that massive, designing and constructing a collider powerful enough to create it in sufficient numbers to detect it stretched the imaginations of machine builders throughout the high-energy physics community. It clearly had to be a multi-TeV collider involving protons and perhaps antiprotons. Because protons and antiprotons are composed of quarks and gluons, collisions between these fundamental constituents occur at energies only about a tenth that of their parent particles, whose energies must thus be ten times higher to compensate (see appendix 1). Given the limitations on the possible strengths of magnetic fields, even with the most advanced superconducting magnets, such a gargantuan machine able to produce particles with masses up to 1 TeV would extend tens of miles or kilometers in circumference and cost billions of scarce US dollars or Swiss francs (SFr, also written CHF). But the most daring leaders of the US high-energy physics community were undaunted by such awesome challenges—whether scientific, technological, economic, or political—as the new decade began.

DREAMS OF A WORLD ACCELERATOR

By the early 1980s, a constellation of accelerator visions had coalesced, partly out of the aspirations of high-minded physicists eager to advance their science while offering a contrasting example to the ideological Cold War confrontation. Since the mid-1950s, they had slowly elaborated the idea of a very large accelerator to serve physicists from countries on both sides of the Iron Curtain, which would share in the huge cost of such a mammoth machine, too expensive for any single nation to build.[19] The organization of CERN earlier that decade served as an important step toward international cooperation in physics, for it demonstrated how individual nations could successfully collaborate on a major scientific project.[20] In 1957 the International Union of Pure and Applied Physics (IUPAP) established a commission to help "encourage international collaboration among the various high-energy laboratories to ensure the best use of these large and expensive installations."[21]

Soon physics meetings throughout the world included discussion of such an ambitious worldwide cooperative accelerator project to serve as a possible model for peaceful international collaboration. In September 1959, USSR Premier Khrushchev's successful summit meeting with US President Eisenhower encouraged dialogue about cooperative scientific projects. Eight months later a delegation of physicists that included Edward Lofgren of Lawrence Berkeley Laboratory (LBL) and Robert R. Wilson of Cornell visited the USSR to explore the joint construction of a large accelerator.[22] But the opportunity for collaboration evaporated after the Soviet Union downed an American U-2 reconnaissance plane over its territory in May 1960.[23]

A small group of idealistic physicists continued to discuss the dream that, as Wilson later wrote, "in building and operating a World Laboratory we would not only be exploring nature, but we also might be exploring some of the ingredients of peace."[24] An August 1960 meeting organized by Wilson to consider 1,000 GeV (or 1 TeV) accelerators in a global context initiated serious discussions on the design of such a large international facility.[25] Its basic technical concept would eventually serve as a starting point for the design of the Superconducting Super Collider.

A window for discussing international cooperation opened during the 1970s period of détente. Nixon and Brezhnev's historic agreement of June 1973 identified basic research on the fundamental properties of matter as one of three areas that were particularly useful for "expanded and strengthened cooperation for mutual benefit, equality and reciprocity between the U.S. and the U.S.S.R."[26] In addition, scientific cooperation and the free flow of information were included as basic human rights in the August 1975 Helsinki Accords. By the mid-1970s, therefore, initial steps toward a World Accelerator had occurred, but there were still no concrete plans for building one; many of the physicists involved in the discussions of international cooperation were beginning to lose patience.[27]

MIT theoretical physicist Victor Weisskopf, formerly the director general of CERN, stimulated the next move by inviting physicists to a seminar on international collaboration in New Orleans in March 1975. Sparks flew at the meeting, igniting, as Wilson recalled, "impassioned speeches to the effect that a world laboratory along the lines of a worldwide CERN would be necessary and desirable if we are to push into the multi-TeV region of proton energy."[28] Leon Lederman, then the director of Columbia University's Nevis Laboratory, dubbed the proposed machine the "Very Big Accelerator," or VBA, proclaiming that "the world community of high-energy physics [should] bite the bullet and organize together to bring this 10 TeV machine to realization."[29] Participants in the meeting recommended the formation of a VBA study group led by Weisskopf. Various designs were discussed by this group at meetings held over the next several years, while the aspirations to use the VBA idea as a model for achieving world peace continued. In late 1975 Wilson suggested that "such an undertaking might well provide some of the experience in international living so necessary for human survival—a candle in the darkness."[30]

At its May 1976 meeting in Serpukhov, Russia, the study group pursued the definition and scale of the proposed VBA. Members conceived it as either a 10–20 TeV proton accelerator or a 200 GeV electron-positron collider.[31] Two months later, at the International Conference on High-Energy Physics in Tbilisi, Georgia, the IUPAP Commission on Particles and Fields agreed to establish a subcommittee, called the International Committee on Future Accelerators, or ICFA, to organize meetings aimed at studying the VBA and future regional facilities and collaborations.[32] By now the physicists' vision was filtering into mainstream discourse, as reflected in an October *New York Times* article about possible construction of a "world machine" that would dwarf any accelerator then in existence.[33] A few months later, Lederman spoke about the VBA at the Particle Accelerator Conference in Chicago.

Suggesting that New York City, then on the brink of bankruptcy, be selected as a potential site, he joked that most of the necessary facilities already existed, including "high-rise international headquarters, educational resources, pre tunneled terrain, and the usual degree of inaccessibility."[34]

But real progress on the VBA was plagued by political and organizational intricacies. While physicists from each region expressed support, the actual funding had to come from national treasuries. And when a choice had to be made between supporting the VBA or one's own national or regional laboratory, the first loyalty of high-energy physicists would usually be to the latter.[35] Meanwhile, a cloud gathered on the horizon, as American high-energy physics advisory panels struggled to come to grips with the fact that leadership in the discipline was shifting to Europe.

The goal of demonstrating national prowess gradually overcame the will to cooperate, as competition began to dominate the planning discussions. A subpanel of the US High-Energy Physics Advisory Panel (HEPAP) led by Samuel Treiman of Princeton expressed concern that without increased funding "the U.S. program will inevitably lose its eminence" and its ability "to compete with Western Europe."[36] Another panel headed by Cornell accelerator physicist Maury Tigner insisted that "we must redouble our efforts to improve the cost effectiveness of our accelerators if the needs of U.S. particle physics are to be met in the resource-limited situation."[37] The US high-energy physics program indeed faced strong competition from CERN, then led by Herwig Schopper, previously director of DESY. Both European labs were aggressively pursuing construction of and plans for new particle colliders.

It was becoming increasingly clear that an even more powerful instrument was needed to investigate the emerging agenda of high-energy physics. In addition to the Higgs boson, theories that had arisen in the wake of the Standard Model—such as grand unification, supersymmetry, and technicolor—predicted that very massive new particles should appear at the TeV energy scale (see appendix 1).[38] Many US physicists had become concerned that Isabelle, the proton collider then under construction at Brookhaven, would not attain sufficient energy to produce such particles. Moreover, the high costs of Isabelle and its superconducting magnet problems (see below) threatened to "bleed the rest of the program," according to Lederman.[39] In January 1982 another HEPAP subpanel led by LBL physicist George Trilling concluded that Isabelle might have to be abandoned if additional funding was not forthcoming. HEPAP recommended that in order to maintain a vigorous US high-energy physics program "another major facility had to be started in the mid-1980s to be available for research by 1990," capable of exploring new frontiers. The completion of Fermilab's Energy Doubler (renamed Energy Saver during the Carter years) was deemed the US high-energy physics community's "highest immediate priority."[40] Meanwhile, the lack of urgency and the absence of serious funding, as well as continuing frustration with the slow pace of international collaboration, hindered participation of US physicists in the VBA planning process.[41]

SUPERCONDUCTING PARTICLE COLLIDERS

"Superconductivity is a magic potion," Wilson proclaimed in 1977, "an elixir to rejuvenate old accelerators and open new vistas for the future."[42] As the 1970s ended with the Standard Model triumphant and the most pressing quarry—the W and Z bosons—to be discovered at higher energies beyond the capacities of existing accelerators, Brookhaven and Fermilab resorted to this intoxicating brew as the best possible way to ensure their continued viability on the research frontiers of high-energy physics. Both laboratories began building large proton rings that relied on hundreds of superconducting magnets to keep energetic particles whirling around on course.[43]

Discovered in 1911 by the Dutch physicist Heike Kamerlingh-Onnes, the property of superconductivity means that at extremely low temperatures near absolute zero electric current flows without resistance in certain metals and alloys. Magnet coils wound with wires made of such alloys as niobium-titanium (NbTi) can generate very high magnetic fields with almost no power consumption. While ordinary magnets manufactured with copper coils can develop nearly 20 kilogauss (for comparison, the Earth's magnetic field is about half a gauss at its surface), the new superconducting magnets promised fields with strengths of up to 100 kilogauss (or 10 tesla) if their daunting technological problems could ever be solved. The most serious of these problems was that a superconducting magnet will "quench" if the temperature of even a tiny part of its coils happens to rise above a certain critical temperature, which is 4.35 degrees Kelvin (4.35 K) in the case of NbTi. The alloy suddenly "goes normal" and returns to its ordinary resistive state of electrical conduction—releasing large quantities of stored energy, with consequent generation of heat that can melt the coil or even lead to a disastrous explosion.

During the 1970s Brookhaven and Fermilab had pursued R&D programs aimed at solving the difficult problems involved in large-scale superconducting magnet systems.[44] Among these were the design and development of quench-protection mechanisms and complicated cryogenic systems to supply the liquid helium needed to cool and maintain magnet coils at 4.3 K. By 1978 these problems appeared to be solved. Both Brookhaven and Fermilab had built large prototype superconducting magnets that could sustain fields of 4–5 tesla (or 4–5 T). In these pioneering efforts the two laboratories were well ahead of high-energy physics laboratories in Europe and Japan.

The apparent success of these R&D programs led to DOE approval of two ambitious projects—to build the Isabelle proton-proton collider at Brookhaven and the Energy Doubler (or Saver) at Fermilab.[45] Using the much higher magnetic fields that are achievable with superconducting magnets, physicists could confine beams of circulating protons (and, at Fermilab, antiprotons) at much higher energies within the existing real estate at the laboratories. Then, using colliding-beam technologies that had been perfected at CERN and SLAC, they planned to make the high-energy beams clash with one another, generating tremendous collision energies that would be sufficient to create the massive W and Z bosons, among other subatomic quarry. Brookhaven's plan followed CERN's Intersecting Storage Rings design,[46] circulating twin beams of 400 GeV protons in a pair of interlaced rings housed within a new 3.8 km tunnel. Fermilab's idea was to add a ring of superconducting magnets inside the existing 6.3 km tunnel that housed its 400 GeV Main Ring; bunches of protons and eventually antiprotons would circulate at nearly 1 TeV in opposite directions in this new storage ring, colliding at two interaction regions.

At Brookhaven, difficult problems in manufacturing the superconducting magnets struck the Isabelle project soon after its construction began in the fall of 1978. An initial set of 12 dipole magnets, whose coils had been manufactured by Westinghouse, could not achieve the design field of 5 T.[47] Instead, as the applied current increased, the magnets quenched at fields of only 3–4 T; the coils apparently were shifting slightly in response to the tremendous magnetic forces upon them, generating heat that made them go normal. After repeated attempts, in one case involving a hundred quenches, the magnets attained fields above 4 T, but none of them could achieve the ambitious design goal of 5 T. Any such lengthy "training" procedure would be absolutely unacceptable in practice for the many hundreds of superconducting dipoles needed for Isabelle.[48]

Brookhaven terminated the Westinghouse contract in 1979 but proceeded with construction of the tunnel and experimental halls. Meanwhile, it began a new R&D program to try to ascertain the causes of the quenching and determine why production magnets could not reach the 5 Tesla fields that had been attained in a 1977 prototype.[49] By the time these thorny problems had been resolved in late 1981, with model magnets finally able to achieve the required fields, construction was essentially finished but the estimated cost of Isabelle had more than doubled from an initial $275 million to about $600 million dollars (in part because of inflation).[50] Project completion was still at least five years in the future—well after CERN was expected to discover the W and Z bosons with its daring proton-antiproton collider. And a Republican administration led by Ronald Reagan had moved into the White House in January 1981. It began taking a fresh look at how US taxpayer dollars were being spent by the Department of Energy, aiming to dismantle the agency and fulfill a Reagan campaign promise.

Since Fermilab's construction had begun in the late 1960s, superconducting magnets were considered the principal option for increasing the machine energy.[51] Wilson insisted that his designers leave room in the Main Ring tunnel to install an additional ring of such magnets at an appropriate time in the future. His primary goal in so doing was to upgrade the proton beam energy to 1 TeV; the idea to use this superconducting ring as a proton-antiproton collider came afterward, in the late 1970s. But an immense amount of research and development had to occur before the lab could be ready to make the nearly one thousand magnets required. In 1975 Fermilab began a crash R&D program that eventually convinced the DOE that the laboratory could succeed in such an effort.[52] (DOE staff had already committed to supporting Isabelle, however, and it took them time to recognize that Fermilab's magnet design was in fact superior.) Fermilab set up an on-site manufacturing line that turned out over a hundred prototype magnets—ranging from small-scale to full-size models—plus another facility where strings of these magnets were tested under exacting conditions. The design, production, and management experience gained from this program resulted in the ability to fabricate the hundreds of 21-foot-long, 4 T dipole magnets required for the Energy Doubler/Saver.

Part of the reason for Fermilab's success was its decision to employ twisted, multi-filament superconducting cable that had been pioneered in the early 1970s by the Rutherford High Energy Laboratory in Great Britain.[53] Brookhaven had instead opted for braided cable that many observers later considered to be deeply flawed.[54] But Fermilab had also based its R&D program on the production of *many* full-size prototypes, making small changes and understanding their effects while solving the difficult manufacturing problems involved. In contrast, Brookhaven researchers had initially concentrated efforts on only a few model magnets; they never completely

FIGURE 1.1 Leon M. Lederman (*left*) and Robert R. Wilson at dedication of Wilson Hall, 1980. Courtesy of Fermilab.

understood why in 1977 one of these prototypes had been able to reach a field of 5 T after only a few quenches.[55] Unfortunately, the Isabelle design approved by DOE in 1978 was based on expectations that hundreds of full-size magnets reaching this design field could be readily manufactured by US industry.[56]

Official DOE authorization of the $47 million Doubler project finally came in July 1979, a month after Lederman had stepped in as Fermilab's director, succeeding Wilson. By that time, the project had been dubbed the "Energy Saver," reflecting the Carter administration's concerns about the energy crisis of the mid-to-late 1970s and the fact that the superconducting ring could save $5 million a year on the lab's power bill.[57] By 1980, after resolving one last design problem, Fermilab was finally ready to begin full-scale magnet production using in-house manufacturing facilities. One sector of its superconducting ring was installed by January 1982; tests in the first half of that year proved completely successful. In June 1982 the lab shut down the Main Ring to begin installation of the rest of the superconducting magnets and other components in preparation for full-scale testing and operations in 1983.[58]

THE 1982 SNOWMASS WORKSHOP AND THE DESERTRON

In the summer of 1982, US high-energy physicists gathered in Snowmass, Colorado, to survey the status of their field and compare ideas for future facilities. Collisions between protons and antiprotons had begun on CERN's bold new proton-antiproton collider, the Sp$\bar{p}$S. It was widely expected that the two large experiments at this facility, including one headed by Harvard physicist Carlo Rubbia, would soon discover the long-sought W and Z particles among the debris. Although the difficult problems of the Isabelle superconducting magnets had apparently been resolved at Brookhaven, it was still going to take several years before this proton-proton collider could begin operations, assuming that the DOE allowed construction to proceed. By then, many US physicists felt, CERN would already have garnered all the important discoveries to be made in this new energy range; Isabelle experiments could at best confirm them at higher energy and collision rates.

To add to these concerns, the European high-energy physics community was moving forward with two new particle colliders scheduled to begin operating later in the decade that would keep its research efforts vigorous well into the 1990s. CERN had just begun tunneling for its huge LEP collider, a 27 km electron-positron storage ring. By far the largest atom-smasher in the world (for comparison, Fermilab's Main Ring has a 6.3 km circumference and SLAC's linear accelerator is only 3 km long), it had been designed to promote detailed studies of the W and Z particles—and to discover other exotic, massive particles (such as the top quark) that might turn up in its energy range. And once the LEP tunnel was completed, CERN physicists could contemplate installing superconducting magnets to circulate and collide protons at total energies above 10 TeV, allowing this facility even more discovery potential.[59]

The 1982 Snowmass gathering brought together representatives of the US high-energy physics community—theorists, experimenters, and machine builders—to make projections for its future collectively.[60] Over the previous three decades, government decisions about whether to build expensive new facilities had generally come in response to proposals from various university and national laboratories, which often had their own parochial interests (such as ensuring the lab's scientific productivity and hence survival) at heart. (One noteworthy exception had been the late-1960s establishment of the National Accelerator Laboratory, or Fermilab, near Chicago after an intense debate and a competitive national site-selection process.)[61] By the early 1980s, only Brookhaven, Fermilab, SLAC (all funded by the DOE), and Cornell University (funded by the National Science Foundation, or NSF) continued to operate accelerators or colliders for high-energy physics research, competing with each other for the increasingly scarce federal construction funds. Prominent figures within the US high-energy community began suggesting that this multilaboratory approach might not be the best way to meet the daunting European challenge.[62]

One such voice was that of Fermilab Director Lederman. "Are we settling into a comfortable secondary role in what used to be an American preserve?" he chided his colleagues at Snowmass. "In the U.S., the problem is that we have, over the past two decades, been reduced to four aging laboratories."[63] What the nation needed instead for the late 1980s and early 1990s was "a very bold advance" into the multi-TeV energy range where rich new physics discoveries were almost certain to occur, thus leapfrogging their European competition. But a collider able to generate such tremendous energies would likely need a very large and flat site many kilometers across that was virtually uninhabited, but close to power lines and a major international airport; it could not be built at one of the existing high-energy physics labs. In his Snowmass talk, Lederman called it the "Machine in the Desert," and it soon became known as the "Desertron."[64]

The energy and design of this multi-TeV collider, with proton beam energies as high as 20 TeV, followed designs that had been promoted for the proton VBA at the ICFA workshops, including possible use of 10 T superconducting magnets.[65] Because superconducting dipole magnets of only 4–5 T had proved successful by that point, however, the Desertron was initially thought to require such a large expanse of real estate that it could be built only in a desert. (As subsequent R&D efforts pushed the attainable magnetic fields higher, however, smaller sites in less remote areas came under consideration.) It was also meant to explore the nearly empty theoretical "desert" that particle physicists had previously expected to occur at the TeV energy scale—but which had recently become populated by a host of possible discoveries, including of the Higgs boson itself (see appendix 1).[66]

Two groups at Snowmass considered how to construct such an enormous collider and addressed the critical question of what it might cost. One pursued a "conventional" approach of scaling up the Fermilab Tevatron design from a total collision energy of 2 TeV to 40 TeV, assuming its superconducting magnetic fields could be raised by a factor of 2.5 to 10 T after a sufficient R&D program. Such a collider, this group reported, would still require a ring 60 km in circumference and "presumably involve the establishment of a new laboratory in the western desert." Depending on its exact design and other factors, its costs were estimated at 2 to 3 billion dollars—much more than any accelerator then built or under construction.[67]

Another group, whose éminence grise was Wilson, considered what measures might be taken to reduce these costs, focusing on "superferric" magnets that include within them sufficient iron to help shape the superconducting magnetic fields. Such magnets would have had substantially lower field strength, but were expected to be easier to manufacture and thus have lower costs. And they would be small enough to be installed inside a three-foot-diameter culvert buried under six feet of earth and serviced by robots. The total cost of building such a "sewer-pipe in the desert" ring, which had to be nearly 200 km in circumference to compensate for its lower fields, was estimated to be 1.5 to 2 billion dollars, including the laboratory infrastructure, inflation, and contingencies.[68]

Despite these differing designs, a consensus began to emerge at Snowmass that the future of US high-energy physics was to lie in building a gargantuan new multibillion-dollar proton collider in the American Southwest with a total energy of at least 20 TeV.[69] (No matter what the design or its underlying assumptions, it was clear that it would cost over a billion dollars.) These hopes were soon bolstered by the 1983 achievement of the first beams in the Fermilab Energy Saver, or Tevatron, the world's first superconducting accelerator.[70] As physicists returned from Snowmass to their universities and national laboratories, they carried with them a renewed spirit that promised to reinvigorate their field. The "spirit of Snowmass" and the dream of building the Desertron soon infused a large fraction of the US high-energy physics community, particularly those experimenters who studied proton collisions.

But taking such an approach to restore US leadership in high-energy physics would necessarily involve abandoning the international VBA project as it had been envisioned by Wilson, Lederman, Weisskopf, and other leaders of the worldwide high-energy physics community during the 1970s. If other nations wished to climb on the Desertron bandwagon—which some physicists advocated at Snowmass—it would have to be as junior partners in

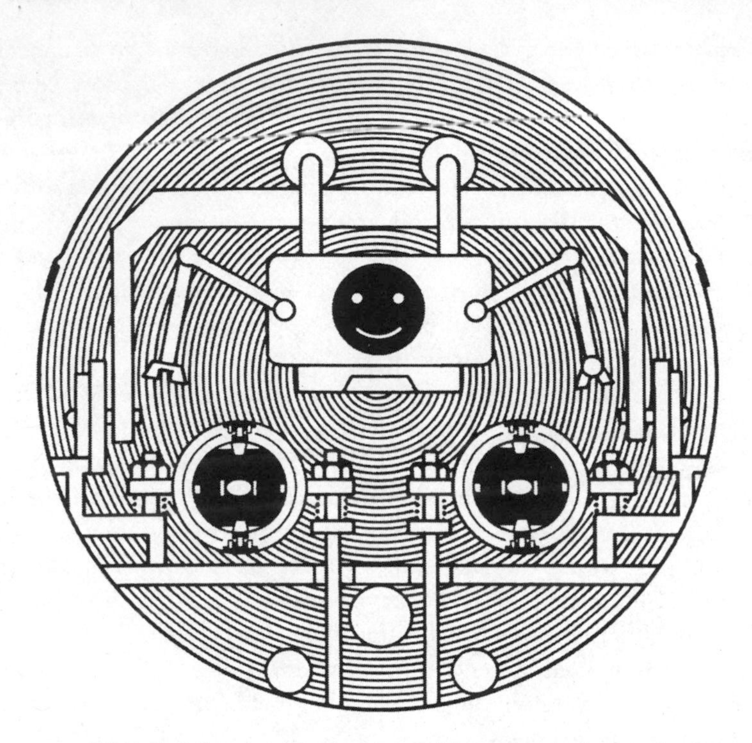

FIGURE 1.2 Whimsical drawing of a magnet-servicing robot in the imagined Desertron tunnel, as it appeared in the *Proceedings of the 1982 DPF Summer Study on Elementary Particle Physics and Future Facilities*. Courtesy of Fermilab.

what was going to be a US-led project. The idealism of the World Accelerator as science for promoting world peace subtly gave way to another, very different, rhetoric that would indelibly stamp the proposed collider as a *national* project whose rationale could be crafted to fit the emerging competitive discourse of the Reagan Revolution.

THE REAGAN ADMINISTRATION AND SCIENCE

There was also a pervasive sense of decline among the American populace during the late 1970s and early 1980s. Long accustomed to considering itself the dominant world power, the United States foundered in the midst of a period of gnawing self-doubt. The Soviet invasion of Afghanistan and the Iran hostage crisis—occurring as they did at the end of an unsettling decade that had witnessed the Watergate scandal, the loss of the Vietnam War, and the Arab oil embargo—had driven Jimmy Carter from the White House and elevated conservative California Governor Ronald Reagan to the office to which he had long aspired.[71]

Reagan's appointees quickly began making deep cuts in social spending, while slashing billions that Carter had devoted to energy demonstration projects. At the same time it was pursuing a dramatic military buildup to counter the perceived Soviet threat, the Reagan administration was cutting personal income taxes to try to stimulate the stagflationary US economy. But tight-money policies instigated by the Federal Reserve Board during 1981–82 to bring inflation under control pitched the country into its deepest recession since the 1930s. Factory output dropped by over 10 percent, and the Dow Jones industrial average plummeted 20 percent from mid-1981 levels. As the Snowmass workshop began in the summer of 1982, nearly a tenth of the US workforce was unemployed, and 16 states had to borrow from the Treasury to cope with jobless payments.[72] More than four million workers, mostly in the industrial Northeast and Midwest, were drawing unemployment benefits, and over a million more had given up trying to find work. With tax payments thus depressed, the US government was headed for its first of many budget deficits in excess of $100 billion—a level it did not drop below again until 1996.[73] The so-called pain index of the inflation rate plus the unemployment rate approached 20 percent in late 1982.

Whole sectors of US industry, especially those that had responded slowly to the exploding energy costs of the 1970s, were on their knees. The auto industry was particularly hard hit, with Chrysler Corporation needing federally backed loans to avoid bankruptcy and Ford in similar straits. Like the steelmaking and consumer-electronics industries, US automakers were fast losing market share to Japanese and Korean manufacturers. And the partial meltdown of a Three Mile Island nuclear reactor in 1979 had dealt a death blow to the US nuclear power industry, which was already staggering because of high capital costs and the effects of inflation. These and other instances of industrial decline, many linked to the economic dislocations of the 1970s, had spawned a crisis of confidence in the nation's technological might.[74]

Soon after stepping in as president in January 1981, Reagan appointed the former South Carolina Governor James Edwards to be his first Secretary of Energy, giving him a mandate to shut down the department that had been created four years earlier by the Carter administration.[75] The DOE and the Office of Management and Budget (OMB) under David Stockman began slashing budgets of demonstration projects for synthetic fuels and solar energy, arguing that commercial applications were the province of private industry, not government. Similar cuts did not apply, however, to nuclear energy.[76]

Nuclear and high-energy physics fared well in this climate because they had long been closely associated with nuclear power and weapons—a relationship that physicists did little to disavow.[77] These fields also benefited from the strong Reagan administration support for basic research, especially in physical sciences, which it viewed as the US government's responsibility. Thus federal funding for high-energy physics remained fairly level in real terms, despite sharp cutbacks in other fields and what *Science* called a "massive shift of scientific and technological resources into the military."[78] Costly space programs, such as the National Aeronautic and Space Administration's (NASA) Space Shuttle, also continued with little disruption during the early years of the Reagan administration.[79]

After months of delay, during which OMB had begun making deep cuts in research funding, Reagan named nuclear physicist George A. Keyworth to be his first science advisor and the director of the Office of Science and Technology Policy (OSTP). Almost completely unknown in Washington circles, Keyworth was then serving as head of the Physics Division at Los Alamos National Laboratory.[80] His appointment had been strongly supported by Livermore physicist Edward Teller, the father of the hydrogen bomb, and Harold Agnew, president of General Atomics—both well known for their hawkish views on national defense. As Reagan had stated that defense would be a priority for his administration, he wanted an advisor who shared his views on the subject.

At a meeting of science-policy analysts in Washington in June 1981 shortly after his appointment, Keyworth stated, "Those areas that are most exciting and those most necessary to the economy or national defense should be supported at a higher level than areas that are dormant."[81] Among the "exciting" fields he had in mind was "weak-interaction physics"—the search for the W and Z particles about to get under way at CERN.[82]

A challenging problem facing Keyworth as he assumed his new post was Isabelle. In July he flew to Brookhaven to meet with its director, Nicholas Samios, and other physicists to find out firsthand what was amiss with the superconducting magnets. What he learned did not please him at all. Not only were the magnets not working as designed, but Brookhaven's managers seemed to lack a coherent plan about how to remedy the problems.[83] "We still don't know the best way to build the accelerator, and the cost is rising very rapidly," Keyworth admitted soon after his return to Washington.[84]

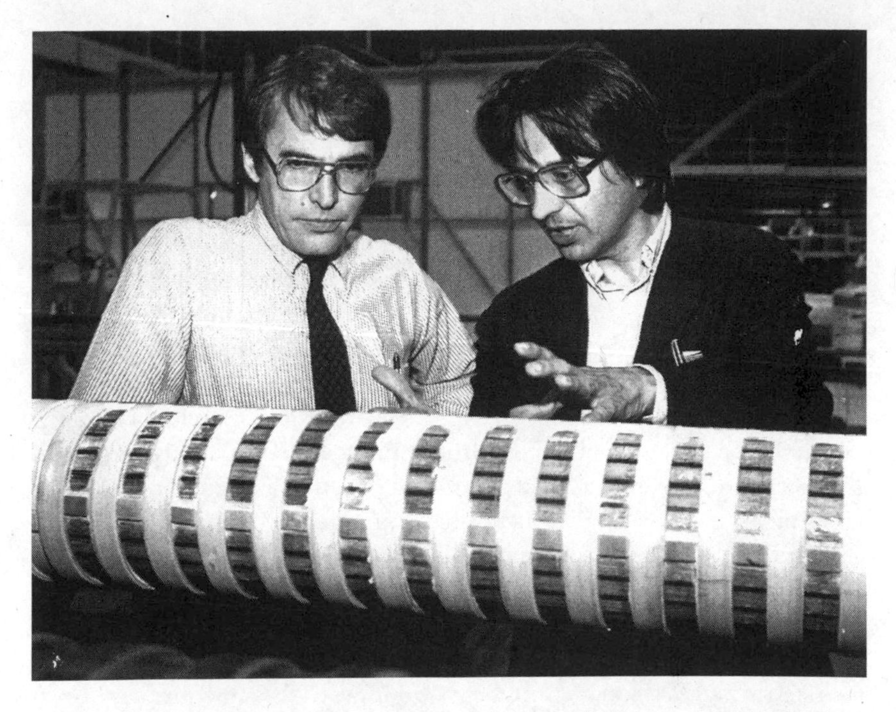

FIGURE 1.3 OSTP Director George A. Keyworth (*left*) and Nick Samios at Brookhaven, examining the core of a superconducting Isabelle dipole magnet, 1981. Courtesy of Brookhaven National Laboratory.

Thus it was difficult to defend high-energy physics from across-the-board 12 percent cuts in R&D spending that Reagan decreed that fall when Congress refused to slash social programs as deeply as he needed to avoid increasing the deficit. With the field already hard hit by Reagan's reductions in the 1981 budget, the further cuts portended major layoffs at Brookhaven, Fermilab, and SLAC, plus reductions in operations of their accelerators to less than 40 percent of available time.[85] In this difficult budget climate, the prospects for Isabelle appeared grim. The Trilling subpanel of HEPAP cautiously recommended the project continue, but *only if* the high-energy physics budget grew to $440 million, an increase of over 20 percent, and remained at that level through completion.[86]

When Reagan's budget request for fiscal 1983 was released the following January, high-energy physics received a mixture of good and bad news. Thanks to last-minute maneuvering by Keyworth, the president's 1983 budget included a big increase of $65 million—or almost 18 percent—to $429 million.[87] But most of the additional money was designated to go toward completion of the Fermilab Energy Saver and to restore accelerator operations at the three major laboratories to more traditional levels of 60–70 percent. All the construction funds for Isabelle had been "zeroed out," although the budget did include $23 million for magnet R&D and studies of other possible accelerator options. According to plasma physicist Alvin W. Trivelpiece, the newly appointed director of the DOE Office of Energy Research, the project had been put "in mothballs."[88]

At the annual meeting of science-policy analysts that summer, Keyworth explained the Reagan administration's reasoning on the surprisingly large increase in funding. "For 10 years support for the three major Department of Energy high-energy physics facilities . . . has been falling behind inflation," he noted. "Today they are starved into a state of near intellectual malnutrition."[89] Given all the problems with the nation's economy and technological infrastructure, however, US taxpayers could not be expected to bail out a faltering "WPA project" that had fallen so far behind schedule that it was about to lose the race with Europe to discover the W and Z particles. As the Trilling panel had suggested, the money would be better spent on trying to revive the rest of the US high-energy physics program.

Two principal goals of the Reagan administration's science policy were to enhance national security and boost US industrial competitiveness, especially in the high-technology sector that many deemed essential to economic recovery.[90] The word "competitiveness" had recently become the new mantra that Keyworth and other Reagan administration officials began sounding in public. As they saw it, the proper role of government was to support basic research in promising new areas that could lead to commercialization and to clear the way for industry to employ this new knowledge in generating innovative products. Thus the funding level for basic research, which had remained relatively constant in real terms during the first two Reagan years while many other budgets were being slashed, was about to rise substantially, especially in the physical sciences.[91]

A major beneficiary of this policy was about to be high-energy physics. With a passion for fundamental research resonating with similar Reagan sentiments, Keyworth said he "wanted exciting projects and quality" to keep America strong and competitive.[92] He later observed that "mediocrity was beginning to creep into our profession." Keyworth had already concluded that Isabelle did "not fill any unique bill," for its energy was too low to address the "really interesting theoretical questions." And it troubled him that Brookhaven leaders were justifying Isabelle mainly on regional grounds, as an East Coast physics laboratory. "When you have facility-based research you are on the road to mediocrity," he said in an interview years later.[93]

Urged on by Lederman, Keyworth and his deputy Douglas Pewitt (who had served from 1978 to 1981 as deputy director of the DOE Office of Energy Research), began maneuvering behind the Washington scenes in 1982 to kill Isabelle. But they also recognized that its termination would leave a huge hole in the US high-energy physics program. So Keyworth and Trivelpiece seized upon building the Desertron—or the Supercollider, as it was soon to become known—as a *national* project to restore US leadership in the discipline. For Keyworth understood, from personal interactions with Reagan and his close friend Attorney General Edwin Meese, that the president inherently favored such large, imaginative, innovative projects.[94]

"A major transformation was taking place in Washington in terms of the attitudes of the nation's policymakers towards science," wrote David Dickson, the US reporter for *Nature* during the early Reagan years.[95] After the socially conscious period of the late 1960s and early 1970s, during which the adverse impacts of science and technology had come under intense scrutiny, "the contributions of science to the competitive strength of American industry and to military technology [had] risen [back] to the top of the priority list."[96] Federal support for research and development grew by 50 percent in real terms between 1980 and 1987, and the military share of this funding rose from 47 to 74 percent during that period.[97] Having abandoned the Nixon-Brezhnev détente, the Reagan administration was determined to fight and win the Cold War, no matter what the fiscal implications of doing so.

Basic research in the physical sciences fared exceedingly well in this climate. The funding for DOE's Office of Energy Research would increase by over 80 percent in constant dollars under Trivelpiece, who served as the OER director from 1981 to 1987.[98] In terms of the renewed interest in science, technology, and their military applications, it was a period that resembled the post-Sputnik years of the late 1950s and the early 1960s. In hindsight, this was an auspicious time to begin planning an audacious new multibillion-dollar, multi-TeV proton collider that could symbolize American resurgence in high-energy physics and science in general.

THE WOJCICKI SUBPANEL'S RECOMMENDATIONS

As 1983 began, physicists around the globe learned of the impressive—but not unexpected—discovery of the W particle on CERN's new proton-antiproton collider, by two teams of mostly European physicists.[99] As had been predicted by the Standard Model, its mass came in at about 85 times that of the proton, a triumph for both theory and experiment. That June, Rubbia's UA1 Collaboration followed up this breakthrough with the discovery of the even heavier Z boson, with almost 100 times the proton mass.[100] These stinging losses to Europe, plus encouragement from the federal government for a major new project, persuaded US physicists to turn their backs on the idealistic World Accelerator project and adopt the outlines of the VBA design (as expressed in the various Desertron designs) as the underlying plan for the Superconducting Super Collider.

High officials within the Reagan administration, principally Keyworth and Trivelpiece, had already decided by late 1982 that US high-energy physics needed to take such a bold new step forward, while other DOE officials continued to hold out hope of resurrecting Isabelle.[101] Having finally resolved its difficult superconducting magnet problems, Brookhaven physicists redesigned and renamed it the Colliding Beam Accelerator, or CBA.[102] They urged DOE to resume construction, but parties to the internal discussions realized that the United States could not do both CBA and the SSC. Counseled by Lederman, who had argued that Isabelle's energy was now too low to make any important discoveries, Keyworth recognized that the faltering Brookhaven project would have to be terminated before he could ever convince the Reagan administration and Congress to support what was obviously destined to become a multibillion-dollar effort.[103]

To try to address these thorny issues, another HEPAP subpanel was formed in early 1983, chaired by Stanford University physicist Stanley Wojcicki.[104] Whereas most of the earlier HEPAP subpanels had usually included leaders from the national laboratories, whose future vitality might well be at stake in the deliberations, this one was distinguished by their *absence*. Instead, its membership included a few relatively young and highly productive university-based experimenters whose future research depended on their access to frontier high-energy physics facilities. Added to the mix to give the subpanel an international flavor were Rubbia and John Adams, CERN's leading accelerator physicist, who had served as its codirector during the late 1970s. In its open-ended charge, the DOE asked the subpanel to make recommendations "for a forefront United States High Energy Physics Program in the next five to ten years" and, notably, to "include a definite recommendation concerning the proposed Colliding Beam Accelerator."[105]

As Wojcicki recalled in a retrospective article published twenty-five years later, this charge was dramatically different from the ones given earlier subpanels in its lack of specific funding constraints.[106] As if to underscore this point, Keyworth appeared before the subpanel on the morning of its initial meeting, on February 25, 1983, in Washington, DC, and told its members that they should "*think big*—that the country was ready for a big new

FIGURE 1.4 Stanford University physics professor Stanley Wojcicki, chairman of the 1983 HEPAP Subpanel on New Facilities for the US High-Energy Physics Program. Courtesy of Fermilab.

initiative."[107] One or two billion dollars was not out of the question, he said, as long as their recommendations were based on scientific merit and restoring US leadership in high-energy physics. In summarizing that initial meeting, Wojcicki stated that the panel "should ignore any political, geographical or institutional issues and concentrate on making a recommendation based on objective scientific issues." And, he noted, "budgetary questions should not drive our recommendations: we should concentrate on what is needed to reestablish American leadership and preeminence in the field during the next decade."[108]

While the Wojcicki subpanel was meeting that spring, workshops at Berkeley and Cornell examined the detector- and accelerator-design issues that would confront a multi-TeV proton collider. Organized by Maury Tigner, the Cornell workshop (held March 28 to April 4, 1983) studied the R&D likely to be needed before building a collider with 20 TeV beams. Extrapolating from the cost of the Fermilab Energy Saver, the participants estimated construction costs at $2.4–3.0 billion (1983 dollars, excluding detectors, equipment, contingency costs, preoperating expenses, and inflation).[109] But they projected that these costs could be reduced to $1.4–2.2 billion after three to four years of accelerator R&D, mainly on its superconducting magnets. Was such a price tag affordable by the United States alone? Could its physicists actually forge ahead with their scientific program without needing the cumbersome international pooling of resources discussed in planning the VBA? If so, construction might possibly be under way by 1987! These enticing speculations came at a moment when US physicists had begun losing hope of ever realizing the World Accelerator.

Wojcicki organized his subpanel into three task forces, one led by Tigner charged with assessing accelerator technical aspects, another to evaluate CBA's physics potential, and the third to examine the overall program, including its future trends and possible risks. Input was solicited from many members of the US high-energy physics community, and the four laboratory directors (Lederman, Samios, Wolfgang Panofsky of SLAC, and Boyce McDaniel of Cornell) made presentations at subpanel meetings. Adams began preparing a document outlining European plans for the next five to fifteen years.

The subpanel also received and considered over two hundred letters from individual high-energy physicists. "Almost without exception, they supported the idea of building a high-energy collider; many of them felt that was essential," wrote Wojcicki. Except for "traditional BNL users," most of them argued against CBA. And with LEP viewed by the subpanel as an exclusively European construction project, members favored the idea that "the US must have its own frontier energy facility."[110]

The full Wojcicki subpanel met three times that spring—at Brookhaven, Fermilab, and SLAC—before gathering at Woods Hole on Cape Cod in early June to make decisions and write its report. The Brookhaven meeting was largely devoted to the question of CBA and the physics that could still be done at this collider. Fermilab leaders presented a proposal to build a 4–5 TeV proton-antiproton collider called the Dedicated Collider that would extend to the limits of the existing site and be able to produce new particles with masses in the hundreds of GeV.[111] And SLAC advocated a TeV-scale electron-positron linear collider that would push the technological limits of the Stanford Linear Collider, about to begin construction at the California lab.

Over the course of that spring, Keyworth became increasingly involved in promoting President Reagan's Strategic Defense Initiative (SDI), which was announced March 23, leaving Lederman and Trivelpiece to lead the campaign for the SSC. Lederman represented the high-energy physics community, while Trivelpiece addressed the politicians, industrialists, other scientists, and reporters interested in the proposal. And thanks in part to Keyworth's advocacy, the high-energy physics program received a big 16 percent boost in the president's budget for the 1984 fiscal year.[112]

The Wojcicki panel deliberated at length at Woods Hole for a week in early June and then at Columbia's Nevis Laboratory from June 30 to July 2. Its members increasingly saw the Desertron (by then effectively renamed the Superconducting Super Collider) as the obvious means for US physicists to respond to CERN's recent achievements (now including discovery of the Z particle, which was heralded in a prominent *New York Times* editorial)[113] with a bold, innovative thrust. As Wojcicki recalled years later, "The perceived need to restore US eminence in HEP through the construction of a frontier collider facility dominated the Subpanel's attitudes toward all the issues in its charge."[114] During the course of its discussions, the subpanel also endorsed Keyworth's judgment that CBA's maximum collision energy of 800 GeV (0.8 TeV) was too low to address important theoretical questions at the TeV scale. Although it would cost less than $1 billion and have a greater "experimental reach" than Brookhaven's CBA, the subpanel concluded, the Fermilab Dedicated Collider would not be able to address physics at this mass scale either. But the Wojcicki panel remained sharply divided on the question of whether to cancel CBA; several straw votes on this issue yielded small majorities in favor of terminating it.[115]

The superconducting dipole magnet technology, by then being proved out at Fermilab, could also reduce costs and allow building the SSC on a smaller, more accessible site if higher magnetic fields could be attained. And—having recently demonstrated the successful operation of a superconducting proton accelerator—Fermilab might even be able to manufacture the thousands of magnets required for the Super Collider.[116]

The Wojcicki subpanel's report unanimously recommended: (1) build an American accelerator called the Superconducting Super Collider, roughly 30 kilometers in diameter, with 10–20 TeV protons in each of its two beams, at a cost then estimated at less than $2 billion (1983 dollars) spread over twelve years; (2) complete both the Tevatron at Fermilab and the Stanford Linear Collider at SLAC, and upgrade the Cornell Electron Storage Ring; and (3) support advanced accelerator technology. By a majority vote, it recommended the CBA project be terminated.[117]

HEPAP met July 11 in Washington and unanimously endorsed this report—including the highly contentious recommendation to cancel Isabelle/CBA. In a letter transmitting the subpanel's report and the panel's decision to the DOE, HEPAP Chairman Jack Sandweiss of Yale stressed that the SSC "would be the forefront high-energy facility of the world and is essential for a strong and highly creative United States high-energy physics program into the next century."[118]

Based on HEPAP's recommendations, DOE officials decided to redirect most of the existing Isabelle/CBA funding toward initiating the Super Collider as a US project. Congress agreed to this request, reprogramming $18 million of the project's $23 million fiscal 1984 allocation to begin R&D on the SSC.[119] Brookhaven supporters were deeply distressed with the decision to cancel Isabelle/CBA, especially since its superconducting magnet problems had apparently been solved.[120] And Brookhaven physicist Robert Palmer pointed out at the HEPAP meeting that the $2 billion cost estimate did not include any detectors or contingencies. "That is a very optimistic number," he argued. "I am appalled at the kind of risk you're recommending."[121] Even Wojcicki was quoted as saying that the path that the community had embarked upon was "bold, risky and perhaps foolhardy."[122]

FOREIGN REACTIONS TO THE US DECISION

The announcement of the Woods Hole recommendation and its acceptance by HEPAP and DOE was not enthusiastically received in Europe. As the CERN Council had recently given its go-ahead on the LEP project and construction began in July 1983, the US proposal to build the SSC during the coming decade posed a threat to CERN's long-range plans. For the 27 km LEP tunnel had been deliberately designed to be roomy enough to accommodate superconducting magnets, which could be added later to convert this electron-positron collider into a multi-TeV proton collider. But the possible proton beam energies ranged from 5 TeV to a maximum 9 TeV—less than half the highest energy contemplated for the SSC.[123] Such a lower-energy collider would be much less attractive if the SSC were to be completed.

Since its establishment in the early 1950s, CERN had been steadily building one accelerator or collider after another as extensions of the existing facilities while growing its other laboratory infrastructure, both physical and human, accordingly. Acceleration of electrons and positrons for the 910 MSFr LEP project, for example, was to be accomplished by converting its existing Proton Synchrotron and Super Proton Synchrotron to serve as injectors.[124] CERN officials hoped to continue this steady expansion of experimental capabilities well into the future. As its director general Schopper stated at the Twelfth International Conference on High-Energy Accelerators at Fermilab in mid-August 1983, "It is the intention to make a further study of the option of a proton ring in the LEP tunnel in the coming years."[125]

High-ranking officials from particle-physics communities in other nations frankly expressed their concerns at an ICFA meeting held during that conference. As recounted in a memo by SLAC Associate Director Burton Richter, an ICFA member, "The Japanese and European delegations expressed considerable unhappiness that the Woods Hole recommendation had been made without any previous discussion in ICFA and without any consultation between the regions. They regarded the Woods Hole recommendation as a 'nationalistic approach' which would be detrimental to the interests of the high-energy physics community as a whole."[126]

Spirited discussions erupted at this meeting about what the proposed SSC project portended for the fervent dreams of the World Accelerator. "The idea of a world laboratory for the largest machines has been discussed for years at meetings and dinner parties, but always in the context that such a laboratory was far away in time and that each region could proceed with its own high-energy physics program without considering activities in other regions," Richter noted. "The Woods Hole panel recommendation that the United States proceed with an accelerator in the multi-billion [dollar] class seems to have made people realize that the time for a world laboratory may be closer than they had thought."[127]

There had already been extensive international collaboration in high-energy physics by the early 1980s, but it was almost entirely restricted to experimental research. Individual nations or regions (such as Europe) would proceed on their own initiatives to design and build new, frontier accelerator or collider facilities—for example, the Tevatron at Fermilab. Scientists from other nations were generally welcome to join the collaborations that were to use these facilities for experiments. They were expected to contribute both manpower and equipment to the detectors needed for these experiments; funding for such foreign contributions had to come from the scientists' resident nations. In a way, this process resembled the foreign involvement in scientific research sponsored by the US National Aeronautics and Space Administration. NASA provided launchers and often satellites, plus the launch and operations facilities, while foreign scientists could become involved in the design, construction, and funding of the experimental payloads.[128]

As the costs of scientific projects began to reach billions of dollars, however, other approaches to international collaboration were required. For such expensive projects, the Group of Seven (G-7) Western nations had started to address the issues of how to organize and fund them internationally at the June 1982 Versailles G-7 summit meeting, continuing the discussion at Williamsburg in 1983. Out of these high-level meetings arose the Summit Working Group on High-Energy Physics, led by Trivelpiece, which first convened in Washington, DC, in October 1983. At this gathering, European and US physicists were already squaring off regarding the SSC versus a proton collider in the LEP tunnel. "If these two efforts go forward, it is not likely that Europe-US cooperation in their construction will occur," stated the draft meeting report. It added that if "after LEP construction is complete, should Europe embark on an effort to put superconducting magnets in the LEP tunnel, European participation in the U.S. SSC construction could not realistically be expected."[129]

SSC advocates typically responded to such criticisms by saying that the new laboratory would still be an international center for high-energy physics—a major step toward the shared vision of a World Accelerator. With the United States taking a forceful lead on such an ambitious, expensive project and thus eliminating the need for lengthy, difficult political negotiations, the timetable for realizing it could be dramatically shortened. High-energy physicists from other nations were welcome to participate according to the existing model of international collaboration, by which they could contribute to the design, construction, and operation of the experimental facilities.[130]

A different approach for obtaining major foreign SSC contributions was the so-called HERA model being pursued by the West German accelerator laboratory DESY on its new $300 million electron-proton collider HERA. According to this approach, one nation takes the lead on a project, while bilateral agreements are signed with other nations to make in-kind contributions to its construction. In this case, France and Italy agreed to manufacture many of the superconducting magnets for HERA, while other nations made smaller contributions. In all, about 20 percent of the accelerator costs and 60 percent of the detector costs came from such foreign contributions.[131] This approach, a departure from previous ways of accelerator and collider construction, was increasingly advocated for the SSC.

With its booming economy and deep pockets, but thus far only moderate contributions to basic research, Japan was considered the most likely major SSC contributor. In the early 1980s, the nation was immersed in the construction of the 60 GeV TRISTAN electron-positron collider at its KEK National Laboratory for High-Energy Physics in Tsukuba.[132] The first world-class particle accelerator built in the country, costing around $400 million, TRISTAN was absorbing essentially all the nation's available high-energy physics funding and manpower. But after it was completed in 1986, the Japanese particle physics community would be seeking to become involved in a major new project elsewhere—most likely in Europe or the United States.

Thus in March 1984, Keyworth led a delegation of US government officials and physicists on a visit to Tokyo to speak with Japanese ministers and scientists about joining the SSC. The goal was to involve Japan in the project early in the planning and design stages. They struck a posture that the United States would proceed with the SSC alone if necessary, but that it would prefer to have Japan on board early as a partner.[133] The delegation visited Japan's Science and Technology Agency, the Ministry of International Trade and Industry, and the Ministry of Education and Culture, Monbusho, which had dominion over high-energy physics. Monbusho officials politely expressed interest in joining the SSC. But because that would require a major commitment of Japan's fairly limited scientific resources, its entire scientific community had to reach consensus, a process that could take a few years.[134] Meanwhile, Japanese physicists would participate in SSC R&D under the existing US-Japan Agreement on Cooperation in High Energy Physics.[135]

That same week, European physicists were gathering in Lausanne and Geneva, Switzerland, for a workshop to study possibilities for building a multi-TeV proton collider at CERN. As the introduction to the workshop report stated, "The installation of a hadron collider in the LEP tunnel, using superconducting magnets, has always been foreseen by ECFA and CERN as the natural long-term extension of the CERN facilities beyond LEP."[136] There were good arguments for doing so, including the extensive existing facilities and infrastructure that could be adapted for use in such a collider, with attendant cost savings. And CERN's long history of building proton colliders since the late 1960s gave it a deep pool of experienced accelerator physicists able to accomplish such a difficult task.[137]

But much depended on the attainable magnetic-field strength. With proven superconducting technology based on niobium-titanium wire, physicists would be able to circulate 5 TeV protons in each beam, for a total collision energy of 10 TeV, only a fourth of that projected for the SSC. If 10 T magnets based on niobium-tin materials operating at superfluid helium

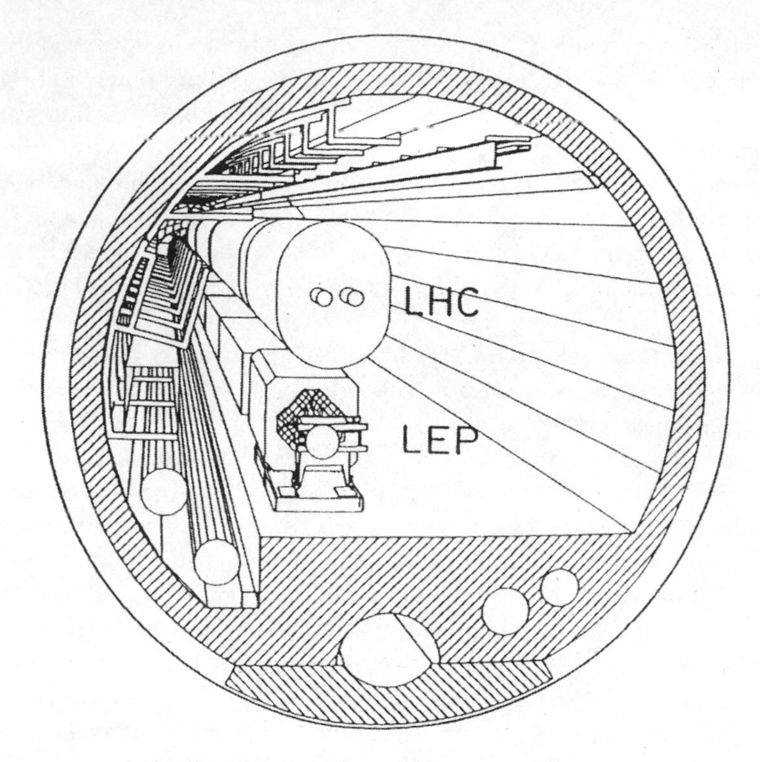

FIGURE 1.5 Early drawing of Large Hadron Collider in the LEP Tunnel, from the cover of the March 1984 workshop report. Courtesy of CERN.

temperatures below 2 K could be achieved after sufficient R&D, 18 TeV collision energies were possible. In either case, some of the quark and gluon constituents of the protons would collide at total energies close to 1 TeV (see appendix 1).[138] The report's conclusion stated that "the facilities of the CERN site form a feasible basis for an economical extension to explore a new energy region . . . where 'new physics' beyond the Standard Model can be expected."[139] But whether pursuing this physics research would be worthwhile if the proposed SSC had already been completed was left unaddressed.

This growing polarization between the two leading high-energy physics communities reached a climax at a May 1984 ICFA workshop at KEK. Convened to compare the major facilities then under construction or in the planning stages, the five-day gathering included dozens of leading high-energy physicists and machine builders from around the world. During the opening ceremonies, former CERN Director-General Weisskopf exhorted them via a video address to stay the course on the long-range ICFA goal of establishing a World Accelerator.[140] But few present were listening, except for Japanese member Yoshio Yamaguchi of the University of Tokyo. He roundly castigated US physicists for pursuing the SSC unilaterally and thereby snubbing the international VBA process.[141]

Speaker after speaker summarized their progress on or plans for national or regional accelerators and colliders, with little discussion of how they might work together or avoid overlaps and competition. Tigner reviewed preliminary designs under consideration for the SSC (see chapter 2). Following him, Giorgio Brianti of CERN outlined the nascent plans to install superconducting magnets in the LEP tunnel and convert it into a hadron collider with total energy up to 18 TeV. In a panel discussion of international collaboration, Lederman made a pitch for following the HERA model on the SSC, while Schopper argued it would be more economical to adapt and expand existing high-energy facilities like CERN's LEP collider, then under construction.

In the final session, a consensus emerged that ICFA should not attempt to arbitrate among various options under consideration but instead could promote greater collaboration on all phases of machine construction.[142] This compromise position represented a critical failure of ICFA on one of its central goals: to avoid overlaps and duplication of high-energy physics facilities among various nations or regions involved. And because of this lapse, two similar, multi-TeV proton colliders were now under serious consideration in Europe and the United States. As these facilities crossed the line into the multibillion-dollar class, leaders of the high-energy physics community were still not ready to set aside national and regional prerogatives in favor of a genuine World Accelerator.

With the Wojcicki subpanel's SSC recommendation and the HEPAP vote endorsing it, US high-energy physicists had effectively abandoned international collaboration in favor of competition. Encouraged by Keyworth, the panel decided to "think big." Given the political climate of the moment, its choice was not surprising. By mid-1983, Reagan administration officials had been sounding the mantra of US competitiveness for more than two years. The Oval Office has tremendous power to set the national agenda, even in the face of congressional opposition, and competing with Europe (and Japan) had become one of its major themes. Little US opposition was anticipated or ever encountered to the SSC being built as a national project.

As Wojcicki noted after a May 1983 discussion with him, Keyworth had "stressed the political difficulties in getting a new project approved . . . he said that the key word is *competition* as the thing the Congress will understand." The SSC "will put us ahead" while Brookhaven's CBA "will only make us equal."[143] Even the *New York Times* (which would turn about-face and oppose the SSC later in that decade) chimed in. "American accelerators should be designed to win or not be built at all," its editors proclaimed. "The 3–0 loss in the boson race cries out for earnest revenge."[144]

"The possibility of international collaboration on the high-energy collider did not play a prominent role in the subpanel's deliberation," observed Wojcicki twenty-five years later, despite inclusion of Adams and Rubbia.[145] According to the then-extant model of international collaboration, the Super Collider was to be built within the United States and almost entirely with US funds; foreign physicists would be welcome to contribute to and participate in the experimental collaborations—as they were at other high-energy physics laboratories. The possibility of the United States taking a leadership role in a truly international project (like the European Union has assumed in the International Thermonuclear Experimental Reactor (ITER) project)[146] apparently was not considered. Pursuit of such an approach, with the United States providing the site for the Super Collider and approximately half its funding, would certainly have been more cumbersome and time-consuming than proceeding with the project on a national basis, but it might have been more sustainable in the long run.[147]

US scientists tried to promote a vision of the SSC as an international center for high-energy physics, with contributions of manpower and resources from other nations. This was the portrait painted at the 1984 London Economic Summit, where Trivelpiece led a working group on high-energy physics.[148] But this argument failed to persuade the other members of ICFA, who recognized their loss of economic influence and thus any serious decision-making voice in a US national project. The original, nationalistic rhetoric of the SSC had become firmly entrenched; it would prove very difficult to undo.

With hindsight, the SSC decision was clearly fueled by extreme optimism about its total costs and the likelihood that they could ultimately be borne by the United States alone. At the time of the decision, SSC costs were estimated at just $2–3 billion—not including detectors, computers, R&D, inflation, and contingencies. But its overall price tag would ultimately exceed $10 billion when these and other costs were eventually included in the early 1990s. "I note with dismay the tendency which has grown up in recent years to err on the side of optimism," Adams observed in a prophetic memo to his fellow subpanel members. "This daring approach can have unfortunate consequences which put at risk the future of large physics communities if the optimism turns out to be unrealistic."[149] In 1982–83, some cost estimates were made based on projections that 10 T superconducting dipoles (and therefore a much smaller collider footprint) would be attainable after sufficient magnet R&D had been done. Over three decades later, high-energy physicists are still struggling to achieve this challenging goal in actual practice. As the costly teething problems of the LHC have demonstrated (see epilogue), achieving even 8 T superconducting magnets was to be an extremely challenging design goal.[150]

With the 1983 HEPAP recommendation, based on the report of the Wojcicki subpanel, the US high-energy physics community embarked on a high-risk venture in a bold attempt to reassert its hegemony in a scientific field it had long dominated. There was no detailed or even conceptual design for a multi-TeV proton collider at the time, only hopeful projections that it could eventually be achieved at attainable cost after suitable R&D on its superconducting magnets. US high-energy physicists abandoned the multilateral, international VBA process in hopes that the federal government could bear these expenses mainly by itself, supplemented by moderate contributions from abroad. They also assumed that such a high level of support could somehow be sustained through the inevitable turnovers in Congress and the White House. Ensuing events over the following decade were to prove that these expectations were wildly optimistic.

CHAPTER TWO

A New Frontier Outpost, 1983–88

Since the time of Lawrence and Wilson, physicists have appreciated the frontier nature of the instruments they needed. When one is pushing the limits, one must expect the unexpected.

—MAURY TIGNER, February 1999

The physicists who conceived and promoted the Superconducting Super Collider frequently referred to it as a "frontier instrument" for exploring exotic new subatomic domains. This imagery, intended in part to appeal to Washington and the US public, evoked familiar historical narratives of the American frontier—a subject that had been eloquently addressed in the 1890s by Frederick Jackson Turner just as the western frontier was closing.[1] The siren call of the frontier was also exploited by Vannevar Bush as World War II ended. He invoked this deeply felt imagery in his influential report *Science: The Endless Frontier,* advocating large-scale government support of scientific research.[2]

Often featured in these frontier narratives is the establishment of an outpost, a temporary fort or launching pad at the outskirts of the unknown territory.[3] During the formative period when the life of the expedition pivots about such an outpost, scouts based there prepare for their ventures of exploration, typically voicing great optimism about and enthusiasm for the project ahead—and taking risks that later, less adventurous settlers are too timid or prudent to attempt. The Central Design Group (CDG) organized in California at the Lawrence Berkeley Laboratory served as the frontier outpost for the SSC design effort. Physicists gathered there from across the United States to develop intellectual as well as physical provisions for the planned foray into new scientific territory. According to two of the leaders involved, the work at CDG was "exciting and hectic" and the intellectual atmosphere there "fantastically exhilarating."[4]

But a frontier outpost is a temporary, fragile institution, and the people working there are in constant danger of losing their positions and control over future events. Its inevitable abandonment or collapse is a critical moment for historians to examine. In the tensions and conflicts over funding and control that occur between the explorers and those who support their efforts, historians can begin to discern the larger forces at play that will govern the evolution of the future settlement.

THE REFERENCE DESIGNS STUDY

Between December 1983 and May 1984, a group of high-energy physicists working in Berkeley as the National SSC Reference Designs Study (RDS) conducted the first substantial work on designing the SSC.[5] Essentially a feasibility study, it had been chartered by the directors of the US high-energy physics labs in cooperation with the Department of Energy. Its primary focus was to estimate the cost of building the SSC.[6] The effort grew out of an enterprising letter written in November 1983 by LBL nuclear physicist Hermann Grunder and several of his colleagues asking the lab directors of Brookhaven, Cornell, Fermilab, and SLAC to contribute to starting the SSC by allowing some of their staff members to conduct early ad hoc design work.[7] The directors agreed, and the DOE approved the plan.[8]

The DOE referred to this first round of planning as the SSC's "Phase 0." It was understood by those involved that their work on the SSC would occur only to the extent allowed by their home laboratories. The funding that DOE then authorized for the RDS over the Christmas holidays of 1983, a total of $19.5 million, was soon spread among the existing high-energy physics labs and the new Texas Accelerator Center (TAC)—a recent entrepreneurial creation of Houston businessman George Mitchell and physicist Peter McIntyre of Texas A&M.[9] Secretary of Energy Donald Hodel, who had succeeded James Edwards in 1982, officially announced the study's launch on January 9, 1984.[10]

Maury Tigner of Cornell, widely viewed at that time as the best of the next generation of accelerator builders, had become deeply involved with planning the SSC after chairing the 20 TeV Hadron Collider Workshop in the spring of 1983 (see chapter 1).[11] Along with some 200 hopeful future SSC "users," he was also an active participant in a discussion group led by Bruce Winstein of the University of Chicago, known as "Physics at the SSC" (PSSC).[12] Other workshops occurred around the country, including an APS (American Physical Society) Division of Particles and Fields gathering at the University of Michigan in December 1983 dealing with superconducting accelerator aspects of the SSC.[13] Tigner was the natural choice to lead the RDS, given that its primary goal was to build on the work of the 1983 Cornell meeting and the PSSC to flesh out the SSC's preliminary design. A 46-year-old former graduate student of Fermilab's founding director, he had been nurtured in the tradition that Wilson established with his colleague Boyce McDaniel of building imaginative but frugally designed accelerators at Cornell.[14] Later, as director of the Laboratory for Nuclear Studies there, Tigner managed its electron machines as well as pursued his own innovative research on superconducting microwave cavities. By 1982 he was a member of the select cadre of physicists who were regularly called upon by the US high-energy physics community to serve in advisory capacities; for example, in 1982 he identified the configuration to be used for Fermilab's antiproton source.

For the SSC Reference Designs Study, Tigner brought together at LBL about 150 scientists and engineers from national laboratories and universities.[15] He selected Berkeley not only because many excellent accelerator physicists were based there, and because California was likely to attract other participants, but also because LBL Director David Shirley had offered suitable working space. Moreover, LBL was a neutral site that, unlike Fermilab and Brookhaven, had not been embroiled in the bitterness caused by the cancellation of Isabelle. Working steadily under Tigner's leadership, the group produced its report, the *SSC Reference Designs Study,* in only three months—in time for discussions at the final PSSC meeting on April 30, 1984.

The design of the SSC superconducting dipole magnets was fundamental and critical, for the maximum strength of the magnetic field at the collider design energy determined the machine's size and thus its cost. The higher the attainable field, the smaller the machine that could be built. And the different magnet concepts under consideration formed the basis of three different collider designs, the circumference of which ranged from 90 to 164 km depending on the magnets used. One design, proposed by teams from Brookhaven and Berkeley, relied on superconducting coils and a surrounding iron yoke to reach the highest field, 6.5 tesla (6.5 T). A second design, put forth by a Fermilab group, was iron-free and would reach 5 T. The third design, proffered by TAC physicists, was a 3.5 T superferric magnet that used a lot more iron than the Brookhaven-Berkeley design; it would in theory produce a more uniform field, but its lower field meant a much larger ring would be required to achieve the same energy as the other two designs. Weighing these parameters, feasibility, and costs for the three designs, the RDS came up with estimates for all three that were roughly the same: $2.72–$3.05 billion, in 1984 dollars.[16] The RDS judged all three magnet types

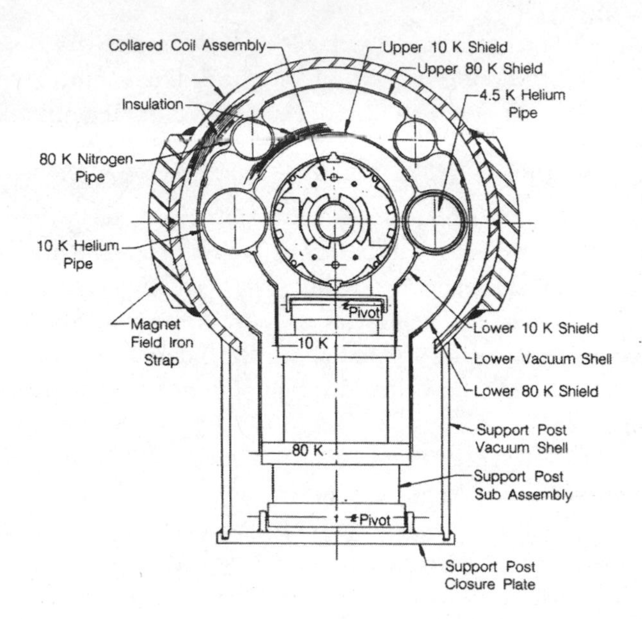

FIGURE 2.1 Cross section of Reference Designs Study superconducting dipole magnet design B, the high-field option. Courtesy of Fermilab Archives, SSC Collection.

feasible but left identifying the best choice for the next round of DOE's project management process, Phase 1, to the Central Design Group.[17] The estimated timeline in the RDS was nine years—three for the R&D phase plus an additional six for construction.[18] Tigner submitted the report *SSC Reference Designs Study* to the DOE on May 8, 1984, one week after its review by the directors of the major participating laboratories.[19]

A DOE review committee, to which the report next proceeded, approved the study on May 18, judging "all three basic SSC designs . . . technically feasible" and recommending increasing the cost estimate by $200 million.[20] The next to receive the report was Energy Secretary Hodel, on June 4. As everyone awaited his approval, *Science* reporter Mitchell Waldrop reflected on the intense year of activity since HEPAP committed America's high-energy physics community to the Super Collider. "Starting from nothing—little more than a community-wide consensus that the SSC is a good idea—they have created a set of reference designs that give a reasonably firm estimate of the cost (about $3 billion over 10 years) . . . put a management structure in place, . . . and . . . honed the design of the SSC in innumerable working groups." DOE officials were "delighted" with the design study, he reported. "Truly outstanding," was how Director of High Energy Physics William

Wallenmeyer described it. "The amount of work and the depth is much better than many things we go forward with construction on."[21]

The reactions on the political front were also encouraging. "Congress has been generally receptive to the SSC, and the Administration has been downright enthusiastic," Waldrop observed. There seemed to be justifiable concern, because "everyone flinches at the price tag." As he noted, "Initial enthusiasm aside, $3 billion takes a lot of justifying—'years of preparation,' as Keyworth warned the physicists, 'not just of the technical design but of the public rationale as well.'"[22]

Still, few were surprised when in August Secretary Hodel approved the RDS and allocated $20 million to begin Phase 1 during fiscal 1985. He commended those who had prepared the report, affirming that "this project is totally in the spirit of this administration's commitment to the advancement of science and technology as an essential ingredient in the achievement of national goals."[23] Many high-energy physicists saw Hodel's approval as a sign that the vision that emerged at Snowmass in 1982 might indeed become reality. In a tone often invoked in these surroundings, Fermilab physicist Paul Mantsch had written:

> The moment was right for the great vision to become reality. The mood of optimism was on the Land. The technical progress that enriches the lives of the People has at its foundation a vital and dynamic program of scientific research. The People know little of quarks and leptons. But they know curiosity. They know asking questions and seeking answers. They share the spirit of pioneers at a new frontier. And they know it is good. For they were great People of a great Land where boldness and innovation is legendary. This, after all, is the Spirit of Snowmass.[24]

EARLY WASHINGTON PERSPECTIVES ON THE SSC

As DOE's transfer of $20 million for the CDG later that summer did not need further congressional approval, the SSC project could advance, maintaining a low profile in Washington.[25] At this point, the project had few prominent Washington supporters besides Hodel and Keyworth. Another was Trivelpiece, who had been a vice president at Maxwell Laboratories and the Science Applications International Corporation (SAIC). Besides serving as director of the DOE Office of Energy Research, he acted as Hodel's scientific advisor. A Reagan appointee from California, Trivelpiece was the principal Washington official who oversaw the SSC project. A thoughtful administrator, he was widely viewed as capable of bridging the most relevant communities: scientists, politicians, policy-makers, industrialists, and the media.

Another crucial SSC supporter in Washington was James Leiss, the DOE associate director for High-Energy and Nuclear Physics, who had authority over funding for high-energy laboratories and other facilities, subject to congressional approval.[26] And despite lingering hard feelings generated by the US high-energy physics community's break from ICFA's fervent hopes to build a World Accelerator, OSTP Director Keyworth began pursuing foreign contributions to the SSC, inviting Japanese participation in the project.[27] Trivelpiece, meanwhile, tried to encourage European contributions to the SSC, continuing his efforts via the Economic Summit process, which began in earnest at the 1983 Williamsburg summit meeting of the G-7 nations (see chapter 1).[28] As yet unrecognized was the inherent contradiction between the SSC project's need for foreign contributions and the Reagan administration's nationalistic rhetoric. Another potential barrier to progress on the SSC was the unresolved question of how to fit the project into Wallenmeyer's ongoing DOE strategy of maintaining mutually supportive relationships among the three major high-energy physics laboratories—Brookhaven, Fermilab, and SLAC, all in states with powerful congressional delegations.

Meanwhile, the topic of managing R&D for the SSC project had come up in Washington at various times. On July 13, 1983, the day after HEPAP recommended the SSC as the high-energy physics community's top priority, Norman Hackerman, chair of the Council of Presidents of the Washington-based consortium Universities Research Association Inc., wrote to URA President H. Guyford Stever, asking him to send "DOE a proposal offering the services of URA in the management of the R&D phase." Consisting at that time of more than 50 research universities across the United States, URA had successfully managed construction of Fermilab, including its Main Ring and Tevatron (then the world's highest-energy accelerator complex). Hackerman noted that the R&D management had "to be an effort entirely separate from the management of Fermilab."[29] Stever promptly sent the requested proposal to Hodel two days later.

Although the fact that URA was managing and operating Fermilab was a compelling qualification, it presented a potential conflict of interest because the lab was naturally interested in hosting the SSC, too. Circumventing any such conflict by separating URA's involvement in SSC R&D from its management of Fermilab was therefore critical. In November 1983, URA decided to create two distinct Boards of Overseers (BOO), one for Fermilab and the other to deal exclusively with matters involving the SSC project.[30] At its first meeting on March 23, 1984, the SSC BOO elected Boyce McDaniel of Cornell to serve as its chairman.[31] Earlier that month, Leiss had asked URA to provide a management plan for the project's initial phase, "to assist the department in the management of the national effort during the R&D and Conceptual Design Phase of the SSC Project." He acknowledged that URA was the natural choice to manage this effort and hoped to have its management plan in place by the time Secretary Hodel approved Phase 1 in August 1984.[32]

ORGANIZING THE CENTRAL DESIGN GROUP

While the possibility of building the SSC was under consideration in Washington, the Reference Designs Study served physicists by addressing some of its technical design questions, providing topics for many additional meetings held during the next several years. The largest gathering was the second Snowmass workshop, where some 250 physicists convened from June 23 to July 13, 1984, to review technological progress made since the 1982 meeting and look to the future.[33] The urgency of building the SSC was further heightened on July 3, when the UA1 collaboration at CERN led by Rubbia announced it had six candidate events for the top quark.[34] Although these results were eventually found to be spurious, they had a strong motivational impact on SSC planning.

An important question to be resolved early in the SSC design process was how high a total collision energy to shoot for. The Wojcicki panel had specified the range 20–40 TeV (see chapter 1), but it mostly favored the higher end of this range due to the added physics potential.[35] At the 1984 Snowmass gathering, preprints of an influential review article titled "Supercollider Physics" circulated widely.[36] It predicted production rates for possible physics processes to be expected at proton colliders with energies from 2 to 100 TeV. For the most interesting cases of TeV-scale physics to be explored (see appendix 1), ratios of signal to noise were substantially better at 40 TeV than at 20 TeV energy. "By the end of that summer, the community consensus for 40 TeV . . . was strong," recalled Chris Quigg, an author of the paper.[37]

On June 20, just before the Snowmass meeting, Stever announced that URA's new SSC BOO had chosen Tigner to lead CDG and develop a detailed conceptual design for the SSC.[38] His selection to direct the CDG efforts was no surprise. Tigner's level-headed effectiveness in leading the RDS had been widely appreciated. *Physics Today* reported that DOE officials considered him to be "ideal as chief designer of the SSC" and expressed the widely held

FIGURE 2.2 Cornell University accelerator physicist Maury Tigner, speaking at the May 1984 ICFA Workshop in Tsukuba, Japan. Credit: Fermi Film Collection, courtesy of American Institute of Physics (AIP) Emilio Segrè Visual Archives, *Physics Today* Collection.

expectation that he would "maintain the head of steam . . . that drove the Reference Designs Study to completion in three months."[39] Congressman William Carney of New York stated that "Tigner is one of the most informed and imperturbable witnesses I've seen on Capitol Hill." Many perceived him as the *only* member of the high-energy physics community who could prevent "a shoot-out among different groups, each claiming the best plan."[40]

At a September 1984 meeting, the SSC Board of Overseers officially endorsed the 40 TeV energy choice, which was by then essentially a foregone conclusion.[41] From that point forward, CDG physicists worked toward achieving 40 TeV in the SSC design process, with little if any consideration of lower-energy options.[42] As European physicists had earlier that year broadcast intentions to proceed eventually with their plans for a proton collider in the LEP tunnel capable of energies up to 18 TeV (see chapter 1), a 20 or 30 TeV US collider could easily have been seen as duplicating possible CERN efforts. The SSC had to offer substantially more.

Tigner proceeded to assemble the CDG staff, drawing upon leading members of various laboratories. As his principal deputies, he chose two he had served with on the Woods Hole subpanel: experimenter Stanley Wojcicki from Stanford as his deputy director for external relations (encompassing relations with the DOE and Congress, foreign institutions, the US science community, and the general public) and theorist J. David Jackson of Berkeley to be his deputy director for operations. Other early appointees included Jim Sanford of Brookhaven to direct conventional facilities planning, Tom Elioff of LBL to lead management and administration, Alex Chao from SLAC as head of accelerator physics, Peter Limon of Fermilab to oversee accelerator systems, Victor Karpenko from Lawrence Livermore Laboratory as the director of superconducting magnet R&D, Robert Matyas of Cornell to serve as management advisor, and Paul Reardon from Brookhaven to lead an external advisory group. Rene Donaldson of Fermilab agreed to serve as technical writer and senior editor. Many others joined later.

As the core group of CDG physicists (which would in time grow to about forty) settled into temporary offices arranged around the "Cockpit," an open conference area near the historic Bevatron, workshops and panels were arranged that brought 250 experts and other short-term visitors to LBL between June 1984 and March 1986.[43] Tigner defined the milestones for Phase 1, among them submission by April 1985 of the site criteria document, the first SSC users meeting in May 1985, magnet selection in July 1985, preparation of the SSC Conceptual Design Report by March 1986, selection of the SSC site by December 1986, and the start of its construction in October 1987.[44] In his many talks over the next few years, not only to physicists but also to DOE and congressional committees, he would typically explain the project status, showing the milestones in overhead slides and inviting comments or advice. Tigner set a tone of purpose, dedication, and excitement that sustained CDG staff through long hours and held their work to his tight schedule. Those serving with him there recalled the CDG atmosphere enthusiastically. One staffer remembered Tigner as a leader who inspired loyalty and devotion by working very hard and expecting his staff to work almost as hard.[45]

Tigner's authority was, however, constrained by the cumbersome financial arrangement DOE set up for disbursing CDG funds. In their original memorandum of understanding, the directors of the existing US high-energy laboratories and DOE agreed that CDG would coordinate the design work of the participating laboratories, but ultimately DOE officials controlled the purse strings.[46] Although he negotiated budgets with each laboratory at the beginning of the fiscal year, Tigner lacked full authority to "harness their abilities and enthusiasms more efficiently."[47]

The SSC's base of support in Washington was shaken in January 1985 when President Reagan named Secretary Hodel to head the Department of Interior. Tigner now had to win support from Hodel's successor, California real-estate developer John S. Herrington, who was persuaded only gradually about the SSC's importance. Meanwhile, URA itself changed when Stever, a former National Science Foundation (NSF) director and science advisor to President Gerald Ford, announced his retirement as president in the spring of 1985. URA trustees sought another president who was well connected in Washington. They selected Edward A. Knapp, a Los Alamos staff physicist since 1958, NSF director from 1982 to 1985, and a member of the SSC Board of Overseers. On August 1, 1985, URA announced that he would serve as the first full-time president of URA.[48] A year later, in October 1986, Knapp hired physicist Ezra D. Heitowit as the new URA vice president. Formerly the staff director of the House Subcommittee on Science, Research and Technology, Heitowit brought extensive congressional experience to his new position.[49]

Obtaining a long-term federal commitment to building the SSC remained a problem. In June 1984, when Tigner was named the project's chief designer, many had been optimistic about the DOE's intention to build the collider. Two months later, when Energy Secretary Hodel's endorsement of the RDS urged SSC advocates to seek foreign contributions, an actual financial endorsement of the project was not part of his decision memorandum. In the fall of 1984, Trivelpiece told *Physics Today* that he anticipated that it would take another three years to accumulate "enough information to make a clear commitment on whether or not to ask Congress for money to build SSC."[50] And in June 1985, DOE officials told *Science*, "Right now we are just trying to keep the idea alive."[51] Mixed messages like these were worrisome to potential users as well as to everyone at CDG designing the machine.[52]

THE SSC CONCEPTUAL DESIGN REPORT

In the fall of 1985, the CDG began to focus on producing a description of the collider's technical footprint on a generic site, known as the "SSC Conceptual Design Report."[53] As the design depended critically on the choice of magnets, Brookhaven, Fermilab, and the Texas Accelerator Center fabricated prototypes to determine the specifications and evaluate the performance and reliability of their proposed magnets. A vital concern was the size of the magnet aperture, the opening through which the proton beam had to pass while it was being accelerated to higher energy. A larger aperture implied a substantial increase in materials, including the amount of superconducting wire in the magnet coils. As material quantities scale roughly as the square of the aperture, a small increase in aperture would seriously impact the cost of the entire project, because thousands of magnets would be required. To keep costs down, the aperture had to be as small as possible, but reducing it inevitably led to deterioration of the magnet field quality where the protons would pass through. Chao, the head of CDG's Accelerator Physics Division, described selecting the aperture as a delicate balancing act with major financial consequences.[54] The painstaking work of deciding on the aperture involved writing and interpreting computer simulation programs, and running repeated simulations that tracked protons for hours as their energies increased from 1 TeV to as high as 20 TeV. It was necessary "to be inventive in our algorithms and evaluation criteria," recalled Chao. "For accelerators before the SSC, one usually adopted a conservative approach. When in doubt, the aperture could be increased without much financial penalty," he added. "This was no longer the case for the SSC."[55]

A task force appointed by Tigner to make the aperture choice worked from October 1984 to July 1985. As Chao recalled, the task force "was a loosely organized group of physicists from various labs who [had] their own work and priorities at their home institutions." Again, prioritizing SSC work against other responsibilities influenced the magnet-development process. Among the relevant issues studied were the physical arrangement of magnets in the collider ring (the "lattice"), schemes to correct field aberrations, extrapolation of existing superconducting-magnet performance to what could be expected in the near term, and computer simulations of the proton orbits in the collider.[56]

To settle the "most difficult technical issue," Chao organized the Magnet Aperture Workshop, held in November 1984, consisting of about sixty participants.[57] "After a countless number of skipped lunches, working into the evenings, and some sleepless nights," they selected an aperture of 4 cm.[58] It was a daring choice, based on Wilson's philosophy of not overdesigning an accelerator beyond the minimum needed.[59] Everyone at CDG recognized that this choice would require correction coils around the beam tube to help cancel out non-uniformities in the magnetic field. The guiding idea, as Tigner recalled, was not to keep "a factor of two in our back pocket, but rather to control costs while meeting the agreed-upon specifications."[60]

Choosing the aperture was only part of the task, however. To help him decide on the specific magnet design, Tigner appointed a Technical Magnet Review Panel chaired by Alvin Tollestrup of Fermilab, who a decade earlier had led development of the superconducting dipole magnets used for the Energy Saver, which eventually became the Tevatron. By the time this panel first met in early July 1985 at the Texas Accelerator Center, CDG had narrowed its magnet choices to two candidates: the "high-field $cos\,\theta$" (also known as the "conductor-dominated") magnet developed at Berkeley and Brookhaven with Fermilab's aid, and the superferric magnet proposed by the Texas Accelerator Center (which was similar to what Wilson had suggested at Snowmass). The first of these corresponded to RDS magnet design B (see fig. 2.1).

The Magnet Type Selection Advisory Committee, chaired by Frank Sciulli of Columbia University, convened in Berkeley in late August 1985, to consider costs, technical merits, and site politics related to the magnet designs.[61] It had less than a week to prepare its report for Tigner, due September 1. The Texas team objected when the panel's selection of the high-field $cos\,\theta$ magnet design was announced on September 18. An important reason for this choice was that this design's projected 6.5 T field led to a collider circumference of about 52 miles (84 kilometers). Using the lower-field superferric magnet, which was arguably cheaper to fabricate, would require a ring about 100 miles in circumference to achieve the same proton energies. But building such a large collider would not only increase the cost of tunneling and microwave systems; it would also drastically limit the number of potential sites for the SSC (eliminating the smaller, more populous Northeast and Midwest states). The choice of the high-field magnet design made sense to all but the Texas group.[62]

The CDG decision about the SSC magnet design was significant enough to attract the attention of the science press. Describing the SSC in a front-page article, *New York Times* reporter William Broad wrote that for the past two years the collider had been "a vague goal" pursued by hundreds of scientists working on feasibility studies. He interpreted selection of the magnet design as a "firm commitment to a particular design for the huge machine," whose footprint would thus encompass an area the size of New York City.[63]

Almost in passing, Broad observed that building the SSC was expected to cost $3–6 billion, expressing concern that the SSC's "huge appetite for funds will squeeze out other science experiments" (see chapter 3). By this time some twenty states were already preparing site proposals for the "multibillion-dollar boon for contractors chosen to dig the tunnel and make the magnets."

Broad also mentioned that while researchers in other fields feared that the SSC's budget would "distort the Government's broad financing of many areas of science," several Nobel laureates, among them James Cronin of the University of Chicago, defended the project as "a scientific imperative." Lederman argued that the fundamental questions of science could not be answered without this "essential tool."[64]

In *Science* magazine, a more ominous assessment of the magnet decision came from Mitchell Waldrop, who had been watching the SSC magnet story closely.[65] He noted that the schedule for construction of the SSC was

> hanging in a state of limbo at the moment. The supercollider has come under attack by scientists in other disciplines who are concerned that huge expenditures on the machine will cut into their own funding. Moreover, a construction start in October 1987 means that the Energy Department has to make an official commitment to the supercollider sometime next year, before the fiscal year 1988 budget is prepared. Given the deficit situation, it is not at all clear that the department will be willing or able to take that step. In fact, some physicists in the project have begun to resign themselves to a delay of several years.[66]

Six weeks later, Waldrop reported that an October hearing about the SSC before a House subcommittee was "rife with concerns about the financial viability of the supercollider." Although the "prestigious group of European and American physicists" who endorsed the project at the hearing heard "praise for the scientific quality of the supercollider concept," there was "skepticism about the viability of the $4-billion project in the face of a mounting federal deficit."[67] Indeed, even as the SSC hearing was in session, the House was debating a major deficit-reduction measure, the Gramm-Rudman-Hollings Balanced Budget and Emergency Deficit Control Act of 1985.[68] This law introduced automatic, across-the-board spending cuts or "budget sequestrations" if the total discretionary appropriations exceeded annual budget targets.

Despite this increasingly pessimistic climate, CDG physicists pressed on in their preparation of the conceptual design report. Serving as the report's editor-in-chief from September 1985 through March 1986, Jackson coordinated the writing assignments, drafting most of the opening chapters himself. It was a heavy burden for him. Hoping to lighten the load with humor, his wife gave him a T-shirt for Christmas depicting a toy monkey on his back. In March, when the advance copy of the four-volume report was completed on schedule, in time for the start of DOE's fiscal 1987 budget cycle, editor Rene Donaldson presented Jackson with a stuffed monkey clutching a miniature of the report.[69] When the 712-page document (over 2,000 pages with additional appendices) was delivered to the DOE on April 1, 1986, it projected a total cost of $3.01 billion in 1986 dollars, including $529 million as a contingency.[70] This estimate did not include many as-yet unknown costs, including essential R&D, commissioning, computers, experimental detectors, and inflation.[71]

While DOE was digesting the massive document, the culmination of two-and-a-half years of R&D efforts by scores of contributors, the three laboratories that had backed the high-field magnet design continued to plan and develop these magnets. LBL agreed to work on the superconducting cable and on the short model magnets; Brookhaven took on building the full-length magnets; and Fermilab assumed the primary responsibility for developing cryogenics systems and for testing all these magnets.[72]

Meanwhile, at the DOE program offices in Germantown, MD, the conceptual SSC design was subjected to an intensive review, which included an independent cost estimate that confirmed the CDG total figure, adding only $5 million.[73] L. Edward Temple, who headed the Facilities Management Division in the DOE Office of Energy Research, chaired the review. Alex Chao recalled Temple's review team as an army of fifty to sixty people (larger than the CDG staff) that "invaded the CDG." Initiated in the early 1980s, such "Temple reviews" of the budget and management systems of DOE projects had earned a reputation for being exacting, comprehensive, and fair. They marked a new level of DOE project management, oversight, and scrutiny of research funding, and quickly became a required step in the long DOE process of bringing a proposed project to formal construction status.[74] Expectation of a Temple review forced laboratory managers to set specific budgets and schedules from the outset, including careful audits and rigid management of programs to meet stated goals.

While this review was under way, on April 7, CDG suddenly and without warning faced what it perceived as "shooting from the left and the right."[75] Russell Huson and Peter McIntyre of TAC protested the choice of the high-field magnet design in a well-publicized letter to Sciulli, who had led the magnet-selection panel. Claiming that the Texas superferric design would be more economical than the high-field choice, they asked HEPAP to reconsider CDG's September 1985 decision.[76] To investigate their claims, Trivelpiece appointed a subpanel chaired by SLAC Director Richter. In a May 15, 1986, meeting at LBL, the Richter subpanel upheld the Sciulli panel's recommendation and CDG's consequent decision.[77]

The Temple review of the SSC Conceptual Design Report, released in May 1986, deemed its scope and cost appropriate, emphasizing that strong management would be essential.[78] But a subcommittee chaired by Robert Siemann of Cornell expressed concern that the 4 cm aperture might be too daring and suggested that perhaps CDG should maintain the option of increasing the aperture to 5 cm as a possibility. After considering this concern, Temple's review committee approved the 4 cm aperture under the condition that, until more was understood about it, "one must hold open the possibility that the field qualities will have to be improved."[79] The CDG was already planning to do just that, reflected Chao.[80] But its sweeping vision of the SSC had finally passed muster at DOE, just over a month after the report had been submitted. The conceptual design report for the SSC was widely regarded as "a model of scientific clarity."[81]

PRESIDENT REAGAN THROWS DEEP

Despite the Temple review panel's positive evaluation of the conceptual design report and increased support from particle physicists, the nation's growing anxiety over federal budgets and spending in the mid-1980s cast a dark shadow over the early optimism about the SSC. Journalists did not hesitate to express the growing worries about the SSC's future.[82] And with congressional attention beginning to focus on reducing the budget deficit, every recipient of federal funding wondered whether, or when, "the automatic Gramm-Rudman-Hollings (GRH) ax" might fall on their work.[83] As a leading statesman of physics, Lederman decided to express the widespread concern about potential Gramm-Rudman-Hollings cuts in a March 1986 letter (unpublished) to the *New York Times*. He feared that the GRH budget process would work to "reduce the deficit even if it destroys the country." He argued that the reduction of research, once triggered, would be substantial, "more precipitous than the retrenchment which began in 1967," and lead to a "chilling and escalating drag on progress and an even greater negative influence on the essential process of recruiting and training." Lederman portrayed the research process as "historically asymmetric—it is very easy to damage, exceedingly difficult to repair."[84]

SSC supporters had initially believed Keyworth's scenario, in which the big project would be funded with "new" money—above the existing "base program" of support for ongoing high-energy physics research at existing laboratories.[85] But as hope for such new money faded and the estimated SSC cost rose, many scientists outside high-energy physics worried that it would crimp growth in other research areas (see chapter 3).[86] Lederman thus suggested building a less-expensive machine that, like the Tevatron, would collide protons with antiprotons in a single ring of magnets. "In my opinion, if we don't drastically reduce costs," he had written URA President Knapp in October 1985, "SSC will die or, worse, drag out to become an unfruitful drain on the rest of the program."[87] But a June 1986 review of this suggestion concluded that such a proton-antiproton supercollider would not be as reliable as a proton-proton machine, have inadequate luminosity for the proposed experiments, and save less than $200 million.[88]

In mid-1986 the immediate problem was that presidential approval, required for the SSC given its multibillion-dollar construction cost, was far from assured.[89] It was unnerving to those working on the SSC design when, after ninety-one congressmen signed a petition urging Reagan to support the project, Energy Secretary Herrington stated in a March 1986 hearing before the House Science and Technology Committee that he "had not yet made a decision" on the SSC, adding that "we certainly don't want to give the impression that we're opposed to the project." But he also remarked, "We don't want to waste further funds at this time until we make a decision." In his summary of SSC deliberations on Capitol Hill, Herrington admitted that although "the technology is there, . . . this is not the year" for new construction funds.[90]

A few days later, in an effort to address the troubling fact that the fiscal 1987 budget contained no SSC funding for the first time in several years, more than eighty SSC supporters from academia, industry, government, and the scientific community attended a strategy session in Washington. It had been called by Illinois Governor James Thompson to keep the SSC project on Washington's mind. Though it is not clear whether this gathering played a significant role, Secretary Herrington had by late 1986 become one of the SSC's major political supporters. On December 3, 1986, he announced he was convinced that the Reagan administration should build the SSC. "I have seen enough that I think we will get a lot of benefit out of the SSC," he said. Herrington now strongly agreed with Trivelpiece that the SSC "is something the country should do." He acknowledged, however, that construction of the huge machine hinged on finding $4–$5 billion to pay for it. A decision about whether to bring the SSC project before President Reagan, with the intention of getting his endorsement for a construction start in fiscal 1988, was not expected until late that month.

The tensions due to delays in construction funding were compounded by another shuffle of Washington personnel that left the SSC with even less political support. At the end of 1985, Keyworth had stepped down as OSTP director; his position remained vacant until William R. Graham took over in

October 1986.[91] A high-level NASA administrator, Graham had no particular allegiance to the SSC. And James C. Miller III, Reagan's second director of the Office of Management and Budget, was no friend of the SSC, either. He continually advocated reducing its budget, claiming that the machine had a high cost and only marginal benefits. As the administration's budget watchdog, it was part of his job to cajole cabinet members to reduce spending on their programs. Miller told Reagan that "if he wanted to invest in science, alternative use of this money would yield better results."[92]

In the same period, James Leiss retired as DOE Associate Director of Energy Research for High-Energy and Nuclear Physics; the position was temporarily filled in an acting capacity by Wallenmeyer, who also continued serving as Director of High Energy Physics. But when Wilmot Hess was appointed in August 1986 to fill this post, he did not share his predecessor's enthusiasm for the SSC (at least not at first). Trained as a nuclear physicist at UC Berkeley, Hess had served in leadership roles at NASA during the 1960s and subsequently at the National Oceanic and Atmospheric Administration and the National Center for Atmospheric Research—three agencies well removed from the culture and concerns of high-energy physics.[93]

From late 1985 through late 1986, Trivelpiece was thus virtually alone as the SSC's only high-level political advocate in Washington. It was his responsibility to translate the SSC conceptual design report into a format suitable for presentation to Reagan. He briefed the White House Domestic Policy Council and Economic Policy Council on the scientific merits of the Super Collider near the end of 1986.[94] Based on these briefings, the Domestic Policy Council authored a memorandum to Reagan with arguments for and against funding construction of the SSC. Estimating its costs at $4.5 billion (which, over a ten-year period, would be $6 billion in as-spent dollars, including computers, detectors, R&D, and inflation), the memo stressed its role in meeting national concerns, including "scientific prestige," "technological competitiveness," maintaining US leadership in high-energy physics research, and serving as a good continuing training ground for American scientists.[95] But because of budget conflicts with other high-profile programs (including the Strategic Defense Initiative, or "Star Wars," the Orient Express space plane, and Space Station Freedom), the council had become deadlocked over the SSC. Expected increases in federal revenues due to supply-side economics had not materialized, and little new funding was going to be available for new projects.[96] Relieving the deadlock thus required a presidential decision.

At the request of Secretary Herrington, Trivelpiece briefed Reagan in the White House at a January 29, 1987, meeting of the Domestic Policy Council.

FIGURE 2.3 DOE Assistant Secretary Alvin Trivelpiece, making presentation about the SSC to President Ronald Reagan and the Domestic Policy Council in the Cabinet Room of the White House, January 29, 1987. To Reagan's right is OMB Director James Miller, and across from him, in glasses, is Vice President George H. W. Bush. Courtesy of Ronald Reagan Library.

Having already presented his talk twice, Trivelpiece gave a polished presentation.[97] "It's about the size of the Washington Beltway," kidded the president during the talk. "Maybe we could just put it around Washington!"[98] After Trivelpiece finished, the former actor and television personality responded with characteristic high drama. Pulling an index card from his pocket, he read a poem by Jack London:[99]

> I would rather be ashes than dust!
> I would rather that my spark should burn out
> in a brilliant blaze than it should be stifled by dry-rot.
> I would rather be a superb meteor, every atom
> of me in magnificent glow, than a sleepy and permanent planet.
> The function of man is to live, not to exist.
> I shall not waste my days trying to prolong them.
> I shall use my time.

When asked what he meant in citing this poem, Reagan summoned a line from his days as a radio sportscaster, before he moved to Hollywood and starred as George Gipp in a 1940 movie about Notre Dame football coach Knute Rockne. He invoked Kenny Stabler, the Oakland Raiders' quarterback who had a reputation for calling risky offensive plays and throwing long, "Hail Mary" passes into the end zone as time expired. Stabler once told him, Reagan explained, that the poem meant one should "Throw deep!"[100] With this cryptic endorsement, he had approved the SSC. And when Secretary Herrington announced the administration's support for the project the next day, he called it the high-energy physics equivalent of "putting a man on the moon."[101]

THE UNSOLICITED URA PROPOSAL TO MANAGE AND OPERATE THE SSC

There were serious differences, however, between the expectations of high-energy physicists and the DOE officials administering major construction projects. This dichotomy became clear after CDG submitted a 14-page URA management proposal (which had not been solicited by the DOE) in March 1987, and it elicited no response. It was written largely by Edwin Goldwasser, a physicist who had served as Wilson's deputy director during Fermilab's first decade—and later as vice chancellor of the University of Illinois.

In September 1986 Goldwasser had joined CDG to work as associate director for development. Soon thereafter, he drafted the proposal, arguing that URA was the natural organization to manage the SSC.[102] In his judgment, everyone would save time, money, and effort by the DOE letting a "sole-source" contract to URA for management and operation of the SSC. Physicists had followed this approach for decades in negotiating government contracts for operating the national laboratories. Goldwasser's proposal explained why URA was "uniquely qualified to provide the broad representation of the nationwide academic community together with the management experience that is essential to the success of the SSC laboratory and its research program."[103] On March 2, 1987, URA President Knapp sent the proposal to Trivelpiece, who, after assessing the legality of letting such an unsolicited contract, forwarded the proposal upstairs to his DOE superiors at the Forrestal Building in downtown Washington.

When the proposal was considered at a critical spring meeting that included Trivelpiece, Herrington, and other top DOE officials, according to Goldwasser, the central issue was whether sole-sourcing the SSC management contract was appropriate.[104] A high government official (reportedly Herrington or Undersecretary of Energy Joseph Salgado) was said to have "had pen in hand ready to sign the unsolicited proposal."[105] But this DOE official asked whether resolving the matter right then was absolutely urgent. Reluctant to claim that it was, Trivelpiece hesitated. The proposal was thus tabled and received no further consideration.[106] This twist of fate was the first clear sign that the SSC would not be judged in the same way as prior high-energy physics facilities. Indeed, new procedures and growing complexities would delay all large DOE scientific projects from then on. Had Trivelpiece replied that approving the URA proposal was of utmost urgency, Goldwasser believes, it might have been accepted in 1987. But others disagree, pointing out that the postwar era of carte blanche physics had ended—and that with its multibillion-dollar price tag, the SSC had crossed an invisible line beyond which sole-sourcing its management contract was politically impossible. According to this viewpoint, the SSC needed to run the customary Washington gauntlet and be subject to a far more elaborate process that included making a competitive response to an official DOE request for proposals, partnering with industrial firms, and following accounting and oversight procedures typically used by military-industrial contractors.[107]

While the unsolicited proposal sat inactive, the SSC project lost the last of its original influential Washington supporters. Having secured Reagan's endorsement for high-energy physics' premier scientific instrument, Trivelpiece left on a high note on April 1, 1987, to become the executive director of the American Association for the Advancement of Science (AAAS). There was speculation that he might eventually be a candidate for the SSC director and that he left DOE to avoid a future conflict of interest.[108]

But because of confirmation delays lasting over a year, it was not until August 1988 that physicist Robert O. Hunter took over as the new DOE director of energy research. The founder and president of Western Research Corporation, a defense-contracting firm in San Diego, he had previously worked under Trivelpiece at Maxwell Laboratories and served as a member of the White House Science Council advising Reagan. In the interim, James Decker, a former Bell Laboratories physicist then serving as deputy director of the Office of Energy Research, filled in for Trivelpiece as acting director. By the time Hunter was confirmed, it had become clear that DOE viewed the SSC as much too large a project for physicists to manage by themselves.[109] Thus the project that Trivelpiece had skillfully shepherded through the Reagan administration chain of command was left without a politically well-connected physicist to champion it in Washington.

SUPERCONDUCTING MAGNET TRIALS AND TRIBULATIONS

The quality of the Central Design Group's continuing work on the SSC design impressed many leading high-energy physicists. SLAC's Wolfgang Panofsky gave CDG staff "the major credit for having managed the transition from the multiple design initiatives . . . generated independently at several US laboratories to a single integrated design concept." Serving as a consultant to the annual DOE review of CDG in January 1987, he had judged its work as an "absolutely first-rate job in the face of unusual, and at times difficult, circumstance," and he recommended that "the CDG team form the core management team of that contracting arrangement."[110]

But the superconducting magnets under development for the SSC were not functioning as well as hoped. While the cable being wound into the magnets (made of niobium-titanium filaments embedded in a copper matrix) seemed to be working well, and the dipole magnet field quality and cooling systems were satisfactory, vexing problems had arisen in scaling the short test-magnet models up to full size, resulting in troublesome electrical shorts and sudden current losses (quenches) before the magnets had attained their full design field of 6.6 T. These quenches were attributed to tiny movements of the coil windings: they shifted slightly when cooled down to the low temperatures needed because the different materials involved have different thermal expansion coefficients. Another likely explanation was that the coils would shift slightly under the strong magnetic forces being exerted upon them by the intense fields being generated. These powerful magnets could literally tear themselves apart if not constrained from so doing. For whatever reason, such coil motion generates heat that can warm the superconducting wire beyond the 4.35 K critical temperature in small, localized regions, especially near the magnet ends, so that it is no longer superconducting there; resistive electrical heating then causes the coil to go normal, quenching the magnet. This was the same kind of problem that had confounded Brookhaven and Fermilab magnet developers in the late 1970s.

Designing appropriate metal "collars" to constrain the coil movement was difficult and frustrating.[111] The challenge was to clamp the coils in place firmly as the magnet was energized. Eight years earlier, Fermilab's superconducting magnets had experienced a similar problem, but it was proving far more difficult to solve in the case of the SSC dipole magnets because of their greater length and stronger magnetic field. And it did not help

FIGURE 2.4 Prototype of SSC superconducting dipole magnet, revealing details of the cold mass at its core. Courtesy of Brookhaven National Laboratory.

that CDG's design efforts and the measurements of the prototype magnets had to occur "in a fish bowl," said Tigner, referring to the close DOE oversight and attention from science reporters. Yet CDG and laboratory physicists gradually succeeded in developing appropriate constraints.[112]

Another issue that CDG had to deal with was developing superconducting wire to be wound into cable used in making the electromagnets. CDG viewed this problem as a research rather than an engineering task. "Since the time of Lawrence and Wilson, physicists have appreciated the frontier nature of the instrumentation they needed and approached its development accordingly, not expecting that the first design of anything would work as it ultimately can be made to," Tigner later explained. "When one is pushing the limits, one must expect the unexpected and not be upset by it, but learn from it and proceed in the light of that learning."[113]

Fostering a cohesive, productive collaboration was challenging, too, given the management structure in which the CDG had to operate. In an attempt to promote cooperation among participating laboratories, DOE officials had decided that LBL, Fermilab, and Brookhaven should work collaboratively on the SSC magnets (see above).[114] But as each lab was compensated for its contributions directly by DOE, CDG leaders did not have full control over the magnet R&D. Moreover, by separating the magnet testing (done at Fermilab) from development (at LBL or Brookhaven), insights from the testing could not be rapidly applied to subsequent magnet design and development.[115] This dichotomy could not be avoided in the SSC magnet R&D, partly because Brookhaven physicists and engineers had difficulty working with their Fermilab counterparts after the contentious Isabelle affair. And Fermilab, at the time fully committed to developing its Tevatron research program, was not in any position to assume additional responsibility.[116] It was becoming clear, however, that maximum cooperation would be needed throughout the magnet-development process.

To try to oversee the magnet R&D program, Tigner appointed a four-member magnet management group, consisting of himself, Goldwasser, Fermilab physicist Peter Limon, then head of the CDG accelerator systems group, and Victor Karpenko, a former Livermore engineer then leading CDG's magnet group. This "gang of four," as the team called itself, visited Brookhaven for two days and then Fermilab for two days every four weeks.[117] Such an unwieldy arrangement might have worked in a wartime context, where all groups share a single urgent defense goal, as was so in the Manhattan Project, but the SSC's magnet goals were not sufficiently unifying.[118] Accomplishing this oversight was "a tricky business," reflected Goldwasser. "We would just be there hovering over what they were doing, meeting at their meetings, and impressing them with the fact that their responsibility was to do what we asked. But that had to be done tactfully enough so that very bright and talented people would still feel that they had sufficient autonomy that their own ideas could be tested and given a fair hearing." This oversight effort became "even trickier because there was a different person there every time—sometimes Tigner, sometimes Limon, sometimes Karpenko, and sometimes Goldwasser."[119] Limon's outspoken manner of expressing his criticisms produced so much resentment at Brookhaven that Tigner decided to relieve him of his responsibility, leaving the remaining three to make the inspection visits. It was challenging enough for physicists working at different laboratories to work together in their technical efforts, but when physicists, engineers, and outside contractors had to collaborate on a common goal, the interaction problems multiplied.

When Karpenko retired as CDG magnet leader in the fall of 1987 after suffering a heart attack, Goldwasser agreed to step in for six months as temporary head. But he felt that John Peoples, then head of Fermilab's Accelerator Division, would be better at this job, for he had been involved in the lab's superconducting magnet development. Aided by several magnet experts, Goldwasser persuaded Peoples to take the position for one year, starting in October 1987.[120] Peoples later reflected on the awkwardness of the experience, saying that upon arrival at CDG, "I was a little bit like the plumber who'd been called in to fix the leaks and the toilets that are overflowing during a dinner party, but I wasn't exactly invited to dinner."[121] Within a month of starting this new assignment, a major accident at Fermilab worked in his favor. A superconducting dipole magnet being tested there failed spectacularly due to a short circuit, as a segment of the coil literally evaporated. The subsequent investigation of this catastrophic failure, which could have destroyed the entire magnet, resulted in a deeper understanding of the underlying physics.[122]

Peoples established an R&D program focused on understanding the causes of the movement of the dipole magnet coils. "We didn't know whether it was linear expansion of the magnets or radial expansion that had to be restrained," confessed Goldwasser. Peoples drew on the experience of Fermilab's earlier development of superconducting magnets during the late 1970s. To isolate the cause of the problem, physicist Alvin Tollestrup had had hundreds of prototype Tevatron magnets built, each differing from the previous one by a single design parameter.[123] Following that approach, Peoples instigated a one-year program of building and testing many SSC prototype magnets, each having slightly different parameters, "with every magnet identified, along with what the changes in design would be, and what tests would be made, and what each would prove," explained Goldwasser.[124] Peoples recalled that this was a "methodical approach" that involved checking each magnet and working closely with each laboratory to develop the best possible magnets.[125]

During the first half of 1988, efforts to scale the model magnets up to the full length of 17 meters led to too many short circuits and quenches to be considered successful. "Despite the organizational challenges and financial obstacles that plague us, we are making good progress," Tigner insisted, "thanks to the skill and devotion of our colleagues across the country."[126] Under the focused leadership of Peoples, the magnet R&D program had begun to bear fruit. Aluminum and stainless steel collars clamped around the windings—keyed and locked to the surrounding iron yoke—helped to constrain and limit coil movements. By August 1988, two full-length prototype dipoles had reached or exceeded the 6.6 T design objective after only one or two quenches.[127] In a letter to *Physics Today* that month, Tigner reported that "a full-length SSC dipole has now operated at 7.6 tesla, 1 tesla above the normal SSC operating point."[128] While such encouraging performance was by no means sufficient to consider development of the SSC dipole magnets complete, these tests suggested that by continuing the magnet R&D program along the lines being followed, CDG would eventually succeed. "Because of the many improvements to all coil and collaring fixtures, we expect that dipoles made starting this fall will either meet or exceed the performance of [the two successful magnets]," Peoples reported in September 1988, just before returning to Fermilab. "We certainly expect them to reach the design field without excessive training."[129]

In working to solve the SSC magnet problems, Peoples also noticed what he called a "philosophical" problem in the CDG program, a misalignment between the management practices of research physicists and those of weapons-lab engineers such as Karpenko.[130] He recognized what Tigner understood but had little power to fix: that physicists could usually work with members of an industrial culture but often clashed with members of a culture "in which military officers call the shots" (as Tigner later reflected).[131] Peoples realized that Karpenko was "used to one kind of management [style] and really not used to trying to cajole people, seduce them into doing things." In the case of physics research, however, "that of course is how you have to do it," he explained.[132] As Tigner elaborated, there was a mismatch in the work of designing and building the SSC between the disciplines of engineering and physics. This mismatch occurred between professional engineers who understand their practice as having "all relevant parameters well within the range of current practice," and scientists, who aim to "achieve some novel purpose or . . . advance in performance or specific cost. It is the latter that 'frontiers' are all about."[133] In the limited context of the CDG magnet program, such mismatches could be recognized and corrected, as occurred when Peoples assumed leadership. In the more complex clash of the scientific and engineering cultures that erupted later at the SSC Laboratory, such problems would eventually prove insurmountable (see chapter 4).[134]

POLITICAL CLOUDS GATHER

The lagging superconducting-magnet program began to exert increasing political pressure on the CDG and DOE. Many recalled the cancelled Isabelle program, which stumbled largely because of lingering superconducting magnet problems; anxious DOE officials were concerned that something similar might be about to occur in the case of the Super Collider. And Huson and McIntyre of the Texas Accelerator Center had not given up on their superferric magnet design. They were calling reporters, influential contacts in Congress, and Hess at the DOE, claiming that theirs was still the better option and that the problems of the CDG high-field *cos θ* approach could not be resolved in a sufficiently short period of time.[135]

CDG did not demonstrate a successful long prototype superconducting dipole magnet until the summer of 1988, and it had only two by that September, when Peoples returned to Fermilab.[136] The 1986 SSC conceptual design report had, however, projected that CDG would freeze the superconducting dipole magnet designs in June 1987. But a year later they were still making small changes to these magnets, trying to obtain reliable prototypes that would operate at the design field of 6.6 T. Unfortunately, however, the funding was not available to do that. It would have taken additional millions to fabricate many full-length prototypes, which could not be accommodated within CDG's $25 million annual budget.[137] And even this figure was not entirely secure, for DOE rescissions to cover congressional earmarks and Gramm-Rudman-Hollings sequestrations could further reduce its budget. Thus the superconducting magnet R&D program was seriously behind schedule, which became a growing source of concern—and conflict—between DOE and CDG.

In Washington, meanwhile, there had been important changes in the DOE program office. In October 1987 William Wallenmeyer, who had directed the High Energy Physics Office since the 1960s, left to become president of the Southeastern Universities Research Association, a Washington-based consortium (like URA) that was organized to represent institutions building a new electron accelerator laboratory in Virginia. Hess assumed his responsibilities while a successor was sought.[138] He also named Robert Diebold, who had come from Argonne National Lab in 1985 to work at DOE, as director of a new Office of the SSC, initially reporting to him.[139]

In a major oversight, DOE had somehow neglected to include funding for the SSC Site Selection Task Force led by Hess and Temple, which was visiting the seven best-qualified sites in the spring of 1988 and making detailed analyses of their merits and demerits as potential sites for the Super Collider (see chapter 3). The DOE was spending millions of dollars for this purpose that had not been allocated in the 1988 budget. The total cost of this effort came to about $8 million by that time.[140] But Hess and other DOE officials were reluctant to return to Congress and either request the money or ask for it to be reprogrammed from other DOE reserve funds that might exist, according to CDG Deputy Director for Operations Chris Quigg, a Fermilab theorist who had succeeded Jackson in 1986.[141]

Instead, Hess hoped to extract $8 million from the CDG budget and use it to cover the task-force expenses. But by the spring of 1988, CDG was already halfway through the 1988 fiscal year, and most of its budget had either been spent or was committed for staff salaries, magnet R&D at the DOE laboratories, or purchasing superconducting wire. If DOE wanted to extract $8 million from the CDG budget at that point, it meant that Tigner would have to lay off most of his staff. He was reluctant to do that because so much work remained to be done, especially on magnet development. And he could not just lay off staff physicists and ask them to come back at the beginning of the next fiscal year in October. As Tigner stalled, the already strained relationship between CDG and the DOE program office deteriorated.

The matter came to a head one particularly "black day," Friday, May 13, 1988, when Hess phoned Tigner and bluntly *ordered* him to prepare a list of people to be laid off. Tigner reluctantly began to comply with this directive, but a last-minute reprieve came through just after lunch.[142] Apparently, the appropriations committee chairmen had finally relented and agreed to let DOE use other funds to cover the site selection task-force expenses. While the immediate crisis had been averted, however, the tensions between CDG physicists and DOE officials remained. This troublesome incident aggravated the worsening relationship between Tigner and Hess, who began to consider Tigner a person whom DOE officials could not trust to follow their orders.[143]

THE DOE REQUEST FOR PROPOSALS TO MANAGE AND OPERATE THE SSC

Meanwhile, the URA's unsolicited proposal remained in limbo at DOE. In an attempt to resuscitate it and stir action on it, URA submitted a slightly revised version in February 1988.[144] But Decker, to whom the proposal was addressed, was serving only as acting director of the Office of Energy Research and thus lacked the clout of a confirmed political appointee like Trivelpiece or (later that year) Hunter.[145] He was unable to push the proposal forward to higher-level DOE officials for their approval as required. On June 1, 1988, SSC BOO Chair Boyce McDaniel wrote Salgado urging DOE to "take immediate action to set up an SSC organization where management responsibility is centralized in some entity drawn from the scientific community." Salgado replied on July 22 that DOE was "now developing a plan of action for establishing a permanent management structure."[146]

In early August 1988 there seemed reason for hope when DOE issued its official request for proposals (RFP) to manage and operate the SSC,[147] with formal responses due by November 4. With only three months—an incredibly short time—to prepare such a major, detailed proposal, the DOE seemed to be targeting its request directly at URA, which could react quickly and represent the interests of the physics community. But the text of the RFP also conveyed the clear message that DOE would accept only responses written in the form of a "standard industrial buy-a-widget sort of contract"—which was alien to academic scientists.[148] The RFP called for a strong construction-management infrastructure, plus a list of the names of the designated SSC director and top-level managers.[149] It marked the formal rejection of the unsolicited URA proposal, and with it ended an era of largely unquestioned federal support that high-energy physicists had enjoyed since the early 1950s.

URA President Knapp, having learned several weeks earlier about the DOE decision to seek competitive bids for the SSC management and operating (M&O) contract, had already discussed how to respond with advisors, including Panofsky. Then an outspoken member of the SSC BOO, Panofsky wrote that "the responsibility for preparing the proposals should be that of the President of URA."[150] Knapp thus decided to respond to the RFP with a Defense Department (DOD)-style proposal that he and others believed DOE expected, written in the style typical of the multibillion-dollar proposals submitted by industrial contractors working on military projects.[151]

Rumors circulated that the huge industrial firm Martin Marietta planned to prepare an M&O proposal in collaboration with MIT professor and Nobel laureate Samuel Ting.[152] While that firm, which also approached URA, never did submit a proposal to DOE, Knapp appears to have been heavily influenced by the possibility. He sought the aid of someone with experience in writing a DOD-style proposal, consulting Hugh Loweth, a longtime OMB staff member,[153] who recommended N. Douglas Pewitt, a physicist who had earlier served as an OMB analyst, deputy director of the DOE Office of Energy Research, and as Keyworth's deputy in the OSTP (see chapter 1). In 1988 Pewitt was working in San Diego for the Washington-based firm Science Applications International Corporation, which provided scientific, technical, and managerial support to government contractors—and did defense contracting itself.

Pewitt understood the political, managerial, and budgetary requirements for a successful proposal of the kind that Knapp and the URA BOO believed the DOE wanted, one that would satisfy both the DOE and Congress, with a fully detailed project management plan. Like Knapp, Pewitt believed that URA, as a physicist-dominated organization, was vulnerable to criticism that scientists were unqualified to manage multibillion-dollar projects;

FIGURE 2.5 OSTP Deputy Director N. Douglas Pewitt, speaking at a physics meeting in October 1982. Courtesy of AIP Emilio Segrè Visual Archives, *Physics Today* Collection.

they would therefore require industrial partners. Thus he viewed the earlier unsolicited URA proposals as inappropriate, especially given his opinion that high-energy physicists refused to recognize that they were spending public funds.[154] So Pewitt cast the new proposal in a style that industrial firms and Washington consultants often used in proposals to build large, multibillion-dollar military systems. It came to look, as Knapp reflected, more "like [a proposal for] building an aircraft carrier than like building a particle accelerator."[155]

Most of the CDG physicists did not like the idea of teaming with industrial partners, which Pewitt included in the proposal.[156] While from the earliest days at CDG it had been assumed that industrial involvement would be needed in building the SSC, the nature of the relationship with industry—in particular how much control companies would have on the project—remained unclear. "There is a great deal of misunderstanding concerning the term 'industrial involvement' in the SSC," wrote Panofsky as far back as January 1987. This misunderstanding encompassed exactly how "the talents of industry can be brought to bear on the solution of problems," and affected as well how the work of the laboratory would transfer the technologies developed to industry.[157]

To deal with the issue of industrial partners, Pewitt drafted a memorandum in late August or early September 1988 intended to set forth URA's position on the matter. "There has developed an understanding on the part of DOE and many others that construction of the SSC, due to its much larger size, requires capabilities not possessed by URA nor, in fact, typically present in a DOE-HEP laboratory or academic community," it stated. Admitting that "the fairness of the perception can be disputed," Pewitt argued that teaming with industrial firms was a key element to include in the proposal by URA to the DOE, essentially to give DOE confidence that the "management system can perform acceptably on a major multibillion-dollar, highly technical project."[158] In this manner, the controversial practice of teaming—in which industrial firms were to be treated as partners on an equal, or almost equal, footing with URA—entered URA's proposal.

URA quickly selected two firms, EG&G Inc., a provider of administrative services, electronics, and scientific instrumentation for DOE and DOD facilities, such as the Idaho Engineering Laboratory and the Nevada Test Site, and Sverdrup Corp., a construction management firm with experience doing major contracting for the armed forces.[159] In particular, Sverdrup had managed construction of the Space Shuttle Complex at Vandenberg Air Force Base and the Fort McHenry Tunnel in Baltimore Harbor. Brief letter contracts summarizing the teaming relationships with the two firms were drafted and signed; they were intended to be expanded into longer and more formal agreements if DOE accepted the proposal. Terms of these relationships—including licensing procedures, press releases, rules against raiding personnel, how to handle proprietary information—were sketched out.[160]

Adapting the style in which DOD proposals were typically written, Pewitt arranged that three different groups would participate in the process. A "red team" would write the first draft, and a "pink team" would then revise it. Finally a "gold team" would critique what had been written, identify what might be missing, find errors, and generally fix whatever the other team had not done to best advantage. The challenge was to prepare a proposal acceptable not only to the DOE but also to the other communities involved by then with the SSC. Pewitt took direct responsibility for Volume I, the technical proposal, while others in the writing group handled Volume II, the

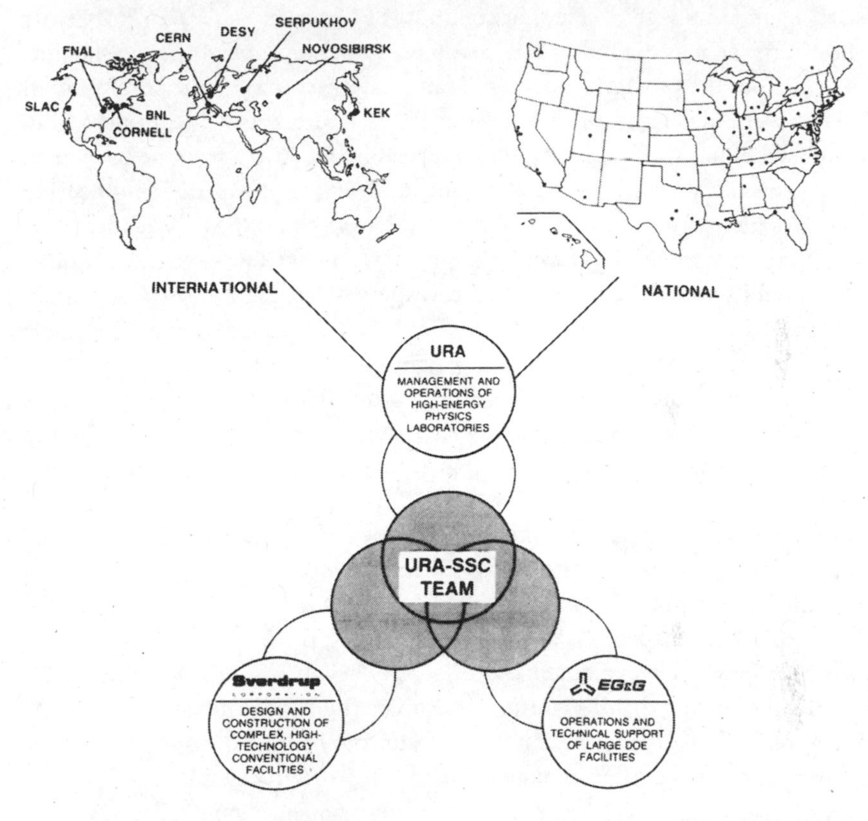

FIGURE 2.6 Proposed teaming relationship among URA, EG&G Inc., and Sverdrup Corp. Courtesy of Fermilab Archives, SSC Collection.

business-management proposal, and Volume III, on SSC costs and budgetary coordination. A Pewitt memorandum to Knapp dated August 24, 1988, laid out the procedures to be followed and specified a strict schedule for responding to the RFP. Once all corrections had been made, the document, in the form of three detailed volumes, would be printed on October 26.[161]

The work of writing the proposal required negotiating among conflicting approaches regarding management and operation of a research project. Several "significant policy-sensitive matters" were mentioned in Pewitt's August 24 memo to Knapp. One concerned naming SSC managers in the proposal, as required by the RFP. As the memo indicated (somewhat cryptically), the question of the experience and qualifications of the "heads of the major organization units," including the lab director, "represents a long standing difference of view on the part of Ch/Ops [DOE Chicago Operations

Office] and the DOE accelerator community." Another sensitive policy issue Pewitt raised was that "the RFP clearly presumes a project organization different from the CDG Management Plan model [which the proposal writing teams were instructed to use]."[162] Finally, he wrote, the red team needed to know whether or not URA would be forced by DOE to "bring on as a 'team member' a systems integrator," a concept familiar in systems engineering but as yet foreign to the high-energy physicists, who opposed the idea.[163]

In contrast to the open, academic environment at CDG in which URA had prepared its unsolicited proposal, this response to the DOE RFP was written behind closed doors. And to avoid what Panofsky considered a potentially disqualifying conflict in which URA might appear to be taking unfair advantage of its intimate relationship with the group, URA chose to exclude CDG from the writing process.[164] URA felt it had to maintain an arm's-length relationship with CDG, which also had to make itself available to any other potential proposers on an equal basis. While it is difficult to assess whether this was really necessary, the exclusion of the group clearly resulted in great mistrust and resentment by the physicists involved in it.

Among the discussions that were thus off-limits to CDG was the all-important one involving selection of the SSC director. An August 1 Panofsky memorandum explained the uncertain legal grounds by which URA justified selecting the director in secret once the RFP had been issued.[165] As he noted, "formally identifying the Director publicly before submittal of the proposal might be interpreted as violating the privileged nature of proposals." Expanding the rationale in a subsequent paper, Panofsky wrote:

> Identifying a Director is tricky business. We must assume that competitive proponents will also attempt to identify a candidate, or may already have a candidate. For this reason and in general because of the competitive nature of the procurement, there cannot be a public search. If there were such a public search, then this would automatically establish a connection among the different proponents and again this would be illegal.[166]

Accepting this logic, the discussions over the next month about selection of the SSC director were private and took the form of communications exchanged via a "buddy system."[167] Among the many candidates under discussion were Lederman, Tigner, Richter, Samios, Peoples, former NASA Director James Beggs, Sam Ting, Steven Weinberg, Charles Baltay of Yale, Wojcicki, DESY Director Bjorn Wiik, Trivelpiece, Trilling, Hermann Grunder of LBL, and Panofsky's top choice by now, physicist Roy Schwitters of Harvard University.[168] According to the URA proposal eventually submitted, planning of the choice of director had begun back in May 1988 with a formulation by the SSC BOO of the attributes the SSC director should have[169] (but little had been done about it since then).

As stated in the minutes of its August 19–20 meeting, the SSC BOO decided to "establish itself as the Search Committee for the SSC Director." Schwitters and Trilling were excused from discussion of the issue, as they were possible candidates, and Jerome Friedman of MIT, Frank Sciulli of Columbia University, and Martin Walt of Lockheed were added.[170] Each committee member agreed to interview at least eight individuals by phone, and the information would then be privately discussed by the BOO. The suggestions made by the committee members varied widely. For instance, Fred Gilman of SLAC, then chair of the American Physical Society's Division of Particles and Fields, suggested selecting a "tycoon" from the industrial/administrative ranks who could serve as "a symbol of unity among small and Big Science," and giving Tigner free rein as SSC technical director.[171] The three leading candidates for director were Lederman, Samios, and Schwitters.[172]

At a meeting held at the O'Hare Hilton in Chicago (its only in-person meeting) on Sunday, August 28, hardly a week after its formation, the URA search committee selected Schwitters as the designated SSC director.[173] Having served as a cospokesman of the large CDF Experiment at Fermilab, which involved hundreds of physicists, Schwitters seemed to bring considerable management experience to the position. Urbane and polished, he had risen quickly through the ranks of particle physics after playing an important role in the 1974 discovery of the psi particle at SLAC, which led to a Nobel Prize for his group leader Burton Richter.[174] A recipient of the NSF's prestigious Alan T. Waterman Award, Schwitters had joined Harvard as a full professor of physics in 1979 before becoming CDF cospokesperson in 1983.

For other leadership positions, the URA proposal listed Maury Tigner as deputy director; Helen Edwards as head of the Accelerator Division, a position she occupied at Fermilab; Bruce Chrisman, then associate director for administration at Fermilab, as director of laboratory administrative services; and Robert Robbins of Sverdrup as the head of the Conventional Construction Division. "Strangely, no current member of the CDG was contacted for suggestions or advice during the process," Wojcicki recalled.[175] This exclusion may have been a consequence of URA policy to avoid any perception of insider trading, by keeping CDG at arm's length during the proposal-writing process. As might be expected, however, it stimulated resentment on the part of CDG physicists and other staffers, who had expected that Tigner would be named SSC director. Having worked closely with him on designing the collider, they recognized his capacity for leadership in accelerator design and admired the intensity with which he pursued

FIGURE 2.7 URA's designated SSC Laboratory director, Harvard University physicist Roy F. Schwitters, standing before the CDF detector at the Fermilab Tevatron. Courtesy of Fermilab.

goals. Indeed, everyone involved agreed that he was excellent technically and managerially.

But some leading members of the high-energy physics community worried that Tigner might project the wrong image in Washington, for he could on occasion come across as guarded or impatient, especially with nonphysicists.[176] Some thought that the image of a "cowboy in the Wilson tradition" might not encourage newer members of Congress to support the SSC.[177] According to SSC BOO Member Martin Blume of Brookhaven, Tigner had been contacted and interviewed by search committee members, and he had been ranked highly on his management abilities, technical skill, and judgment, as well as on his ability to attract excellent staff. But whether he possessed the "ability to promote and defend the SSC in all the appropriate forums, and to work effectively with the DOE"—one of the attributes the search committee deemed critical for the director[178]—was questioned. There were also pervasive rumors that important DOE officials were opposed to Tigner being named the SSC director.[179] As Blume recalled, the SSC needed "someone who could present an external face in addition to [being] the tough man."[180] In short, the search committee considered Tigner the perfect builder of the SSC, but hoped to find another individual such as Lederman, Peoples, Samios, or Schwitters who could work with Tigner and represent the SSC to the outside world.

It was now the delicate task of those involved to inform Tigner of the URA decision. McDaniel and Panofsky agreed that the latter should be the one to do so, and he told Tigner on Sunday, September 4.[181] When URA leaders visited LBL for a mid-September 1988 meeting, they faced the uncomfortable job of explaining to the CDG staff what was already known by then, that URA had designated Schwitters as the SSC director. The staff responded negatively to this choice, for they considered him an accomplished physicist who had been involved only peripherally in the five-year SSC design effort, as a member of the SSC BOO. And it seemed a huge leap from being a co-spokesperson for Fermilab's CDF Experiment to becoming the director of the SSC Laboratory. But as Knapp reflected, there were only a few top-notch candidates then in the younger generation of physicists—and none had the large-project management skills that Washington considered necessary to lead such an enormous project.[182]

URA Board of Trustees Chairman John Marburger accompanied Knapp and BOO Chairman McDaniel on this Berkeley visit. They had hoped to mend fences, but it was too late. The meeting was awkward for all. As Quigg recalled, Knapp and McDaniel shifted back and forth uncomfortably at the podium.[183] CDG physicists were embittered by their exclusion from the process of making this crucial selection, particularly because many others in the high-energy physics community *had* been consulted. When some of them later spoke with Schwitters, "they were concerned with my viability and that their input simply didn't get into this process," he recalled. "The way they put it to me was that they could have accepted any decision as long as they had had the chance to get their input in."[184]

With the director selection settled, Pewitt took the first steps toward writing the URA management and operations proposal in Berkeley, and the red team began work in a planning meeting there on September 14, 1988.[185] Pewitt recalled receiving little cooperation from CDG despite his overtures,[186] so he soon relocated the writing effort to rented office space in an industrial park in the Chicago suburb of St. Charles, near Fermilab, where the assembled group of about twenty individuals met periodically over the next month and a half.[187] CDG physicists were asked to visit St. Charles to provide information, but they were otherwise excluded from the proposal-writing process.[188] Blume, as chairman of the SSC BOO Proposal Oversight Committee, monitored the hurried affair; the URA proposal eventually became a six-inch-thick document in three volumes. Encompassing a myriad of technical and practical details, including the many aspects of staffing, daily operations, interface with the other national labs, construction, magnet industrialization, corporate partnering, wages, salaries, and fringe benefits, costing procedures, and property management, this proposal was very different from the simple 14-page unsolicited URA proposal of 1987. But, as Knapp reflected years later, "Bureaucratically it was gorgeous."[189] URA submitted it to DOE on November 4, 1988—just before the US national elections.

In requiring Tigner to report to Schwitters, URA officials had expected the two to work together and let them negotiate their respective roles. But it became clear over the next few months that they could not find a mutually acceptable way to do so, for they had too many differences. For example, Tigner strongly opposed the degree of industrial teaming described in the proposal, while Schwitters viewed it as "an opportunity" for physicists to focus on physics rather than all the mundane details of building the machine.[190] Tigner requested, but was not granted, full control over any decisions regarding accelerator matters and the selection of people who would report to him as project manager and deputy director.[191] He attempted to cooperate with Schwitters in an "uneasy alliance" for a brief period, but, according to Wojcicki, "the discussions with Roy did not go at all well."[192] Schwitters recalled many visits and discussions with Tigner between September 1988 and the following January:

> We discussed project manager. We discussed deputy. We discussed all kinds of things. I thought Maury was very difficult to talk to. Clearly, he appeared to be very disappointed after not being chosen [director]. The Board was urging me to find a role [for him] and [we] were having lots of discussions. . . . And it never really converged on something that Maury felt gave him what he wanted. . . . I had the uncomfortable feeling though that if we couldn't have a heart-to-heart conversation that it could cause some uncomfortable feeling down the line.[193]

Schwitters later described his differences with Tigner as a "classic conflict of styles." Tigner's was the pioneering, maverick style of Wilson, the

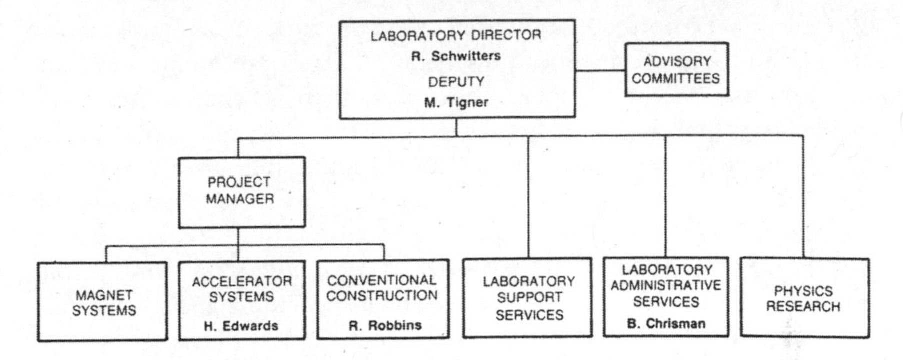

FIGURE 2.8 Laboratory management team, as set forth in the URA proposal for management and operation of the SSC Laboratory, November 1988. Courtesy of Fermilab Archives, SSC Collection.

independent, self-taught tinkerer working on a shoestring. Schwitters was more aligned with what he and URA construed as the new bureaucratic Washington regime.[194] But as Robert Diebold, who headed the DOE Office of the SSC in 1988, recalled, "Tigner tried to cooperate with us the best he could. He made reasonable arguments. We would try to listen to him. . . . He did a good job. Roy came in, and he wanted to put his own stamp on it. He didn't want the CDG folks," certainly not in top leadership roles.[195]

In October Tigner agreed to let his name be included on the URA proposal as SSC deputy director, but even then he wavered. The final proposal included an organization chart showing Tigner in this position, second in command, reporting to Schwitters.[196] The chart also included a box for the position of project manager, but it was empty, although Tigner seemed the obvious choice for the position.[197]

As DOE received no other proposal, it accepted URA's and began to draft a formal contract with the consortium to construct, manage, and operate the SSC Laboratory. Knapp recalled these negotiations, with a DOE team led by attorney Ed Cumesty of its Chicago Operations Office, as "one of the more painful experiences I have ever had," for the DOE officials "wanted everything . . . control over personnel policies, control over raise policies, control over vacation policies, control over rules and regulations." It was clear that the DOE did not trust URA physicists to manage such a huge, multibillion-dollar enterprise. But URA could (and did) take advantage of Reagan's impending January 1989 exit from office—and of Herrington's wish to announce the formal launch of the SSC project before that—to stall and obtain important concessions (see chapter 4).[198]

On November 10, 1988, just two days after George H. W. Bush was elected to succeed Reagan as president, the national site competition ended with the news that Waxahachie, Texas, was the winner (see chapter 3). The announcement came in a DOE press conference at the Forrestal Building with the theme "Supercollider: Opening a New Frontier into Inner Space." But the Washington political celebrations had little impact in Berkeley. Few members of the Central Design Group were asked to go to Texas, and even fewer elected to sign up for the project.[199] It was a crucial break in the continuity of the SSC project, with the unfortunate result that many important CDG design and planning efforts were not transferred to Texas.[200]

As Mike Davis argued in his history of urban Los Angeles, *City of Quartz*, there comes a point in the history of frontier ventures when the pioneers must confront the reality of change.[201] For the SSC's frontier outpost, the Central Design Group, that stark realization finally hit home late in the summer of 1988, when CDG physicists were largely excluded from the URA proposal-writing process and Tigner was passed over as a possible SSC director. Until that moment, they had continued to believe that CDG controlled the direction of and the work on the project, despite the growing indications that DOE officials wanted to be in charge. Subtle signals suggested the widening chasm between Berkeley and Washington, but CDG staff did not grasp their full significance until September 1988. Thus they were caught by surprise when DOE began requiring a new style of laboratory management that involved partnering with industrial firms in the construction and administration of the SSC Laboratory.

The context for doing large-scale physics research had been changing slowly and subtly during the Cold War.[202] While the old AEC, working under the aegis of the Congressional Joint Committee on Atomic Energy, looked upon physicists as the heroes who had helped win World War II and had then contributed to maintaining America's nuclear supremacy, the new DOE could hardly remember their heroic contributions. The idea that basic research in physics could serve national defense had lost the political traction it enjoyed in the postwar years after Vannevar Bush wrote *Science: The Endless Frontier.* To many in Washington, high-energy physicists had become "just another group of constituents" who were reluctant to comply with the requirements expected of those who received major federal funding.[203] And to Norman Ramsey, a seasoned veteran and URA president from the AEC period, it seemed that the new generation of physicists did not work hard enough to make friends in Washington.[204]

The SSC project was also far larger than any previous project physicists had attempted, and its estimated costs were growing steadily. From the $2 billion figure bandied about at Snowmass in 1982, its price tag had grown to $3.0 billion when the SSC Conceptual Design Report was released in 1986 and to $4.5 billion in January 1987 when President Reagan decided to "throw deep."[205] These increases occurred against a backdrop of increasing US fiscal austerity. As the anticipated benefits of Reagan's earlier supply-side economics failed to materialize while the costs of his defense buildup mounted, annual budget deficits exceeded $200 billion in the mid-1980s. His cabinet was deeply divided about whether or not to proceed with the SSC project, with OMB Director Miller an influential voice of opposition. It took the executive decision of the world's most powerful leader to resolve this impasse in favor of building the Super Collider, which accorded well with his administration's promotion of US competitiveness and his personal love of large high-tech projects. But SSC advocates did not fully comprehend how close a decision this had been.

Such multibillion-dollar funding almost always comes with strings attached, however. In hindsight, CDG physicists were politically naïve to expect that the DOE could sole-source a contract entailing some $5 billion to a not-for-profit corporation representing only a few thousand physicists. What might have been acceptable below the billion-dollar level was no longer politically feasible.[206] This was a *huge* amount of public money that would have to be closely accounted for—to both the DOE and Congress. In addition to their bureaucratic expertise, the political clout of industrial firms representing tens of thousands of workers was going to be required to help the SSC project navigate the annual congressional gauntlet. Such industry involvement *might* however have been included as an explicit part of the unsolicited URA proposal if commitments could have been made when it was submitted, for example, to firms interested in manufacturing the SSC magnets.

But the superconducting dipole magnets, then estimated to represent about a quarter of the SSC total project cost, were dangerously behind schedule by early 1988, and no such commitments could in fact be made. Remembering the Isabelle fiasco, DOE officials were understandably anxious. As Pewitt argued, the problems in that instance were not just technical; they involved the failure of the Brookhaven management to deal aggressively with their superconducting magnet schedule slippages.[207] From this perspective, one can more easily understand why DOE officials might want to bring industrial managers into the SSC leadership mix.

CDG's management of the SSC magnet R&D was, however, hindered by the funding structure imposed upon it by DOE. Tigner did not have sufficient control over the money allocated for this purpose, which was funneled directly to the three labs involved and thus difficult to hold up or redirect. And what funding he did have was not enough to support a truly vigorous magnet R&D program. CDG magnet managers had to cajole the labs to do as requested or promised—which Peoples understood but Karpenko did not. In any case, the DOE kept control of the purse strings, to the ultimate disadvantage of SSC magnet R&D. Had these difficulties been recognized at the time, they might have served as warnings that attempting to manage the SSC project from Washington was not such a good idea, after all.

All these events played out in the mid-1980s against the backdrop of the emerging nuclear-waste problems at the DOE weapons labs and production sites, a Cold War legacy that would eventually cost over a hundred billion dollars to clean up. Although leaders of high-energy physics labs could (and did) argue that they were different and did not suffer from these problems, their protestations fell largely on deaf ears in Washington, where all of the DOE labs were often lumped together as one. So the increasingly strident congressional demands for closer, stricter DOE management of its far-flung facilities also applied to basic-science labs such as Brookhaven, Fermilab, and SLAC. And they applied in particular to a new DOE high-energy physics lab in Texas about to cost taxpayers over $5 billion.

The early 1980s had witnessed hopes of uniting the US high-energy physics community behind the SSC as a way to restore its competitive edge in the field. But by the late 1980s, rifts were forming within the community as it tried to respond to political realities. What had earlier worked as a partnership between physicists and the AEC was gradually being replaced by a more hierarchical management structure reminiscent of military-industry relationships, with the DOE in the driver's seat and the physicists expected to comply with its directives. The SSC project would become more like building an aircraft carrier than a high-energy physics laboratory.

The style and commitments of Tigner, who epitomized Wilson's frugal, independent, self-taught, tinkering approach, contrasted with the more bureaucratic, professional approach of Schwitters in trying to respond to what he and others construed as the new DOE mandates. The clearest indication of the new style came in August 1988 in the terms of DOE's request for proposals calling for competitive bidding of the SSC's management and operations contract. It set forth a daunting variety of bureaucratic requirements to be met. Whereas CDG had hoped that the SSC's management philosophy would follow the more traditional Fermilab model, as expressed in URA's unsolicited proposal, DOE envisioned a partnership between high-energy physicists and military-industrial engineers that could somehow be managed from Washington.

Strongly promoted by Panofsky, the choice of Schwitters as SSC director can also be viewed as part of a general shift in "big physics" during the CDG years—from the frugal but riskier accelerator-building philosophy characteristic of Wilson and Cornell to the corporate, conservative (and therefore more expensive) approach that was eventually adopted at the SSC Laboratory (see chapter 4). Teaming with industrial partners was an important element in this approach, as well as an attempt to mollify Washington critics, who did not believe that "long-haired academics" could ever manage such an enormous project effectively. Many of the CDG physicists, including most crucially its leadership, disagreed with this approach. As CDG Deputy Director Wojcicki noted in his personal memoir of those years, this partnering with industry was "the beginning of a drift away from the historical HEP culture in laboratory management."[208]

After Tigner departed, the entire top leadership of the Central Design Group subsequently returned to home laboratories and universities in 1989 while the new SSC Laboratory began to take shape in Texas (see chapter 4). And many in its next leadership tier followed their example. Thus the SSC suffered a dramatic loss of continuity, especially in its institutional memory. As the project's frontier outpost lost its relevance in late 1988, the path-breaking high-energy physicists who had conceived the SSC—and put flesh on the buoyant Snowmass visions by developing a detailed conceptual design—realized that the golden ring of opportunity, which had earlier shone so brightly, was becoming tarnished by political compromise.

CHAPTER THREE

Selling the Super Collider, 1983–88

When a community of fifteen hundred scientists is proposing to ask the public for billions of dollars for a major research project . . . they'd better be telling the American people why this is a good investment.

—GEORGE KEYWORTH, July 1985

Although the SSC project enjoyed the support of the Reagan administration, including the president himself, it was still necessary to convince Congress and by extension the general public to approve building it. If its advocates wished to spend billions of taxpayer dollars on the huge project, they would have to convince their fellow US citizens why it was so needed. But most people—including other scientists, politicians, and molders of public policy—viewed the project in a much different light than did the physicists promoting it. Almost all lacked sufficient scientific understanding to grasp their excitement about its fundamental physics goals—for example, why discovering the Higgs boson might be important in learning how particle mass originates, or why this mass is intriguing in and of itself. Essentially all the public understanding of the Super Collider came through articles written by journalists in nonscientific terms. Often uncomfortable themselves with the SSC physics goals, they focused instead on aspects that their readers could hope to comprehend, such as the project's enormous size, extreme cost, or difficult management problems.

This tendency was not as true of other large scientific projects, such as the Hubble Space Telescope, the International Space Station, or the Human Genome Project, where the topics involved were less abstract and much more accessible to average readers than high-energy physics.[1] For nonphysicists, it was hard to grasp how a gargantuan colliding-beam machine that created fundamental but esoteric subatomic particles could have any impact on their livelihood other than to cost huge sums that might otherwise be spent in ways more closely aligned with more pressing human needs. Therefore, despite strenuous efforts of several well-known physicists such as Nobel laureates Sheldon Glashow, Leon Lederman, and Steven Weinberg, plus a few talented science writers who tried to relate the physics of the SSC to everyday issues, the general public was not widely drawn to the news of the SSC project.[2]

On the other hand, almost everyone could readily grasp that constructing an expensive new facility within their state would enhance local prosperity by creating thousands of high-tech jobs. Hence there was considerable interest in the nationwide contest to select a site for the SSC. But once the actual site had been chosen, it was going to be very difficult to maintain this interest in the states that had been passed over. And this approach ran the risk that the SSC might become fixed in the public mind as a costly public-works project rather than a scientific undertaking to push back the frontiers of human knowledge and reestablish US leadership in high-energy physics.

In another sense, the SSC project had to be "sold" to the foreign governments that might become involved, especially their science agencies that would eventually be asked to pay for portions of the detectors or even the collider itself. What were the likely benefits to another country of contributing to the project and supporting its scientists who were interested in SSC research? Of course, there were the obvious benefits of their participation in likely discoveries such as the Higgs boson, but that was not sufficient to justify sending tens or possibly hundreds of millions of dollars' worth of components overseas. For this was funding that could otherwise be spent on indigenous research that would almost always be of more immediate benefit to the individual nations involved. And the credit for any SSC discoveries would still redound largely to US researchers, as the laboratory was to be sited on US soil, and not to foreign scientists. For other nations to contribute significantly to the SSC project, there had to be a deeper, stronger rationale.

THE SUPER COLLIDER AND THE AMERICAN PRESS

Before the selection of the SSC's site, the Central Design Group coordinated public relations efforts under the leadership of physicist Stanley Wojcicki, the CDG deputy director for external relations.[3] Although he had little prior experience in public relations, Wojcicki communicated well in the important Washington circles, especially with congressional staffers. Supporting his efforts were technical writer and senior editor Rene Donaldson and her staff, who attempted to convey the SSC's scientific purpose and importance to the administration and congressional staffers, science reporters, and other scientists. Wojcicki had a budget (augmented by URA in areas where it might be questionable to spend federal funds) to support communication and public outreach. Donaldson and coworkers produced colorful, informative booklets and pamphlets that explained the project in understandable and accessible formats. As she later observed, "Selling the cost of the SSC was part of my job."[4]

Wojcicki brought these glossy, well-written publications with him on his frequent trips to Washington, distributing them strategically to government officials and bureaucrats who might be in positions to influence SSC decisions. CDG also had a professional clipping service that monitored press coverage of the SSC. Whenever any confusion, misperception, or injurious information appeared, Wojcicki or one of the other CDG leaders attempted to correct it. He also hired K. C. Cole and Kate Metropolis, both accomplished science writers, to do freelance writing projects related to the Super Collider. These outreach efforts resulted in a steady stream of articles about the project that appeared in science-focused magazines—for example, *Discover, Nature, Physics Today, Science,* and *Scientific American.* These publications proved particularly effective in bearing the SSC messages to scientists in other disciplines or physics subfields, and to congressional staffers and others working in science policy.[5]

But Wojcicki and his team had far less impact on SSC coverage in general-interest publications. For one thing, the SSC was not covered as often in the major newspapers and magazines as in science magazines.[6] And when the popular press did report on the SSC, the emphasis was typically not on its scientific goals, which physicists found so exciting, but instead on topics such as staff appointments, the enormous collider ring, or the high cost of building the machine.[7] Even in science magazines, articles frequently emphasized the machine's unprecedented size and cost.[8] Such coverage did not generate much excitement or support for the SSC.[9]

As the SSC meant different things to different groups, explaining its science in appealing terms in newspapers and magazines was a continuing challenge. The uneven state of science literacy in the United States made this effort more difficult. One rhetorical strategy employed to interest people in the Reagan administration or Congress was to describe the SSC as an exploration vessel for entering or advancing "bold" new frontiers.[10] Comparison with Columbus setting out across the Atlantic in search of a better route to India and discovering the New World was a favorite approach. Other strategies included emphasizing that the SSC would be built using state-of-the-art technologies, or that its technological spin-offs would enhance industry, culture, or society—just as had previous scientific advances, such as the discovery of electromagnetism or X-rays. Or advocates argued that it would bolster the US competitive edge in the international scientific arena by helping its physicists recapture their earlier lead over European colleagues in scientific discovery.[11] Transatlantic rivalry was a useful, recurring theme in rhetoric aimed at the White House, for which the huge project could serve to symbolize President Reagan's efforts to reassert America's role as the dominant global superpower.

The SSC's high cost was a primary concern to people in Washington, and for wider audiences as well; its price tag remained the theme reported most often in the news. But different audiences saw different implications. For many scientists working outside of high-energy physics, the Super Collider loomed as a potential drain on their tightening research budgets. "Scientists in other fields, such as biotechnology, are worried that their federal funding may be reduced to help pay for the SSC," reported the *Chicago Tribune* at the end of 1984, when the collider's projected cost hovered around $3 billion.[12]

Members of Congress viewed the SSC, at least at first, as a multibillion-dollar chance to "bring home the bacon" and bestow thousands of jobs upon constituents. But it gradually dawned on individual groups of these constituents that the chances of loss might be equal to or even greater than their chances of gain. For regions that had already lost another major facility or a site contest, it was especially obvious that only the winner of the SSC site-selection contest (see below) would reap the benefits of what was becoming known as the "Big Pour"—a massive, long-term construction project that would enrich one US regional economy with federal largesse.

And Brookhaven's disastrous experience with Isabelle remained a sensitive subject. Representing the Long Island district that had just lost that project, Congressman William Carney convened an October 1983 hearing to investigate the HEPAP recommendation to cancel Isabelle and pursue the SSC. For Carney, these hearings were not merely an exercise to reassure troubled constituents. By asking questions that ranged well beyond Isabelle's cancellation, he probed the DOE by identifying two sensitive core issues: Do the HEPAP recommendations advance not only the interests of a scientific community but also those of taxpayers? And what responsibility does this community have to Congress for developing a consistent scientific policy? "How do I go back to the people [in my district] and tell them it's prudent to abandon a $200 million concrete doughnut in their midst to move on to a machine I can't even describe?" Carney asked. "I don't know if it will cost $2 billion or $8 billion; I don't know its site or who heads its team of designers or operators or how long it will take to complete."[13]

Even Keyworth, who in February 1983 had urged the Wojcicki subpanel to "think big" (see chapter 1), warned his listeners at the April 1983

American Physical Society meeting that they should not assume taxpayers would pay for expensive new atom-smashers without a clearer statement of their priorities.[14] In an era of growing budget deficits and increasingly scarce research funds, scientific priority setting was becoming mandatory.

Taking their case directly to the American public had not been necessary for high-energy physicists in funding previous accelerators and colliders. Unlike NASA, for which public education about outer space is a core part of its mission, the AEC and DOE paid only lip service to promoting new projects and disseminating their findings to the public. Especially in AEC days, during the depths of the Cold War, the approach to public communication could be described as part of a "culture of secrecy."[15] Prior generations of physicists had been able to obtain federal funds for their projects and research, and to publish the results in scientific journals without communicating their research plans to the public, in part because scientific budgets had not been as large or as visible, and public trust in the value of scientific research had been greater. If anything, high-energy physicists had only to make a good case for a project before the largely supportive Congressional Joint Committee on Atomic Energy.[16] Scientists and laboratory staffers who *did* try to communicate high-energy physics to the public were often held in low esteem by this community, so few were very good at it. The resulting dearth of good communicators became one of the factors that later worked against positive reception of the SSC by the US public.

By the mid-1980s, SSC advocates had to go beyond traditional scientific and technical justifications for their machine to try to allay emerging worries about its high cost and develop a convincing public rationale for it.[17] High-energy physicists had of course been among the first to attribute meaning to the SSC, relating their excitement to reporters following its endorsement by the DOE. One of the first to write about the project for general audiences was *New York Times* reporter William Broad, who remarked in September 1983, "Put simply, the project would be the biggest endeavor in the history of pure science, a colossus that would rival the building of the pyramids and the construction of the Panama Canal." While his primary focus was on the coming SSC site competition, Broad was also struck by the great sense of wonder expressed by physicists he interviewed. "The magnitude of the envisioned project—almost its sheer audacity—has . . . produced a spirit of reverential awe and cooperation that transcends the politics of site selection."[18] Among the glowing phrases used by *Chicago Tribune* reporter Ronald Kotulak in a December 1984 article were "ultimate machine," "boggle the mind," and "most fantastic." Like Broad, Kotulak compared the SSC to other supreme engineering achievements, such as the pyramids and the Apollo project. He noted that "the precision in building the machine will be so great that it would be equivalent to aiming a rifle at an insect sitting on the moon's surface and hitting it in the eye."[19]

Over the next few years, the SSC continued to be compared to some of the world's most famous structures. In 1985, for example, *Physics Today* reporter Irwin Goodwin predicted that the SSC would likely be "one of the Wonders of the World—certainly as impressive as the pyramids, Hoover Dam or Sears Tower, each in its own way unequalled."[20] But once its site had been selected in 1988, the press no longer built on this initially positive imagery about the Super Collider (see chapter 5).[21] Instead, articles explaining the project's feasibility and arguing the case for building it usually appeared in science magazines. These articles rarely reached general audiences, however. Diebold's 1983 *Science* article, like others of its kind, was aimed at readers with a good technical understanding of accelerator physics or engineering, which few members of the American public possessed.

A feature article on the SSC by science writer Gary Taubes in the July 1985 issue of *Discover* included material on its physics goals but, like others, focused on the project's huge size, costs, and the competition with Europe. The Super Collider "will, its supporters claim, fire the imagination of American youth," he wrote.[22] "It will restore America's dominance in the flashy and profoundly intellectual endeavor known as high energy physics." Taubes quoted Keyworth, who was becoming a good spokesman for the SSC, prominently in the article—especially on the need to restore US leadership in high-energy physics. But he also stressed the need to tell the general public why the United States should build the facility: "When a community of fifteen hundred people is proposing to ask the public for billions of dollars for a major research project, then I think they'd better be telling the American people why this is a good investment."[23]

Among high-energy physicists, Fermilab's Leon Lederman was perhaps the most successful in bringing the SSC into the public eye. In the mid-1980s he traveled far and wide to explain the machine's intellectual significance in popular talks and interviews, many of them broadcast on radio or TV, and often reported on in the press.[24] Taking ample time to inform reporters, Lederman helped to promote a popular discourse supportive of the SSC. Among his points was that the research planned at the SSC was necessary to advance both the pursuit of the fundamental building blocks of nature and to aid efforts to understand the forces that hold the universe together. Famous for his teaching skills, he made the physics accessible to general audiences. By appealing to humor and not bound by scholarly accuracy, he offered an account of the history of particle physics that traced the search for the Higgs boson back to the teachings of the ancient Greek philosopher Democritus. By doing so, Lederman attracted large audiences to the subject of the SSC. He employed breakthroughs from the history of science to underscore how basic research had brought useful advances to civilization—for instance, electricity, X-ray diagnostics, transistors, computers, and television. Lederman noted that physical discoveries, dating back to James Clerk Maxwell's unification of electric and magnetic forces in the 1860s, form the foundation for most of the modern technological world. "Probably half of our gross national product today" is based on Maxwell's theory of electromagnetism, he claimed.[25] He remained among the few physicists who never lost hope in getting the public to understand the merits of building the SSC.

Lederman also took the opportunity to comment on science-policy issues of the day, using the SSC as a foil. At the annual meeting of the American Association for the Advancement of Science in February 1985, by which time the estimated cost of the collider had increased to roughly $4 billion, he framed the growing opposition among scientists to the SSC as symptomatic of a larger problem in federal funding of research.[26] "I don't detect opposition to SSC so much as concern," Lederman told the *Chicago Tribune*. He suggested that "the decision on the SSC provides a good time for the government to reexamine how we support science in the United States."[27] But his early assessments of scientists' (and politicians') attitudes toward the SSC would be increasingly challenged in the press.

While the speeches and writings of Lederman and other SSC supporters informed and appealed to many, a far more comprehensive effort was needed to reverse the ebbing tide of public interest and turn it back in favor of the project. For countering the attempts of SSC supporters to communicate its importance was a growing crescendo of opposing voices of scientists in other disciplines, who feared the expensive project would undercut their research funding.

CONFLICTING PERCEPTIONS OF THE SUPER COLLIDER

During the mid-1980s many leading high-energy physicists, including Jackson, Quigg, Tigner, and Wojcicki of CDG, as well as Lederman, Chicago's James Cronin, Harvard theorist Sheldon Glashow, and Schwitters, continued their efforts to promote the Super Collider. Attempting to convince their fellow scientists that the SSC was indeed a worthy investment, they published articles in science magazines such as *Physics Today, Science,* and *Scientific American.*[28] An article by Schwitters published in *American Politics* was thought to have influenced Secretary Herrington to back the SSC.[29] In March 1985, Glashow and Lederman coauthored a lengthy *Physics Today* article discussing the merits, feasibility, and affordability of the SSC, inviting "our fellow physicists to join us, however vicariously, in a very great adventure." They cited four prominent reasons why America needed the SSC: the challenge, its spin-offs, national pride, and sense of duty. Glashow and Lederman declared that the "track record of high-energy physics construction projects is very good." They predicted that cost overruns with the SSC would be "extremely unlikely," and noted that $4 billion was "not out of line" for such a sophisticated project. Their article ended by presenting the SSC as a stark choice between preserving the "scientific vitality [and] validity" of high-energy physics, and terminating "the 3,000-year-old quest for a comprehension of the architecture of the subnuclear world."[30]

Following publication of this article, both critics and champions of the SSC, from within as well as outside high-energy physics, spoke out in letters to *Physics Today.*[31] One of the more spirited exchanges occurred in the December 1986 issue, when three scientists—Charles Hailey, Gordon Freedman and Pedro Echenique—debated the pros and cons of building the SSC with Glashow.[32] "There is nothing essentially wrong with constructing the SSC," argued Hailey. "What is wrong is that the costs and benefits of SSC have not been compared with an alternative." Though fairly mild, these letters represented an unusual challenge. Before the SSC came along, physicists in other subfields might grumble quietly among themselves about "those greedy high-energy physicists," but few had come out in the open and published letters or articles voicing their discontent.[33]

Just weeks later, the physics community worldwide was both excited and puzzled by the discovery by J. D. Bednorz and K. A. Muller at IBM Zurich of high-temperature superconductivity. The materials involved (such as lanthanum-barium-copper oxide and other "perovskites") began to be superconductors at much higher temperatures (above 30 K) than NbTi and Nb3Sn (niobium-tin).[34] Condensed-matter physicists soon discovered other such compounds that became superconductors at even higher temperatures above that of liquid nitrogen (77 K or −136 °C). Opponents of the SSC seized upon these revolutionary materials, claiming that their possible future use in powerful superconducting magnets argued for delaying the project. If these new materials could indeed be used, could the extremely expensive and often dangerous cooling processes needed for such magnets perhaps be circumvented? Could even higher fields be generated? Could the tremendous cost of the

FIGURE 3.1 Imaginary footprint of the SSC, plus LEP and the Tevatron (*smaller circles*), superimposed on a Landsat image of the Washington, DC, area. Courtesy of *Physics Today*.

SSC be reduced? As such questions could not be answered immediately, the discovery stirred debate about whether the SSC should be delayed while the new materials were tested and developed for use in superconducting magnets in the hope of eventually achieving significant cost savings.

Among the most vocal scientists calling for a slowdown were SSC critics in the condensed-matter physics community, particularly Nobel laureate Philip W. Anderson of Princeton and the American Physical Society Vice President James A. Krumhansl of Cornell. Writing about the new superconductors in a February 19, 1987, letter to Energy Secretary Herrington, hardly three weeks after Reagan's decision to support the SSC, Krumhansl exuded:

> *The implications are vast! These materials are inexpensive; they can be easily made, and can be refrigerated by a variety of widely available cheap methods. They unquestionably have the potential to save billions of dollars in construction and operation of particle accelerators like the SSC.* Because they are easily fabricated, I have little hesitation in predicting that they will be brought to technological usability in three to five years, if materials research is supported adequately.
>
> By contrast with particle physics, I can assure you that this discovery is so important that it will find its way into almost every area of materials, energy, electronic and military technologies.[35]

Krumhansl subsequently testified before Congress that the new superconducting materials might reduce the diameter of the SSC ring from 50 to 10 miles, saving billions of dollars in construction. Anderson suggested that it might even be possible to eliminate helium cooling or reduce the size of the ring in two or three years.[36]

Two groups of scientists and engineers studied the question—a National Academy of Sciences panel headed by John Hulm of Westinghouse, a leader in the superconducting materials industry, and a DOE Basic Energy Sciences Advisory Committee, chaired by Albert Narath of Bell Labs. Both panels concluded that at least four, and possibly eight or more, years would be required to develop the materials for practical applications. The consensus was to continue SSC design work using conventional superconductors.[37]

One outcome of the discussion was to raise questions in Washington about the SSC's proposed fast-track schedule.[38] The critical discord over the SSC in the scientific community had become sufficiently contentious by the spring of 1988 for president of National Academies of Science Frank Press to warn that "this sniping and carping . . . is disturbing and destructive."[39] Irwin Goodwin of *Physics Today* characterized this war of words among scientists as "mainly over priorities: big science or little science, crisis response or attention to long-term research, national preeminence or international collaboration."[40] The public heard little of this debate because it occurred largely among scientists, but it did resonate within the halls of Congress.

The SSC cost continued to be chronicled in the press as the site-selection process heated up during 1987–88 (see below). Cost had become the dominant issue for many SSC critics, even for some of its supporters. "These devices are becoming so expensive, and what they're trying to find is so obscure, that we may be at the point where scientists can no longer justify the cost," a Federation of American Scientists official told the *New York Times*.[41]

At the same time, SSC advocates were touting the project as good for business and the US economy. At a DOE site-proposal conference in Washington in April 1987, for example, the conveners explained that the promise of jobs and economic spin-offs would transcend the benefits that industry gained through the support of science. A conference called the International Industrial Symposium on the Super Collider (IISSC) grew out of a Colorado meeting held on December 3–4, 1987, in Denver, where physicists met with industrial leaders and government officials in an effort to involve them in the project.[42] (Between 1987 and 1993, the IISSC meeting would bring together physicists and industry representatives to interest companies in the SSC components and the spin-offs of SSC research, which would eventually bring economic benefits.) Speaking before this audience, Keyworth—now a private consultant and chairman of the Council on Superconductivity for American Competitiveness[43]—voiced some of the reasons he so favored the SSC. Beyond the original primary objective that the United States could "retain preeminence in science," he now stressed "economic growth . . . by applying the kinds of national competitive advantages that others envy the US for. I'm referring, of course, to our science and technology, as well as to the entrepreneurial environment that enables us to take quick advantage of new opportunities. That's what I would call the basis for a national competitiveness policy, and, ultimately that's the broad force behind the SSC."[44]

For the general public, however, the science mission and broader economic implications of the SSC were upstaged by hopes of "the most sought-after scientific plum of the century" to provide some 4,500 new construction jobs over eight years. And with an estimated annual operating budget of $270 million, the SSC Laboratory would employ 2,500 permanent workers.[45] This was the principal message preached by each state's team of SSC advocates. The promise of America's resurgence in high-energy physics, so dominant in SSC discussions before President Reagan first gave the SSC his approval, was becoming eclipsed by the quest for federal dollars in 1987. "There aren't very many big federal projects that are going on these days," said one Texas official with regard to his state's hunger for the SSC.[46]

In the spring of 1987, the Illinois Department of Energy and Natural Resources, partnering with a state booster group called "SSC for Illinois," conducted a public opinion study through the Center for Governmental Studies at Northern Illinois University. Feasibility and economic-impact studies had been completed for an SSC site adjacent to Fermilab. Entitled "An Overview of Citizen Reactions to the Proposed Superconducting Super Collider," the report explained that 600 residents in the six counties that might be affected by the SSC's location there had been asked four questions:

do you know about the SSC project; do you support the SSC being in Illinois; do you want the state to provide any financial incentives to help it win the site competition; and are you concerned about tunneling during construction? The results of the poll indicated that there was "a strong base of support for construction of the SSC near Fermilab, that the citizens were supportive of state actions to pursue the project for Illinois, that as local residents were given specific information about tunneling near their homes for the SSC their concerns were ameliorated, and that even among those who knew little of the SSC, support for the project was high."[47] These results encouraged state officials to pursue the project for Illinois.

In less than a year, however, an opposing view surfaced among people living in a few states vying for the big project, who organized under the umbrella of Citizens Against The Collider Here, or CATCH. Their goal was not so much to kill the SSC as to drive it away from their neighborhoods.[48] For them, the SSC was a threatening, potentially disruptive, intruder. At public hearings and in local newspapers, they voiced attitudes objecting to potential loss of homes and land, fearful of not receiving fair-market value for their properties if forced to relocate, and generally demonstrating their resentment of state agencies and other pro-SSC groups that they perceived, in some cases, to be forcing the Super Collider on local residents. Some also worried that the machine might release nuclear radiation in the event of a mishap, while a few even compared the Super Collider to a "Chernobyl waiting to happen."[49] The SSC began to become a symbol of Big Government not to be trusted.

By July 1988 the general enthusiasm for the glowing SSC vision launched five years earlier had given way to more skeptical representations of it. The increasing opposition to the project became blatantly evident in August at the Democratic and Republican national conventions, when televised images of a hand-held sign proclaiming "Super Collider, Super Mistake" were broadcast into the homes of millions of Americans during roll calls for the presidential nominations.[50]

In the same month, Lederman and Quigg attempted to counter the growing negative portrayals of the SSC with a broader, national communications effort. Following the model of a 1965 campaign in which famous physicists—including Hans Bethe, T. D. Lee, C. N. Yang, Julian Schwinger, and Victor Weisskopf—contributed to a collection titled *The Nature of Matter* that argued for building the accelerator that eventually became Fermilab,[51] Lederman and Quigg now gathered 58 endorsements in a compendium, *Appraising the Ring: Statements in Support of the Superconducting Super Collider.*[52] Its many contributors—including government officials, university presidents, leaders of industry, and the usual Nobel laureates—framed the

SSC in terms of various themes including scientific duty, voyages of discovery, and international competition. Some of the writing was beautifully crafted, eloquent, and even inspirational.

Quigg, for example, wrote that the cost of *not* building the SSC would mean "turning our back on a fundamental scientific field in which the United States has traditionally held a position of leadership, . . . abdicating our responsibilities to the future as a wealthy and enlightened society on which fortune has smiled." Nobel laureate Val Fitch of Princeton evoked the great exploration voyages of Columbus and Magellan, suggesting that "a lot of the spirit of the country could be restored if as large a fraction of the gross national product were devoted to basic research in the United States as in Western Europe and Japan."[53] University presidents Harold Shapiro of Princeton and John Marburger of Stony Brook added their enthusiastic support, as did industry executives like Samuel D. Bechtel of the Bechtel Group, J. Fred Bucy of Texas Instruments, Robert W. Galvin of Motorola, John K. Hulm of Westinghouse, James J. O'Connor of Commonwealth Edison, and George Pake of Xerox Palo Alto Research Center. *Appraising the Ring* was published by URA in July 1988 and mailed to a long list of public and private officials who might become advocates and lend their support to the SSC project.

On Capitol Hill, where the SSC's fate would ultimately be decided, Congress had at least concluded the Super Collider would deliver good—if not necessarily affordable—science. At the other end of Pennsylvania Avenue, the future of the project looked bright on March 30, 1988, when Reagan offered a strong endorsement of the SSC at a public White House ceremony in the Rose Garden. Stressing the adventurous themes of research to an audience of the leading high-school science students from almost every state and four Nobel laureates in physics (James Cronin, Burton Richter, Samuel Ting, and Steven Weinberg), he explained his rationale for building the SSC:

> I know that some people may question the practical applications of the Superconducting Super Collider. The strange world of subatomic particles, they may think, will never be more than an arcane interest to a few highly specialized scientists. But the truth is, the practical applications of this new technology are already changing the way we live. . . .
>
> . . . Every time someone turns on his desk computer, makes a phone call or plays a video game, he is plugging into that mysterious world of quantum physics. . . . The Superconducting Super Collider is the doorway to that new world of quantum change, of quantum progress for science and our economy.[54]

According to Quigg, "They had kids [there] from every state." Each child received a SuperCollider T-shirt emblazoned with a colorful image of particle

FIGURE 3.2 Energy Secretary John Herrington (*at lectern*) speaking in the White House Rose Garden, March 30, 1988. Next to him is President Reagan; Nobel laureates are (*l. to r.*) Samuel Ting of MIT, Steven Weinberg of the University of Texas, and SLAC Director Burton Richter. Courtesy of Ronald Reagan Library.

tracks exploding forth from a collision, and two of the children presented a T-shirt to the president. Quigg remembered being "very moved by the notion that our equations and ideas had, for 15 minutes, gotten the attention of the most powerful leader on Earth."[55]

In reality, beyond the Rose Garden and Beltway, the political context for the SSC was anything but rosy. The sudden US stock market crash in October 1987 had revealed an American economy that was weaker than most believed. Instead of the economic recovery forecast by the Reagan administration, the federal budget deficit was steadily worsening. These developments reinforced congressional perception of the SSC as a "budget bomb."[56] No foreign contributions to the project had yet materialized. Of some two-dozen news articles about the SSC appearing in *Science* during 1983–88, all but five focused on its financial aspects—funding problems, the search for foreign partners, the total estimated cost, or cost overruns.

Reagan had requested $363 million to begin construction of the SSC in his fiscal 1989 budget, but the request faced a stiff uphill battle in Congress. Quigg remembered frequent calls to CDG leaders from DOE about funding uncertainties.[57]

The tenuous state of the project was reflected in the SSC coverage in *Science,* with the reported debates rarely focused on the science but instead on the project's high cost, DOE and congressional indecision on whether to commit, and on the implications of magnet R&D for its total estimated cost. In both *Science* and *Physics Today,* the core issue had shifted from whether the SSC was going to do good science to whether or not that science was affordable.[58] Addressing the latter question was not getting any easier, for as the SSC design became more specific, its cost estimates doubled—to about $6 billion by late 1988.[59]

The project remained in serious financial doubt through 1988, as the nation went through the presidential campaign and November elections. The biggest communication challenge facing high-energy physicists was how to persuade Congress that the country could afford the SSC despite a worsening US economic and political climate. That challenge had not eased since the buoyant days of 1983.

EARLY DOE PURSUIT OF FOREIGN CONTRIBUTIONS

Informal government-to-government contacts like Keyworth's March 1984 Japan visit, usually between DOE or OSTP officials and those in equivalent foreign agencies, continued to occur in 1985 and 1986 while CDG was designing the SSC. By the time of Reagan's January 1987 decision to "throw deep" on the SSC (see chapter 2), these visits had ascertained that foreign contributions totaling about $1 billion might be expected on the project. Aside from Japan, Canada and Italy were most often cited as likely to be willing to make large in-kind contributions to the construction of the accelerator and its detectors—both on the order of $100 million dollars.[60] At that time, the total cost of the SSC was estimated at about $4 billion in 1988 dollars, including the detectors but not contingency or inflation.[61] Thus at most 25 percent of the projected construction costs were expected to come from foreign sources when Reagan made the decision to proceed with the SSC.[62] It was generally assumed that these foreign contributions would be made according to the HERA model of international collaboration. The United States would supposedly sign one-on-one agreements with contributor nations—generally between the Department of Energy and the foreign agencies responsible for high-energy physics.

The crucial decision facing the US Domestic Policy Council that January was whether to proceed with the SSC project or defer making the decision for another year. SSC advocates argued, among other things, that proceeding now would put the United States "in a better bargaining position to obtain cost-sharing commitments from other parties."[63] But delaying the decision, say for another year, might lead to a loss of US leadership in high-energy physics in the 1990s. According to a memorandum that the council sent Reagan, "European governments are awaiting our decision on the SSC and may go ahead with their own equivalent at CERN, Switzerland, if we delay."[64] It was a difficult quandary that Reagan finally resolved with his January 29, 1987, decision. And he apparently made it with the expectation that US allies would eventually fall into line, as they were doing on other US initiatives like Space Station Freedom.

In his memorandum on the SSC decision, Council Chairman (and close Reagan ally) Edwin W. Meese, then the US Attorney General, wrote that the United States would proceed with the SSC in fiscal 1988 "using the maximum possible cost-sharing from other countries, private industry, and state and local governments."[65] It designated the Secretary of Energy as the government official responsible "to seek Congressional authorization, select a site, and obtain cost-sharing commitments." That same day, Energy Secretary John Herrington held a press conference in which he called the president's decision "a momentous leap forward for America and science and technology."[66] The DOE press release did not mention any foreign participation in the project, but his prepared statement did. "The United States has a long history of collaborative research efforts with other friendly nations," he told reporters. "It is also our intention to seek maximum cost-sharing funding from other countries." The train was about to leave the station, that is, but it was as yet unclear who would be aboard. In response to a reporter asking whether the United States was prepared to go it alone, Herrington replied, "I don't think it will be necessary, from the information that I have, in talking to various foreign-interest parties. But this is an American project, American leadership. We are going forward with it."[67]

When Trivelpiece stepped down as OER director three months later, in April 1987, however, it was a major setback in the efforts to obtain foreign SSC partners. For he had chaired the Summit Working Group on High Energy Physics and was the project's principal representative to foreign governments. This responsibility was assumed temporarily by OER Acting

Director James Decker, a career DOE civil servant who lacked the extensive political connections and clout of a Reagan appointee like Trivelpiece.

Nevertheless, efforts to obtain foreign contributions proceeded despite the DOE change in command. First an interagency working group was established that included representatives from the DOE, OMB, OSTP, the National Security Council, and the Departments of Commerce, Defense, and State; it would ensure that foreign commitments made would be consistent with US economic and national security policies. It was headed by George J. Bradley Jr., the DOE principal deputy assistant secretary for international affairs, who reported to Undersecretary Salgado.[68] After a series of meetings to discuss the ground rules for foreign participation, Secretary Herrington sent letters in December 1987 to his counterparts in science ministries of the larger countries in the Organization for Economic Cooperation and Development (OECD), inviting them to participate in the SSC project. Although the terms of such participation (or partnership) were still vague, the invitations met with expressions of willingness to confer with DOE representatives to discuss further the possibilities of collaboration on the SSC.[69]

In January 1988 a DOE team headed by Bradley visited Rome, Paris, London, Bonn, and Geneva, while another led by Decker went to Ottawa and Tokyo. Foreign officials they met with expressed strong interest in the SSC as a facility where their physicists could do research in the future, but Bradley and Decker returned without any significant financial commitments. Nevertheless, optimism within DOE remained high. In testimony before a House Appropriations Subcommittee that March, Herrington said the prospects for foreign participation were "not only good, they are excellent," suggesting that he expected "45 to 50 percent" of the project costs (then estimated at $4.4 billion in 1988 dollars) to come from foreign sources.[70] This was exactly what Congress wanted to hear, for members then faced the difficult prospect of having to impose severe cutbacks in the funding for other science programs due to Gramm-Rudman budget reduction targets if the SSC were approved. But if foreign contributions of 50 percent could be expected, remarked the House Energy and Water Appropriations Subcommittee Chairman Tom Bevill (D-AL), "then it's a different ballgame."[71]

Reality, however, was far different from the Reagan administration rhetoric. Until that time, the only foreign contributions to US high-energy physics projects had been in-kind contributions to the big experimental particle detectors at Fermilab and SLAC, representing typically 25 percent or less of their total cost. Even if similar contributions could increase to 50 percent on the SSC detectors, they would amount to only about $360 million, or less than 10 percent of the SSC's total price tag (at the time). To obtain anything like the suggested 50 percent foreign contributions overall, the partner nations would have to either produce essentially all of the superconducting magnets or pay up in cash—both beyond political possibility.

Europe in particular was strapped for cash, what with the expensive LEP and HERA projects under construction at CERN and DESY. Wealthier European OECD nations—France, Italy,[72] the United Kingdom,[73] and West Germany—were fully committed to seeing these projects to successful completion before they could ever begin to think about joining a major new US effort. Large European contributions to the SSC project could not occur before the mid-1990s, and then only if the proposed LHC project was not pursued at CERN. Canada, with a relatively small contingent of perhaps a hundred high-energy physicists and annual funding in just the tens of millions of dollars, was planning a US$400 million upgrade of its TRIUMF meson-physics facility in Vancouver. And DOE's summary rejection of a proposed cross-border site between northern New York and southern Quebec (see below, esp. n. 112) cooled Canadian interest in the SSC—and called into question how serious DOE officials really were about international partners.[74] That left Japan as the only major OECD nation that might be willing and able to make a major contribution to the project.

Laying the groundwork for a large foreign contribution generally involved negotiations at different levels. At the scientist-to-scientist level, physicists from a partner nation had to work in close cooperation with US physicists to establish the scientific and technological needs to be met by in-kind contributions; this work usually required R&D efforts on both sides, with possible industry involvement, too.[75] After suitable component designs or performance characteristics were achieved, formal contracts had to be enacted between the DOE and the appropriate science agency or ministry to deliver the goods on an agreeable schedule. For in-kind contributions to particle detectors that might be worth only several million dollars, no further agreements were necessary, as long as physicists remained within budgets and technologically sensitive items did not cross US borders. But if the value of the foreign contributions increased into the tens or hundreds of millions of dollars, as anticipated for the SSC, other ministries and foreign agents would inevitably have to become involved in the interactions—including heads of state when the numbers got truly large. And as the size of the contributions reached such levels, foreign governments would naturally insist on being treated as *equal partners* in a laboratory's overall scientific enterprise, involved in its design, operation, and management. This was the kind of egalitarian organizational structure that had existed for decades at CERN and been advocated as part of the VBA process. But the United States had little or no previous experience with such an organization or process, at either the DOE or NASA.

Since 1979 the United States and Japan had been pursuing joint R&D projects in high-energy physics, including work on superconducting wire, cable, and magnets at KEK.[76] And Japanese physicists were participating in CDG efforts focused on the design and development of particle-detection components able to function under the extreme operating conditions anticipated at the SSC. Thus much of the preliminary physicist-to-physicist groundwork had been laid by the late 1980s, with Japanese physicists becoming involved in the SSC design effort. And after commissioning its TRISTAN collider in 1986,[77] Japan was beginning to cast about for its next major accelerator project. And unlike their European counterparts, Japanese high-energy physicists were no longer burdened by prior commitments to a project still under construction. So SSC advocates naturally viewed Japan as the foreign country most likely to make significant contributions potentially worth hundreds of millions of dollars. With the nation enjoying a booming economy in the late 1980s—and the fact that it had not supported basic scientific research at levels similar to what the other OECD nations had done—Japan supposedly had the required deep pockets to afford such a major commitment.

But getting another nation involved in the *construction* phase of a scientific project (as had been done at HERA) was uncharted territory for the United States, and not without significant difficulties. Uppermost among them was the need to transfer superconducting technology to Japanese companies—or at least work closely with them on joint R&D projects. Keyworth and other Reaganites had touted the SSC as a means for high-energy physicists to help the United States reestablish its industrial competitiveness. Large technology-oriented companies such as General Dynamics, Grumman, and Westinghouse were lining up to work with CDG (and after it the SSC lab) on the superconducting magnets, and thereby gain valuable expertise in the technology as well as lucrative contracts to supply them. To send a portion of this technology overseas to their feared Japanese competitors flew directly in the face of such seemingly laudable intentions.[78] Another problem was that the CDG magnet R&D was well behind schedule in 1988 (see chapter 2). No full-length prototype dipoles had been successfully tested by early 1988, so they were not yet ready to be transferred to industry, whether in Japan or the United States.

FIGURE 3.3 US and Japanese physicists about to go hiking near Nikko, Japan, during a break in May 1984 workshop on US-Japan cooperation in high-energy physics: (*l. to r.*) Leon Lederman, J. David Jackson, Tuneyoshi Kamae, and Roy Schwitters. Credit: AIP Emilio Segrè Visual Archives, J. D. Jackson Collection.

Financial benefits to the host country of in-kind foreign contributions may have been overrated, too, as Panofsky observed.[79] The added managerial complexity of having to coordinate design details across borders and in different cultures, plus the attendant lack of flexibility in choosing the manufacturers, often meant that any cost-avoidance factors due to such contributions came in well below 100 percent. For example, he noted that "foreign participation in HERA has indeed been a positive step in terms of securing international recognition and commitment, but it may not make a significant reduction in the construction cost borne by the Germans."[80] In any event, the United States could not begin to make formal arrangements with Japan, or with other potential contributor nations, before January 1989 because an M&O contractor for the SSC project had not been selected.[81] The Central Design Group did work closely with Japanese physicists and institutions on R&D efforts, but it could not as yet enter into any formal contracts to produce and manufacture actual components for the project.

And always looming in the background to these discussions of international contributions was the possibility of CERN building the LHC as an alternative to the SSC, with US participation in that project. In fact, CERN Director Herwig Schopper had suggested the possibility of US involvement in April 1987, in testimony before the House Science Committee, as did Carlo Rubbia, then directing CERN's Long-Range Planning Committee.[82] An October 1988 report by the Congressional Budget Office in fact suggested joining CERN in building the LHC as a viable alternative to the SSC project. Although the report gained substantial attention in Washington, it was roundly criticized by DOE officials, mainly concerning this recommendation.[83]

To congressional appropriators increasingly pressed by ballooning US budget deficits and the Gramm-Rudman process, the suggestion to join CERN was especially attractive, for it was a way to keep US high-energy physicists intellectually active at a fraction of the SSC construction costs. As then proposed, the LHC might be able to discover the Higgs boson up to a mass of about 1 TeV (see appendix 1); the SSC could be delayed and later built as a truly international project if higher energies turned out to be needed. But a principal flaw in this argument was that designs for the LHC were still sketchy in the late 1980s and cost estimates correspondingly unreliable.[84] And CERN was already fully committed to running its new LEP collider through the 1990s, which meant that the LHC could not come on line until the next century at the earliest. Finally, a US commitment to participate in the LHC project would potentially leave the nation without a frontier high-energy physics facility while Europe had three.

By early 1988, funding for SSC construction was caught in a classic "chicken and egg dilemma," observed the Republican Technical Consultant Todd Schultz.[85] "Congress would like to see foreign contributions to the project before undertaking SSC construction, but foreign commitments cannot be ascertained until a signal of Congressional intent to fund the project is given to the international community." Efforts to appropriate construction funds had failed in 1987, and would fail once again in 1988. Although President Reagan had decided to proceed, Congress was inclined to bide its time until an SSC site had been chosen, a contract to manage and operate the laboratory enacted, and major foreign contributions committed. But foreign science ministers were equally cautious. They much preferred to see strong congressional backing of the project before committing any scarce funds to it.

Although the Reagan administration had requested $363 million to begin SSC construction in FY1989, Congress deleted construction funds and appropriated just $98 million to continue project R&D.[86] Still, that was almost triple the previous year's appropriation; it would permit the CDG to do more intensive magnet development. The DOE budget request for FY1990 assumed that a third of the SSC project costs, or $1.8 billion, would eventually come from nonfederal sources, as the OMB was recommending.[87] But at this juncture, observed the Congressional Budget Office report, "DOE has yet to produce other evidence to support the claim that foreign and state donations are likely to cover $1.8 billion of SSC costs."[88]

THE SITE-SELECTION PROCESS BEGINS

As the pursuit of major nonfederal contributions started to play out behind the scenes in foreign capitals, a very public site-selection process began in the United States. By early 1987 a national competition seemed to many the most logical way to determine the SSC site, for DOE officials widely believed that the 1965–1966 process used to select the Fermilab site (then called the National Accelerator Laboratory, or NAL) had unified national support for that project—especially among the nation's scientific, industrial, and political communities.[89]

William Wallenmeyer, who directed the DOE Office of High Energy Physics, was an early proponent of an SSC site contest. Having served as an administrator with the AEC during the NAL site contest, he had an insider's view of this process. Wallenmeyer encouraged his new boss Wilmot Hess, director of the Office of High Energy and Nuclear Physics, to support a similar site contest for the SSC. Persuaded by Hess about the merits of a competition, Trivelpiece believed it would help ease collaboration problems among the different labs and states interested in the SSC. He then convinced the rest of the DOE hierarchy that such a site contest was desirable, despite the risk that most of Congress might lose interest after a specific site had been designated.[90] Little, if any, consideration was given to needs of foreign nations and physicists during this site-selection process.

The issue then arose: should the site for the SSC be a completely new greenfield site, or should it be built as an add-on to an existing accelerator laboratory? Europeans favored the approach of building new facilities at existing laboratories, in the belief that reusing the existing human and physical infrastructure led to greater cost-effectiveness.[91] At Snowmass in 1982 Lederman had suggested building the huge machine at a remote desert location (see chapter 1), as he believed that the new collider under discussion would be so large that it would have to be built on a vast, uninhabited site in the American Southwest.[92] But a year later, after Fermilab had successfully demonstrated the Tevatron based on superconducting dipole magnets, Lederman and others there recognized that an Illinois "Prairietron" with a smaller main ring might be possible if superconducting magnets with higher fields could be developed. They began to emphasize that DOE would instead realize major cost savings by building the SSC adjacent to Fermilab, using the Tevatron as a proton injector.[93]

CDG leaders initially supported an SSC site contest, too, with the implicit goal of finding a greenfield site for the collider.[94] According to Jim Sanford, who headed a group working on the required site characteristics, Tigner did not encourage the proposal of adding the SSC to an existing accelerator laboratory. He in fact steered Sanford's team away from that approach for fear that they "would appear to have been in effect choosing a site."[95] When Illinois advocated building the SSC adjacent to Fermilab, the suggestion challenged CDG efforts to develop criteria for "fair and equal" evaluation of the site proposals. Tigner worried that any appearance of undue bias toward Fermilab might jeopardize the credibility of the site-selection process and possibly threaten national support for the SSC itself.[96]

Senior DOE officials were reluctant to consider the Fermilab site proposal seriously because they felt that congressional support might evaporate if even a hint of bias or impropriety infected the site-selection process. As Trivelpiece commented, the competition had to be "clean and unassailable," so that the winning site would be "perceived to be a level-playing-field selection, and that the selection . . . could be audited, explored, the people who made it interrogated, and in no sense would there ever be criticism that there [was] improper behavior . . . because that would have been fatal."[97]

Many DOE officials favored building the SSC at a new site. Trivelpiece, for example, thought that option would actually be less costly due to the additional expense of converting and upgrading existing components before building the new machine. As he later explained, "a greenfield start is usually cheaper than going back and taking a place like Fermilab and upgrading it to the point [of making it into] a suitable injector."[98] In support of this viewpoint, Ed Temple, who had conducted dozens of reviews at DOE laboratories, noted that institutions atrophy, becoming more bureaucratic and less efficient over time. He considered Fermilab an already "aging, beginning to be aged, institution."[99] On the other hand, many others argued that building up the required physical and human infrastructure—such as offices, shops, a research library, and business-administration services like procurement—at a greenfield site could prove unduly expensive.

Senior DOE officials also had strong political reasons to favor an SSC site competition. Given that its unprecedented cost was often cited in the press, the officials involved with planning the site competition worried about maintaining congressional support after the site was chosen.[100] As Temple recalled, Hess often reminded people that the SSC was the first machine that had to be sold to Congress and the American people. As a nation-wide site competition would entail extensive promotion of the SSC in the individual states, DOE managers came to regard this as a good way to sell the project to Congress as well. Federal lawmakers, they figured, would more willingly support what was billed as the world's largest public-works project if their respective states were given a fair chance at the 4,500 construction jobs, its 2,500 permanent positions and $270 million annual operating budget, as well as the prestige that would come with hosting the "crown jewel of particle physics."[101]

Not surprisingly, individual states endorsed the idea of a national SSC site competition. Each state studied and then emphasized what it had to offer. Texas took the unique approach of inviting proposals from communities throughout the state before selecting its best sites to put forth.[102] So great was the prize represented by the SSC that some states spent millions of dollars, conducted countless studies, and drafted dozens of reports. A *Chronicle of Higher Education* survey in April 1987 identified 14 states that had already spent at least $1 million in pursuit of their proposals.[103] Other states soon crossed this threshold.

Sanford's CDG group began by developing the SSC site parameters and evaluation criteria, starting with an approach similar to what had been used two decades earlier in selecting the NAL site. The team submitted its report to DOE on February 10, 1987.[104] Temple's group had worked closely with this team to facilitate and refine the work, turning the recommendations into the official DOE request for SSC site proposals. It required delicate diplomatic work because, as Sanford sensed, the DOE bureaucrats didn't trust the physicists to manage the SSC project on their own. He tried to instill their confidence in the CDG efforts, and by the spring of 1987 thought Temple had finally come to regard their work as trustworthy.[105]

The site-selection process began officially on April 1, 1987 (the same day that Trivelpiece left office as OER Director). DOE issued an "Invitation for Site Proposals for the Superconducting Super Collider." Responses were due by August 3, 1987, but this deadline was later extended to September 2. As the invitation explained, the agency planned to hold a "fair and open" national competition with three separate rounds of evaluation. At first, there would be an internal DOE task-force screening of all the proposals against a set of minimum qualifications. Proposals that survived this cut would then be reviewed by a panel of experts selected by the National Academies of Science and Engineering, who were to announce the finalist sites by the end of 1987. After that a second DOE task force, to be headed by Temple, would have another six months to examine and evaluate the finalist sites much more deeply before making its recommendation of the preferred SSC site to the secretary of energy, who would announce the final decision. If all went according to this plan, the decision could be announced months before President Reagan left office in January 1989.[106]

As described in the April 1987 invitation, the DOE plan was to use two sets of criteria to winnow out less-qualified candidate sites. In the first step, all proposals had to meet five minimum qualifying criteria, that the site:

- be entirely located within the United States
- cover at least 16,000 acres, able to encompass a 52-mile collider ring
- be transferred to DOE at no cost
- be free of environmental problems that could not be mitigated
- have access to 200 megawatts of electrical power and 2,200 gallons per minute of industrial-quality water.[107]

These criteria were relatively clear-cut and easily applied. The DOE did so on its own, completing this step two weeks after the September 2 (extended) deadline.

The second set of criteria was more open to interpretation. DOE listed them in descending order of importance. The highest priority was "geology

& tunneling." A flat, stable, and dry site suitable for tunneling would be essential in building the SSC. Second on this list was "regional resources." It was important that there be good transportation, housing, schools, and hospitals available—as well as institutional support from local communities, state and local governments, and universities. The ideal site would lie within 20 miles of a vibrant urban area, close to a major airport. The DOE envisioned that at least 80 percent of the physicists working at the SSC lab at any moment would probably be visiting from universities and other laboratories around the world. The third criterion concerned the "environment" more generally, including environmental impacts, regulatory compliance, and the extent to which any adverse SSC impacts could be mitigated. Issues of land availability, including flexibility of the site to accommodate relocations and other changes, fell under the heading of the fourth criterion in the set, vaguely referred to as "setting." Bidders would also have to document items relating to criterion five, "regional conditions," a catch-all phrase for obstacles, both natural and man-made, like vibration, climate, and noise. The sixth and final criterion, called "utilities," referred primarily to the availability of power and water.[108] Some of these criteria were clearly "soft"—and allowed for considerable discretion in the evaluations. Given that its high cost was so prominent in press reports about the project, it is puzzling that the site-specific SSC cost was not considered very important by the second DOE task force.[109]

By September 2, 1987, DOE had received 43 bids from half of the states. They included proposals from governments, commissions, organizations, and even private individuals.[110] Within two weeks, DOE announced the first round of cuts in proposals being considered, rejecting seven because of failure to meet minimum specifications. One of the rejected sites was a cross-border New York/Canada site, extending from Franklin County in northern New York to Huntingdon County in Quebec, Canada. The only international site proposed, it was rejected outright by Secretary Herrington because it did not meet the first criterion of being located entirely within the United States.[111] The site offered considerable cost savings, including lower costs of electricity (because low-cost Canadian hydropower would be available.) And by straddling the border between the United States and Canada (comparable to CERN straddling the Swiss-French border), it would reflect a genuine attempt to internationalize the SSC Laboratory. But the DOE condition that any proposed site had to be *entirely within* the United States automatically disqualified this site; it was therefore summarily dropped from consideration. This decision stimulated substantial concern among the

New York congressional delegation, especially Representative Sherwood L. Boehlert (R-NY) from the upstate Utica district.[112]

The other sites rejected in this initial review phase were in Texas, Utah, and Washington state—and even one in outer space![113] Of the 36 sites that survived, 25 were located west of the Mississippi River. Three proposals from New York were the only successful ones in the Northeast. Boehlert and others in Congress wanted cost savings to be included as a site-selection factor, but DOE, particularly Temple's task force, did not consider cost savings relevant for those sites that would proceed to the final round.[114]

The next stage of the review process involved an independent panel named by the National Academies. Chaired by Edward Frieman, director of Scripps Institution of Oceanography—and formerly a professor of astrophysics at Princeton and the director of the DOE Office of Energy Research in 1980–81—the panel included 21 experts in high-energy physics, accelerator physics, engineering geology, economics, environment, procurement, large construction, and large-facility management. Their task was to select an unranked list of sites known as the Best Qualified List. Tensions between DOE and the National Academies emerged even before the panel convened to examine the proposals. Concerned with keeping on schedule, Temple and Richard Nolan, a rising DOE project manager, wanted to establish firm guidelines for the panel. But they were rebuffed by the NAS staffers led by Raphael Kasper, who argued that panel members were sufficiently qualified to establish their own procedures. Temple accepted a vague three-page work statement, the essence of which was that the panel's evaluations of the sites were "to be developed in the light of past experience with large science research laboratories, [and] will stress those items within the DOE-announced criteria and subcriteria . . . that are likely to be most critical in determining scientific productivity of the SSC laboratory."[115] While the basis for the panel's assessments continued to be the DOE evaluation criteria, as given by Temple's task force, exactly how panel members interpreted those criteria was left to their own discretion.

The panelists formed working groups based on the six technical evaluation criteria—geology and tunneling, regional resources, environment, setting, regional conditions, and utilities—plus a seventh criterion dealing with SSC life-cycle costs. Agreeing that a rigid set of factors to allow mathematical ranking of the proposals had potential practical flaws and was probably impossible, the panel decided to provide an initial evaluation (good, satisfactory, questionable) of each proposal on each subcriterion.[116] By mid-November 1987, they had selected eight sites for the Best Qualified List:

Arizona (Maricopa), Colorado (Ft. Morgan), Illinois (Fermilab), Michigan (Stockbridge), New York (Rochester), North Carolina (Durham), Tennessee (Murfreesboro), and Texas (Dallas-Fort Worth).[117] No ranking of these sites was ever made public, but panelists recall Texas and Illinois as the top two.[118] According to Kasper, the panel had principally used the leading criteria of geology & tunneling and regional resources to distinguish among the many proposals.[119]

When the Best Qualified List was revealed at the end of December 1987, reactions were mixed, with some observers claiming few surprises and others grumbling about bias, presidential influence, or snobbery.[120] An obvious counter to the charges of presidential influence was the absence on the list of Reagan's—and Energy Secretary Herrington's—home state of California, which was eliminated largely because of the unstable geology of its two sites.[121] The consensus, by and large, held that the recommendations of the NAS/NAE panel were fair, and based on technical rather than political criteria.[122] But DOE's insistence that the selection process was apolitical still met with skepticism. Nationally syndicated humorist Dave Barry mocked that tale by predicting that the winning state would be selected "based on how well the governor cleaned the shoes of key federal officials with his tongue."[123] Beltway pundits and the media both characterized the process as having *two* distinct threads—technical and political.[124] And during the next nine months, swarms of lobbyists continued visiting the Forrestal Building and Capitol Hill on behalf of their states' proposals. Much of this politicking was, however, aimed at maintaining congressional support for the SSC.

Broad support for the project weakened in the wake of the Best Qualified List announcement, which left far fewer states in the running. For despite all the rhetoric about advancing frontiers of science and increasing national prestige, many states regarded the SSC primarily as a job-creation machine. And once that opportunity had evaporated, many politicians began to question the intrinsic scientific merits of the Super Collider—so many, in fact, that promoting the project as a whole came to be more important than promoting any of the individual sites. "There was no point in focusing on the site," said an Arizona lobbyist: "The real challenge was to save the program."[125]

Within days of the Best Qualified List announcement, New York Governor Mario Cuomo withdrew the Rochester, NY, site because of strong local opposition.[126] That left seven finalist sites. The fact that most of the others—except for Illinois—were located in rural settings reduced the potential disruptions of adverse local reactions. As had been anticipated, state enthusiasm for the project diminished whenever that state's proposal lost one of the site-selection rounds. In the same way that DOE's nationalistic rhetoric lessened the project's appeal to potential foreign contributors, the economic rationale originally given for states to host the SSC reduced its luster as the selection process lowered the number of potential sites.

At the same time as mainstream America was finally becoming aware of the SSC by virtue of press coverage of the site race, news stories took an increasingly cynical view of the project. "Now that the Department of Energy has eliminated all but seven states from the site selection process," declared Congressman Don Ritter (R-PA) soon after the finalist states were announced, "we will begin to see that the elected representatives from districts no longer eligible for this science gravy train will join a number of Members of this House who oppose SSC in recognizing the Superconducting Super Collider for what it is—a $6 billion assault on the nation's science priorities and science budget." His harsh but catchy phrase "quark barrel politics" became a familiar refrain.[127]

The seven finalist sites had various individual advantages. It was estimated, for example, that only four families would have to be uprooted by the selection of Arizona's site circling the Maricopa Mountains in the desert 35 miles southwest of Phoenix. Colorado's site, offering exceptional geology because much of the site sat atop a shale formation that would allow faster and cheaper tunneling, was similarly remote, in the hilly grasslands sixty-five miles northeast of Denver, near Fort Morgan. The geology of the Illinois site was good, because the limestone-dolomite formation underlying it was excellent material for tunneling. Moreover, the northern Illinois area around Fermilab offered outstanding regional resources, including Chicago's cultural institutions and major research universities, transportation infrastructure, a large skilled labor force, and ample housing. Michigan also boasted an impressive industrial base in the vicinity of its Stockbridge site, between East Lansing and Ann Arbor, in addition to their major research universities. North Carolina's proposed site, ten miles north of Durham, showcased the state's world-renowned Research Triangle, as well as the area's general desirability. Tennessee's Murfreesboro site, 30 miles southeast of Nashville, offered excellent limestone geology, a major reason its construction costs were among the lowest. Tunneling costs would also be low at the Texas site in Waxahachie, 30 miles south of Dallas, due to a chalk formation through which boring machines could tunnel relatively quickly. The broad technological base and an array of important regional resources (such as a major international airport) in the Dallas-Fort Worth area lent the Texas site great appeal.[128]

The Illinois site stood out in its strengths and advantages, especially as the US government had already made a major investment there in Fermilab's infrastructure and Tevatron, then the world's most powerful particle accelerator.[129] The Frieman panel cited its existence as an important asset, saying that the centerpiece of Illinois's bid was "an existing infrastructure built upon almost two decades of experience in operating an outstanding high-energy physics research laboratory."[130] Illinois site boosters emphasized the cost savings that this infrastructure and experience would offer. One study unveiled by the powerful Illinois congressman Dan Rostenkowski (D-IL), chairman of the House Ways and Means Committee, estimated that the roads, buildings, and especially the experienced personnel already in place at Fermilab could save at least $3 billion for the SSC, and reduce its construction time. "For the government not to build [the SSC at Fermilab] and reinvest in another area is folly," Rostenkowski declared.[131]

DOE OFFICIALS DETERMINE THE SSC SITE

The dominant mindset within the DOE (or at least among its site task-force members) held that cost should be a less-important factor in the SSC site selection process than the technical evaluation criteria. According to Temple, the answer was ultimately determined by the agency's desire for a fair and open site competition and, according to Sanford, DOE's desire to establish a new national laboratory.[132] If monetary concern were a key site criterion, argued Temple, "the Illinois proposal would be way ahead . . . [of] all sites except [states that] were willing to put up large sums of money."[133] The competition would then turn into a bidding war among a handful of rich states, dashing DOE hopes of maintaining interest in the SSC among a larger number of states in order to sustain annual appropriations for the project. DOE's public stance on matters of cost, as set forth in its original invitation document, was that while cost considerations were "significant" in the site-selection process, the "primary emphasis" would be on the technical criteria. The agency specified no mechanism by which costs would be considered in the evaluations, however, other than to state that life-cycle costs would be calculated for each site proposal.[134]

But the DOE position did not prevent a bidding war. By the summer of 1987, it was no secret that certain states were preparing to offer substantial financial incentives with their bids. Texas in particular attracted major attention when its state legislature approved $500 million in revenue bonds for the SSC (contingent on winning the site competition), including money for site improvements, university research, and low-cost electricity. This extraordinary incentive, coordinated by political consultant Karl Rove, was part of an all-out effort led by Republican Governor William Clements to bring the SSC to the Lone Star State. In 1986 he had appointed prominent Texans, including geologist Peter Flawn, formerly president of the University of Texas in Austin, and J. Fred Bucy, former CEO of Texas Instruments, to a state commission, the Texas National Research Laboratory Commission, whose aim was to prepare a strong proposal and win the site competition.[135] The commission received approval from Texans in the fall 1987 elections—several weeks after the proposal deadline—for an additional $500 million in general obligation bonds targeted for the SSC, thus sweetening the pot with a tantalizing billion dollars in potential cost savings.

The Texas grandstanding prompted a coalition of smaller states led by Senator Peter Domenici (R-NM) to try to "level the playing field" by passing a bill prohibiting the DOE from considering any financial inducements in its selection process. States had to submit incentives in sealed envelopes that the DOE could not open until it had finally selected the SSC site.[136] But there was no way to keep secret offers of such magnitude—California had offered a $1.2 billion package, while Mississippi's incentive approached $2.1 billion—especially since many required legislative or voter approval, and local politicians took every opportunity to hype them in the media. Among finalist states, Tennessee offered to "forgive" a $2 billion electric bill owed by the DOE to the Tennessee Valley Authority, while North Carolina's $1 billion incentive included $300 million over thirty years for new university faculty positions. Illinois promised $570 million in all, including funding for tunneling and land acquisition. While members of Temple's task force acknowledged hearing such reports and rumors, they all insisted that the sealed envelopes remained in a locked safe for the duration of the selection process, and that their evaluations and final recommendations were never influenced by the incentive packages.[137]

Nevertheless, the task force attempted to evaluate Fermilab's infrastructure. Primary responsibility for determining life-cycle costs for the Best Qualified List of sites rested with Daniel Lehman, Temple's trusted deputy in the DOE Construction Management Support Division. A civil engineer with many years of experience in construction management, Lehman had worked closely with Sanford's CDG group on an SSC layout and other foundational work, and helped draft the site-selection timeline. Now he headed the task force subcommittee assessing the most important evaluation criterion, geology & tunneling. But it was the analyses of life-cycle costs that proved most challenging. Lehman's group prepared two sets of cost estimates, one for the 36 initial proposals and a more refined subset for the best-qualified sites. The initial cost estimates fell within such a narrow range that the NAS/NAE panel assigned them a "minor" role in its evaluations.[138]

The first round of estimates left unresolved how much to credit Illinois for its proposed use of the Fermilab facilities in its SSC site. After stating in its report that credit for Fermilab would only make the Illinois site "one of the less costly" sites, the NAS/NAE panel questioned the reliability of the Tevatron for use as the injector for the SSC. Several panelists expressed a concern shared by other physicists that some or all of the Tevatron's aging equipment might need to be replaced long before SSC equipment wore out.[139] Their concerns were subsequently confirmed by a special DOE-convened panel of accelerator physicists headed by Robert Siemann, a SLAC accelerator physicist (by way of Cornell). Its conclusions essentially nullified any chance of cost savings becoming the trump card for Illinois.[140]

Based on the findings of the Siemann panel, Lehman concluded that the state should be credited with between $495 million and $1,033 million for the Fermilab infrastructure. Of that, $240–312 million was for the construction costs saved by using the Tevatron (officially, it was $361 million, but $60–140 million was deducted for changes necessary to upgrade the Tevatron for use as the SSC proton injector).[141] The largest single credit granted was $233–699 million for the operating costs to be saved during the remaining five to fifteen years that the Tevatron was expected to continue operating for physics research (almost $47 million per year).[142] All told, the estimated cost savings were, at best, a third of what Illinois had claimed. In Lehman's thinking, the SSC cost driver was not its laboratory facilities but the superconducting magnets and to a lesser extent labor and utilities. Essentially nothing was allowed in the Siemann report for the value of the human infrastructure already in place at Fermilab.

The Illinois proposal ultimately suffered from a three-pronged mandate from DOE under the guise of a fair and equal consideration of all sites. First, neither Lederman nor other employees of Fermilab, the only existing national laboratory involved in the site competition, could speak out on its behalf, according to a DOE memo from Undersecretary Salgado. Illinois also could not use the total value of the federal government's 20-year investment in Fermilab to represent a cost savings for this site. And building the SSC at Fermilab should not interfere with its ongoing research.[143]

More than two decades later, Lehman and his team still expressed confidence in their estimate of the value of Fermilab's physical infrastructure. But hindsight has helped them recognize errors in the credit given for its *human* infrastructure, including expert managers, planners, engineers, technicians, and procurers. DOE's technical criteria emphasized factors important to both construction and operation. Curiously, during the site-evaluation process in 1988, existing management infrastructure was a factor that Lehman's team did not really consider. "We were just looking at getting a piece of land," he recalled. "We weren't into the next step: how do you build it? That wasn't part of our criteria."[144] Similarly, the NAS/NAE panel had earlier stated its intent to focus on "technical requirements and such other factors . . . *as would assure scientific productivity*."[145] Of course, estimating the value of existing human infrastructure for construction and operation of a future facility is elusive. There can be several widely varying estimates, each reasonably justified. But the DOE task force's decision to *ignore* completely the human element in its Fermilab life-cycle cost estimate can be questioned in hindsight, as Temple himself later acknowledged.[146]

Temple's task force was meanwhile evaluating a host of specific concerns that the NAS/NAE panel had noted among the best-qualified sites. These issues ranged from geological flaws to groundwater issues to distance from a major city. Typically spending four days at each site on alternating weeks, the team and its support staff resolved many known concerns and uncovered others not apparent in documents supplied by the states. These visits usually began with closed-door sessions in which each state's representatives discussed their proposals and answered questions. Air and ground tours were usually scheduled for the second day, while the third and final days were filled with meetings with the leaders of government, business, labor, and industry so that the task force could get a sense of what it might be like to do business in the state. At the end of each day, the various task-force subgroups convened for debriefing sessions. Members also had the benefit of additional data collected by consultant engineers.[147]

The task force conducted additional site visits between April and July 1988. Temple characterized site fly-overs in planes and helicopters as "a fantastic way to get a feel and a handle for what the site offered in the way of communities as natural kinds of landmarks, such as mountains, lakes, waterskiing, and skyscrapers."[148] In Illinois, SSC proponents sponsored a gala banquet on the fifty-seventh floor of Chicago's First National Bank building. Guests included Governor James Thompson, Congressman Rostenkowski, the conductor of the Chicago Symphony, and celebrity athletes.[149] During the Texas site visit, the task force enjoyed a parade in support of the SSC followed by a dinner hosted by billionaire industrialist H. Ross Perot. Wilmot

Hess, chairman of the task force in attendance for this visit, was duly impressed. Texas also staged a well-supported media effort to publicize the SSC and raise awareness of the benefits of bringing the project to their state. Magazine ads and articles—and even television commercials featuring local beauty queens—helped excite Texas citizens.[150]

In some states, however, the site visits sparked another kind of reception, rooted in a vision of the SSC very different from the image of a squeaky-clean high-tech industry promoted by its boosters. In these cases, grassroots protesters issued dire warnings about a project that would displace hundreds of families, create few enduring jobs for community members, and possibly produce radioactive waste—then a deepening concern because of growing revelations at DOE nuclear-weapons labs. The effectiveness of these local opponents varied from state to state. On the Illinois site visit, for example, the DOE task force encountered angry protestors with signs at the Fermilab entrance and cars driven by hostile residents following its motorcade.[151] The contrast with the Texas welcoming parade was striking.

Town meetings and public hearings held in February 1988 and again that fall at each of the seven finalist sites indicate how local residents reacted to the SSC. In Illinois, public opposition surfaced after the state's Fermilab site was announced as a finalist, when nightly informational meetings began to occur in communities around the lab. State and local officials attended these meetings of over a hundred mostly fearful residents of small communities in DuPage and Kane Counties. Emotions raged as fears were voiced of traffic, radiation, dynamite blasting, lowered property values, loss of family homes and farms, and disruption of quiet country lifestyles. At least one representative of the group SSC for Illinois or from the state's Geological Survey or Department of Energy and Natural Resources was usually present to try to answer questions with scientific facts and provide state policy information.

At a public hearing held at Fermilab on February 18, turnout was heavy. Besides complaining about the SSC itself, opponents repeatedly accused Illinois officials of being secretive, arrogant, and insensitive. Craig Jones, a member of the opposition group CATCH–Illinois, charged that the state's team had known the location of the proposed SSC tunnel since at least early 1987, and certainly since formally submitting its bid to DOE in September of that year, but that officials chose not to reveal it to local politicians and residents until a January 1988 town board meeting.[152] Also in attendance, Governor Thompson assured SSC opponents of his support for "any reasonable effort by the legislature to compensate property owners for declines in property values." He acknowledged that local tax values would not be what they are today without Fermilab. Warrenville Mayor Vivian Lund rose in support of the project, saying the fears sounded familiar, for she had heard it all twenty years before, when Illinois won the site competition for Fermilab. "None of these [fears] have materialized," she observed. "None of us glow in the dark, and our hair is falling out only because we are getting older."[153]

Similar uncertainties swirled about the proposed North Carolina site, which might have required more than 600 property owners to sell homes, land, or mineral rights.[154] Some objected that the state had developed its SSC bid without consulting local officials. As in Illinois, SSC proponents in the Tarheel State were left struggling to overcome mistrust and anger.[155]

In sharp contrast were the much more supportive reactions among audiences in Colorado, Michigan, and Texas—even though the latter two states were both proposing to relocate hundreds of people.[156] A multitude of factors undoubtedly contributed to this response. Michigan had initiated social-impact assessments back in 1986, hoping to assuage residents' concerns and curb opposition before it could gather momentum. Colorado's SSC team took similar care during the early stages of proposal preparation, working with the state Department of Local Affairs to form committees with residents from the Fort Morgan area. Not only did this approach help to include local perspectives on problematic issues and improve the state's submissions to DOE; it also reassured residents about both the SSC and the state's intentions. Texas proponents held early informational meetings throughout the state and selected Waxahachie (and another site) through an intrastate competition in which several localities submitted site bids.[157] Texas was the only state to hold such an internal site competition, which helped generate a widespread base of popular support for the proposed project.[158]

Fermilab had worked to create a similar sense of community since the early days of establishing the laboratory. After two decades, the facility was regarded as a good neighbor and its resident scientists considered model citizens. Neighbors could freely wander about Fermilab's scenic campus, which included an award-winning restored prairie, a buffalo herd, and a magnificent sixteen-story central laboratory building modeled on the thirteenth-century Beauvais Cathedral in France.[159] Joggers and bicyclists coursed its pathways, while foreign films, lectures, and other cultural activities attracted thousands of visitors annually. Groups from rival states, seeking to gain an idea about how the SSC might look, came away impressed.[160] But suspicions of local real-estate interests, along with a not-in-my-backyard attitude, agitation by members of CATCH, and the uncertainty about the exact location kept

Illinois from capitalizing on Fermilab's good reputation. As the site competition neared its end, the Illinois team submitted answers to DOE questions and began drafting a list of things to be done if the state won.[161]

While many members of the US high-energy physics community held strong opinions about the preferred SSC site, they did not publicly question the DOE site-selection process. But in late August 1988, as the task force was about to wrap up site evaluations, University of Chicago physicist and Nobel laureate James Cronin weighed in and called upon his colleagues to do likewise. As a member of CERN's Scientific Policy Committee, he appreciated that European physicists enjoyed a unique advantage in this research because their nations had to support only two major facilities—CERN and DESY. Cronin concluded that it would be a grave mistake to start yet another US high-energy physics laboratory. He expressed this opinion in a letter to URA Chairman John Marburger: "This experience has led me to believe most strongly that the SSC be located at Fermilab," he stated. "If the site is chosen elsewhere, we will be starting a fifth high-energy laboratory which will spread even more thinly the human resources and scarce operating funds."[162] Taking a firm stand in favor of the Illinois site adjacent to Fermilab, Cronin resigned from the URA Board of Trustees and tried to solicit the support of the high-energy physics community via an open letter to his colleagues urging them to write Hess to endorse the proposed Illinois site, stressing the need to concentrate limited human and financial resources at Fermilab.[163] His request stimulated over thirty letters from community leaders; almost two-thirds of them agreed with his viewpoint.

But these letters did not seem to have much, if any, impact on the DOE site-selection process. In mid-September, having studied the best-qualified sites for nine months, Temple's task force left its Germantown offices and gathered in a Frederick, Maryland, hotel for a week-long retreat to rate the individual sites.[164] As the discussions progressed, a draft of what would become known as the "Green Book" after the color of its cover, began to take shape with assessments of the strengths and weaknesses of each site. For each of the six evaluation criteria, the seven sites were graded as outstanding, good, satisfactory, poor, or unsatisfactory. During this week, final site rankings based on individual state scores emerged.[165]

As predicted by many site-competition participants and pundits alike, the strongest candidates proved to be Illinois and Texas.[166] Both states received "outstanding" grades for the geology of their respective sites, and similarly for their regional resources. But although these were the two most important criteria for the evaluations, they were not the *only* criteria, and Illinois lost points in the other four categories. Despite the many environmental awards Fermilab had won, Illinois received only a "good" on its environment evaluation, compared to the "outstanding" awarded Texas. But it was the fourth criterion, setting, for which Illinois was most penalized; the state was rated "poor"—the only such grade issued to any state on any of the criteria. Disturbed by the local opposition, the task force also cast doubt about whether the state Department of Energy and Natural Resources could handle the land acquisition in a timely manner.[167] For the CATCH-Illinois group threatened to "tie up this project in the courts for 35 years."[168] Another contributing factor was the task-force members' impression that having to use the Tevatron as the SSC injector permitted very limited flexibility in ring placement. Other experts insist, however, that had the DOE really wanted to use its investment in Fermilab, it could have found a way.[169]

The Waxahachie site had few flaws, according to the task-force evaluations. Environment subgroup leader Warren Black figured that putting the Super Collider there would facilitate efforts to restore land modified by decades of agriculture that was not being subdivided and developed as were the counties around Fermilab.[170] As for its setting, Texas plans for land acquisition and relocations were considered feasible and sensitive to landowners, while only minor limitations compromised its flexibility in the placement of the collider in the unused cotton fields south of Dallas. Texas was the only site to receive four "outstanding" ratings.[171]

The task force made its decision based on the site scores and wrote its report in September 1988, ignoring the fact that the seven finalist states were still working on their responses on the Super Collider's environmental impacts, which were due on October 17.[172] On November 7, 1988, three days after the SSC Laboratory M&O proposals had been due, the final report was submitted to the DOE administration. The following morning—on the day George H. W. Bush was elected forty-first president of the United States—the task force presented its findings to top-level DOE officials on its Energy System Acquisition Advisory Board, which included Secretary Herrington, who reportedly made the final decision.[173]

In a press conference on November 10, Herrington announced Waxahachie as the winning site (see fig. 3.5). For a competition that DOE had launched twenty months earlier with the proclamation it would be fair and open, such timing could hardly have been less appropriate. The losing states inevitably linked the choice of Waxahachie with Bush's victory.[174] As requested by Congress, the US Government Accounting Office launched an

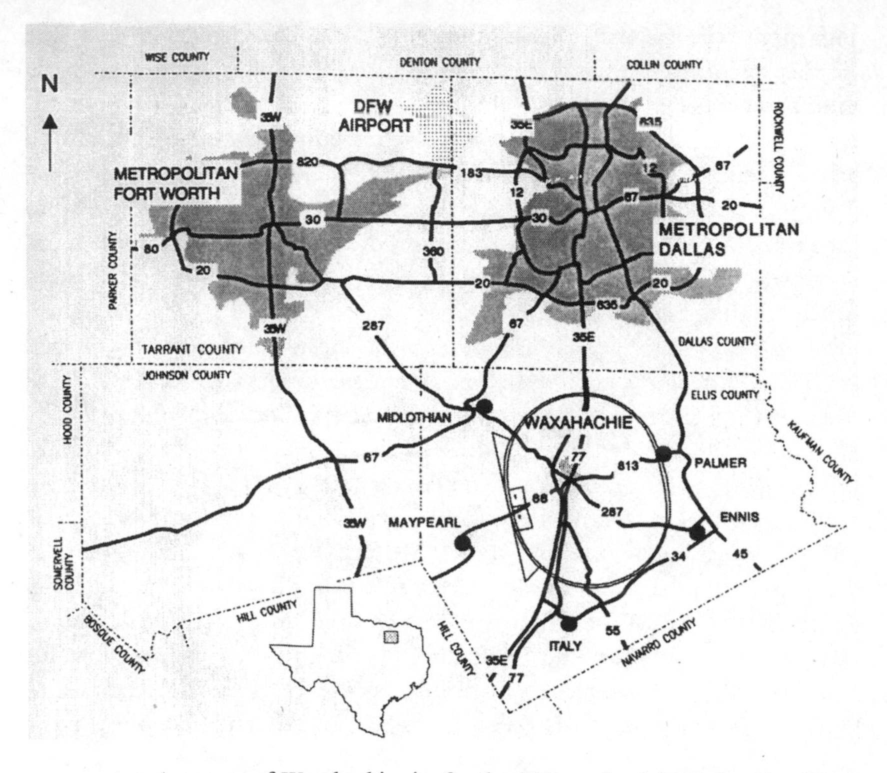

FIGURE 3.4 Area map of Waxahachie site for the SSC, south of the Dallas-Fort Worth metroplex. Courtesy of Fermilab Archives, SSC Collection.

investigation; the watchdog agency uncovered nothing untoward in four investigations of the site-selection process.[175] But knowledgeable Washington observers probably recognized that the powerful Texas congressional delegation, plus the fact that the new president was an adopted Texan, must have played some role—consciously or not—in the DOE decision.[176] It certainly helped that the Waxahachie site had come in ahead of the other sites in the task-force rankings.

Eager to emerge from the difficult economic times owing to the faltering oil business, savings-and-loan failures, and rising unemployment, Texas officials were jubilant over the victory, which capped a bipartisan effort involving dozens of state and federal lawmakers, faculty and administrators at four Texas universities, Washington insiders, and many private individuals and interests.[177] "We pulled a lot of different constituencies together," boasted Governor Clements on the day after the DOE site announcement.

"We hitched all these people to the same wagon, and those horses really performed."[178] In an editorial headlined "Thanks for the Extraordinary Sales Job," the *Dallas Morning News* was equally effusive about the "remarkable team effort, the likes of which this area never has seen before."[179]

US science had not seen a site competition quite like the one for the SSC, either. *Life* magazine soon featured an article about Waxahachie, with photographs of "Miss Waxahachie" as well as glowing references to the Super Collider.[180] The political stars seemed in perfect alignment over the Lone Star State. A Texan was president-elect, Texan Lloyd Bentsen—who had just lost the election as the vice presidential candidate on the Democratic ticket—still chaired the powerful Senate Finance Committee, and Fort Worth Congressman Jim Wright was the Speaker of the House. "We've got 27 Congressmen, two Senators, and a President," said Representative Joe Barton of the Ennis congressional district in which the Super Collider was to be built. "If any state can do it, Texas can do it."[181]

FIGURE 3.5 DOE Secretary John Herrington announcing selection of Waxahachie as the SSC site, on November 10, 1988. Seated at the table are (*l. to r.*) OER Director Robert Hunter, Wilmot Hess, and L. Edward Temple. Credit: AIP Emilio Segrè Visual Archives, *Physics Today* Collection.

The joy in Texas was matched by dismay and disbelief in the other finalist states, especially in Illinois and at Fermilab, as most of the physicists there had been expecting to win. Only a few correctly predicted the outcome, among them Richard Lundy, who in late October had asked a colleague to record three predictions: the SSC would be sited in Texas; URA would be the M&O contractor; and the project would fold in five years or less.[182]

News of the Waxahachie site selection had leaked to the press and reached the Chicago area early on November 10. Many Fermilab employees heard about it on the radio as they drove to work that morning. Small groups gathered in corridors and corners to share their great disappointment and ponder the future of the lab and their jobs. Just before noon, nearly a thousand members of the Fermilab community poured into the main auditorium in Wilson Hall to hear what Lederman—who a month earlier had been awarded the Nobel Prize in physics—had to say about the SSC site selection and what it meant for the lab.[183]

He strode to the podium promptly at noon, having prepared a surprise he hoped would relieve the community's distress. From under the podium, Lederman pulled out a large Stetson hat and put it on his head as the surprised audience burst into laughter. After waiting patiently for it to subside, he removed the hat and joked, "I don't know if we have such a thing as a prairie hat." Again the audience roared with laughter. Then he began to read his prepared statement, one of two he had written the night before for the two possible outcomes.

"As you know, the Department of Energy selected Texas as the site for the SSC," Lederman remarked. "I personally believe that it would have been much easier to build the machine in Illinois and harder to build it in Texas." He tried to assure listeners that Fermilab would not close. He spoke briefly about the success of the Tevatron and the need for its research program to continue as "the prime tool for keeping physics in the United States going over at least the next decade, if not the next fifteen years," noting that the SSC would not be ready to publish any research until roughly 2000.

"Is there any chance that Fermilab would be put in a box and shipped to Texas?" someone asked. "Move the apparatus?" Lederman reacted. "We won't let 'em." Another inquired whether he might be headed for the SSC, but he sidestepped the question. "I think we'll all be involved in it," he said. "Fermilab has a major role to play in the R&D program." Unsure whether he believed what he was saying, yet certain he needed to say it, he forecast a likely buildup at the lab to provide facilities to train the next generation and do some of the work that the SSC would require.[184]

After responding to a few more questions, he put the Stetson back on. "Call me Tex," he drawled hoarsely, and soon stepped from the podium as the audience cheered, although downcast.

As 1988 came to an end, SSC advocates could look back with pride on their achievements of the past four-and-one-half years. They had taken what was largely a speculative idea in July 1983 and developed a detailed conceptual design for the Super Collider that won accolades throughout the high-energy physics community. They had secured the support of the Reagan administration—including of President Reagan himself—for the multibillion-dollar project. And after a highly public nationwide competition, DOE officials decided that the SSC was to be sited near Dallas, Texas, with URA serving as the management and operations contractor responsible for transforming this dream of a world-class scientific laboratory into reality.

But the project still faced major hurdles. Most importantly, it needed the support of Congress—and by extension the American public—if it were ever to be built. For in the US government, Congress controls the purse strings, not the president. And there were many other interests competing for a share of the federal budget during the late 1980s and early 1990s, when it was being battered by burgeoning demands and a weakening US economy.

Congress was hesitant to proceed with a project that would suddenly need hundreds of millions of dollars annually and continue to have such requirements for the next decade. Before fiscal year 1989, the SSC R&D budget had been only $25 to $35 million, an amount that was easily accommodated. But Congress balked at the Reagan request for $363 million to begin SSC construction that year, allowing the project only $95 million to continue doing R&D.[185] Part of the problem was the lack of any firm, significant foreign commitments.

Much more than any other major US accelerator project, the Super Collider had to be sold to the American public. But the Department of Energy and the high-energy physics community did a lackluster job of taking the scientific case for the SSC to this audience. As there had never been much need to do so on prior projects, the necessary talent and expertise had not been cultivated in either the agency or the community. The Central Design Group made valiant efforts, giving government and media relations high priority and placing good communicators into positions of authority. But actually communicating the exceedingly abstract physics goals of the Super Collider was a difficult task, even for the best science writers. Many reporters fell back on the project's enormity, its cost, the site competition, and the competition with Europe as major themes in their SSC coverage. These themes gradually began to dominate the public discourse on the Super Collider, to its detriment.

High-energy physicists failed even to convince many other physicists, who probably understood the scientific arguments for building the Super Collider,[186] of the importance of the frontier research it would enable. Concerned about the possible impact of the SSC's multibillion-dollar cost on their own research, given the growing budget pressures, "small scientists"—especially condensed-matter physicists—began voicing serious objections in Congress and the press. Such explicit, vocal opposition to another physics project was unprecedented. It weakened the scientific case for the SSC and gave congressional skeptics seemingly valid reasons to vote against it.

In a similar vein, DOE had difficulty convincing foreign science agencies that a major commitment of funds should be made to an overseas laboratory being billed as the means to restore US preeminence in high-energy physics. These funds could be spent much more effectively on their own research programs or to support high-energy physics research at CERN and DESY. Smaller amounts might be committed to support some of their physicists who wished to work on SSC research, but such amounts were likely to be tiny compared to the multibillion-dollar total project cost. Only Japan seemed at all likely to make a major foreign contribution to the SSC project.

Taking their cue from the successful site contest for Fermilab two decades earlier, DOE leaders had calculated that a fair and open competition for the SSC site would help generate widespread support for the project. It indeed stimulated interest in the project among the US public, which might not otherwise have paid as much attention. But this was the wrong kind of attention, for the Super Collider subsequently became identified primarily as a public-works project and secondarily as an important science project required to reestablish US leadership in high-energy physics. When the final decision was eventually made to build the SSC in one state, the congressional support that had accumulated was likely to wither. As a senator who supported the SSC told Steven Weinberg, "A hundred Senators will favor it, but once the site is chosen, the number decreases to two."[187]

DOE officials strove to hold an equitable site competition and largely succeeded. But the criteria that they established focused too much on the physical features of the sites and on the social and cultural characteristics of the surrounding areas. The physical infrastructure at Fermilab was consequently evaluated at between $500 million and $1 billion. But the *human* infrastructure of this successful, world-class high-energy physics laboratory—with thriving relationships already established among the diverse individuals who had to interact within it—was almost completely ignored in this evaluation.[188] In retrospect, the *true* value of the existing Fermilab infrastructure, both physical *and* human, probably should have been set somewhere in the range between $1 billion and $2 billion.

On the specified DOE criteria, Waxahachie probably was the leader among the seven best-qualified sites. The "poor" rating on setting assigned to the Illinois site because of the minority but vocal public opposition to the SSC seriously damaged its overall standing and gave the edge to Texas. But had the Fermilab human infrastructure been properly included in the equation—and sufficient priority given to the cost-effectiveness of the finalist sites—the decision might easily have gone the other way. As discussed in subsequent chapters, the project costs could have been controlled more easily and effectively in an existing laboratory with strong management structures already in place.

URA now faced the truly daunting task of recruiting top-notch physicists, engineers, technicians, planners, designers, managers, procurers, and a myriad of other specialists needed to a region that many high-energy physicists regarded as an intellectual backwater, at least from a scientific perspective. SSC leaders had to begin fusing all these disparate agents and actors into a world-class high-energy physics laboratory—a new secular citadel of science—on the windswept plains south of Dallas. And many likely recruits still remembered Dallas as the "City of Hate," where twenty-five years earlier that November President John F. Kennedy had been felled by an assassin's bullet.[189]

CHAPTER FOUR

Settling in Texas, 1989–91

You don't start out a project by asking for more dough; that's the way to kill a project.

—ADMIRAL JAMES D. WATKINS, December 1989

One of the first high-energy physicists to visit the newly designated SSC site in rural Ellis County was Dennis Theriot of Fermilab, who had served as a deputy to Schwitters on the CDF Experiment. He arrived there to survey SSC prospects in November 1988, just days after Texas had been declared the winner. With its graceful 1895 red granite-and-sandstone courthouse at the center of town, and scattered Victorian houses dating back to its turn-of-the-century cotton-farming roots, Waxahachie had been the setting for several films—including *Tender Mercies, Places in the Heart,* and *The Trip to Bountiful.* But the town's historic charm belied a world very different from what Theriot knew in the busy suburbs west of Chicago, as he and his wife discovered in meeting for lunch with the two Texans who were to escort them around:

> We walked into this cafe and the hostess asked us, "Upstairs or downstairs?" One of the Texans said "Upstairs." So we went upstairs and sat down. When the waitress came up to take our order, they all ordered Coors. I knew from reading that this area of Texas was dry, and I was somewhat puzzled how you could get beer in a dry county. The waitress looked at us and said, "Are you boys members here?" One of the Texans said, "[Waxahachie businessman] Buck Jordan sent us." She said, "Well, if Buck sent you, that's OK." And so she took our beer order and then I ordered . . . chili to go with it. But I knew I was in a different atmosphere than I had ever been in before.[1]

Culture shock was hardly the only challenge facing Theriot and the other pioneer builders of the SSC Laboratory. After lunch his hosts showed him the first parcel of land that Texas would be donating, for the laboratory's west campus. Part of his assignment was to locate a suitable place for a high-tech laboratory and office trailers, with sufficient power, water, and sewage facilities for as many as 300 people. The site-specific design for the SSC would have to be developed there. But an international scientific laboratory seemed inconceivable in this location. Theriot found himself surrounded by fallow cotton fields that were "flat as a pancake" and devoid of any vegetation except brown grass. A single string of electric utility poles carried only three wires, and the nearest major power lines were more than three miles away.[2] In town and south of Dallas, foreclosure signs were everywhere.[3]

When told that local residents used septic systems to treat sewage, Theriot asked how large a drain field might be needed for 300 people. This question drew laughter from one of his hosts, Edward C. Bingler,[4] a geologist who said it would cover the entire county, because only three feet beneath the surface was hard rock. "I decided then and there that it would take some time developing an infrastructure before the lab could actually be at the site in Waxahachie," recalled Theriot.[5] Instead he chose a suburban industrial park south of Dallas as the initial home for the SSC design team. His decision was reminiscent of Robert Wilson's 1967 selection of Oak Brook Executive Plaza as the first home for his NAL team, before its 1968 move to Weston.[6]

As planned, the SSC Laboratory (SSCL) was to encompass over fifty miles of tunnels filled with over ten thousand sophisticated superconducting magnets and cost some $5 billion in 1988 dollars. To achieve this goal, the prime contractor URA had deliberately teamed with two subcontractors, EG&G and Sverdrup, with extensive experience working with the US armed forces. While this unwieldy arrangement helped allay DOE concerns that the management of this enormous project be in capable professional hands, it required the delicate merging of two different and potentially adversarial cultures—academic-scientific and military-industrial—to collaborate rather seamlessly on a common goal. Forging this alliance was to prove far more difficult than most anticipated.[7]

FORMING A MANAGEMENT TEAM

As 1989 began, URA and the DOE were negotiating the M&O contract that would specify their interactions over the SSC lab. Known as "government-owned, contractor-operated" (GOCO) institutions, all DOE laboratories are governed by such contracts.[8] The principal point in contention was a paragraph allowing DOE to order URA to use specific subcontractors for portions of the work. To the URA negotiators, this was a serious problem.[9] It left the door open to political manipulation of DOE by corporate lobbyists, and it could make URA responsible for work of subcontractors that it would not otherwise have selected or approved. Nothing like it had been included in previous DOE (or AEC) contracts for Brookhaven, SLAC, or Fermilab. But DOE negotiators led by attorney Edward Cumesty from its Chicago Operations Office would not yield on this paragraph, so the two parties finally "surrounded it with paper"—setting up a process whereby any invocation of the provision could be appealed all the way up to the Secretary.[10]

Wolfgang K. H. Panofsky, the founding director of SLAC then serving as vice chair of the SSC Board of Overseers, remarked at the time that the DOE was treating the project "more as a procurement" than as the traditional "partnership" between government and academia that had existed since postwar years.[11] Instead of URA, the DOE leadership would have preferred that a major corporation with extensive experience on billion-dollar projects, such as Lockheed or Martin Marietta, be the entity primarily responsible for building the SSC Laboratory, which would then be handed over to "us longhairs" to operate, Panofsky said. But that would have been a big mistake, he argued, for building a successful scientific laboratory involves much more than constructing roads, buildings, and accelerators. An intellectual edifice also has to be erected during construction, and that requires the intimate involvement of its eventual scientific users in the building process.

The SSC M&O contract between URA and the DOE was finalized and inked on January 18, 1989, days before Ronald Reagan stepped down as president and George H. W. Bush assumed office. Enacting it before Reagan departed was a pressing goal of Energy Secretary Herrington. In fact, the press release announcing the signing suggested the lab be named the "Ronald Reagan Center for High Energy Physics," in honor of the president who had pushed hard for its establishment.[12] For over a month afterward, physicists joked that the SSC might become known as the "Gippertron," reflecting Reagan's acting role as Notre Dame quarterback George Gipp. But the incoming Bush administration did not force this issue upon the laboratory, and the idea gradually died away—to the relief of many in the high-energy physics community who considered such a politically motivated name to be inappropriate for what they hoped would be an international scientific laboratory.[13]

On the same day the M&O contract was signed, Maury Tigner resigned as the deputy director designate, saying he wished to return to Cornell. The timing of his resignation suggests he had already made up his mind, well before that.[14] Schwitters recalled that he had had long discussions with Tigner, dating back to the previous September, about the appropriate roles that he might play at the SSC Laboratory,[15] but that they could not reach a clear understanding about these matters. According to Schwitters, Tigner was requesting too much authority: "It was clear he wanted total authority on lots of the detailed stuff that couldn't be granted," recalled Schwitters. "He had a very specific list of people he demanded in key roles," namely his CDG associates and colleagues.[16] It was crucial for the transition to Texas that the SSC design team that had worked together well under Tigner be folded seamlessly into the emerging SSCL management structure. But after their leader had departed, many CDG members followed his example and returned to their former universities and laboratories.[17] A number of leading CDG physicists indeed went to Texas to work full time on building the SSC, but they were sprinkled throughout a complex management structure in midlevel positions that lacked the authority they had formerly enjoyed at CDG.

Curiously, Tigner had been listed only as deputy director designate on the URA M&O proposal, and *not* as project manager—the position many thought ideal for him. In fact, the DOE request for proposals did not require that the individual filling this crucial position be explicitly named, and thus nobody was. Panofsky and Schwitters have speculated that this omission was deliberate: that the DOE expected to fill this role from within its own ranks.[18] Tigner, for whom the transition from CDG to the SSC lab remained a sensitive topic years later, allowed that "the DOE had indicated that I would not be acceptable as project manager."[19] Whatever the truth, one of the most capable accelerator physicists in the world, who had been central to the SSC design effort for half a decade, and many of the people loyal to him, were effectively lost to the project. Consequently, the minimalist Wilsonian accelerator-building philosophy that had pervaded the CDG design effort did not find its way to Waxahachie.

Tigner's departure left a gaping hole in the SSC management structure that was among the director's topmost priorities to fill. While Helen Edwards—one of four physicists who had led the successful Fermilab project of building the Tevatron—was already on board as technical director, and Fermilab business manager Bruce Chrisman had agreed to become associate director for administration, the SSC Laboratory needed a strong leader to shoulder the day-to-day responsibilities of getting the Super Collider built. That would entail establishing a project-management team and setting up a cost-and-schedule control system (often referred to by the shorthand CS^2) that was required to track construction activities and expenditures.[20] Doug Pewitt, whom Schwitters had earlier hired as the associate director for integration and management, responsible for liaison with Washington, attempted to cover some of these responsibilities while the search began for a person to fill the position permanently. With others, Pewitt began trying to set up the cost-and-schedule control system but encountered difficulty in getting physicists to buy into using the system, which was foreign to their normal operating procedures.[21]

The search soon focused on Richard Briggs, an associate director for beam research and magnetic-fusion energy programs at Lawrence Livermore Laboratory, who had led its induction accelerator project.[22] In April, Schwitters offered him the position of deputy director and project manager, second-in-command at the SSC Laboratory. Briggs accepted, but his appointment was blocked by Robert Hunter, the new director of the DOE Office of Energy Research, who held the authority to approve candidates for the major SSC positions. Hunter admitted in a later interview that he thought Briggs lacked the large-project management experience needed to cope with such a huge, multibillion-dollar effort.[23] Others have echoed similar sentiments, saying that Briggs was more a program manager than a true project manager.[24]

Concerned about the SSC Laboratory's inexperience with large, billion-dollar construction projects, Hunter insisted that *someone* with such experience be included in its top management ranks.[25] Schwitters had served only as the cospokesman for the CDF Experiment at Fermilab, roughly a $50 million project at that point in its evolution, while Briggs had comparable experience at Livermore. The inclusion of Sverdrup in the SSC M&O contract was supposed to remedy the problem, but this teaming arrangement was not working out as planned.[26] Sverdrup executives began sending their third-rate employees to Texas—people who probably would otherwise have been laid off; they proved incapable of launching a project of such magnitude and complexity. The other teaming partner, EG&G, was responsible for technical and administrative services. As yet, however, nobody on the SSC management team had anything close to the needed project-management expertise.

Hunter had also decided that the SSC was too politically visible and costly a project to be "managed" in the usual DOE (or AEC) manner—through the program office in nearby Germantown, Maryland, and the responsible DOE field office, the Chicago Office at Argonne National Laboratory.[27] In anticipation of the project becoming a reality in 1988, the DOE Office of High Energy and Nuclear Physics had already set up a separate SSC division in Germantown, headed by former Argonne high-energy physicist Robert Diebold. But after the M&O contract was signed in January 1989, Hunter began reorganizing the Office of the SSC (OSSC) within the Forrestal Building.

Louis Ianniello, a DOE career civil servant, became the acting associate director of this office and began formulating plans for the new organization with Diebold and other DOE officials. As this plan emerged in the spring of 1989, it soon became clear that the OSSC was to be no small operation, for it entailed over 200 full-time employees serving either in Washington or in a site office at the SSC Laboratory.[28] Hunter wanted the DOE to take direct responsibility for managing construction of the laboratory. According to his plan, SSC physicists were supposed to design and specify the required facilities, while DOE officials would contract out the actual construction process and manage its implementation.

This was hardly the first attempt at setting up such a relationship between the DOE (or AEC) and its laboratory contractors. A long history stretching back to the establishment of Brookhaven National Laboratory and including the negotiations leading to construction of the Stanford Linear Accelerator Center reveals instances where the agency assumed or tried to assume direct control over subcontractors doing the actual construction.[29] The argument usually made was that university professors lacked the expertise required to manage such large-scale construction projects. Physicists countered that establishing a successful scientific laboratory entails a lot more than erecting buildings and associated experimental facilities. As the individuals involved in this process are also laying the groundwork for a new laboratory culture, the control of the project should rest in their hands.[30] Scientists should be the ones to manage the project, according to this view, while the federal agency's proper role was to "oversee" their activities.

High-energy physicists could—and often did—point to their successful construction of SLAC and Fermilab during the 1960s and 1970s, by teams led by Panofsky and Wilson.[31] Each project had cost well over $100 million (the equivalent of about $400 and $750 million in 1990s dollars) and had come in on schedule and at or below budget.[32] In both cases, scientists had effectively managed the construction process, while the AEC had overseen their activities. But DOE skeptics could refer to the recent examples of Isabelle at Brookhaven and the Tevatron at Fermilab, which involved superconducting magnets and encountered long delays and large cost overruns.[33] Both were managed by high-energy physicists, with DOE serving in an oversight capacity.

Panofsky and Wilson were already seasoned accelerator builders when they built SLAC and NAL, with professional roots and connections reaching all the way back to the Manhattan Project. They were prominent members of an elite group of scientific "master builders," who emerged in the postwar years, able to navigate the corridors of government (and military) power and build large projects or national laboratories.[34] By comparison, Schwitters and Briggs were relative neophytes with little experience on such major projects.[35] And the SSC was at least an *order of magnitude* larger and more complex than anything yet attempted by US physicists. Hunter had good reason to be concerned about the emerging SSC management team, but he may have underestimated the skills of the physicists from DOE laboratories.

In addition, the Energy Department was experiencing severe criticism because of the recent public outrage over pollution at its nuclear-weapons labs—a problem unrelated to its high-energy physics labs. In January 1989 President Bush had asked Admiral James D. Watkins, a mechanical engineer (and a protégé of nuclear submarine pioneer Admiral Hyman G. Rickover) who had served as chief of naval operations in the Reagan administration, to step in as energy secretary and address this problem. As a former commander of a nuclear submarine and cruiser, Watkins had immense respect for physicists, who had provided an almost limitless energy source for naval propulsion. But he was determined to bring military discipline to bear on what many Washington insiders considered a rather aimless, dysfunctional agency.[36] His ascendance only heightened the growing DOE determination to exert stronger control over its sprawling network of national laboratories.

With his primary focus that first year on cleaning up the nuclear-weapons labs, Watkins allowed Hunter fairly free rein in his dealings with the SSC project. As spring turned into summer, it became increasingly obvious to URA that the DOE wanted to manage the project by itself through the new Office of the SSC. According to Schwitters, the DOE "had a philosophy that . . . the physicists should specify [the SSC design] and hand it over to what they called an industrial contractor," which would then "execute the design, build it, and then hand it over to the physicists to operate."[37] In this "military procurement" approach, the DOE Office of the SSC would directly manage the individual contractors through completion of each task. Schwitters called it an "unworkable" approach, completely unacceptable to URA.

The dispute reached a climax in August 1989, when it emerged that the DOE management plans included use of Washington-area subcontractors (often dubbed "Beltway bandits") to help the Office of the SSC manage the project, replacing URA in some of its contract responsibilities; this was effectively invoking the contentious paragraph in the M&O contract. URA appealed to Secretary Watkins, in a meeting at the Forrestal Building that included Hunter, Panofsky, Schwitters, and Thomas Luce, the chairman of the Texas National Research Laboratory Commission (see below). An experienced lawyer and Washington lobbyist, Luce pointed out that exactly the same verbiage had been used to describe the contracted work expected of both URA and these subcontractors. Watkins admitted that this was absurd.[38] In a September 7 memo approving the DOE's management plan, he scaled back the responsibilities of the Office of the SSC and significantly reduced its possible staffing levels, to at most thirty people at headquarters and another sixty at the Texas site office.[39] He also gave URA full responsibility for all aspects of SSC management, technical direction, design, and construction.

This was a major victory for URA, or at least it seemed like one. Briggs was soon approved as the SSC deputy director and project manager. In mid-October, Hunter resigned under pressure and returned to Southern California.[40] For a brief, shining moment, it seemed as if URA and the SSC Laboratory had won the management battle.

THE SITE-SPECIFIC SSC DESIGN

Among the first activities of the new SSC Laboratory, even before moving to its temporary Dallas headquarters, was to convene a panel of expert physicists and industry representatives to review CDG superconducting magnet R&D activities. As discussed in chapter 2, CDG had experienced significant management problems and difficulties in fabricating prototype magnets able to attain the 6.6 tesla magnetic fields required for full 20 TeV operation. On March 21, 1989, Schwitters delivered a charge to the Collider Dipole Review Panel, requesting that it "carefully review, discuss, and report to me on the present status and future prospects of the laboratory-based SSC collider dipole R&D development program and on the current best full-length designs coming out of that program."[41] Led by Gustav Voss of DESY, who directed its HERA project, and CDG magnet chief Tom Kirk, about two dozen physicists and engineers gathered at LBL for a week in late April to review the SSC magnet R&D program in detail.

In its report, the panel highlighted several areas of concern, including "the insufficient operating field margin for meeting the machine's specifications at the highest operating energy."[42] A few prototype magnets had indeed reached the fields required, but only one had succeeded at the 4.35 K design temperature. For trouble-free operation under ordinary SSC conditions, the report continued, "a 10 percent margin in the operating field before quench should be achieved." Prototype magnets had sufficient margin for SSC operation at 18 TeV per beam, that is, but not 20 TeV. The report also voiced concerns about the quality of the magnetic fields during the injection process, which was intended to occur at low energies close to 1 TeV. Here, stray "persistent currents" in the superconducting NbTi materials might induce erratic, time-varying fields that would enlarge the low-energy beams after proton injection, potentially leading to serious beam losses.[43] Means to correct for these fields needed to be designed, tested, and implemented, which would involve additional time, effort, and expense.

Not included in the panel's original charge was the question of the dipole magnet aperture, the inner diameter of the beam tube, which CDG had set at 4 cm after extensive analysis (see chapter 2). A belated request came in—likely from Helen Edwards or Schwitters—to consider the possibility of enlarging the aperture to 5 cm and to assess the impacts of so doing. As the report stated in its introduction,

> After receiving the formal charge from Dr. Schwitters, the panel's scope was broadened by the addition of one very significant topic. This was the question of increasing the aperture of the dipole from its current 4 cm value to a new value of 5 cm. There are a number of beneficial technical reasons for considering such a change, but it is anticipated that this change would likely result in cost increases and schedule delays.[44]

Among the major benefits would be reducing proton beam losses at injection, but the "cost increases and schedule delays" mentioned could be significant indeed. "If a redesigned magnet is to be thoroughly tested prior to the start of the [magnet industrialization program], it is likely that *a delay of two years could occur*," the report continued. "Although this appears on the surface to be the least risky program for changing the aperture, the loss of momentum from the delay in starting transition to the industrial team would be very costly."[45] Now that the SSC project was about to become a reality—and in part because Tigner was no longer involved—serious questions about excessive design risk began to surface. And design conservatism had started to creep in, with all its attendant costs.

During the CDG design process, the question of magnet aperture had been examined extensively. It was obvious by 1985 that a 5 cm aperture would result in better magnets, with a larger cross-sectional area of uniform fields at their core. But it also meant that more iron, superconducting wire, and other materials would be needed to fabricate these magnets, driving up their cost. Thus Tigner decided they would design for 4 cm apertures, which meant that correction-magnet coils *had* to be added just outside the beam tube—an approach then being pursued on HERA.[46] Operators could adjust the currents in these correction coils independent of those in the main

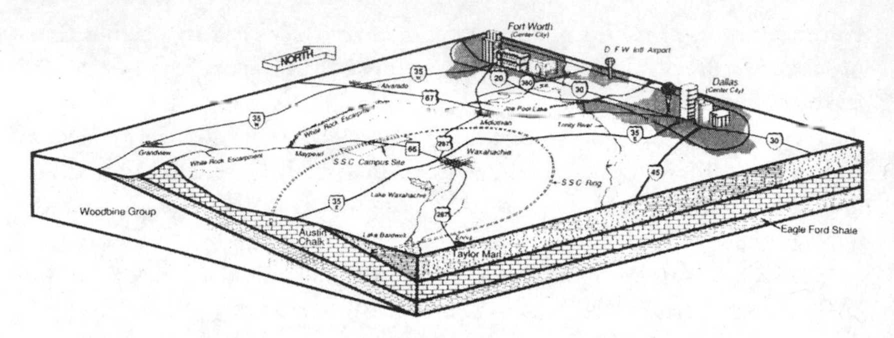

FIGURE 4.1 Artist's conception of the Waxahachie, TX, site, showing the underlying geological layers of Taylor Marl, Austin Chalk, and Eagle Ford Shale. Courtesy of Fermilab Archives, SSC Collection.

superconducting coils to try to reduce or eliminate unwanted stray fields and achieve the required field uniformity at the center of the tube where the beam passed through it. But insufficient R&D had been done on these correction coils by the time of the review.[47] Although successful tests had been performed at LBL with shorter-length model magnets, tests of full-length, 17-meter magnets with correction coils installed in them had yet to begin.

By late spring of 1989, the nucleus of the SSC Laboratory technical staff had formed at an office park on Beckleymeade Avenue in South Dallas rented by EG&G as the temporary headquarters. Led by Edwards, a group of physicists was in the midst of adapting the generic SSC design, developed by the CDG to be applicable to *any* potential site, to the peculiar characteristics of the Waxahachie site. This "site-specific design," which had to conform to specifics of local geology and land-use patterns, would normally have involved only minor changes to the original CDG design.[48] For example, a question arose about whether to tilt the plane of the main tunnel slightly to keep it entirely within the more stable Austin Chalk formation and avoid the unstable, lower-lying Eagle Ford Shale; a decision was subsequently made not to do so.

As an outcome of the magnet review, however, the group began to revisit the principal SSC design parameters such as the aperture and proton injection energy. Prodded by Edwards, whose design philosophy of "greater margin" contrasted with Tigner's minimalist approach, this group was much more inclined to build in additional safety margins by increasing the aperture and injection energy, despite the extra costs.[49] Their thinking had been buttressed by new, sophisticated supercomputer simulations of beam performance done by physicists David Ritson of Stanford and Rae Stiening of the SSC lab, which indicated that significant proton losses would occur at injection due to stray fields.[50] These losses could be substantially reduced by increasing the magnet aperture or the energy at which protons would be injected into the two main collider rings. In late June, Edwards decided to proceed with the SSC design *as if* its aperture would be increased to 5 cm *and* the injection energy to 2 TeV.[51] But Schwitters had made no formal decision yet to do so.

Management problems related to the difficulty of merging the academic and military-industrial cultures at the fledgling laboratory had also begun to emerge.[52] In a confidential June 26 memo to the members of the SSC Board of Overseers, chairman Boyce McDaniel of Cornell reported that EG&G and Sverdrup were "flooding the place with people," most of them "incompetent" or poorly suited to the tasks they were being assigned.[53] Designated project manager Briggs was on-site, but he lacked the authority to exert managerial control, with Hunter blocking his appointment. In summarizing the overall situation, McDaniel wrote:

> The situation is not quite as bad as I had feared, but it is still very serious. . . . I am worried about appearances and the fact that EGG, Sverdrup and the large industrial complex syndrome is calling the shots.
>
> Money is being spent on all kinds of personnel that probably are not really contributing to the project. Roy has committed the project to big operators who are from outside of the community. It isn't clear that the project belongs to "us" anymore. *We are in danger of the cost going up by a sizable factor.*[54]

In early October, the SSC Laboratory held its first workshop to discuss the experiments being planned for the new collider, inviting high-energy physicists from across the United States and around the world to attend.[55] The project had just received congressional approval of its first major appropriation for construction funding during the 1990 fiscal year (see chapter 5), and spirits were high. In addition, DOE officials had finally approved Briggs as the SSC deputy director and project manager, removing a major roadblock.

By that time, the decision of the accelerator physicists under Edwards had essentially been made.[56] As she explained in a workshop lecture, they would be moving forward with a new, less risky SSC design based on a 5 cm magnet aperture, doubling the proton injection energy, and incorporating stronger focusing into the "lattice" of dipole and quadrupole magnets that keep protons on track and confined within the beam tubes.[57] Although the changes would cause delays and increase the costs, Edwards considered

FIGURE 4.2 Helen Edwards, SSC Laboratory associate director for accelerator systems, lecturing about the Tevatron in 1983. Courtesy of Fermilab.

them necessary to reduce the risks in the original CDG design, which could lead to serious problems and delays in commissioning the SSC once built. In an interview years later, Schwitters likened the aperture increase to the larger Prince tennis rackets that had become popular in the 1980s: they had a much larger "sweet spot" at the center of the racket, thus greatly reducing one's chances of error when hitting the ball.[58] The new 5 cm aperture would also eliminate the need for beam-tube correction magnets, which were absolutely necessary in the earlier 4 cm design.[59]

On November 1, Secretary Watkins visited the South Dallas offices of the SSC lab and was given a personal tour of the Waxahachie site by lab officials. Schwitters told him that they were in the midst of the site-specific design process and several design changes might be in the offing. Watkins told Schwitters that the total project cost should not exceed the $5.9 billion price tag that Congress had just approved that September when it appropriated $225 million to begin construction.[60]

But the design changes under contemplation by Edwards and her team were extensive indeed, and they were going to cost plenty. In addition to increasing the dipole magnet aperture from 4 cm to 5 cm, SSC designers intended to shorten them from 17.35 m to 15.85 m and to use 8,800 of them rather than 8,600. To focus the circulating proton beams more strongly, there would be one quadrupole magnet for every five (instead of six) dipoles, and to accommodate the additional magnets, the circumference of the main ring had to increase from 52 to 54 miles. And the injection energy would increase from 1 TeV to 2 TeV. Although poorly understood in the fall of 1989, the total project cost was likely to increase by at least a billion dollars.[61]

On a November 9 Washington visit (the same day the Berlin Wall crumbled), Schwitters communicated these design changes to Jim Decker, who had stepped in again as acting director while a new OER director was being sought to replace Hunter. Decker dutifully sent the information up the DOE line to Deputy Secretary Henson Moore, a former Louisiana congressman who was second in command to Watkins. Moore was not pleased by the news. According to a story that soon broke in the *Washington Post,* he insisted that "either we're going to build it for the figure of 5.9 [billion dollars] . . . or we're not going to build it at all."[62] When Watkins learned of the design changes and associated cost increases, he "went non-linear," according to an unnamed source in a *Physics Today* article.[63] Another DOE official (who requested anonymity) recalled that "Watkins had to be scraped off the ceiling" when he heard the bad news. In a mid-November meeting in the Forrestal Building, Schwitters said, "Watkins went berserk and started screaming . . . and swearing" at him about the likely cost overrun.[64] According to the *Post* article, the SSC cost might grow by as much as $2 billion over the estimated $5.9 billion total project cost. The brief honeymoon between URA and the DOE was suddenly over.

In early December, the SSC Machine Advisory Committee (MAC) endorsed the proposed design changes. This group of about a dozen of the world's foremost accelerator builders gathered at the SSC lab to hear presentations from Edwards, Chao, and other members of her team about the risks inherent in the original CDG design and the changes being proposed to address them.[65] Of central concern was the likelihood of significant losses of protons during the nearly hour-long injection process of filling the beams at low energies, when they would be fatter. In addition, the use of beam-tube correction coils with the 4 cm dipole aperture promised a long and potentially difficult "commissioning" process—getting the SSC to operate as designed after its construction was completed. Given the extremely high political visibility of the SSC, Schwitters and Edwards wished to avoid the kinds of problems that had hobbled Fermilab during its difficult teething process in the early 1970s.[66] Urged on by Schwitters, MAC members concurred, stating in their report that the 4 cm aperture "now presents a significant risk to timely commissioning and specified performance level."[67] And because of the potential difficulties of serious beam losses owing to uncontrollable persistent currents during the low-energy injection process, they recommended that "the collider injection energy be increased in the range up to 2 TeV."

In retrospect, it was probably the correct decision—at least technically.[68] Although Tigner might have been able to build and operate the collider designed under his leadership, it likely would have endured a long, difficult commissioning, possibly taking years to reach full design performance. He acknowledged that it was an "aggressive" design, but he strongly believed that it entailed "acceptable risk" and that he could have successfully built such a supercollider.[69] It would have been unrealistic, however, to require a different group of accelerator physicists to build a machine that they sincerely believed had such a high level of risk. Tigner's decision not to go to Texas as deputy director or project manager meant that the SSC lab would now pursue a more conservative—and thus significantly more expensive—collider design. And have to endure the bruising political consequences of so doing.

A $2 BILLION COST OVERRUN

On November 21, two days after the *Post* article appeared, the SSC Board of Overseers hastily convened an emergency meeting of its executive committee at the SSC offices to address the issues raised by the magnet problems and machine redesign. Schwitters, Briggs, and Pewitt briefed BOO members on the likely cost increases and DOE reactions. At that moment, increasing the injection energy to 2 TeV seemed a foregone conclusion, but URA physicists were still debating whether to use a 5 cm aperture, in part because of the costly schedule delays that would be required to redesign the magnets and test them.[70] And successes had recently been reported in testing the existing 17-meter-long, 4 cm aperture magnets at Fermilab.

According to a memorandum of this meeting by Panofsky, SSC lab staff had conducted a "bottoms-up cost review" of the entire project, including the costs of a "wishing list" of additional features that the high-energy physics community had suggested, such as fixed-target beam lines and extra experimental areas. With those items included, the total costs would balloon to an estimated $8.4 billion (in as-spent dollars); without them, it was still over $7.0 billion, including about $800 million for the new, more conservative features designed to reduce project risk.[71] Pewitt said that this cost estimate had already been "scrubbed" to rid it of unnecessary items. "The only way to get these costs down further would be a reduction in scope" of the collider. "Reduction to 17 TeV [per beam] would give a cost of $6.552 billion and reduction to 15 TeV or lower would be required" to get back down to $5.9 billion.

Such scope reductions required redesigning and relocating the tunnel to a smaller "footprint" less than fifty miles in circumference and installing fewer magnets. The lower proton energies that would be available initially could be compensated for later by increasing the beam intensities, and hence the frequency of collisions (or "luminosity"), or by reducing the helium temperatures below the planned 4.35 K, which would allow higher magnetic fields and energies.[72] But there were potential downside risks involved, too, such as possibly having to reopen the site-selection process. Perhaps prodded by Panofsky, who strongly advocated such a "descoping" effort, the committee concluded: "Balancing these risks the Executive Committee felt strongly that a very good case could be made for the 'reoptimization' of parameters including an energy of 17 TeV/beam and that one should emphasize the overriding need for a confident cost estimate and operational reliability."[73]

Following that meeting, Panofsky began another memo on the SSC's cost growth; he circulated drafts of it in early December—at about the time of the MAC meeting at the SSC.[74] In it he advocated implementing the more conservative design features being contemplated to minimize commissioning and operational risks, plus reducing the energy to 17 TeV per beam to try to hold the line on SSC cost increases. But by then the MAC had weighed in strongly in favor of raising the dipole magnet aperture to 5 cm, which would require a redesign and testing phase, adding about a year to the construction schedule and another $600 million to its price tag.[75] The total estimated cost of the SSC could well approach $8 billion if all these features were to be implemented and its beam energy remained 20 TeV. Lowering this energy to 17 TeV would reduce the time required for tunneling and magnet manufacturing (and the costs thereof), thus compensating for the extra time needed to redesign and test the superconducting magnets with 5 cm apertures.

It was a period of severe stress on the federal budget, caused by the savings-and-loan bailouts, environmental remediation at the DOE weapons labs, and other pressing national priorities. Annual deficits were again growing alarmingly, into the hundreds of billions of dollars. Having moved in Washington circles since the late 1950s, Panofsky recognized the broader political context in which the SSC project had to operate—and how that could affect the laboratory's longer-term viability. "The total pattern demands that the SSC demonstrate to all constituencies that within this context its scope is not a 'sacred cow,'" he warned his URA readers. "Its predicted performance must [not] be held inviolate at all costs while other needs of the nation . . . are under severe budgetary pressure."[76]

As Panofsky correctly noted, there was nothing sacred about 20 TeV, despite what many theorists might have been saying; the SSC's "discovery potential" would not change much by reducing the energy from 20 to 17 TeV per beam (and that could be raised later). Lowering it much further, say, to 14 or 15 TeV to try to return to the original $5.9 billion cost figure, however, *would* begin to hurt its planned physics program.[77] And still haunting these discussions was the LHC possibility increasingly being championed by CERN (see chapter 3). Panofsky acknowledged that "the gap between the SSC and the energy attainable by a potential LHC installed in the LEP tunnel narrows to an unfortunate extent."[78] This was probably the primary unspoken consideration when physicists contemplated lowering the SSC energy.

Another study done at the same time indicated that a 3 TeV reduction in the beam energy would not seriously impact the SSC discovery potential, also called its "experimental reach."[79] Problems might occur if the Higgs boson (or whatever else might be responsible for the electroweak symmetry-breaking process that imbues particles with mass) were to come in above 1 TeV, which wasn't considered likely (see appendix 1), but they could be addressed by running longer, raising the SSC luminosity, or increasing the magnetic fields by lowering the helium temperature.[80] And the more robust magnet designs then being contemplated for the SSC would make it easier to boost its energy or luminosity in later years. The report concluded that an "SSC total energy anywhere in the range from 30 to 40 TeV would provide an enormous increase in physics capability over existing machines."[81]

On December 19, URA convened an extraordinary one-day joint meeting of its board of trustees and the SSC Board of Overseers to address deepening concerns about the site-specific design and the cost overruns.[82] Schwitters arrived from Dallas to address the meeting, held in Crystal City, Virginia, near Washington National Airport. Most of the BOO members were present, including McDaniel and Panofsky, the new URA President John Toll,[83] and about half of the URA trustees, including board chairman John H. Marburger III. Although the meeting was billed as being "informational" and "consultative," its recommendations would clearly carry major weight in decisions soon to be reached. Schwitters outlined the situation in Texas and made the case for redesigning the SSC to reduce its operating risks, despite all the additional costs. Although estimates were still in flux, he told his audience that a "minimum-cost configuration" with 5 cm magnet apertures would cost $7.4 billion.[84]

The discussions inevitably turned to the question of whether to reduce the beam energy to 17 TeV and try to hold the line on total costs, or to maintain the 20 TeV design energy. Panofsky presented his arguments to reduce the beam energy, stating that the cost growth under contemplation "will drive a wedge between the SSC and all other science in DOE."[85] But he found himself "a minority of one" in advocating such an alternative.[86] Others pointed out that the cost savings of $500–700 million would not allay SSC critics very much, and that backing off from the original design energy might have other, negative, political consequences. As the meeting ended, the members of both boards agreed to issue a "sense of the meeting statement" expressing "their support for retaining the SSC performance parameters . . . including the beam energy of 20 TeV."[87]

In retrospect, the estimated cost overrun of about $1.5 billion, or 25 percent of the original $5.9 billion total cost, was modest in comparison with other large US projects at the time—in particular, the many billions in overruns being experienced then on the Space Station.[88] And the increased costs could be credibly justified by the need to reduce apparent risks in commissioning and operating the supercollider and to promote subsequent upgrades of its performance. Such considerations no doubt swayed the thinking of SSC officials and the members of the URA boards. But such an attitude ignored how anxiously and enviously the project was being viewed by the rest of the US scientific community, for which only a *million* dollars was an awful lot of money.[89] And it assumed that the US high-energy physics community still enjoyed the strong political support experienced earlier during the height of the Cold War.

URA leaders may also have been influenced by a moderation in the Energy Department's public position on potential SSC cost overruns. Speaking to reporters the week before the joint board meeting, Secretary Watkins had opened the door to entertaining a possible increase after his tempestuous meeting with Schwitters to discuss the need for the design changes. But he indicated that the DOE remained opposed to any major SSC cost increases, insisting, "You don't start out a program by asking for more dough; that's the way to kill a project."[90]

What the DOE needed to help build political support for a projected overrun of more than a billion dollars was a report from a blue-ribbon panel of scientists telling the agency that the design changes causing it were

FIGURE 4.3 DOE Secretary James D. Watkins (*left*) with the first three Fermilab directors (*l. to r.*), Robert R. Wilson, Leon Lederman, and John Peoples, September 1992. Courtesy of Fermilab.

absolutely necessary to the success of the SSC project. On December 21, 1989, Decker sent a letter to HEPAP Chairman Francis Low of MIT, requesting that he convene a subpanel to answer the central question, "Are there some lower bounds on the technical parameters below which the physics research to be undertaken would be scientifically unimportant, marginally productive, or *duplicative of ongoing or proposed work elsewhere*?"[91] HEPAP quickly organized a highly distinguished subpanel of physicists (including five Nobel laureates) chaired by theorist Sidney Drell of SLAC to address the issue.[92] They met with lab officials and scientists at the SSCL in early January and prepared a report in record time to be presented to a HEPAP meeting in Washington on January 12, 1990. It completely supported the SSC lab's redesign proposals, including the 2 TeV injector *and* the 5 cm dipole magnet aperture, and opposed reducing beam energies below 20 TeV. In his letter transmitting the report to the DOE, Low noted, "The Subpanel has concluded on physics research grounds that any substantial reduction in the energy of the SSC would compromise our ability to elucidate the nature of electroweak symmetry breaking," the uppermost goal of the project.[93]

Another rationale for the recommendation, left largely unspoken at the time, was the existence of the proposed LHC project then taking shape at CERN, with expectations that it would eventually reach 8 TeV beam energies. Any substantial reduction in the SSC energy might induce Congress to revisit the question, "Why don't we just join the LHC project and save billions of dollars, which could then be devoted to other priorities?"[94] Given early rhetoric advocating the SSC as the ideal way to restore US leadership in high-energy physics—effectively by leapfrogging European efforts—such an eventuality had to be avoided at all costs. "The SSC should provide a substantial new physics capability when it comes on line," Decker had told HEPAP on January 12. "The United States should not spend this large amount of money for a facility that would be only marginally better than some other existing facility."[95] There were good *political*—as well as scientific and technical—reasons not to lower the SSC beam energy.

With the Drell subpanel recommendations in hand, Watkins had ammunition to go back to Congress and request more money. But exactly how much to ask for was fairly ambiguous in early 1990. The site-specific design still needed to be refined and its costs determined before any DOE trips to the Hill could occur. And Watkins had learned his lesson from this troubling affair—not to trust URA physicists to manage such a difficult, expensive construction project. That kind of responsibility, he figured, was better given to military men. He began seeking capable managers from among friends and colleagues in the US Navy,[96] whom he trusted better to oversee and manage a complex, multibillion-dollar project.

After the SSC site-specific design was finished in June 1990, the DOE held a four-day conceptual design review at the lab offices on June 25–28. Expert staff such as Ed Temple, aided by consultants from other laboratories and industry, pored over the myriad details of the new design, generally giving it high marks.[97] URA officials estimated the total project cost at $7.84 billion, assuming its completion in 1998 and including a contingency of $753 million plus inflation costs of $1.23 billion over the duration of construction. That total represented a growth of almost $2 billion—or about 33 percent—above the original $5.9-billion figure DOE had given Congress a year earlier, exactly as the *Washington Post* had forecast in November 1989. The DOE review team put the total project cost more than a half a billion dollars higher than the URA estimate, at $8.39 billion.[98]

Not content to believe URA or SSC physicists or its own review team this time around, however, the DOE commissioned two independent analyses of the SSC project costs. The first was done by a HEPAP subpanel chaired by John W. Townsend, director of the Goddard Space Flight Center in Greenbelt, Maryland.[99] After about a month's work, this group concluded that DOE should add another billion dollars to the URA figure, pushing it up to as much as $8.9 billion. It questioned the estimates for the superconducting magnets, calling them "the most significant risk," and suggested adding another $300 million for the experimental program, so that the SSC could have two large detectors at the outset.[100] And this panel thought the project schedule was far too tight. "The biggest problem, to put it bluntly, was a lack of confidence about their scheduling," said Townsend.[101] His panel recommended adding another six to twelve months to the schedule, completing construction in 1999, which would raise the SSC cost by $300 million.

But the biggest arguments were occurring behind closed doors that summer. Rumors flew that DOE's Independent Cost Estimating group (dubbed ICE because its assessments would often help to freeze out hopeful projects) had come up with a projected SSC cost close to $12 billion.[102] Leaked copies of the ICE report surfaced in September, confirming these rumors.[103] The report made similar criticisms as the Townsend panel but put higher numbers on the increased costs. ICE considered the 8,800 superconducting dipoles to be "the greatest technical challenge and the greatest technical risk in the design of the SSC."[104] URA estimators had not allowed manufacturers sufficient mark-up and profits in their calculations; thus ICE put the cost of these magnets at $171,400 apiece—or 22 percent above the $140,700 URA figure. In addition, ICE added a large contingency owing to the high technical risk of these magnets. And it included the full cost of two experimental detectors (plus a contingency of $507 million, versus *none* in the URA estimates), assuming no foreign contributions. Finally, ICE observed that the projected SSC funding profile would peak well above $1 billion per year in the mid-1990s, and that no guarantees could be given that Congress would even appropriate such huge annual sums. Thus it suggested that one to two years be added to the schedule, at an extra cost of $476 million.

All told, the ICE figure came in at a staggering $11.8 billion—$4 billion, or 34 percent, higher than the SSC lab's total project cost of $7.8 billion, and *double* the original $5.9 billion. While there might have been some exaggeration in the report, its authors thought otherwise. "The ICE staff strongly believes that the SSCL cost and schedule estimate for the SSC is both unrealistic and unachievable," they wrote in their conclusions. "*This* [independent cost estimate] *should not be interpreted as a worst-case scenario.*"[105]

But the ICE report got little play in the press, as the DOE leadership managed to squelch it, focusing media attention on the other three estimates. In doing so, the political appointees in the Forrestal Building won out over the career bureaucrats in Germantown, for Watkins and Moore well knew that they had to return to Congress with a politically palatable cost estimate. By October 1990, the DOE had reportedly settled on a compromise figure of $8.25 billion, close to the lower end of the range of the four estimates. The Office of Management and Budget held up release of the official figure, however, maintaining that it should be set closer to the ICE number.[106] In part, the OMB objection stemmed from what it considered a low inflation rate used by the DOE, but there were other, more substantive factors, too.

Mainly as a result of this roadblock, the new estimate of SSC total project cost did not surface publicly for another four months. In early February 1991, the DOE finally submitted its detailed, updated SSC budget, based on the site-specific design with a 5 cm dipole magnet aperture, 2 TeV injector, and other improvements (see chapter 5).[107] The DOE projected the total cost of the Super Collider at $8.25 billion in as-spent dollars, spread over eight years, with construction to be completed in 1999.[108] The increased price tag, which represented a 40 percent growth over the original $5.9 billion figure, was due not only to these upgrades but also to other items omitted in the 1989 estimate—for example, costs of magnet development, laboratory infrastructure, and a one-year schedule delay. But the cost of the superconducting dipole magnets had nearly *doubled* in real terms from what the Central Design Group had estimated in 1986. And many of the major problems highlighted by the ICE and Townsend reports had been largely overlooked in reaching this $8.25 million figure.

THE EMERGING MILITARY-INDUSTRIAL LABORATORY CULTURE

The year 1990 had been a time of consolidation for the SSC Laboratory. While the new, more conservative site-specific design was completed and published in a detailed report,[109] development and testing of the 5 cm aperture superconducting dipole magnets began at Brookhaven, Fermilab, and LBL. The lab management was reorganized, with troublesome gaps in its leadership filled by reassignments or new recruits. The DOE site office began taking shape under a newly appointed director and its staffing grew accordingly. The SSC Laboratory contracted with an architect-engineer/construction manager (AE/CM), which soon developed plans for and began erecting the first buildings on the site. As the total laboratory staff more than doubled in 1990, from 400 to almost 900 employees, high-energy physicists from around the United States and the world began forming large scientific collaborations to propose and design experiments for the mammoth high-energy physics facility.

The most glaring problem in early 1990 was the absence of a seasoned project manager able to cope with such an enormously complex undertaking. As Hunter had argued before leaving DOE, Briggs was not up to the task—although Hunter may have hobbled his effectiveness at first by holding up his appointment. In his capacity as the director of integration and management, Pewitt struggled to fill the gap and then help Briggs try to fulfill the role after his appointment was approved in October 1989, but it was just not working out.[110] Schwitters finally recognized the problem in March 1990, issuing a memorandum that acknowledged "the very large and growing amount of work that needs to be done to manage an enterprise of this magnitude" and acknowledged "the perception within DOE that the Laboratory's project management needs strengthening."[111] He named Pewitt as acting project manager and initiated an extensive search for a capable person to fill this role permanently. Briggs began devoting full time to his responsibilities as deputy director, a position for which he was better suited.

In part because of this gap in the SSC management structure, an essential management tool got short shrift: the project-management control system or cost-and-schedule control system. It is usually a complex computer model of the project whose inputs include estimated costs of the various project components and their expected completion dates. Using it properly, managers can tell quickly whether a project is on schedule and within budget, and see what "critical path" items are lagging dangerously. It also allows them to calculate an "earned value" for items completed and compare it with amounts spent thus far—and to communicate this information to others who might need to know, such as in the government agencies and Congress.[112] Industrial firms (e.g., SAIC and Western Research) had become accustomed to using these systems in their government-contracting work, but they were relatively foreign to high-energy physics in the late 1980s.[113]

The SSC M&O contract specifically *required* such a system be implemented early in the life of the project, and the DOE—especially Hunter—had urged the lab to start setting one up.[114] Without a project manager in place, however, it was not clear to whom this responsibility should fall. Tom Elioff had developed one at the Central Design Group based on personal computers, but it was not transferred to Texas.[115] Top SSC managers had little sense of urgency about implementing such a system; they regarded it as yet another bureaucratic DOE requirement getting in the way of doing their jobs. According to Tim Toohig, a CDG physicist who moved to Texas and joined the civil construction division, EG&G tried to implement another system based on a mainframe computer, but it apparently went nowhere.[116] From his government-contracting experience at SAIC, Pewitt understood the pressing need for a cost-and-schedule control system and had begun trying to implement one back in the spring of 1989.[117] But in the absence of a strong project manager to lead such an attempt, and lacking top-management buy-in, his efforts were doomed to failure like the others.[118]

In fairness, the SSCL management had to deal with so many other pressing problems requiring immediate attention in the early years that the lack of a cost-and-schedule control system might have been viewed as one to be resolved later, *after* they had a good project manager firmly in place. And the SSC design was still in flux in 1989–90, too; that would encourage managers to wait until they had reliable costs and schedules to put into the system. But such systems are notoriously difficult to implement once a project is under way; they really need to be in place at the outset, before spending serious money on project components. Two years later, the SSC Laboratory had dozens of people feverishly attempting to establish such a system after the House Science Committee and General Accounting Office criticized the lab for not having one already working (see below). This problem might have been avoided had the transition from CDG to the SSCL gone smoothly, too, with its design and management philosophies transferred largely intact from Berkeley to Texas—instead of its leadership dissipating back into the US high-energy physics community.

One of the most serious problems in the early days occurred in the teaming relationship with Sverdrup, which was supposed to take responsibility for conventional construction—paving roads, erecting buildings, digging tunnels. The company apparently did not take the job very seriously. According to former CDG Deputy Director Wojcicki, who served as a consultant to the SSC lab, Sverdrup had essentially no experience in tunneling, despite being involved in the Baltimore Harbor Tunnel project.[119] Robert Robbins, the Sverdrup executive whose name was listed on URA's M&O proposal as director of conventional construction, had to be relieved within months and replaced by Lieu Smith, who lasted only a bit longer. Schwitters recalled that the company "was going from a real engineering company with a great tradition . . . into essentially a strip-mall management firm."[120] URA terminated its contract with Sverdrup within the first year, to which DOE officials did not object. Cornell physicist Robert Matyas took over briefly as acting head of conventional construction, aided by Jim Sanford of Brookhaven and CDG, until a permanent associate director could be named for this division.

The other teaming partnership, with EG&G, fared better but offered its own peculiar array of management challenges. The firm had expeditiously rented office space for the nascent lab, but as Boyce McDaniel noted after his June 1989 visit, EG&G was flooding the place with employees of questionable ability (see above). Stanford's David Ritson later likened the situation to a "CIA sting operation, where a large dummy corporation is created out of thin air."[121] Under Jack Story of EG&G, who served as the SSC associate director for technical services, the company was building up a division providing services such as design, drafting, computing, and engineering support. This partnership helped the SSC Laboratory to staff up rapidly in a location where high-energy physicists had few connections and little community interaction. But after the SSC was redesigned in 1989–90, leading to significant scheduling delays, the presence of these EG&G employees constituted "a standing army marching in place" that needed to be paid salaries, wages, and benefits, adding substantially to the growing SSC cost overruns.[122] To exacerbate the problem, the administrative services director, Bruce Chrisman, one of the four SSC officials originally designated by URA in its M&O proposal, left the project and returned to Fermilab in May 1989, before many of the responsibilities could be sorted out and capable individuals hired to address them.

Pewitt lasted just two months in the role of acting project manager. After a dispute about the SSC magnet R&D program with Tom Bush, the director of the magnet systems division hired from industry (see below), he resigned abruptly in May 1990 and was replaced by LBL engineer Ted Kozman, director of accelerator systems.[123] The search for a permanent project manager continued; candidates from within and outside high-energy physics were interviewed. John Casani of NASA's Jet Propulsion Laboratory and Harold Forsen of Bechtel Corporation both turned the position down when offered it.[124] In August the search focused on Paul Reardon, a 59-year-old physicist who had served as Brookhaven's associate director for high-energy physics facilities during the 1980s; before that, he had been involved in building the Tevatron at Fermilab and managed construction of the Tokamak Fusion Test Reactor at Princeton. To many high-energy physicists, he seemed an excellent choice. Reardon had been instrumental in developing superconducting NbTi wire, contracting with industrial firms to manufacture the large quantities of it required.[125] He was also an experienced business manager who had worked for a time at SAIC (like Bush, Pewitt, and Trivelpiece, as an SAIC vice president), and enjoyed interacting with both engineers and scientists.

The SSC BOO approved Reardon's appointment in August 1990; in early September Schwitters met with Deputy Secretary Moore in Washington to inform him of the choice and request DOE approval. But it was not forthcoming. Instead, about a week later Admiral Watkins asked URA to consider another candidate, Edward Siskin (fig. 4.5), an engineer who had served in the nuclear Navy and more recently been executive director of Stone and Webster Engineering Corporation.[126] Invited to Texas for interviews, Siskin visited the SSC offices in late September to meet with management and staff, generally earning high marks. Out of those gatherings, a proposal emerged to establish a new position for him of general manager, reporting directly to Schwitters; Reardon would then serve as the project manager, reporting to Siskin. It met with mixed reactions, as recorded in a Panofsky memo.[127] The addition of Reardon and Siskin would bring in much-needed management expertise and help allay DOE concerns about the lab's weakness in this regard, he wrote. But the new position would add yet another layer of bureaucracy to an already complicated management structure, separating technical staff from the director and further impeding information flows among them. And because of Siskin's relationship with Secretary Watkins, Panofsky observed, "The access available to Ed Siskin to top DOE management could be abused so that de facto the General Manager receives instructions from DOE bypassing the Director."[128] Despite these reservations, the BOO approved the two appointments and the new SSCL management structure in early October, as did the DOE later that month.

In an interview nearly a decade later, Watkins acknowledged that he had recruited Siskin in a deliberate effort to gain better control of the SSC project, which he viewed with growing alarm—particularly after the $2 billion cost overrun looming in 1990.[129] One more such cost increase, he felt, and Congress would kill the project. And he did not trust academic scientists to manage such a mammoth project. Instead, he turned to people from the Defense Department and Navy, with whom he was more familiar and comfortable, and began inserting them into key SSC lab positions.[130]

Another DOD manager recruited to the SSC project was Joseph Cipriano, who had managed multibillion-dollar contracts at the Pentagon. When contacted by Watkins in the spring of 1990, he was serving as executive director, weapons and combat systems, in the US Naval Sea Systems Command, overseeing as much as $12 billion annually in the research, development, and production of advanced weapons systems.[131] He had known Watkins when the latter served as the chief of naval operations during the 1980s. Watkins checked out Cipriano's record with Navy managers and engineers, who gave him "great kudos" on his management abilities.[132] At a meeting in the Forrestal Building, Watkins told Cipriano that he figured there was less than a 50 percent chance of the SSC project being completed before Congress killed it, especially if any more large cost increases occurred. But because it was a presidential priority, he wanted to give it his best shot. Watkins also told Cipriano that there was "a bunch of physicists down there in Texas who didn't understand the sensitivity of keeping costs under control."[133] If he accepted the position to direct the DOE site office at the laboratory, getting control of these costs would be his principal responsibility.

A Texas native with a degree in physics from Baylor and family living there, Cipriano eagerly accepted the position and its associated challenges. In April 1990, the DOE announced his appointment as "project manager" in charge of the DOE site office at the SSC Laboratory.[134] With it came another appointment as an associate director in the Office of Energy Research, heading the Office of the SSC. Cipriano nominally reported to the director of Energy Research, but he in fact had a separate reporting channel directly to the energy secretary, which lent him great power and influence. As Watkins recalled, "He was to go down there and generally be coequal with the scientific leadership within URA and balance it out from a programmatic viewpoint."[135] This kind of authority, reaching well beyond the normal oversight role that the high-energy physics community was accustomed to dealing with in the head of a DOE site office, was guaranteed to create tremendous friction.

Arriving at the SSC on May 21, Cipriano quickly began taking charge of the site office, hiring staff members from the Pentagon, other government agencies, and industry. Few had prior experience or interactions with high-energy physics; their focus was largely on civil construction and project management. Cipriano started writing biweekly memoranda addressed to Secretary Watkins, informing him about problems encountered and the halting progress being made at the lab.[136] Obtained from Congressman Boehlert's office, these memos provide a remarkable record of the interactions between Cipriano, Watkins, and other DOE leaders such as Moore, for they also contain the handwritten comments of these officials, revealing their reactions to the Texas goings-on plus their individual recommendations for actions to be taken by the SSC site office. The communications were obviously being used as a management tool to try to gain control of operations at the lab.

On September 14, 1990, for example, just as URA was beginning to grapple with the question of Siskin's becoming general manager, Cipriano wrote Watkins that Texas National Research Laboratory Commission Chairman Morton Myerson "agrees with our assessment that the Project Manager needs to be elevated in status at the Lab and we have to do something to get our arms around the conventional construction problem quickly."[137] Earlier that year the lab had signed the AE/CM contract with the consortium Parsons-Brinckerhoff/Morrison-Knudsen (PB/MK) to provide needed architectural, engineering, and construction management services.[138] In keeping with the Secretary's directives, Cipriano asked to step in and manage this contract directly, as well as any additional subcontracts that stemmed from it. But Schwitters and the URA leaders rejected this DOE request. "The URA position at the meeting was that it was unacceptable to them to have the AE/CM contract broken out and that while they admitted there were problems, they saw no advantage to changing their current management structure," Cipriano's memorandum continued. "They were very firm in their position that all aspects of the Lab had to report to the Laboratory Director." (In this copy of the memo, these sentences were underlined by Watkins in a heavy hand, with the two comments "NUTS" and "NO" scrawled in the margin next to them.)[139]

Through Cipriano and the DOE site office, Watkins was effectively trying to build a parallel organization to grab the reins of the SSC project—or at least of its collider and civil construction parts—away from the URA physicists. He wanted to put them in the hands of managers from the military-industrial complex, whom he trusted better to control SSC costs. Watkins adapted the OSSC framework originally established by Ianniello under Hunter (see above) but required that the bulk of its more than one hundred employees work right at the SSC site, not in Washington. "I wanted the real work to be done in the field, in the vicinity of the scientists," he insisted.[140] Their task was not to be only that of overseeing the URA work and signing off on its decisions. Instead, Watkins tasked Cipriano to *manage* the interactions with key subcontractors, which would effectively circumvent the M&O contract that had conferred this responsibility upon URA. (But there was always that troublesome clause regarding subcontracts, which had never been resolved—only "surrounded with paper"—during the 1988–89 negotiations.) This effort would also effectively bypass the DOE's own Office of Energy Research and its Construction Management Division, as well as the Chicago Operations Office, which had built up considerable expertise in overseeing AEC and DOE projects.[141] In addition, career civil servants such as Decker and Temple had established trusting (if not always harmonious) relationships with high-energy physicists; they were more sensitive to the needs and practices of this distinctive scientific culture.[142] But they were effectively shunted aside in the new SSC organizational structure.

This "militarization" of the emerging SSC Laboratory culture could be felt at multiple levels of its organization, not just in the general manager's office and the DOE site office. In early 1989 Schwitters had named Tom Bush to direct the SSCL Magnet Systems Division, instead of Tom Kirk, who was then leading the CDG magnet R&D program. Holding a PhD in engineering physics from the US Naval Postgraduate School in Monterey, California, Bush had extensive experience in managing billion-dollar projects and procurements as strategic systems project officer in the Navy's Trident submarine program. More recently he had been working at SAIC on nuclear-attack drills and studies of the Soviet missile program.[143] Roger Coombes, a high-energy physicist from SLAC and CDG, served as Bush's deputy. Schwitters recognized that over a billion dollars' worth of superconducting magnets would eventually have to be purchased from corporations such as General Dynamics and Westinghouse; thus he wanted someone with deep procurement experience at the helm of this division.[144] Nobody within the US high-energy physics community, which had previously built almost all its magnets at the national labs, had anything close to this kind of experience.[145]

But Bush erected walls of secrecy around his magnet division, not allowing members of other divisions to attend its meetings and forbidding his own people to exchange information with outsiders. According to one SSC physicist, his attitude was, "You tell us how you want the magnets built, and we'll get them built to your specifications."[146] This secretive approach flew in the face of the physicist culture, which values highly the free exchange of information across bureaucratic divisions, but it was typical of how business was done in the military-industrial complex.[147] Many Bush critics at the SSC lab cited his frequent slips of tongue, referring to the sleek 16-m-long cylindrical metallic objects his division was developing as "missiles" instead of "magnets." He was viewed very differently by Cipriano, who wrote to Watkins that Bush "is the best engineer they have. He is not trusted by the physicists because he is an engineer but continues to plug away."[148]

The deepening military-industrial culture at the SSC Laboratory could also be felt in the Conventional Construction Division. The original turmoil experienced in this division, caused by the failure of Sverdrup to meet expectations, was partially resolved in mid-1990 when John Ives became associate director. A retired Navy admiral like Watkins, Ives brought added large-project experience to the SSC lab. CDG physicist Tim Toohig continued serving at a lower level in that division. Working with PB/MK, the division made detailed plans for—and that fall began—constructing the Magnet Development Laboratory in the northwest corner of the Waxahachie site, the first SSC building to appear there.

Given the sheer magnitude of the SSC project, it may have been inevitable that a military-industrial culture would emerge, at least in the construction phase. High-energy physicists did not have this kind of billion-dollar project management or procurement experience.[149] If anywhere, it had to come from the military and its contractors—or perhaps from NASA and the aerospace companies that worked with it. And as the Cold War wound down during the late 1980s and early 1990s, a large number of capable engineers from the military-industrial complex were seeking work at their talent levels. Exactly *how* these people were to be integrated into the laboratory's organizational structure, however, might have been handled differently. Schwitters had expected Sverdrup, EG&G, and individual engineers to come in, help build the lab facilities, and then disappear, leaving laboratory operations and management to the physicists.[150] There was plenty of routine construction work to be done, after all, and not much in the way of advanced accelerator physics after the designs had been established. This kind of tedious work was unlikely to promote a physicist's career.

But control of the project had largely been ceded to the military-industrial engineers by late 1990. They had come to form the bones and sinews of the SSC Laboratory organization, while high-energy physicists—including experimenters, theorists, and accelerator physicists—constituted its flesh and blood. Members of this proud, accomplished scientific community increasingly perceived that an alien, military-industrial culture was emerging at the new laboratory. Thus many high-energy physicists decided to remain at their home universities and laboratories rather than risk making permanent, full-time commitments. Some, like Ritson of Stanford, came to Texas on one-year or two-year temporary contracts; others like Wojcicki served as consultants. But the primary task of building a new scientific culture was left largely to younger, less-well-established physicists.

STRONG TEXAS SUPPORT FOR THE LABORATORY

The debate about the growing SSC price tag was being closely monitored by the Texas National Research Laboratory Commission (TNRLC). Made up largely of politically connected business leaders characterized as the

"Texas collider cartel,"[151] the commission had been established by Governor Clements to bring what they called the "Texatron" to the state (see chapter 3).[152] Having achieved that, the commission's goal as 1989 began was to win what it called the "Battle Within the Beltway"—the looming fight on Capitol Hill for the billions of dollars needed to build the SSC. It would also assist Schwitters and his staff in their efforts to build a world-class laboratory and foster a thriving scientific community.

Ed Bingler, the geologist who accompanied Dennis Theriot in November 1988, had been serving as executive director of the TNRLC since its inception. The former deputy director of the Texas Bureau of Economic Geology, he served as the commission's director during its ten-year life—through the tenures of several chairmen. Throughout its existence, Bingler provided the SSC lab invaluable technical and logistical support.[153]

Bingler also envisioned a political role for the TNRLC. From its early days, he had urged the commission to extend its sphere of operations beyond Texas and onto the national stage. With strong support from Governor Clements, it established a Washington liaison team to support SSC-related activities of the Texas Office of State-Federal Relations. A key adviser for the political campaign was Robert Strauss, a Texan who once served as national chairman of the Democratic Party. Strauss provided gratis the services of his prestigious law firm—Akin Gump, one of the world's largest law firms—for help lobbying Congress and other assistance.[154]

Texas arguably wielded greater political clout in the nation's capital than any other state, thanks to its potent congressional delegation. The Texas delegation had a long and enviable history of working closely together for common purposes. It was the only state delegation to hold weekly luncheon meetings attended by both Democratic and Republican members. During recent decades, the delegation had helped Texas land the Johnson Space Center, a Navy base, and SEMATECH, the semiconductor R&D center in Austin.[155]

But only *after* Texas won the site competition did its political power really become important for the SSC. Once Waxahachie was selected, the SSC gradually became recognized outside the Lone Star State as a "Texas pork" project—and thus more vulnerable on Capitol Hill. A sizable portion of the Super Collider's growing budget would undoubtedly have to be spent in other states, as is typical for federal projects of such magnitude. But the Texas victory left members of Congress from other states, especially those that had entered the site competition, with greatly diminished political incentive to support the multibillion-dollar project.[156]

The Texas cartel fully recognized this political vulnerability. Among them was Tom Luce, who in early 1989 became Bingler's next boss as the TNRLC chairman. A corporate attorney with extensive experience in negotiating federal contracts for clients that included Ross Perot, he also understood the workings of Congress. Luce's initial goal was to persuade Congress to commit funds to begin SSC construction. He figured that once construction was under way, stopping the project would be more difficult. "It's like building a house," he claimed. "If you get halfway through, it's pretty hard to stop."[157]

Luce and the TNRLC faced a tough task on Capitol Hill, partly because of widespread concern about the project's escalating cost. In October 1988 the nonpartisan Congressional Budget Office (CBO) had warned that the Super Collider's price tag might soar by as much as 46 percent, based on the DOE track record with recent accelerator projects.[158] Two weeks later, amid the furor over Texas winning the site competition, a *New York Times* editorial urged that SSC construction be delayed to save money for other scientific fields. "The power of the Texas delegation in Congress may offer the best hope of funding the $5 billion machine," it declared, "but the deeper question, given little attention by the Energy Department, is how paying for the project will affect other science."[159]

A powerful tool at the disposal of the commission in trying to blunt this criticism and broaden the base of congressional support for the SSC was the $1 billion in state funds that Texans had committed to the project (see chapter 3). The TNRLC, which was authorized by act of the Texas legislature to disburse these funds, decided to gamble.[160] In March 1989, days after the CBO had suggested SSC postponement or termination as two options for reducing the federal budget deficit,[161] Luce testified before the House Subcommittee on International Scientific Cooperation. Rather than allocate his state's billion-dollar contribution at a rate proportional to the federal investment—roughly 1:5—Luce promised Texas money up front to help get the SSC rolling.[162] Two months later, he put his state's offer in writing. In a letter to the powerful chairman of the Senate Appropriations Subcommittee on Energy and Water Development, Louisiana Senator J. Bennett Johnston, Luce stressed "our willingness to 'front-load' more than our pro-rata share of available State funds during the first two years of construction."[163] For fiscal year 1990, contingent on Congress appropriating $250 million to begin construction, the commission would commit $100 million toward the SSC project, and a similar percentage in fiscal 1991.

In addition, wrote Luce, the TNRLC would set aside $100 million from the $1 billion total "for the purpose of establishing an ongoing research fund to support external research programs."[164] This funding would go to universities and national laboratories around the country for such things as generic R&D needed for the SSC detectors. Proposals would be peer reviewed by a panel of experts who would make recommendations to the TNRLC, which had final say in the decision. This funding proved a boon to the SSC Laboratory, helping to jump-start detector R&D without having to wait for Congress to appropriate funding. A stream of proposals soon flooded in from all across the country, many of them from URA institutions.[165]

TNRLC staff had even drawn up a map identifying potential subcontractors, with a view to funneling dollars to them so as to increase political support from the members of Congress representing those states. DOE and the national labs had been utilizing a similar policy for years.[166] The Texas commissioners hoped that once SSC operations gathered momentum after their pump-priming efforts, Congress would feel obligated to start committing federal dollars to what was, after all, a federal project. "I felt like that was a helluva step by us," Luce later said, reflecting on his state's offer to Congress.[167]

Finally, on September 7, 1989, after a summer of "Texas-type wheeling and dealing" among federal lawmakers, Texas got its wish. A House-Senate conference committee approved construction of the Super Collider (see chapter 5). Of the $225 million allocated for the project in the fiscal year beginning in October, $135 million was specifically set aside for construction. Congress had officially committed itself—and the American taxpayer—to building the SSC.[168]

Thirteen hundred miles away in Waxahachie, the TNRLC was honoring its commitment to the SSC, working closely with DOE and the rapidly growing SSC Laboratory to recruit scientists and establish priorities for design and construction of the laboratory. Although functioning mainly as a funding agency, the commission's activities extended well beyond writing checks. TNRLC staffers collaborated with representatives from DOE and the SSC lab on at least eight working groups that handled various aspects of the project: real estate/conventional construction, socioeconomics, survey and site modeling, physics research, policy, outreach, education, and private inducements. The commission served as an on-site facilitator whose local knowledge and network of contacts proved essential in helping the newly arrived physicists, engineers, and federal bureaucrats settle into the Texas surroundings.

In December 1989, Luce stepped down as TNRLC chairman, succeeded by another cartel member, J. Frederick Bucy, a solid-state physicist and retired CEO of Texas Instruments Inc.[169] Many of the projects Luce had set in motion during his chairmanship came to fruition during Bucy's tenure, including finalization of the footprint for the SSC ring, laboratory, and support facilities. Until this issue was settled, the commission could not begin acquiring the 16,550 acres of land required for the SSC Laboratory. TNRLC staff assisted SSCL staff in developing the site-specific Super Collider design, adapted to the local geology, hydrology, and other Waxahachie site parameters. In March 1990, DOE announced the final SSC footprint. The news prompted Ellis County farmers and ranchers to flood Texas land offices to learn whether its tunnel would run under their lands.[170]

Two months later, Texas sold its first $250 million in state bonds for the project,[171] for use partly in land acquisition. For half of the sprawling SSC site, the state had to purchase just subsurface rights, since surface ownership was needed only at experimental areas (5,000 acres), campus areas (3,500 acres), and selected points around the ring where entry halls were to go. The land-acquisition process officially began in July 1990, when the commission made its first small purchase. A *Washington Post* story spotlighted two longtime landowners who were reluctant to be displaced. But most county residents seemed to have "super-collider fever," as evidenced by comments from Waxahachie Mayor Jordan. "The SSC means so much not only to Dallas-Fort Worth, not only to Ellis County and Waxahachie, but to the world scientific community," he explained. "This is the biggest thing, literally and figuratively, that's ever happened to us."[172] Among other affirmations of local support was an agreement between Ellis County and five adjacent counties to raise vehicle license taxes to pay for new roads to the SSC site.[173]

During the fall of 1990, at the N15 site in the northwest corner of the SSC site, construction began on the $12.5 million Magnet Development Laboratory. This 110,000-square-foot structure was the first major construction project for the SSC. Its completion a year later meant that SSC employees could officially begin working on-site. By the summer of 1992, the N15 site would also become home to the smaller Magnet Test Laboratory and the associated Accelerator Systems String Test facilities. These three buildings encompassed fully equipped work areas totaling 200,000 square feet of floor space, with facilities capable of fabricating twenty-five superconducting magnets per year and testing ten magnets simultaneously.[174]

Meanwhile, on February 1, 1991, Secretary Watkins signed an official Record of Decision, the final step in the prolonged environmental review process for the SSC. With that, the TNRLC could begin land acquisition in earnest.[175] An even stronger indicator of the project's gathering momentum that spring was the size of the SSCL workforce. Thanks in part to lower mortgage rates and other generous relocation incentives negotiated by the commission, over 1,000 scientists, engineers, and other employees had been hired.[176] Hundreds of physicists were descending on the gestating laboratory to participate in its activities. Many returned to their home institutions with R&D grants from the TNRLC.[177] And in 1989–90, the commission awarded fellowships to a select group of young physicists to spend a year working at the laboratory.

TNRLC staff also helped the laboratory find temporary workspace near Waxahachie for its mushrooming staff. The Beckleymeade Avenue office park had quickly become inadequate for its staff and advanced technical activities. In early 1991, at the request of Siskin and the DOE site office, the TNRLC acquired a 500,000-square-foot warehouse near Waxahachie that was quickly converted into offices, laboratories, a library, and a cafeteria. The huge building became known as the Central Facility. By renovating an existing building rather than erecting a new one, the commission reportedly helped save the project $13 million and allowed certain R&D activities to proceed two years ahead of schedule.[178]

Thus, thirty months after Texas had won the site competition, the SSC was becoming reality—physically, technically, and intellectually. Such was the happy picture being circulated by the TNRLC's public-affairs office to the world beyond Waxahachie. But in the background lurked a worrisome story that had begun surfacing in November 1989. The extensive SSC redesign (mentioned above) had added over $2 billion to the project's estimated cost, raising its official price tag to $8.25 billion. TNRLC members and staff had done everything possible to help transform SSC visions into reality on the windswept plains south of Dallas. But the transformation would cost billions of taxpayer dollars, and by the summer of 1991 the project's increasing cost estimates were causing alarm—in Washington as well as in Texas. In May 1991, an amendment to kill the SSC was offered on the House floor to the 1992 Energy and Water Appropriations Bill (see chapter 5). Although defeated by a solid margin, it could only disturb Texas supporters.

On July 5, 1991, Bucy suddenly resigned as TNRLC chairman in an effort to underscore the worsening situation. Citing the growing concern on Capitol Hill about the collider's cost, he said it was time to face the issue of foreign support, "because next year the SSC will need even more money, and Federal funds will be scarcer." He doubted contributions from abroad would total even $500 million—a shortfall that he feared might lead

Congress to kill the SSC.[179] "It's not beyond salvation," Bucy insisted, "but it's getting close."[180]

Such words of warning should have been a rallying cry for everyone involved in the project, particularly the physicists. But in the eyes of the TNRLC, Schwitters and the SSC staff were focused on building a particle physics laboratory, largely oblivious to the political storm gathering in Washington.[181]

BUILDING A DIVIDED LABORATORY

In early 1991 the SSC Laboratory had finally begun to emerge. Contractors were erecting the Magnet Development Laboratory, with occupation scheduled for that spring. Plans for other buildings were nearly finished. And General Dynamics and Westinghouse had signed contracts to develop and eventually manufacture the thousands of superconducting dipole magnets needed for the main collider rings.[182]

Physicists from universities and other laboratories, both US and foreign, increasingly visited the laboratory to interact with this buildup and participate in its many review panels and planning committees, including the Machine Advisory Committee (mentioned above), the Science Policy Committee (SPC), and the Program Advisory Committee (PAC). Chaired by MIT physicist Jerome Friedman, who shared the 1990 Nobel Prize in physics for experiments leading to the discovery of quarks, the SPC advised SSC Director Roy Schwitters on overall policies to follow in establishing the lab's experimental program. In December 1989, for example, it had advised against lowering the maximum proton beam energy from 20 TeV.[183] And in the following meeting, the SPC recommended that the lab pursue design and construction of two large general-purpose detectors to be ready to begin taking experimental data shortly after the first proton collisions occurred.[184]

The PAC, led by former HEPAP Chair Jack Sandweiss of Yale, advised Schwitters on the primary details involved in implementing these policies—for example, the specific experiments to approve and fund using the available allocation. After a July 1990 meeting in Snowmass, by then a favorite summer rendezvous, he invited high-energy physicists worldwide to begin submitting letters of intent describing the two large detectors.[185] Each of these two detectors, weighing many thousands of tons, was to rest in an enormous cavern at one of the SSC's interaction regions where the two proton beams crossed paths, surrounding the collision point and observing the subatomic debris that emanated from it. Of the $8.25 billion total estimated cost (as of 1991) of the SSC project, DOE allocated $760 million (plus

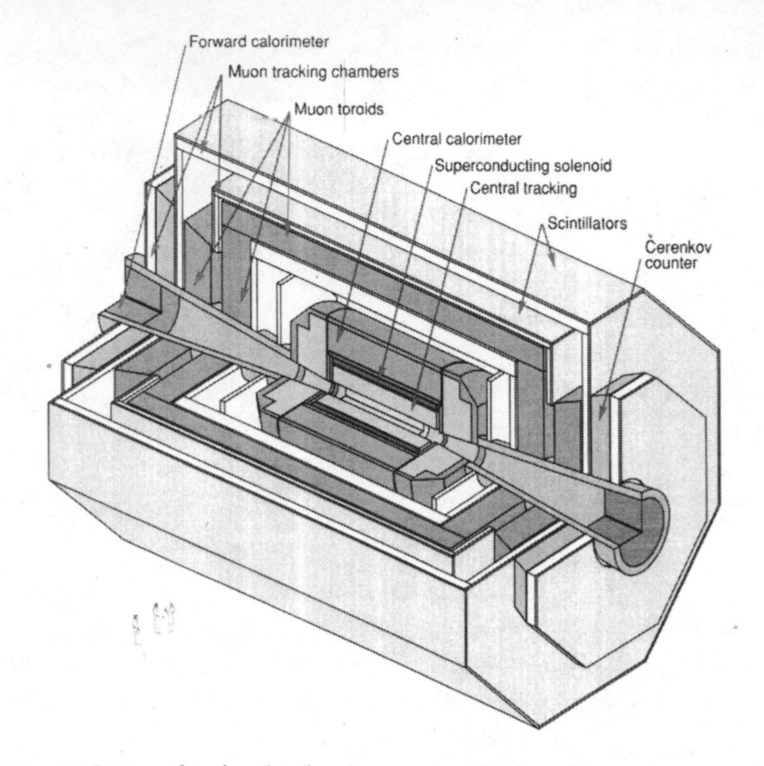

FIGURE 4.4 Cutaway drawing showing the conceptual design and internal structure of the SDC detector. Courtesy of Fermilab Archives, SSC Collection.

20 percent inflation) for experimental equipment and computers; almost three-quarters of this subtotal, or $550 million, was designated for the two large detectors.[186]

In December 1990 Schwitters received letters of intent for three such general-purpose detectors from three large scientific collaborations, each of them encompassing several hundred physicists from dozens of institutions.[187] The Solenoidal Detector Collaboration (SDC), led by the avuncular LBL physicist George Trilling, was soon approved and given funding to develop a formal conceptual design.[188] SDC was a truly international collaboration including many established groups of high-energy physicists from Europe, Japan, and the United States. At the core of its design was a cylindrical superconducting coil that would generate strong magnetic fields to bend the paths of charged particles emerging from the central collision point. Around and inside this coil would go detectors to record trajectories and energy deposits of the emerging subatomic debris. The CDF collaboration at Fermilab had built a smaller detector along similar lines for the Tevatron (see fig. 2.7). SDC used a conservative, established, mainstream design whose intellectual heritage dated back to the collider detector for the SPEAR facility at SLAC—on which Schwitters and Trilling had served together in important roles during the mid-1970s.[189]

A second detector idea was proposed by another large group of physicists from nearly one hundred institutions around the world, many of them in foreign countries, led by the autocratic MIT professor and Nobel laureate Samuel Ting.[190] Dubbed L* (L-Star, or as some SSC watchers liked to joke, "Lone Star"), it downplayed the use of a magnetic field and instead surrounded the collision point with detectors known as calorimeters, designed to measure the energies of emerging particles precisely. It was a good design backed by a strong international collaboration, but the PAC expressed concern about its cost and potential overruns.[191] Another concern was the dominance of foreign physicists in the hierarchical L* management structure, who reported solely to Ting. After an attempt to address this latter concern by including more US physicist groups and establishing a management council headed by Ting and Caltech physicist Barry Barish, Schwitters rejected the proposal in May 1991, citing unresolved concerns about potential cost overruns and the L* management structure.[192] This decision unfortunately meant that a large, influential contingent of foreign physicists (many of whom owed allegiance to Ting) would not be doing research at the Super Collider; with them would go potential contributions from their government agencies.

Some of the L* groups then joined with others from the third collaboration (called EMPACT/TEXAS, which had been turned down in January) to develop a successful proposal for the second general-purpose SSC detector, called GEM, an acronym for "Gammas, Electrons, and Muons"—the subatomic particles it would be optimized to detect.[193] Like L*, its design focused on the use of calorimeters to measure particle energies. Led by Barish and William Willis of Columbia University, the GEM collaboration was dominated by US physicists, although it included major contingents from China and Taiwan.[194]

The two successful detector collaborations were large, extended scientific bureaucracies, each with a physicist spokesman or spokesmen, a governing board, project manager, advisory councils, finance committee, and networks of committees and subcommittees devoted to specific detector elements. High-energy physicists controlled these collaborations; the engineers involved reported ultimately to the physicists in charge. At over half a billion dollars apiece (including the hoped-for foreign contributions), SDC and

GEM each had a budget about equal in constant dollars to the entire cost of building SLAC in the mid-1960s.

Schwitters appointed SLAC theorist Frederick Gilman, whom he knew well from when both had worked at SLAC in the 1970s, as associate director of the SSC Physics Research Division in charge of the laboratory's experimental program. Gilman replaced Murdock Gilchriese of Cornell and CDG, who had filled this position on an acting basis in 1989. To some SSC observers, it was a curious choice. A theorist, Gilman had little experience organizing and managing large groups of (often querulous) physicists—and essentially none in the experimental arena. Nor had he dealt with anything close to the budgets in the hundreds of millions of dollars that would be involved once the detectors went into construction. But Gilman was well liked and respected throughout the US high-energy physics community.[195] He functioned effectively in this managerial role, which also involved planning the experimental facilities needed to house and support the detectors, plus oversight of the detector R&D program.

Schwitters and the laboratory directorate devoted their time increasingly to the welfare of these detector collaborations, leaving to Siskin the details of managing accelerator building and technical services. Schwitters perceived his role as SSC Laboratory director not just as the builder of a high-energy physics research facility, but also as the developer of a thriving laboratory community that would attract first-class scientists from around the world.[196] To achieve this goal, he needed to provide the best possible experimental apparatus and allow the physicists to control their research environment. The latter would prove much easier to realize in the familiar domain of building the particle detectors than in constructing the gargantuan proton collider needed to generate the events they would examine and record.

The culture of the emerging laboratory continued bifurcating into two distinct subcultures that coexisted uneasily. Engineers, many with backgrounds in the armed forces or their industrial contractors, dominated the building of the collider and its chain of injectors. Thus Reardon reported to Siskin, not vice versa. But authority over the detector projects remained firmly in the hands of the high-energy physicists; they reported to Schwitters through a chain of command that included only scientists.

During his first year at the lab, Siskin put primary emphasis on trying to get a fully computerized cost-and-schedule control system up and working, throwing more than seventy employees into the effort.[197] With the project well under way, spending hundreds of millions per year, this was an extremely difficult proposition—somewhat akin to repairing a jet plane

FIGURE 4.5 SSC Laboratory General Manager Edward J. Siskin. Courtesy of Fermilab Archives, SSC Collection.

in flight. For a while, Cipriano considered halting the entire project in its tracks to fix the system, but the project's "burn rate" of over $1 million per day made it too expensive to stop for any significant amount of time.[198] "It will be a year before the Lab has a certified cost [and] schedule control system in place, so there is a lot of manual effort required to determine status," he wrote Watkins in January 1991. "On the other hand, when you have to do it manually, you really understand it when you are done."[199]

The continued lack of buy-in from high-energy physicists involved in supercollider construction to what they considered a military-industrial project-management tool did not help. According to Siskin, the typical reaction was, "Do you want me to do my job, or do you want me to crank out all this paper?"[200] John Rees, an accelerator builder from SLAC who eventually became SSC project manager in January 1992 (see chapter 6), recalled that Reardon dragged his feet on implementing the system, much to Siskin's chagrin, and the physicists designing the detectors were largely reluctant to use it.[201] Helen Edwards was even more defiant than Reardon, according to Siskin: "Those estimates were on Helen's wall, and she'd walk over and change them," he recalled. "And not tell anybody, and then get up in the AD [associate director] meeting and say, 'I'm on schedule.'"[202]

Extreme tensions resulted between physicists and engineers. On April 1, 1991, Edwards finally resigned after a clash with Siskin and Cipriano over the design of the superconducting quadrupole magnets for the main collider rings. In what must have been the worst lapse in the SSC design process, their aperture had remained at 4 cm after the dipole aperture was increased to 5 cm in 1990. Not only would this have resulted in a beam pipe with a variable diameter and consequent complexities in collider construction and operation, it would also have seriously complicated and restricted future installation of a beam-tube liner, if needed to deal with radiation emitted by the proton beams at higher energies and intensities.[203] With the support of the Machine Advisory Committee, Edwards began insisting that the quadrupole aperture *had* to be increased to coincide with the dipole aperture.[204] But Siskin and Cipriano, backed by Watkins, refused to approve this design change, which would have added another $50 million to the total project cost and given congressional SSC opponents evidence that the project was out of control.[205] After Schwitters failed to overrule them, Edwards decided that the time had come to return to Fermilab.[206] By April 1991, therefore, little more than two years into the project, *all* the key SSC managers designated in the URA proposal except Schwitters had departed—in order, Tigner, Chrisman, Sverdrup's Robert Robbins, and Edwards.

Schwitters continued struggling with the DOE site office, which had swelled to over one hundred members by late 1991. He thought Cipriano was trying to micromanage the project rather than serve in the traditional oversight capacity DOE usually played on high-energy physics projects.[207] And he was in fact correct. In early 1991, Cipriano had established a Change Control Board, with members from both DOE and the SSC lab, which reviewed all major design changes plus some minor ones.[208] Before any contingency funds (controlled by DOE) could be applied to pay for these changes, they had to be approved by the board. This procedure, by then common in military procurements but not in high-energy physics, shackled the accelerator physicists in their customary practice of making quick trade-offs as they attempted to refine the site-specific SSC design. In effect, it froze the SSC design at the established baseline parameters and stifled innovation.

These kinds of conflicts occurred because engineers need to freeze the designs of components and systems early for construction or manufacturing development, while physicists prefer instead to keep choices fluid until the last possible moment in order to benefit from the latest technological advances.[209] But doing so can easily lead to cost overruns, as the case of the quadrupole aperture helps illustrate. To deal with such overruns and attempt to keep costs under control, physicists would often make trade-offs with other aspects of the machine design, which requires free and open information exchange. The SSC Change Control Board restricted that process and put the balance of decision-making power in the hands of engineers. The clash between them and the physicists at the SSCL became a classic confrontation between scientific and technological innovation on the one hand, and efficient, rationalized production on the other.

At the SSC Laboratory, this problem was exacerbated by the fact that the engineers involved, coming largely from the military-industrial complex, had little to no experience working with high-energy physicists. There are plenty of highly experienced engineers working at DOE national laboratories who have established effective working relationships with the physicists. The two cultures recognize their differences and the unique contributions each can make to common goals. This familiarity reduces conflict and enhances cooperation between them. But in trying to set up the SSC Laboratory at a greenfield site, its engineering culture had to be built anew from available human resources. Because few engineers went to Texas from the US national laboratories, their roles and responsibilities had to be assumed by engineers with little, if any, familiarity with the culture of high-energy physics.[210]

Establishing a new laboratory infrastructure and scientific culture south of Dallas proved much more difficult than anticipated. The most important missing element, at least in the early going, was a strong project manager to implement the grand plans of high-energy physicists. Maury Tigner *might* have functioned well in such a capacity, given his success in organizing and managing the Central Design Group. Had he been willing do so—and had DOE approved his appointment—more of the well-oiled CDG management team would surely have gone to Texas with him. And he would likely have pursued the original CDG design, with minor modifications. The SSC almost certainly would have experienced cost overruns, but probably not the billions of dollars that eventually would have been spent had it survived.

DOE officials correctly insisted that *someone* with large-project management experience be included as part of the SSC's leadership. It was folly to think that a project that rivaled the Manhattan Project and Panama Canal in scope could ever be successfully managed by physicists from universities and US national laboratories. Project management talent of the highest caliber—someone like a General Groves—was required, if only to direct and control the many corporations and outside entities involved. Unlike Brookhaven, SLAC, and Fermilab, which were built largely by in-house operations, the SSC required major industry involvement, adding a new and unruly dimension to the project that had to be forcefully managed.

But DOE officials, especially Robert Hunter, attempted to manage the SSC project from day one—never trusting that URA physicists could organize a strong management team. Instead, DOE began assembling an unwieldy and most likely unworkable bureaucracy based in Washington, using Beltway contractors to fulfill the roles contractually assigned to URA. It is doubtful that such an ad hoc management team, lacking deep experience with high-energy physics projects, could ever have succeeded, either in Washington or Texas. Reacting sharply against this dangerous salient, and with the aid of TNRLC leaders, URA won the opening battle with DOE. But it was soon outflanked and began to lose the war.[211]

After Helen Edwards and her cohorts substituted a more conservative site-specific SSC design for the generic CDG one, the URA victory proved short lived. The new design team naturally wanted to reduce the risks inherent in the design it had inherited, but the costs of doing so added at least $1 billion to the SSC price tag. Another $1 billion or more arose from resulting schedule delays and the need to add infrastructure items that were not included in the original CDG estimates or the 1989 submittal to Congress. Secretary of Energy Watkins concluded from this experience that he could not trust academic physicists to manage such a challenging project. Deepening the divide between them, he turned to men he knew from his military background to step in and seize the reins of the SSC lab's construction. Its project management was therefore imposed from without.

And at a critical juncture in the project's evolution, in late 1989 and early 1990, the high-energy physics community refused to reduce the project scope, or to take a phased approach to reaching the design energy of 40 TeV. Despite Panofsky's urgings, this community insisted that it had to have it all, and all at once. While the rest of the US scientific community was coping with funding shortages and pleading for financial relief, the Drell subpanel and HEPAP voted unanimously for a conservative design that could be rapidly commissioned at full energy—adding billions to the projected SSC cost at a time of ballooning federal deficits. It proved a huge mistake.

This first big cost overrun bolstered the case of SSC critics trying to arouse congressional opposition to the project. Henceforth, DOE officials would struggle mightily to assure Congress that the complex project was under control—in part by imposing their own selected managers on the laboratory, especially in the collider construction area. But continuing cost increases undermined their efforts. SSC costs quoted in the nation's capital would soon rise toward $10 billion (see chapter 6), well beyond the official $8.25 billion figure, further weakening its congressional support despite the efforts of the powerful Texas delegation to defend it.

And the imposition of what physicists viewed as an alien, military-industrial culture upon the existing scientific one exacerbated the already difficult problem of building a harmonious, effective laboratory culture. One can only guess at the added costs incurred due to all the infighting and disruptions that ensued, but they were not trifling. Had the SSC been sited instead at Fermilab, project leaders could have built on the strong human infrastructure already in place, in which engineers reported to physicists and worked much more cooperatively with them.[212]

Perhaps in such a context—and absent the many other disruptions involved in organizing a new laboratory infrastructure and scientific culture—the physicists involved might not have been averse to using a computerized project-management control system. These systems are now employed routinely throughout high-energy physics, and for far smaller projects than the SSC.[213] They are essential tools in sound project management. It is difficult to fathom why SSC physicists were so opposed to using such a system on what was then the largest basic scientific project ever attempted.

After 1991, the year the Cold War formally ended, an uneasy truce settled upon the SSC Laboratory. Engineers from the military-industrial complex firmly controlled construction of the collider and much of the lab's conventional facilities. High-energy physicists still controlled the design and building of two gargantuan particle detectors, SDC and GEM, which together were estimated to cost over $1 billion. The SSC Laboratory had evolved into a dysfunctional bipolar culture, making it much harder to recruit top high-energy physicists to come to Waxahachie. By and large, the best and brightest of this proud scientific community failed to show up there in sufficient numbers.

By contrast, the example of CERN's Large Hadron Collider project is telling. Built at a thriving international high-energy physics laboratory, with well-integrated physicist and engineering cultures employing a sophisticated project-management control system, this multibillion-dollar collider project endured a number of troublesome management problems and finally began to take data in 2010 at 7 TeV, albeit after a major mishap, long delays, and at low luminosity (see epilogue). It surely helped that CERN had initiated this project in the 1990s with most of the human infrastructure already in place, as well as an existing 27 km tunnel and proton injector chain. And it had the benefit of having a *single* project manager throughout the project life—Lyndon Evans—a physicist who had worked on proton colliders at CERN since the mid-1970s.[214] The social cohesion of a strong, existing laboratory culture, with a long tradition of diverse disciplines working successfully toward common goals, helped overcome the difficult problems that arose on the LHC project.[215]

CHAPTER FIVE

Washington and the World, 1989–92

> The United States has a long history of collaborative research efforts with other friendly nations. But this is an American Project, American leadership. . . . We are going forward with it.
>
> —ENERGY SECRETARY JOHN HERRINGTON, January 1987

While the SSC Laboratory was putting down roots in Texas soil, political sands were shifting dramatically in Washington and the wider world. With the fall of the Berlin Wall in November 1989 followed by the collapse of the Soviet Union in 1991, the bipolar Cold War confrontation that had governed international relations since World War II was coming to a sudden end, to be replaced by a "new world order."[1] High-energy physicists were slow to grasp what this global transformation might mean for their costly scientific field, which was almost completely dependent on robust government funding to build and operate its frontier research facilities. Assumptions that might have been the norm a decade or two earlier were no longer valid. At the scale of the SSC, such miscalculations could have dire consequences.

As George H. W. Bush became president in January 1989, he envisioned a kinder, gentler administration that would redress the imbalance between defense and domestic spending that had emerged under Reagan. But the federal deficits that arose in tandem with the Reagan military buildup continued unabated. Major new expenses, such as for the savings-and-loan bailouts and for cleaning up the nation's nuclear-weapons complex, put heavy demands on available funds. Federal deficits almost doubled under Bush, from $153 billion in fiscal 1989 to $290 billion in fiscal 1992. Finding the money to support a multibillion-dollar physics project in Texas would prove challenging, but Bush made the SSC a "presidential priority" in early 1989. That helped galvanize support for the project within his administration and serve as fair warning to members of Congress seeking vulnerable budgets to slash.

With prior service as ambassador to the United Nations and director of the Central Intelligence Agency, Bush preferred to focus on international affairs. Thus SSC advocates were planning to seek his aid in securing the needed foreign contributions. But Reagan administration rhetoric about building the SSC to restore US leadership in high-energy physics would make this task more difficult than many had assumed.

THE EMERGING CONGRESSIONAL DEBATE

As the fiscal 1990 budget battle began in 1989, DOE and URA had a powerful new ally in Congress: the Texas delegation (see chapter 4). Speaker of the House Jim Wright had great influence over other members by virtue of his power to appoint them to key committee positions.[2] He had named his fellow Texas Democrat Jim Chapman to the all-important House Appropriations Subcommittee on Energy and Water Development, which would put forward the House figure for the SSC budget every year for the next five years.[3] Democrat Martin Frost of Dallas served on the House Rules Committee, which had the power to limit floor debates on the SSC and amendments to bills that might affect the project. And Republicans Ralph Hall and Joe Barton, who represented the Ellis County district, served on the House Committee on Science, Space and Technology, where issues of science policy are often debated. In the Senate, Texas had Democrat Lloyd Bentsen chairing the Finance Committee and Republican Phil Gramm, an author of the Gramm-Rudman-Hollings Budget Act. And in the background stood President Bush, wielding the power to veto any rash congressional bill that might prove unfavorable to the SSC.

In early 1989, the Bush administration requested a modest allocation of $250 million for the SSC project, including $160 million for a construction start in the 1990 fiscal year, to begin on October 1, 1989.[4] It seemed a trifling amount compared to the total of $67.3 billion requested for R&D funding—mostly for military R&D—and especially compared with the $2.05 billion requested for the Space Station and $5.9 billion for the Strategic Defense Initiative (SDI), two other Reagan administration legacies.[5] But granting such a new construction start meant making a commitment to funding the project in future years, when its annual appropriation would increase by hundreds of millions of dollars, a stealthy process known in Washington circles as "getting the camel's nose under the tent." With the annual federal budget deficit attaining levels unseen since World War II, and threats of Gramm-Rudman-Hollings sequestrations looming, congressional appropriators were leery of adding yet another costly gigaproject to the growing US debt load—especially without firm foreign commitments to contribute and participate.

When the 1990 Energy and Water Development Bill emerged from Alabama Congressman Tom Bevill's Energy and Water Appropriations Subcommittee in June, the proposed SSC budget had survived but was reduced to $200 million, with only $110 million for construction. But even that came up for debate when the bill reached the House floor on June 28. Citing, among other things, the fact that the DOE had no firm commitments yet for foreign contributions, Congressmen Boehlert, Dennis Eckart (D-OH), David Obey (D-WI), and Howard Wolpe (D-MI) offered an amendment to delete all the construction funds for the project, leaving only $90 million for ongoing R&D and other preparatory work.[6] Rallying to defend the SSC, Bevill claimed that "we are well over $1 billion in commitments already, and by this time next year the Secretary of Energy says we will have more commitments if we go ahead and approve funding for this project today."[7]

He was referring not only to the $1 billion Texas commitment (see chapter 4), but to DOE claims that other countries were lining up to contribute. In fact, he had received a letter from Secretary Watkins the day before stating that "we are highly confident that we can achieve the $1.8 billion Office of Management and Budget target" from nonfederal sources, or 30 percent of the estimated $5.9 billion SSC cost. "I am confident," Watkins wrote, "that we can count on the Japanese to contribute significantly to this undertaking, we would hope on the order of $0.5–$1.0 billion." In addition to a $50 million in-kind SSC commitment from India DOE already had in hand, there were also "preliminary indications from Italy, Canada, Spain, Finland, and Switzerland" to participate in the construction phase, plus another expected $200 million that would come from abroad to help in building the SSC detectors.[8]

This statement, plus the support of Bevill and the powerful Texas delegation, helped sway the House and defeat the amendment handily, by a surprisingly wide margin of 331 to 92.[9] This resounding defeat for SSC opponents most likely reflected Bevill's support. His subcommittee's appropriations bill, which he liked to call the "All-American Bill," included energy research and water development projects in many congressional districts throughout the nation.[10] It was unwise for members who had or wanted such projects for their districts to oppose the powerful chairman's will.

When it reached the Senate, the bill encountered little resistance because Bentsen and Gramm had already lined up sixty senators in support, thus ensuring its passage. And it didn't hurt that Louisiana Senator Johnston, chairman of the Senate Appropriations Subcommittee on Energy and Water

Development, had become an ardent SSC supporter after General Dynamics, then vying for one of the contracts to manufacture the SSC's superconducting dipole magnets, told him it would build its plant in Louisiana if victorious.[11] A more collegial body than the House, the Senate was much less concerned about foreign contributions. It also takes a longer view; senators more often vote for what they view as the good of the nation as a whole, rather than just that of individual districts or states. When the Energy and Water Appropriations Act of 1990 emerged from the House-Senate conference committee meeting on September 7, it included a compromise figure of $225 million for the SSC project, containing $135 million for construction—which could begin the very next month, if required.[12] The SSC had finally surmounted a major hurdle. The "chicken-and-egg dilemma" referred to earlier (see chapter 3), had apparently been resolved by the United States taking the first major step forward.

Though soundly defeated, the House amendment led by Boehlert, Eckart, Obey, and Wolpe reflected a growing polarization between two geographical regions of the United States that had been at odds for over a decade. In 1976 congressmen from the Northeast and Midwest had formed the Northeast-Midwest Congressional Coalition in the conviction that their districts and states were being shortchanged by federal policies catering to the interests of the South and West.[13] Reagan's military buildup of the 1980s deepened their sense of disenfranchisement, for its financial benefits accrued more to high-tech Sun Belt companies than to the ailing smokestack industries of what had become recognized as the Rust Belt.[14] And the savings-and-loan bailouts of the late 1980s and early 1990s represented yet another economic transfer from the Northeast to the Southwest.[15] In 1990 the chair of the Northeast-Midwest Coalition was the unabashed liberal Michigan Congressman and SSC critic Howard Wolpe, who had earlier served as a professor of political science at Western Michigan University in Kalamazoo.

Particularly galling to these Rust Belt legislators, whose states had been wracked by the recessions of the 1970s and what they perceived as concomitant job losses to Sun Belt states, was the success of Texans in garnering national laboratories and other federally supported facilities. The Lone Star State appeared to be securing more than its proportionate share of not only military bases and NASA facilities but also public-private consortia such as SEMATECH. A partnership between the US government (funded through the Defense Department) and US microchip makers, this institution was established to develop and promote advanced semiconductor manufacturing techniques to enhance US competitiveness in this crucial industry. Many thought SEMATECH should have been located in California, close to Silicon Valley, but the influential Texas delegation helped bring it to Austin in 1988.[16] Siting the SSC in Waxahachie was another big, shining pearl in this long string of Texas successes.

In the fall of 1989, SSC opponents in Congress were licking their wounds but girding for a long war. And hardly two months passed before the SSC lab offered them potent new ammunition. On November 19, 1989, a front-page article in the *Washington Post* suggested that the costs of the SSC project could balloon by as much as $2 billion beyond the $5.9 billion estimate the DOE had given during the FY1990 appropriations process.[17] As events transpired over the following year, that figure proved accurate, if a little low, as the official SSC price tag rose to $8.25 billion in early 1991 (see chapter 4). The need for foreign SSC contributions became even more pressing—and their required size increased accordingly.

And in Eastern Europe that November, historic events were culminating in the collapse of the Iron Curtain, symbolized most dramatically by German citizens tearing down the hated Berlin Wall.[18] Although it had been hovering on the horizon for months, if not years, this pivotal event in world history marked the beginning of the end of the Cold War. With the collapse of the Soviet Union in late 1991, the Cold War was over, and with it ended the bipolar confrontation with the United States that had served for decades as the basis of international relations.[19] As the leader of the Free World, the United States had habitually taken the lead on major scientific projects, assuming that other Western nations would follow with sizable contributions. Now this Cold War model of international scientific collaboration had lost its raison d'être. In fact, it had been steadily eroding throughout the 1980s.

In the same vein, a key rationale governing the federal funding process was evaporating. No longer could supplicants argue, explicitly or implicitly, that the need to maintain world leadership in science, space, and technology justified their funding requests. Projects now had to be justified more on their own intrinsic merits. In the dawning post–Cold War budget climate, many scientists hoped to see a "peace dividend" resulting from major reductions in spending for the military-industrial complex. But it never materialized—at least not for the physical sciences. The SDI budget leveled off below $4 billion, while Space Station funding continued growing inexorably.[20] And the costs of cleaning up and modifying the nuclear-weapons complex, a Cold War legacy, sopped up an increasing portion of any hoped-for peace dividend, too.[21] As 1990 dawned, the federal R&D budget remained as tight as ever, or even tighter. The overall federal deficit for fiscal 1990 was headed well north of $200 billion, and the threat of Gramm-Rudman-Hollings sequestrations and their associated across-the-board cutbacks loomed large.

But most high-energy physicists seemed oblivious to the hand-wringing then going on in the nation's capital (see chapter 4). URA decided to proceed with a more conservative, less risky Super Collider design that was going to cost an additional $2 billion, rather than reduce the project scope (as NASA did with the Space Station) to hold its costs down. In January 1990, HEPAP convened the blue-ribbon Drell panel to rubber-stamp this decision, providing political cover for the DOE to go along with it.[22] To the surprise of many present, Texas Senator Gramm appeared at the January 12 HEPAP meeting that endorsed the panel's recommendation and cautioned them that the redesigned SSC would not have an easy time competing with all the other projects vying for scarce federal funding. "It's important that we not end up with a Cadillac but with a Chevrolet," he told the assembled physicists and DOE officials. "I'll help Congress get over the sticker shock, and you'll have to make sure of a sturdy and reliable Chevy."[23]

That spring, responding in part to the growing cost of the SSC, the House of Representatives decided to consider a bill to authorize construction. Authorization bills differ from the annual appropriations bills, which specify how much money to allocate each fiscal year on a project and in what ways—e.g., for R&D, construction, and operations. They are instead statements of overall policy and guidance, to be followed by the administration and Congress in subsequent actions or bills. The Superconducting Super Collider Authorization Act of 1990 (House Resolution 4380), which emerged in late March from the House Committee on Science, Space and Technology, established a cap of $5 billion on federal expenditures for the project. It also required that Texas actually contribute its promised $1 billion, and that between one-fifth and one-third of the total project costs had to "be raised from international participants."[24] In addition, it specified that no more than half of any major system or component (e.g., the superconducting magnets) could be produced by foreign firms, and it established the DOE position of director of the SSC Office (about to be occupied by Cipriano) with the "responsibility of *managing* the overall project."[25]

Had the DOE and URA been able to hold the line on costs—to, say, $7.4 billion—they could have met these stipulations with a 20 percent contribution from foreign sources, or $1.5 billion,[26] which might have been attainable. But as the project costs escalated in the ensuing years, additional foreign funding was going to be needed if the US portion was truly capped at $5 billion and the Texans didn't budge. Sensing that SSC costs would probably rise even further beyond the figures being bandied about in early 1990, Boehlert and other SSC critics planned to offer an amendment during the House floor debate "that foreign sources provide at least 25 percent of the total cost" of the SSC instead of the 20 percent figure stipulated in the bill.[27] According to Boehlert legislative aide David Goldston, a major rationale for this figure was to align the resolution more closely with what DOE was claiming.[28]

New Jersey Democrat Robert Roe, chairman of the House Committee on Science, Space and Technology, managed the floor debate on May 2. There was little opposition to the bill itself, which had emerged from his committee on a bipartisan vote of 32 in favor and 9 against.[29] Even Boehlert supported it in speaking from the floor: "I agree . . . that this sort of research becomes a great nation, that unforeseen benefits will flow from it, and that this bill includes many safeguards that have been absent from previous legislation."[30] But, he cautioned, "What will kill the SSC in the long run is not the attacks by its opponents, but the bloated claims of its strongest supporters, which will leave the Congress feeling misled and betrayed." In the end, both H.R. 4380 and Boehlert's amendment passed by comfortable margins.

No further votes specifically addressing the Super Collider occurred in either the House or Senate that year. Nobody offered an amendment to kill the project in the two appropriations bills that shortly made their way through Congress. After the president's budget requested $318 million for the SSC in fiscal 1991 (beginning in October 1990), the House and Senate agreed to a compromise figure of $243 million later that year, including $94 million for construction (not much of which was ready to occur yet, due in part to the superconducting dipole magnet redesign). Texas continued to front-load its contribution, ponying up $149 million to bring the total SSC funding that fiscal year to $392 million.

But House Resolution 4380 authorizing the SSC never became law because the Senate did not take it up, letting it languish. The Senate Committee on Energy and Natural Resources, which had jurisdiction, was chaired by Bennett Johnston, too; he preferred to exert his authority and power though his chairmanship of the Appropriations Subcommittee on Energy and Water Development. He stood in a unique and powerful position by virtue of this dual chairmanship, which gave him extraordinary influence in the Senate (and which he would later use to support the SSC). Nevertheless, House members—especially the growing SSC opposition—viewed H.R. 4380 as "the will of the House" despite the fact that it did not become "the law of the land."[31] The DOE and URA could ignore it only at the SSC's peril.

FIGURE 5.1 Senator J. Bennett Johnston (*center*) on a visit to SLAC, conversing with its Deputy Director Sidney Drell (*left*) and Director Emeritus Wolfgang K. H. Panofsky. Credit: Ed Souza, Stanford News Service.

Congress was not of a single mind on this question of foreign contributions. For a project that had been touted as the way to revive US preeminence in high-energy physics and enhance national competitiveness, letting other countries manufacture its highest-technology components made little sense. But at the billion-dollar level, there was no other option; other nations would never agree to send that much cash. Powerful figures in Congress, such as Johnston and Mississippi Democratic Congressman Jamie Whitten, the octogenarian chairman of the House Appropriations Committee, maintained that the United States should just build the SSC itself and pay the full costs of doing so. But as these costs ballooned in the 1990s and the nation transitioned from the Reagan to the Bush years, realistic politicians and government officials recognized that major foreign participation was going to be mandatory.

THE BUSH ADMINISTRATION BEGINS TO SEEK FOREIGN PARTNERS

After a year's hiatus, the Department of Energy had again begun addressing the crucial question of foreign contributions to the SSC in 1990, a year after Carlo Rubbia became CERN director general, replacing Schopper. Apart from leading the European laboratory in successfully commissioning its new LEP collider, Rubbia initiated a spirited campaign to promote the LHC project as CERN's next machine. He suggested that with a vigorous, fast-track effort, the LHC could be installed and running in the LEP tunnel by 1996 or 1997, well *before* the SSC and therefore in a position to scoop the SSC Laboratory on some of its most important physics goals.[32] For veteran US CERN watchers, it was too much. The lab was then burdened with debts it had incurred to complete LEP on schedule and in no position to incur any more—nor to go hat in hand to ask European governments for additional funding. But Rubbia's clever campaign helped make the LHC project appear as a *possible* reality in the eyes of European physicists and science agencies. Future European funds for high-energy physics, which might otherwise have been committed to the SSC, say, for its detectors, remained east of the Atlantic for the seemingly increasing likelihood they would be needed for the LHC.[33]

DOE efforts in early 1990 were focused on developing overall guidance for international contributions to the SSC. This work had to occur at different levels within the department—and in coordination with other federal agencies, such as the OMB, OSTP, and State Department. These interagency efforts began to coalesce that spring in a draft internal document entitled "Superconducting Super Collider: Framework for International Participation." Its introduction states, "This document represents the distillation of that work, and the starting point for discussions of international collaboration on the SSC."[34] It establishes policies and protocols to follow in pursuit of foreign contributions, but it is vague—perhaps intentionally—about what other nations might expect in return. On the one hand, the document states, "DOE will own and operate the SSC for the benefit of all qualified scientific users, foreign or domestic."[35] According to this document, the DOE would attempt to enter into bilateral agreements with other countries to provide cash or in-kind contributions; it would abide by US procurement laws such as the Buy American Act but try to limit or bypass import controls on in-kind contributions. Only at the end does it attempt to address the question of foreign involvement in the SSC Laboratory management, and then only very vaguely: "For substantial contributions to the SSC, the DOE is considering, and open to, various options in the organization and management of the SSC program which could help stimulate greater foreign involvement."[36]

Reading this document years later, one wonders what possible motivations foreign agencies might have had to contribute hundreds of millions of dollars to this overseas scientific facility. According to traditional practices long established by the International Union of Pure and Applied Physics and ICFA, their physicists would have been able to participate in SSC collaborations or experiments *without* having to make any such major contribution. Much smaller in-kind contributions to its experiments, worth a few million dollars, were all that might be necessary.[37] Foreign physicists would naturally have had a correspondingly important role in the management of an experimental collaboration through its various councils and committees. They already participated in SSC advisory committees such as the Science Policy Committee and the Program Advisory Committee. Egregiously absent from the DOE plan was any concrete suggestion for an international body like the CERN Council to oversee the budget, management, and operations of the SSC Laboratory.

Japanese physicists, for example, were then deeply involved in planning the SDC experiment and were also represented on these two committees. Since the 1980s a large group of them, led by Kunitaka Kondo of the University of Tsukuba, had worked in the CDF collaboration at Fermilab, designing and fabricating the large superconducting magnet at the core of its detector with Hitachi Industries.[38] Out of this collaboration emerged an even bigger group that became engaged in CDG's detector R&D during the late 1980s, including high-energy physicists from the KEK laboratory and the universities of Hiroshima, Kyoto, Niigata, Osaka, Tohoku, Tokyo, and Tsukuba.[39] In all, there were perhaps one hundred Japanese physicists interested in becoming involved in SSC research—almost all of them in the SDC collaboration.

That many participating scientists could justify in-kind contributions to the SDC detector worth about $100 million, distributed over the decade required to design and build it. But even that would be difficult for Monbusho, the Japanese Ministry of Education, Science, and Culture, to accommodate. During the previous decade, under an international agreement with the DOE, it had allocated nearly that much money to support the US contributions of Japanese physicists at Brookhaven, Fermilab, and SLAC.[40] Expanding this program to accommodate their research at the SSC Laboratory would put pressure on Japanese high-energy physicists doing research at the other US labs, as well as those working on the home turf at KEK, where most of them were involved in experiments using its TRISTAN electron-positron collider.

The Bush administration, however, was preparing to request a much larger contribution from the Japanese government, worth a billion dollars or more.[41] With the likelihood of obtaining large contributions from European nations diminishing, this option was increasingly viewed as the only way to achieve the $1.7 billion worth of foreign contributions required to meet the stated goal of one-third nonfederal funding. Such a request could not be accommodated by Monbusho alone, and probably not by its teaming up with the Science and Technology Agency (STA) responsible for Japanese big-science projects, either. It could not be based solely on the scientific rationale for the project—especially with such a tiny portion of Japan's scientific community involved. It had to be an inherently *political* request, made at the highest levels of the two governments, from President Bush to Prime Minister Toshiki Kaifu.

For there were powerful ministries and groups within the Japanese government, especially its Finance Ministry, and the nation's scientific community that staunchly opposed such a large transfer of its wealth across the Pacific. Given the consensual nature of the Japanese decision-making process, a request for a billion dollars had to come down from the top, not percolate up from the bottom, to have any chance of success.[42]

Thus, after detailed preparations and an advance-team visit, a US delegation led by DOE Deputy Secretary Henson Moore departed Washington for Tokyo on May 28, 1990. He brought a letter from Bush to Kaifu inviting Japan "to help make the Superconducting Super Collider a reality, a great scientific success, and a model of international collaboration from which all nations can learn and benefit."[43] Included in the delegation were SSCL Director Schwitters, Acting OER Director Jim Decker, the DOE Assistant Secretary for International Affairs John Easton, and officials from Congress, the Commerce and State Departments, and OSTP. The delegation carried with it a list of possible contributions worth a total of more than $2 billion that Japan would be in a good position to make, given its booming economy and high-tech industries.[44] During the past few months, Bush administrations officials (especially his Science Advisor D. Allan Bromley) had been considering the idea of a "partnership" between the United States and Japan, in which the two nations would share the tasks of not only constructing but also operating the SSC lab.[45] The exact nature of this proposed partnership, however, had not yet been clearly defined.

The US delegation met with its counterparts in the Japanese ministries of Monbusho, STA, and MITI, as well as with the Ministry of Foreign Affairs, members of the Japanese Diet, and the Prime Minister's Council on Science and Technology. (Moore did not meet with Kaifu, as that would have been premature.) The ministers expressed guarded interest in the idea of a partnership, and questioned exactly what was implied by this term.[46] They also seemed "intensely interested in the prospects of sustained congressional support for the SSC, and asked a number of questions about the funding and authorization process and the significance of negative votes in the House."[47] In all, the Japanese ministers appeared excited about the scientific prospects offered by the SSC, but said they would need at least a year to discuss the invitation and reach consensus on how to reply.[48]

On June 4–6, the US delegation visited Seoul, carrying a similar letter from President Bush and meeting with Korean ministers, primarily Minister of Science and Technology Chun Kun-Mo, a physicist and American citizen. Based on the work of the DOE advance team, they were expecting a tepid reception. But the delegation was offered a list of proposed contributions Korea could make, and Chun was ready to sign a statement of positive interest in the SSC.[49] It was a pleasant surprise. "He is a very capable fellow and knows this project," said Moore almost a week later. "The Koreans also would like to be the first country in the world to join us as a partner."[50]

The trip was at best a mixed success. Not meeting with Prime Minister Kaifu or anyone close enough to him, Moore left President Bush's letter of invitation to be delivered by US Ambassador Michael Armacost. Officials in the Energy and State Departments recognized that such a big request, an order of magnitude above what could be justified based on a scientific rationale, would require endorsement at the highest levels of the Japanese government. Otherwise, consensus would be hard to achieve. Fortunately, Kaifu was well informed about high-energy physics because he had headed Monbusho before stepping up as prime minister in 1989.[51] In this earlier capacity, he had even visited the KEK laboratory. SSC advocates hoped that Bush would ask Kaifu about joining such a partnership when the two met in July 1990 at the Houston Economic Summit, but that apparently never happened.

In a meeting that July with DOE officials, Toichi Sakata, a rising star in the STA, provided revealing insights about the thinking within Japanese government circles and the obstacles to be overcome for that nation to make such a large commitment. Paramount among them were perceptions of the SSC Laboratory. According to a memo written by the DOE's Japan Desk officer, Sakata said there was no reason for Japan to become involved at that level "unless the project is truly international."[52] According to this document, he elaborated further on their reasoning:

> For cooperation to take place, the US needs to fully resolve the nature of the SSC as an international facility, and the management structure is the key to that resolution. . . . While they heard and appreciated the Deputy Secretary's message in June about the SSC as an international facility, they remember the SSC being sold to Congress as a tool to improve US competitiveness and international prestige. They are not convinced this is a truly international project.

The original rhetoric of the Reagan administration to "throw deep" and attempt to reestablish US preeminence in high-energy physics was coming back to haunt the project as the Bush administration changed course and tried to internationalize the SSC Laboratory to lessen its financial impact on US taxpayers. It would be an uphill battle.

And at the billion-dollar level of the US request, political considerations were to enter into the deliberations that had nothing to do with science. It would be weighed against and have to compete with other major items in the relationship between the two governments. Japan's multibillion-dollar contribution to the Persian Gulf War about to begin, made in lieu of a military involvement, was one such item that year.[53] This telling reality underscores how the SSC project's fate now depended on high-level governmental interactions that extended far beyond the influence or control of high-energy physicists.

That September another DOE advance team headed by John Easton visited Bonn, London, Paris, and Rome to evaluate European interest in SSC partnerships, but the reception was even cooler than experienced in Japan and Korea. Moore wanted to follow up with another high-level delegation accompanying him, but he was finally convinced it would be a fruitless, and possibly embarrassing, trip. "Henson learned it, to his credit," recalled John Metzler, who handled international affairs for the Office of the SSC. "I think he had his heart set on going to Europe, but it wasn't there."[54]

"We don't have the money," said Metzler, paraphrasing replies from the European science ministries.[55] "We'd be happy to work on the detectors but in the traditional way." Such contributions would be much smaller, however, worth in the tens of millions of dollars at most, and they would occur much later in the project—likely in the late 1990s. Already overburdened by the costs of and commitments to HERA and LEP, European nations were in hardly any position to make large financial commitments to the construction of a US high-energy physics project—especially not one that earlier had been touted as a way to reestablish American preeminence in the field! If they existed at all, spare funds would be held in reserve awaiting a decision during the next few years on whether CERN would proceed with the LHC project. If it came to pass, *that* would become their next priority.

According to a closely held report of October 1990, DOE officials concluded "that we are unlikely to meet the Administration's goal for non-Federal participation (one-third of the total project cost) in the foreseeable future."[56] The only remaining hope of attaining that goal, which would entail a total of $1.7 billion (at that point), was for Japan to make contributions worth at least $1 billion. But the "SSC does not have the complete support of the Japanese physics community," the report noted, "and the Japanese government has stated that any decision will take at least a year to make, and likely require a new budget category and possibly some realignments among their science agencies."[57]

The report also stated that, despite economic decline and recent political upheavals in Eastern Europe, the Soviet Union was the next-best possibility for a major SSC contribution. "The Soviet Union is clearly interested, but has worrisome domestic and infrastructure problems," observed the executive summary. "SSC cooperation needs to be part of the Bush-Gorbachev summit process."[58] Preliminary discussions along these lines had been in the works behind the scenes for almost a year. SSC leaders were talking with the Institute for Nuclear Physics in the Siberian science city of Novosibirsk about developing and manufacturing the magnets and other components for the Low-Energy Booster and Medium-Energy Booster (which did not employ superconducting magnets).[59] A figure of $200 million was commonly discussed. But this arrangement would require spending US funds to purchase components at prices well below what they would cost in America and crediting the cost-savings as an actual Soviet "contribution" to the SSC. Congress might well balk at thus sending US jobs to an old Cold War adversary via such a questionable exchange.

Canada was also cited as a potential contributor in the report, but only on a limited basis. The Canadian government was evaluating whether to make a major commitment to KAON, a proposed meson-physics facility in Vancouver, but doing so would severely constrain what it might contribute to the SSC.[60] President Bush's science advisor, Yale University nuclear physicist D. Allan Bromley (who was born and raised in Ontario) was promoting the idea of a US contribution to that project in return for a major contribution to the SSC, but it was not gaining traction on either side of the border.[61] Earlier DOE rejection of the cross-border site may also have cooled Canadian interest. The October 1990 report concluded that "substantive funding for [SSC] construction is not anticipated because of the relatively modest size of their overall HEP program."[62] A small contribution to the SSC detectors was much more likely. And even smaller contributions might be expected from Brazil and Mexico, according to the report.

TWO CONGRESSIONAL SHOTS ACROSS THE BOW

As 1991 began, with Iraqi forces entrenched in Kuwait and soon to face a calamitous confrontation with US and British armed forces massing in Saudi Arabia, business proceeded as usual in Washington, as if the nation were not about to go to war. Copies of the president's budget for fiscal 1992 reached Congress on February 4; staffers quickly thumbed through hundreds of pages to discover what the Bush administration wanted to spend on their favorite projects and programs. The SSC came in with a surprising $534 million, a 120 percent increase that included $374 million for construction, which would allow tunneling to begin as early as October.[63]

Knowledgeable insiders said that this largesse resulted from behind-the-scenes wrangling going on at the DOE and OMB, with the Texas lobbyists pulling out all the stops. Originally OMB had allocated $325 million to the project, but word came down from the White House in December to bump this figure up above $500 million after TNRLC officials threatened to reduce its FY1991 contribution.[64] But the extra funding had to come from elsewhere in the Energy Department, which began seeking projects to cut in other parts of its budget due to new constraints on the budgeting process.[65] An OMB attempt to lop out Fermilab's next big project, the Main Injector, met with concerted opposition from the Illinois delegation, led by House Minority Leader Robert Michel (R-IL). After that funding had been restored, the Bush request came to $666 million for the base high-energy physics program.[66] Adding in the proposed SSC funding, the 1992 Bush budget included well over a billion dollars for high-energy physics—a new watershed for the field. Crossing that threshold did not go unnoticed by opponents in Congress and other struggling scientific fields dependent on DOE funding.

The following day, February 5, the Industry Association for the SSC (IASSC) sponsored a public gathering in the Hart Senate Office Building on Capitol Hill. The Texans were present in force, including TNRLC Chairman Bucy, who reminded listeners that commissions like his wielded the real decision-making power in the Lone Star State. Following brief talks by Texas Congressmen Barton and Chapman, Senator Gramm stood up before the audience. "We've finally got a total cost figure for ya," he drawled in his thick Texas accent. "I'm gonna swallow hard and give it rawht to ya; it's eight billion, two hundred and forty-nayn million dollars."[67] Then he skillfully dodged questions from reporters, including one about how Congress might react to Japan being allowed to contribute superconducting magnets. "Of course we'd prefer that they just give us the money," Gramm replied. "But I don't know of a country or even a good businessman who does business that way."[68]

Anticipation grew during a break, as people waited impatiently for Deputy Energy Secretary Moore to show up and deliver a formal DOE presentation on the SSC costs, which had been delayed for months. He finally arrived and quickly took charge of the meeting in a direct, almost combative style. After extensive review by three different panels, he said, the DOE had arrived at its best estimate of the total project cost: $8,249 million (see chapter 4). Detailed breakdowns of this total, plus justifications, were included in a green, spiral-bound report his DOE assistants passed out at the meeting, which thereafter became known as the "Green Book."[69] Again and again, Moore stressed that the SSC project was moving forward, that it now had good management, and that things were under control. In concluding, he said, "We're gonna bring the SSC in on time and under budget!"[70]

Following his presentation, Moore was besieged by a cross fire of questions from reporters. Many bore down on the lack of solid foreign commitments. But he nimbly dodged their bullets, saying that DOE had a number of "irons in the fire just waiting to get hot." All told, the agency expected $1.6 billion from foreign partners in the SSC project, he said, but it would naturally take time for formal commitments to be made.[71] As the meeting ended, Moore was still on his feet, the jabs of the press having occasionally thrown him off balance but never sending him to the mat. As he walked down the aisle toward the exit, obviously pleased with his performance, he tapped Ed Siskin (sitting next to Cipriano) on the shoulder and gave him a knowing wink.[72]

The IASSC was just one of the lobbying organizations that had sprouted seemingly overnight to support funding and building of the Super Collider. Others present at the February 5 gathering included Americans for the SSC, headed by Texas lobbyist Henry Gandy, and the National Superconducting Super Collider Coalition, under Hollye Doane. At that meeting, these impromptu organizations handed out long lists of the companies interested in bidding on SSC contracts, ranging from industrial giants such as General Dynamics and Martin Marietta to tiny Insulfab Plastics, Inc., of Spartanburg, South Carolina.[73] Both lobbying groups were funded by Texas but operated as though they were national coalitions. Gandy's organization distributed a US map sporting colored dots representing companies and universities interested in bidding for, or that had already garnered, SSC contracts (see fig. 5.2). According to an attached sheet, over 8,000 contracts totaling nearly $185 million had already been committed.[74] Some of the contracts with universities had been funded using the $100 million the TNRLC had allocated specifically for generic R&D.[75]

According to the *Congressional Quarterly*, Doane's coalition was "selling the collider to Congress much as the Department of Defense sells expensive weapons programs."[76] The DOE and TNRLC were spreading contracts over as many states as possible in order to build a broad network of congressional support for the project. Texas lobbyists descended on key members of the appropriations, budget, and science committees with tailored packages to make sure they recognized that SSC dollars were flowing to their states.[77]

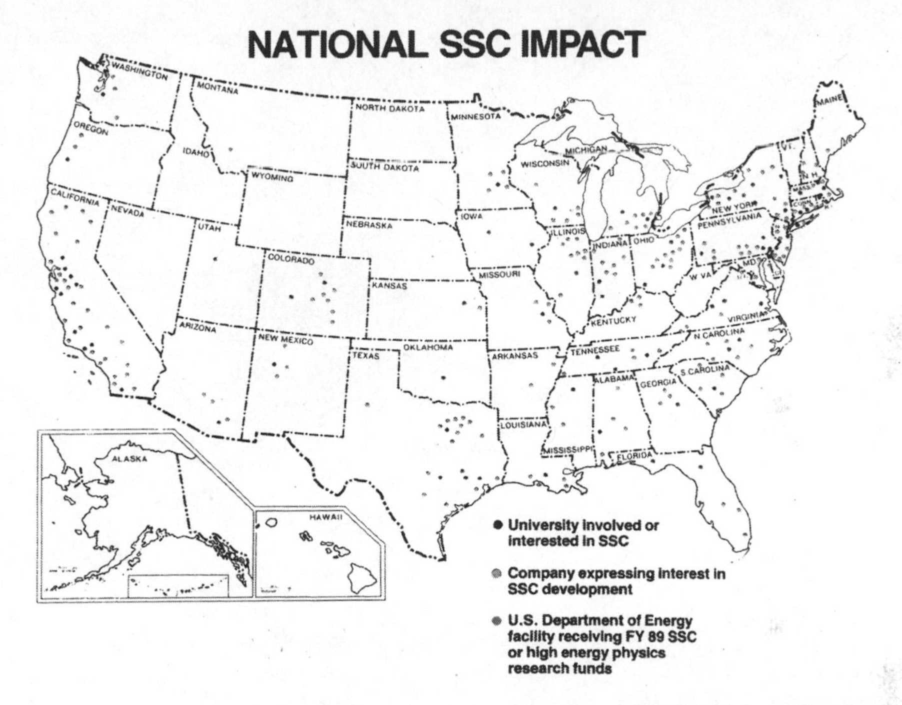

FIGURE 5.2 National SSC impact map of states receiving or expecting SSC Laboratory subcontracts. (Note: This black-and-white rendition of the color original does not distinguish among the types of contract recipients.) Courtesy of Fermilab Archives, SSC Collection.

To California Congressman Leon E. Panetta, then the influential chairman of the House Budget Committee, this resembled the selling of the B-1 bomber. "Once you start passing out contracts around the country on various pieces of [the SSC project]," he noted, "you have a very tough job trying to get everyone to analyze whether that is where we want to commit funds."[78]

But such political maneuvering ignored a new dynamic in the annual budget cycle that had emerged the previous November because of the Omnibus Budget Enforcement Act of 1990, which replaced the Gramm-Rudman-Hollings process.[79] This bill erected a firewall between domestic discretionary spending (such as for R&D) and defense spending, so that decreases in the latter could not be used to pay for increases in the former. There would be no peace dividend. And the growth of domestic spending was limited to the rate of inflation. Thus, increases in one subfield, such as high-energy physics, had to come from decreases in others, like condensed-matter physics—or from beyond physics entirely. To pay Peter, that is, you now had to rob Paul.

SSC Baseline Funding Profile
(Millions of As-Spent Dollars)

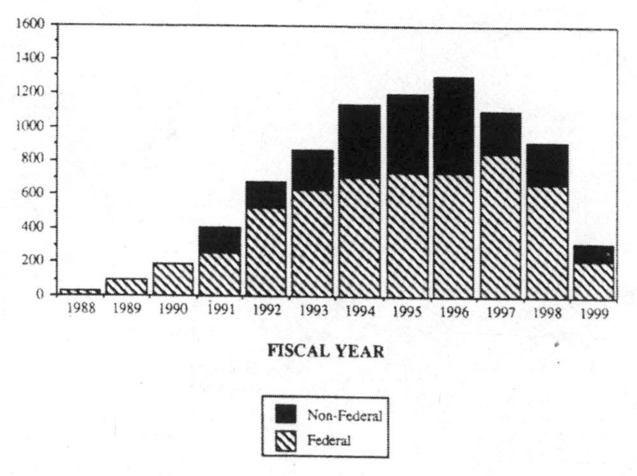

FIGURE 5.3 Expected baseline funding for the SSC project, *Report of the Superconducting Super Collider Cost and Schedule Baseline*. Source: US Department of Energy, Report No. DOE/ER-0468P.

Thus, projects such as the SSC that had to grow substantially during the next few years would inevitably clash with other domestic programs, especially those that came under jurisdiction of the cognizant appropriations subcommittees.[80] With its annual budget ballooning to over a billion dollars in 1994–97, according to the DOE report just released (see fig. 5.3), the SSC project would eventually begin to exert tremendous pressure on other energy and water projects, not just high-energy physics projects, unless foreign contributions somehow arrived to relieve it.

The day of reckoning arrived that spring, sooner than almost everyone had expected. On May 15, 1991, Bevill surreptitiously called a closed-door meeting of his appropriations subcommittee to finalize its energy and water bill for fiscal 1992. When they came out of session later that day, $100 million had been lopped off the Bush budget for the SSC. And Fermilab's Main Injector project, the key to sustaining that lab's vitality through the 1990s during SSC construction, had been "zeroed out" of the budget. Bevill's markup reflected the fact that his subcommittee's allocation could not satisfy all the demands being placed on the budget, so he had established an overall policy of "no new starts."[81] "It's just not possible to do all the things we want to do," Bevill apologized to the lobbyists and congressional staff—many of them from Illinois and Texas—milling about in the hallway outside the hearing room in the Rayburn House Office Building.[82] But, he acknowledged, "There'll probably be some changes before this bill reaches the President's desk."

These actions raised the specter of warfare breaking out between the Illinois and Texas delegations. Sure enough, when the bill was taken up by the full House Appropriations Committee, Illinois Congressman Sid Yates offered an amendment that sliced another $40 million off the SSC budget and diverted it to start the Main Injector. It was handily defeated, but the damage had been done. Washington observers could sense a widening gulf between these two delegations, among the most powerful in Congress. An uneasy truce that had existed since November 1988, when Waxahachie was announced as the chosen SSC site, seemed about to collapse.

When the Energy and Water Appropriations Bill reached the House floor on May 29, Boehlert, Eckart, Wolpe, and a new ally, Democrat Jim Slattery of Kansas, were ready with another amendment to kill the SSC project. Offered by Slattery, it would eliminate all funding for the SSC, leaving $43.5 million to begin construction of the Main Injector at Fermilab. "We were told at its inception that the SSC would be built with 'new money,' that it would not force reductions in the rest of science," said Boehlert in promoting the amendment. "Ask the folks at Fermilab, who are seeing their bid for a new injector come to naught, if the SSC has had an impact on other science funding."[83] Once again, however, the amendment was beaten back, by a seemingly comfortable margin of 251–165—although there had been a substantial erosion of support since the corresponding 1989 vote, 331–92.[84] The SSC project's budget emerged from the House, bent but not broken, at $434 million. And a closer look at the voting patterns revealed a worrisome polarization growing between Rust Belt and Sun Belt states over this gargantuan project.[85]

In the Senate, the power of Bennett Johnston was much in evidence. In mid-June the corresponding energy and water bill emerged from the Appropriations Committee with $75 million tacked back on, at $509 million—just $25 million less than the president's budget request. But when the bill came to the floor, it faced for the first time a Senate amendment to kill the SSC, offered by folksy Arkansas Democrat Dale Bumpers. "You will find that the Superconducting Super Collider cures cancer and earaches and gives you an appetite if you are not hungry," he railed, chastising SSC advocates for their bloated claims of technological spin-offs. Bumpers argued that "we cannot finance the Supercollider and still finance Fermilab, Brookhaven, and SLAC."[86] When the vote was taken, the SSC survived by a seemingly comfortable margin of 62–37 against the amendment. But given the political power of Johnston, Bentsen, and Gramm in the Senate, it was unnervingly close for SSC supporters.

On July 30, a House-Senate conference committee split the difference and compromised at $484 billion for the FY1992 SSC budget, $50 million below Bush's original figure.[87] He signed the bill before the August recess. SSC advocates could breathe easier until the next budget cycle began in early 1992. But with the projected SSC spending growing by hundreds of millions a year until 1996, pressures on other scientific research would only increase unless foreign contributions began to flow.

JAPAN OFFERS A POLITE "MAYBE" ON SSC PARTNERSHIP

By 1991 Japan had become the focus of DOE efforts to secure a major foreign SSC commitment. Of paramount importance in prodding its slow decision-making process forward was a formal, face-to-face request from the US president and its acceptance by the Japanese prime minister at a bilateral summit meeting—what eventually came to be dubbed "the golden handshake" (but never occurred). Given the woeful state of Japan's funding of scientific research at the time, however, its scientists in other fields were ill inclined to support a contribution worth a billion dollars (then about 135 billion yen) to a US project; funding of this magnitude was sorely needed at home.[88] Such a scientific consensus could never be achieved from below. "We have been told by Japanese officials that raising the SSC in discussions between President Bush and Prime Minister Kaifu is a prerequisite to a positive Japanese response, since the Japanese decision to participate is driven much more by political considerations than by a consensus of scientists," the October 1990 internal DOE report had stated. "The Japanese have noted that the SSC was not raised in the two most recent summit meetings."[89]

Nor did Bush raise the issue in their next two meetings—at Newport Beach, California, in April 1991, and Kennebunkport, Maine, that July. Questioned by a reporter about Japanese participation in the SSC at a news conference in Kennebunkport, Kaifu replied, "We did not discuss the issue of the superconducting collider today, but in the past, I received explanation from President Bush about the US position on this."[90] Then he continued: "There is growing awareness in Japan that this sort of thing . . . is important for science and technology. And researchers in Japan are studying what sort of cooperation would be possible. However, I am not prepared today, here, to say what sort of financial cooperation is possible. And I might add that scientific and technological research in Japan is being carried [out] under difficult financial conditions as well."[91] Bush interrupted at that moment, commenting, "But let me say on the supercollider, we only got this far through our talking points and we've got this far to go—the supercollider is in here."[92] If they discussed the question at all thereafter, it was only briefly, and nothing came of it. More important to the Bush administration at this summit were an additional payment of $500 million toward Gulf War expenses and removal of the barriers to large Japanese imports of rice from US farmers.[93]

Bush and Kaifu agreed to address the SSC partnership question more fully at their next summit, scheduled to occur in Tokyo that November. In preparation for that meeting, Bush sent Kaifu an October 1991 letter that stated: "The Super Collider, which we have discussed before in a preliminary fashion, is a critical element in advancing human understanding of the fundamental structure of matter. It will constitute a bridge linking the world's leading scientists in one of the great intellectual ventures of human history. The project requires strong international leadership and commitment, which a partnership between the United States and Japan can uniquely provide."[94] But that summit meeting never occurred. Later that month Kaifu was forced out of office by three powerful factions of the ruling Liberal Democratic Party. Top-level discussion of a possible Japanese partnership in the SSC would have to wait two more months until another summit could occur between Bush and Kiichi Miyazawa, who became prime minister on November 5, 1991.

Japanese support of basic scientific research had never been strong. In fact, the nation was often accused of "coat-tailing" on the scientific output of the other Western powers, doing the applied research and development that could lead to the high-tech products and services produced by its booming, export-driven economy.[95] Japanese scientists recognized this problem, but the solution they envisioned was to boost R&D spending at its major research universities, such as in Kyoto, Tokyo, and Tsukuba—not to send billions of yen abroad to a US supercollider project. Akito Arima, a physicist who had served as president of the University of Tokyo and head of Monbusho before being elected to the Diet, recalled that he'd become "notorious" for pressing Japan's government to increase funding of basic

FIGURE 5.4 DOE Deputy Secretary Henson Moore (*left*) and OSTP Director D. Allan Bromley with President George H. W. Bush aboard Air Force One, 1992. On the opposite side of the table are Texas Congressman Joe Barton and DOE Deputy Undersecretary Linda Stuntz. Courtesy of AIP Emilio Segrè Visual Library, Bromley Collection.

research, especially in the universities.[96] Any Bush administration initiative to get the nation involved as a partner in the SSC Laboratory had to recognize and deal with this troubling reality if it hoped to see a consensus on the question emerge in Japan's scientific community.

Allan Bromley thought he had a solution. In October 1991 he flew to Tokyo with Henson Moore to suggest that the Japanese government make a 500-billion-yen commitment toward "establishing the world leadership in *science and scientific culture* that the Japanese already enjoyed in technology."[97] Of that amount, 200 billion yen—or about $1.5 billion—would be invested in an SSC partnership. They delivered this message to university presidents, members of the Diet, and the Science Council of Japan, which advises the prime minister on science-policy matters. Following that presentation, recalled Bromley, the council voted unanimously to support the idea. "I came back from Japan convinced that the President had only to ask the Prime Minister and the SSC participation at roughly $1.5 billion would be forthcoming," he later wrote in his memoirs.[98]

Bromley and Moore's visit was just one in a succession of high-level US missions to Japan that fall, leading up to a summit meeting between

President Bush and the prime minister (first Kaifu, then, after he resigned, Miyazawa) at which the hoped-for agreement would occur. In September 1991 William Happer, a Princeton laser physicist who had stepped in early that year as director of the DOE Office of Energy Research, led a delegation that included Schwitters, HEPAP Chair Wojcicki, and Nobel laureates Jerome Friedman of MIT and Steven Weinberg of the University of Texas. They met with peers in Japanese universities and science ministries, urging them to support the idea of an SSC partnership.[99]

In November Secretary of State James Baker visited Japan, followed by Energy Secretary Watkins in early December, meeting with counterparts at high levels in the Japanese government.[100] They had other important matters to discuss, of course, but the proposed SSC partnership was high on their lists of talking points, especially for Watkins, whose briefing book was subtitled "Global Partnership for the Superconducting Super Collider and Energy Cooperation."[101] He had hoped to meet with Miyazawa (the briefing book had a sheet of talking points for a possible meeting) but did not, although he did confer with Kaifu, the foreign minister, and the powerful finance minister. None of them gave Watkins much encouragement that Japan might be willing to make the kind of billion-dollar commitment he was seeking. The steadily growing drumbeat of US requests and admonitions that the nation do more for basic research was gradually beginning to "wear thin."[102]

Everything now hinged on President Bush's visit to Tokyo, which had been pushed back into early January—much to the chagrin of his Japanese hosts—because of his domestic political problems. With the United States still struggling through a nagging recession, to which Bush seemed oblivious, his popularity had plummeted from its Gulf War high just months before. As reelection prospects dimmed, White House Chief of Staff John Sununu resigned in early December. He was replaced by Transportation Secretary Samuel Skinner, who was much less favorably disposed toward scientific research and more interested in business, commerce, and trade.[103]

Leading up to such a summit meeting is a choreography by the president's top aides and assistants, who determine the items that appear on his list of talking points and assign them priorities. According to Bromley and others, science took a back seat to trade in this process. Quotas for Japanese manufacturers' purchase of US auto parts was to become the primary US focus at the January summit, "best typified by the overt presence of the chief executive officers of the Big Three Detroit automakers."[104] Yet again, other pressing political issues had pushed the SSC down the list of agenda items for a crucial summit meeting where the official request had to be made. Originally billed as a reaffirmation of friendship between the world's two leading nations on the fiftieth anniversary of Pearl Harbor (which eventually occurred just two weeks after the Soviet Union had formally dissolved), the summit meeting devolved into a discussion of import quotas—a topic the Japanese hated.

But the formal request did not occur that January, as had been hoped and planned by SSC advocates. President Bush fainted and vomited at a private dinner in his honor, due likely to improper medication levels for his Graves' disease and the stress of travel. Items lower on the list of talking points, such as the SSC, were subsequently sacrificed in favor of the topmost US priorities, which all had to do with trade.[105] US and Japanese aides did manage to agree behind the scenes to pursue the SSC question formally, establishing a US/Japan working group to address unresolved issues. To observers familiar with the Japanese government and culture, this was an important signal that the nation was indeed interested in pursuing an SSC partnership, for Japan would never have embarked on the process if the eventual answer was going to be no.[106] But the all-important agreement between the two leaders and the formal Japanese acceptance of the US president's invitation did not occur in Tokyo. It would prove to be a huge loss for SSC advocates.

PUBLIC PERCEPTIONS OF THE SUPER COLLIDER

Through its first few years of existence, the SSC Laboratory had kept a low public profile, primarily reacting to media interest rather than trying to stimulate it. The activism of its CDG origins, during which External Affairs Director Wojcicki served as a member of its top management and carried word about the Super Collider to Washington, New York, and foreign capitals (see chapter 3), was long gone. SSC Laboratory public affairs fell under Assistant Director Raphael ("Rafe") Kasper, who had little experience in this area. Having earned a doctorate in nuclear engineering from MIT, he had been staff director on the NAS panel that winnowed the forty-three candidate SSC sites down to the eight best-qualified ones. Having worked with him well in that effort, Schwitters named Kasper as his chief of staff.

Wojcicki continued to interact with the SSC Laboratory on a consulting basis, participating in such high-profile missions as the September 1991 Japan visit led by Happer. CDG Editor Rene Donaldson and her husband, an engineer, considered SSC positions in 1989; but they decided instead to remain in the San Francisco area and took jobs at SLAC.[107] Later that year Kasper named Russ Wylie of the Houston Area Research Center (HARC) to head the SSC Office of Public Affairs. Hailing from a background in auto-industry public relations, he knew almost nothing about high-energy

FIGURE 5.5 Princeton University physics professor and Nobel laureate Philip Anderson, a leading critic of the SSC project. Credit: AIP Emilio Segrè Visual Library, *Physics Today* Collection.

physics and its culture. A capable but uninspiring media spokesperson, Wylie and his staff began to issue a growing stream of press releases about SSC appointments and contracts signed with industrial firms around the nation. But this approach only reinforced the public image of the SSC as a Texas pork-barrel project.

The barrage of criticism from the "small-science" community continued unabated as the SSC lab struggled through its first year. James Krumhansl, now the American Physical Society president, had to watch his tongue and pen, lest his words be misconstrued as official APS policy. But Philip Anderson was under no such constraint and continued expressing opinions against the SSC. In his February 24, 1989, testimony before the Senate

Committee on Energy and Natural Resources, he lamented, "Scientists like myself in the fields of condensed-matter physics . . . are caught between the Scylla of the glamorous big science projects like the SSC, the genome, and the Space Station, and the Charybdis of programmed research with 'deliverables' aimed at some misunderstood view of 'competitiveness' or at some unrealistically short-term goal."[108] Testifying at the same hearing, Schwitters could only suffer in silence.

Emboldened by the example of Anderson, who then stood at the pinnacle of the condensed-matter physics community, other "small" scientists increasingly broke ranks over the next few years and criticized the SSC project, often in congressional hearings. They included Nobel laureates Nicolaas Bloembergen of Harvard and J. Robert Schrieffer of Florida State University, as well as Theodore Geballe, an expert on superconductivity at Stanford, and Rustum Roy, a materials scientist from Penn State University who soon became the SSC's most outspoken critic. "Big science has gone berserk," Roy told a *New York Times* reporter. "Good minds and a lot of money are going into areas that are not relevant to American competitiveness, American technological health, or even the balanced development of American science."[109] The principled opposition to the project was broad and deep—and was countered ineffectively by Wylie and other SSC spokespersons. There seemed to be an internal assumption among those at the laboratory that once launched, the project could not be halted, that it was not very important to keep reminding the American public and its representatives in Congress about the physics reasons for building it.

Congressional opponents of the SSC pointed increasingly to this criticism of the project by other physicists concerned about its impact on their research funding. In his opening statement before a May 9, 1991, House Science Subcommittee on Investigations and Oversight hearing on the SSC, for example, Michigan Democrat Howard Wolpe observed:

> First, last fall's bipartisan budget summit agreement has changed the rules of the game. Now all non-defense discretionary spending programs must compete for scarce Federal resources. . . .
>
> Second, the new budget process has created the same competition among Federal science and technology programs. There is growing concern that valuable "small science" activities will be severely curtailed by a few capital-intensive "big science" programs.[110]

His partner in opposition to the SSC, Sherwood Boehlert, echoed these sentiments: "The project will swallow up the Nation's already limited science budget, forcing a round of 'beggar-thy-neighbor's scientific discipline.'"[111] Both statements reflected the new reality that, after the Budget Enforcement Act of October 1990, it was going to be essentially impossible to move money out of defense spending to cover growth in big-science projects. An expanding SSC budget therefore portended less money for other sciences, especially other physical sciences funded through the DOE.

The vocal opposition of the condensed-matter physics community provided congressional SSC critics like Boehlert and Wolpe a solid rationale and gave them political cover to oppose the project. As explained by Boehlert's then legislative director David Goldston,[112] Congress was at the time generally favorable to science, which it considered a cost-effective investment that would result in better health, more jobs, and higher profits. It was thus important for individual members not to be viewed by constituents as being *against science,* per se. It was much better for them to be perceived instead as setting priorities among conflicting disciplines or projects: if scientists could not do this on their own, then Congress had to step in, however reluctantly, and do the thankless task on their behalf. And in this case, it also helped to be able to draw the distinction between a basic-science project that *might possibly* make contributions someday to human welfare, versus more practical applied sciences, whose impacts would (more probably) occur in the immediate future. This contrast was particularly stark in an anxious political climate that stressed bolstering US industrial competitiveness over trying to understand nature.

This David and Goliath metaphor of small science in combat with Big Science reverberated throughout the 1991 political season, in articles, editorials, and opinion pieces, as well as congressional testimony.[113] The Super Collider was not the only target of this criticism; so also was the much costlier (and politically more powerful) Space Station.[114] On the same day that Bevill's subcommittee sliced $100 million out of Bush's fiscal 1992 budget for the SSC, another appropriations subcommittee under Robert Traxler (D-MI) zeroed out the Space Station, allowing NASA just $100 million to shut the costly project down. "We simply can no longer afford huge new projects, with huge price tags, while trying to maintain services that the American people expect to be provided," argued Traxler.[115] Tennessee Democrat James Sasser, chairman of the Senate Budget Committee, weighed in on similar themes in a *New York Times* opinion piece:

> If we keep pouring money into the $30 billion [Space] Station, we'll doom other necessary space-science projects to extinction.

> Similarly, the Superconducting Super Collider will grow from $535 million in fiscal 1992 to its overall price tag of about $11 billion. Sustaining this program would relegate smaller-science programs, the heart of America's technological capability, to a budgetary no-man's land.[116]

It could not help the SSC to be compared explicitly with the Space Station, then hemorrhaging cash and requiring annual infusions upwards of $2 billion. At least the Super Collider was truly a scientific project—albeit one that the American public did not understand. Although NASA claimed that the Space Station would enable important scientific research, on such subjects as the long-term effects of human weightlessness and growing crystals in microgravity environments, it was essentially a large-scale engineering project intended to maintain and extend the human presence in space. And since public relations was part of its mission, NASA could enrapture the public and Congress with colorful, elaborate illustrations and simulations of the Space Station orbiting Earth, with Space Shuttles and astronauts hovering about. "Journalists regularly lumped the SSC together with the Space Station as horrible examples of big science, despite the fact that the Space Station is not a scientific project," observed Steven Weinberg. "Arguing about big science versus small science is a good way to avoid thinking about the value of individual projects."[117] But US lawmakers are ill equipped to do such thinking—and beset by political forces that have little to do with such value.

American physicists, however, did not do much better in this regard. The question of the Super Collider's impact on the rest of the US physics program had been highly contentious ever since Reagan chose to endorse it in 1987 (see chapter 3). It was the most divisive issue ever encountered by the American Physical Society, recalled *Physics Today* reporter Irwin Goodwin, who had great access to the internal disputes going on.[118] The condensed-matter community even threatened to leave the society because of its position on the SSC. APS leaders finally hammered out an uneasy compromise in late 1990, publishing a January 1991 statement of the society's official position on the SSC: "The SSC should be built in a timely fashion, but the funding required to achieve this goal must not be at the expense of the broadly based scientific research program of the United States."[119]

APS President Nicolaas Bloembergen reiterated this position that April before the Senate Energy Research and Development Subcommittee chaired by Wendell Ford (D-KY). He went on to advocate that funding to support individual scientists be taken instead from the Space Station—which would not represent a setback for science.[120] He also rebutted the assertions made in the same hearing by DOE Deputy Secretary Henson Moore, who included magnetic-resonance imaging among the technological spin-offs from high-energy physics. Bloembergen, a Nobel laureate who had worked on nuclear magnetic resonance at Harvard, reminded listeners that research by individual investigators—and not by high-energy physicists—had led to MRI technology, buttressing the case for strong federal support of small science.[121]

Robert Park, a solid-state physicist from the University of Maryland who headed the APS office of Public Affairs in Washington, kept drumming its official position that year. He steadily needled Big Science projects—especially the Space Station—in his weekly e-mail newsletter "What's New," which was widely read in science-policy circles. He called the May 1991 vote of Traxler's subcommittee to kill the Space Station a "bold action" and "a pleasant surprise [that] certainly benefited little science."[122] Given his position as an APS spokesman, Park had to toe the line and hew to its official statement on the SSC, but his humorous, tongue-in-cheek comments left little doubt about where he personally stood on the project.[123]

Park also led the counterattack against overreaching claims of potential SSC spin-off technologies, such as in stimulating MRI technology and magnetic-levitation for high-speed trains. SSC advocates in the DOE and Congress increasingly turned to these boasts as a means to help justify the billions of taxpayer dollars about to be devoted to the project. In December 1991, the SSC lab published a brochure titled "Not for Scientists Only: Technology Spin-offs from High Energy Physics and the Super Collider."[124] But touting spin-offs was a dangerous move that often angered scientists (like Bloembergen) who had done the original basic research underlying these technologies. Such ploys earned the SSC more enemies than friends. "Most spinoffs don't come from high-energy physics," said Daniel Pearson, a staff member of the House Science Committee's investigations and oversight subcommittee.[125] "They come from small researchers, usually doing materials research or applied physics."

Claims of potential SSC spin-offs were nonetheless a prominent subject in the May 29, 1991, House debate on the House amendment to kill the SSC. Supporters of the SSC project cited this rationale more often than any other—including the need to maintain US leadership in science and technology.[126] The impact of high-energy physics on medical-imaging technology and possible cancer cures were favorite examples. But congressional SSC opponents retaliated with irony. "We have heard the proponents tell us that the Superconducting Super Collider will cure everything except the heartbreak of psoriasis," ridiculed Ohio Democrat Dennis Eckart. "The fact of the matter is that [the SSC] will not make one person well in this country."[127]

The intensifying public discourse pitting Big Science against small science was a prominent factor in the 1991 House debate, coming in second only to concerns about its excessive costs among the arguments SSC opponents made against it.[128] "This project has become an incredible dollar gobbler which will squeeze out not only taxpayers' dollars but will squeeze out an awful lot of good science around the country, unless we make a tough-minded decision to close it down now," remarked Wisconsin Democrat David Obey, a powerful leader of the House Appropriations Committee. There was little discussion of the scientific goals of the Super Collider, perhaps because few members really understood them in any depth. The only entry of significance on this topic was the Senate testimony of Leon Lederman earlier that month, inserted into the *Congressional Record* by Ralph Hall (D-TX).[129] Even Science Committee Chairman George Brown, who held a degree in industrial physics, had little to say on SSC research, confining his remarks to general observations that "there is no doubt in the scientific community of the value of this project."[130] But, he acknowledged, there was "some division within the scientific community as to the apportionment of funds between large science projects such as this and smaller projects which are in the province of individual investigators or small teams."

SSC advocates belatedly began to realize that they had not made nearly enough efforts to tell the public and Congress about the scientific reasons to build the laboratory.[131] In the first few years after the project went to Texas, only a few newspaper or magazine articles appeared on this subject—as compared with *dozens* on its growing costs, mismanagement, and other problems. Such a bias in SSC press coverage could only create a skewed, negative perception of it as a public-works project, not a science project, which mainly benefited contractors and Texans.

One striking exception was a feature article on particle colliders that made the cover of *Time* in mid-April 1990, two weeks before the House vote on Robert Roe's SSC authorization bill (see above and n. 30). Titled "The Ultimate Quest," it opened with, "Armed with giant machines and grand ambitions, physicists spend billions in the race to discover the building blocks of matter."[132] Author Michael Lemonick focused on recent discoveries at Fermilab, CERN, and SLAC that had filled out details of the Standard Model, plus the ongoing search for the top quark. But the final third of the seven-page article turned to the SSC and its physics, especially the Higgs search. "According to some theories, the Higgs boson is what gives all particles their mass," it stated.[133] "The idea is that everything in the universe is awash in a sea of Higgs bosons, and particles acquire their mass by swimming through this 'molasses.'" A photograph of Schwitters, standing in a muddy cotton field holding an armful of rolled-up blueprints, accompanied a drawing comparing the sizes of the SSC, LEP, the Tevatron, and the SLC at SLAC. The article cast a tale of the worldwide clash of headstrong lab directors—including Peoples, Richter, and Rubbia—and other egotistical physicists vying to be among the first to unearth massive new particles and unlock the secrets of the universe. By doing so, it succeeded in capturing wide public interest.

A much more sedate but accurate and penetrating article on potential SSC physics appeared in *Science* eighteen months later. "The SSC: Radical Therapy for Physics" delved well beyond the nostrums of the *Time* article to describe the deeper scientific motivations for building the multibillion-dollar collider.[134] "In making their case to the taxpaying public and to legions of skeptical scientists from other disciplines," it read, "promoters of the SSC have been uniting under the banner of as-yet-unseen particles, especially an elusive creature by the name of the Higgs particle." That was an easily marketable goal of the Super Collider—a distinct, tangible target that could (usually) be explained to the broader public. Researchers expressed their hopes of discovering completely unexpected results that might prod high-energy physics in fresh new directions. During the past fifteen years, in fact, research in the discipline had become a rather mundane affair of verifying one Standard Model prediction after another, each time only reinforcing the dominant paradigm.

Others sought inspiration in the heavens. "What physicists are staking their hopes on is the Standard Model's prediction that SSC energies will open a realm of physics characteristic of an earlier period in the Universe," observed *Science*.[135] "Physicists often equate higher accelerator energies with the first instants after the Big Bang, when particles crashed in a sort of multibillion-degree primordial soup." Cosmology and particle physics had recently been coming together—based on the growing recognition that particle colliders recreated conditions that had previously existed only during the first few microseconds of existence. Astronomers scanning the heavens found conclusive evidence for the existence of invisible "dark matter" that could not possibly be accounted for by Standard Model entities.[136] Thus there had to be something *else* "beyond the Standard Model" to explain this mysterious matter. Exotic, massive particles such as those predicted by fashionable "supersymmetric" theories (see appendix 1) would serve as ideal objects to seek at the Super Collider.

Elders of the particle physics community finally began to recognize the need to tell the general public why researchers wanted to spend something close to $10 billion worth of taxpayer funds to build the SSC. In 1992 and 1993, Nobel laureates Steven Weinberg and Leon Lederman published popular books closely related to its physics goals; both efforts were long overdue. In his customary elegant, high-minded prose, Weinberg told readers of his *Dreams of a Final Theory* about his lifelong search for unity, symmetry, and beauty in nature—and his ultimate hope of explaining all particles and forces by a "theory of everything" that came into play at the highest energies imaginable.[137] The Super Collider had a crucial role in this grand physics quest. "The urgency of our desire to see the SSC completed comes from a sense that without it we may not be able to continue with the great intellectual adventure of discovering the final laws of nature."[138]

Ever the comedian, Lederman pursued a humorous tack in *The God Particle,* aided by coauthor Dick Teresi, former editor of *Omni*, a general-interest magazine on science and culture.[139] Where Weinberg took the theoretical high road, appealing to readers' intellects, Lederman brought them down into the experimental trenches where the real action of high-energy physics occurs—at the particle accelerators, colliders, and detectors used by intrepid scientists to pry hard-won secrets out of a reluctant nature. Casting his narrative as a dialog with the pre-Socratic philosopher Democritus, who is thought to have coined the word "atom" from the Greek for "uncuttable," Lederman tried to entice readers to enjoy the sheer adventure of the hunt for the Higgs boson. He wanted them to *care* about it, to *feel* why high-energy physicists were so intent on finding it. "There is, we believe, a wraithlike presence throughout the Universe that is keeping us from understanding the true nature of matter," he wrote. "It's as if something, or someone, wants to prevent us from attaining the ultimate knowledge."[140]

Both Weinberg and Lederman were avowed "reductionists" who subscribed to a philosophy of science in which complexity at one level of matter is understood by appealing to a simpler set of more fundamental entities and forces at the next, deeper level.[141] The diversity of atoms and molecules, for example, can be explained by the various assemblages of electrons, protons, and neutrons. This reductionist philosophy naturally puts high-energy physicists at the pinnacle of a scientific hierarchy, for their research results could supposedly be used by others—including chemists and biologists—to deduce the more macroscopic features of nature. If true, that noble status would confer great value on the research to be done at the Super Collider.

For more than two decades, however, a subtle struggle had been occurring between representatives of this reductionist philosophy and another, prominently espoused by Philip Anderson, who argued that nature was not hierarchical but more like an ecological system in which the various individual actors are equal participants.[142] Solid-state physicists, that is, do not depend for their intellectual livelihood on the results of high-energy physicists; they have their own unique, independent observations to make. At every level of matter, Anderson claimed, new and different features emerge that are *not* consequences of the behavior of more "fundamental" entities at deeper levels. Buttressed by such an underlying philosophy, he could and did challenge the fact that so much public money was being spent on the SSC, whose research output would likely benefit only a small community of a few thousand high-energy physicists.[143]

More generally, the revolt of small scientists against the SSC was only the latest and most public instance of the battle against Big Science that had been occurring ever since nuclear physicist Alvin Weinberg's 1961 treatise on the subject.[144] Plenty of grumbling about high-energy physics had been happening for years behind the scenes, but—at least until the SSC came along to require billions of taxpayer dollars—it had almost always occurred in meetings and documents to which Congress and the general public paid little heed. The enormity of the SSC and its likely impact on other sciences were bringing this era of assumed scientific privilege to an end.

In the last analysis, the Super Collider struggled to overcome the founding rhetoric of the Reagan administration, as encapsulated by Herrington's statement quoted as the epigram of this chapter. This was to be an American project—led by US physicists and built on US soil. Its primary goal was to reestablish US hegemony in high-energy physics by discovering the Higgs boson or whatever else might be responsible for endowing elementary particles with mass; foreign physicists would certainly be welcome to participate, as followers. But when SSC costs started to soar after 1988, the Bush administration belatedly began to try to "internationalize" it and seek major foreign partners, with little success.

Congress was not of one mind on this matter. As expressed by H.R. 4380 in 1990, the House wished to cap the US contribution at $5 billion and be certain that a fifth to a quarter of the total SSC cost be borne by foreign partners. This would have helped limit the impact of the project on other sciences and congressional districts. But doing so would also have meant that important superconducting components and associated technologies—a fair part of the original justification for the SSC—had to be developed and built by other countries. Led by Bennett Johnston,[145] who declined to take up this bill in the Energy and Natural Resources Committee that he chaired, senators largely favored maintaining the SSC as a predominantly US project.

And Congress was trying to grapple with the consequences of the Budget Enforcement Act of 1990, which prohibited diverting savings from the post–Cold War military build-down for domestic spending. Passed just before the DOE was to release the increased SSC cost of $8.25 billion, this measure triggered a zero-sum game in which the big project competed nose-to-nose with other energy and water projects for a pot of money that could increase only at the rate of inflation. That was guaranteed to earn it many political enemies as the annual SSC budget soared to at least one billion dollars in the mid-1990s. And it meant that small scientists, especially those who might be funded through the DOE Office of Energy Research, would also be impacted by the project, then a presidential priority.

Ever present in the background of these debates was the possibility of CERN building a lower-energy but more cost-effective and truly international supercollider project in the LEP tunnel. As early as 1987, Schopper and Rubbia had in fact testified before Congress that US physicists would be welcome to join Europeans as partners in this major effort. And the Congressional Budget Office suggested that this option might provide an alternative means to contain costs while ensuring that US physicists would be able to pursue similar research.[146] Such a possibility was part of the reason why it was so difficult for SSC advocates to agree to lower its beam energy when the costs began to surge in 1989. Doing so could have reopened the door to Congress insisting that the LHC option be reconsidered.

Europe was also trying to deal more effectively (and seriously) with impacts a multi-TeV proton collider project would inevitably have upon the other sciences.[147] Putting superconducting magnets in an existing tunnel was a far more cost-effective way to open this new energy frontier while minimizing impacts upon other sciences. European high-energy physicists were leagues ahead of their US counterparts in confronting this problem. West of the Atlantic, the clamorous objections of small scientists to the SSC were a discomfiting nuisance that its proponents never dealt with effectively or seriously—hoping in vain that this vocal opposition would die away of its own accord. It never did.

After the 1988 selection of the Waxahachie site, the SSC was increasingly viewed by the US citizenry and, importantly, by Congress as a Texas public-works project—not a science project benefiting the nation as a whole. This perspective was especially valid for members of Congress in the US Midwest and Northeast. And as the SSC price tag soared toward $10 billion, its inevitable impacts on other sciences (including other high-energy physics labs in California, Illinois, and New York) began to dominate the public discourse about the project, to the exclusion of its promising research. SSC advocates made far too little effort to tell the American public about the scientific rationale for building the collider.

Never good at taking its case to the public, the Department of Energy did not help the project in this regard, and probably hurt it. Nor was it very adept at seeking large foreign contributions—unlike NASA, which had signed up Canada, Europe, and Japan as major Space Station partners.[148] By most accounts, however, Japan was (somewhat reluctantly) ready to become a partner in the SSC by early 1992, needing only a formal request from the US president to agree to sign on.

Steven Weinberg summarized the precarious situation in late 1991 when he sent his manuscript for *Dreams of a Final Theory* to the publisher:

> The future of the Super Collider would be assured if it received appreciable foreign support, but so far that has not been forthcoming. As matters stand, even though funding for the Super Collider has survived in Congress this year, it faces the possibility of cancellation by Congress next year, and in each year until the project is completed. It may be that the closing years of the twentieth century will see the epochal search for the foundations of physical science come to a stop, perhaps only to be resumed many years later.[149]

His words were to prove prophetic.

CHAPTER SIX

The Demise of the SSC, 1991–94

Congressional support for the Super Collider is a mile wide and an inch deep.

—CONGRESSMAN SHERWOOD L. BOEHLERT

In late September 1991, as the Soviet Union teetered toward dissolution, an obscure but important meeting occurred in Room 1E-245 of the Forrestal Building. Confronted by budget requests from DOE national laboratories that could never be met in any reasonable funding scenario, given the Budget Enforcement Act of 1990, Office of Energy Research Director Happer called together a panel of eminent scientists to help him decide which of its projects to push forward with and which to delay or terminate.[1] Chaired by Nobel laureate Charles Townes of Berkeley, highly regarded for his research on masers and lasers, the panel included Stanley Wojcicki of Stanford, Bell Labs research director William Brinkman, and theoretical nuclear physicists Herman Feshbach of MIT and Steven Koonin of Caltech. Prefacing the two-day gathering, Secretary of Energy Watkins told the panelists, "There is no way DOE can do everything that everyone wants us to do."[2] It faced difficult decisions, on which the agency needed their input. But he warned them that, as a presidential priority, the SSC was off limits. Any cuts had to occur elsewhere.

That decision officially made OER funding a zero-sum game—as many had already conceded. The OER budget was projected to rise to about $3 billion in fiscal 1993, to begin a year later, and then remain flat at that level through the mid-1990s as mandated by the 1990 budget act. With the SSC an untouchable priority and its costs ballooning toward $1 billion annually, it would inevitably exert tremendous—and steadily growing—pressure on other DOE labs and projects. The Reagan administration's injunction that the SSC had to be funded by "new money" was finally exposed as nothing but a subtle charade.

DOE laboratories have life cycles that depend on building new facilities to renew their research vitality and give them fresh new scientific domains to explore.[3] Otherwise, they would lapse into repeating previous measurements in ever-finer detail. According to a long-range plan developed in the 1980s under OER director Alvin Trivelpiece, DOE labs were supposed to enjoy a steady succession of construction projects that would ensure their vitality well into the 1990s.[4] Brookhaven was to convert its cancelled Isabelle project into the Relativistic Heavy-Ion Collider (RHIC), which would smash together gold and other heavy nuclei instead of protons, thus enabling research in nuclear physics rather than high-energy physics. The new Continuous Electron Beam Accelerator Facility in Newport News, Virginia, was to supply electron beams for nuclear-physics research, as SLAC was doing for high-energy physics. Argonne National Laboratory south of Chicago was building the Advanced Photon Source to generate X-rays for research in materials science. And so on and so forth. According to this plan, the DOE labs could remain one happy family—as long as the funding envelope could continue growing to accommodate all the new projects. That indeed occurred during the 1980s Reagan buildup, when the physical sciences and the DOE labs pursuing them widely benefited.[5] But the Trivelpiece plan, and the interlab peace and prosperity it promoted, came crashing down in late 1991 as the DOE began to confront budget austerity and face the harsh truth about the SSC.

The DOE Task Force on Energy Research Priorities, known as the Townes Task Force, first recommended that the $1.4 billion Burning Plasma Experiment at Princeton Plasma Physics Laboratory be canceled in favor of a smaller plasma experiment plus US participation in the International Thermonuclear Experimental Reactor (ITER), then in the early planning stages.[6] And Brookhaven would have to proceed more slowly with RHIC. The biggest clash occurred in high-energy physics, after Wilmot Hess revealed an annual base budget that would exceed $800 million in the mid-1990s, with the Main Injector and a new B Factory project to be built at Fermilab and SLAC.[7] Combined with a similar figure for the SSC (which assumed that foreign funds would flow as hoped), this amount meant that high-energy physics needed to consume more than *half* of the entire annual OER budget by mid-decade.

It could not stand. Wrapping up the last session on the second day, Townes abruptly stated, "Well, I guess we've agreed to defer the Main Injector and refer it back to HEPAP for further consideration." Jaws dropped. There had been no such discussion in public. The panel in fact recommended that *both* new projects be referred back to HEPAP to determine whether either could be included in a declining budget for what had recently become known as the high-energy physics "base program."[8]

Fermilab deemed the planned Main Injector crucial for the continued vitality of its research program.[9] And the majority of US high-energy physicists who worked with proton beams considered it essential for doing forefront research and training new physicists during the 1990s, while the SSC was to be under construction. A new feeder ring designed to allow accumulation of intense proton beams before injection into the Tevatron, the injector would help boost its collision rate by about an order of magnitude. Such an improvement, Fermilab officials claimed, would effectively extend the machine's discovery range for rare processes and particles, such as the long-sought top quark that theorists insisted had to exist to fill out the predicted sextet of quarks.[10] Thus the Main Injector upgrade would ensure that US high-energy physicists had interesting research to pursue while awaiting completion of the SSC.

But among the minority of high-energy physicists working with electron and positron beams, the Main Injector was viewed as an expensive luxury. The proton-smashing community had, after all, received approval of its SSC, whose inexorable demands for increasingly scarce dollars during the early 1990s were beginning to exert unbearable pressure on the rest of the high-energy physics base program. Although HEPAP had recommended the Main Injector upgrade as the next-highest priority after the SSC, the electron-physics community chafed that its experimental needs were being crowded out by the requirements of the proton-smashers, in what appeared to them as a "tyranny of the majority." Led by its director, Nobel laureate Burton Richter, SLAC had recently proposed the B Factory, a new electron-positron collider designed to smash electrons into positrons at 10 GeV to produce billions of subatomic particles called B mesons. HEPAP had approved this project, too, as long as added funding could be found to build it.[11]

Hess and program officers in the DOE Office of High Energy Physics strove to accommodate these two principal constituencies by adding both projects to their proposed long-range budget, which also had to accommodate Brookhaven and the many university groups pursuing high-energy physics research. It was a tall order, given the pressures the SSC was already imposing on the rest of the OER budget. Reporters present suggested the Townes panel had been summoned to help DOE put "those greedy high-energy physicists" back in their place and force them to face the hard choices other fields like fusion and materials science were encountering.[12] Happer, for one, was unimpressed by the Main Injector and appeared to favor the SLAC B Factory as a better alternative, because it would open a whole new research domain rather than just extend another lab's discovery reach.[13]

It was a thorny question to resolve, both financially and scientifically. US high-energy physicists of course had to continue pursuing forefront research during the decade that the SSC was under construction if they were going to be well positioned to use this costly machine for research in the next century. They could not spend an entire decade just designing and building SSC detectors. And Brookhaven, Fermilab, and SLAC had to be able to renew their research programs if they ever hoped to continue in existence as productive laboratories. All of this expansion had to happen, too, under the unprecedented budget pressures imposed by fiscal austerity and the inexorably growing costs of the SSC. The parochial interests of individual mission-oriented labs vying to remain productive in research had to be weighed against the longer-range interests of the US high-energy physics community at large. Painful choices had to be made.

As the congressional agent of Fermilab's interests, the powerful Illinois delegation soon weighed in on the Townes panel recommendations. In an October 17, 1991, letter signed by all its twenty-four members, it stated: "What we find so disturbing is that although the Main Injector was given 'highest priority next to the SSC,' . . . the DOE Task Force recommends not proceeding with its construction."[14] That language was mild, however, compared to the bare-knuckled political hardball implied in another letter sent to Watkins that day by House Minority Leader Robert Michel (R-IL) and J. Dennis Hastert, the Republican Congressman from Fermilab's district:[15]

> We object to the manner in which the Department arrived at its recommendations during its recent priority-setting exercise. Because consideration of the SSC was off the table, the hands of the task force were tied from the outset. We find it highly objectionable that the Department has refused to include the SSC in its priority-setting exercise and chosen instead to cut funding in other areas. . . .
>
> Your lack of review leaves Congress with no other responsible alternative but to conduct such a priority-setting exercise itself. We strongly urge the Department to reconsider its position.[16]

The gauntlet was down. Previously the Illinois delegation had given the SSC tepid support in House votes, conditioned on the continued strong administration funding for Fermilab, although that support was beginning to wear thin.[17] In return, the Texas delegation had supported Fermilab initiatives. But if DOE officials delayed the Main Injector, all bets were off. A chastened Watkins apparently heeded these warnings. In public comments before he left for Japan in late November (see chapter 5), he seemed to back off of projected DOE cuts of almost 10 percent from the high-energy physics base budget. "I will be pushing very hard to jack up these numbers," Watkins said, in reference to behind-the-scenes negotiations between DOE and the OMB.[18] Any talk of freezing the $15 million in existing construction funds for the Main Injector was subdued, if not completely silenced. Reporting the negotiations, *Inside Energy* noted that the DOE "has apparently thought better of arousing the ire of the Illinois Congressional delegation."[19]

When the Bush budget for fiscal 1993 became public in February 1992, it included $30 million for continuing the construction of the Main Injector, with the $15 million in fiscal 1992 funding still in place, plus a healthy 5 percent increase in Fermilab operating funds to $142 million.[20] The overall OER budget had increased a healthy 10 percent to $3,070 million, with high-energy physics getting a 15 percent increase to $1,281 million; almost all the increase went to the SSC, which rose by 34 percent to $650 million. After all the *Sturm und Drang*, it was another superb president's budget for high-energy physics. In reporting on the budget for fiscal 1993, however, *Physics Today*'s Goodwin noted, "The combination of the most obdurate recession since the 1930s, the severe limits the White House and Congress imposed on discretionary spending, and the inability to stanch the flow of red ink now suggest that a budget debacle is imminent."[21]

SSC CONSTRUCTION GAINS STEAM

As 1992 began, significant progress was occurring on the construction of the SSC Laboratory under General Manager Siskin and a new project manager, John Rees, who probably had the deepest project-management experience, apart from Panofsky and Wilson, of anybody in the high-energy physics community. At SLAC he had served as the project manager (or its equivalent) on the SPEAR, PEP, and SLC projects—the latter two in the $100 million category—completing them on or under budget. He was a member of the Science Policy Committee before joining the SSC Laboratory in 1991 as the physicist in overall charge of its "warm" accelerators, reporting to the then project manager, Paul Reardon.[22] But Reardon and Siskin could not work together effectively; according to Rees, they had "swords drawn" for much of 1991. Then struggling to implement the long-delayed cost-and-schedule control system, Siskin felt Reardon was "dragging his feet" on the

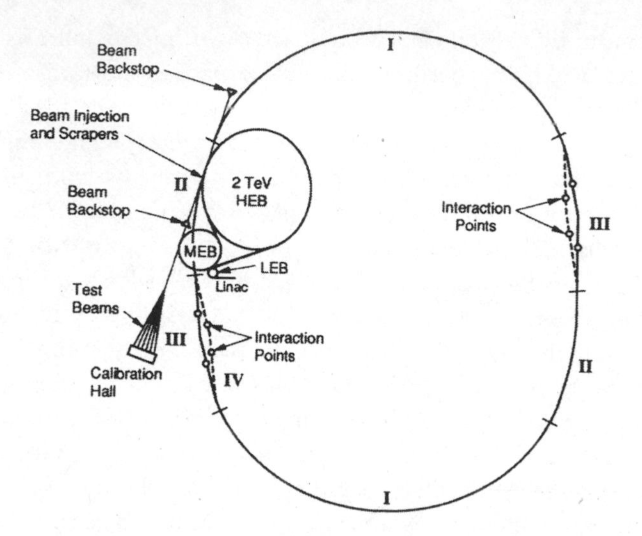

FIGURE 6.1 Schematic layout of SSC, showing main ring and injector chain, plus interaction points and possible location of test beams. Courtesy of Fermilab Archives, SSC Collection.

system, refusing to buy into its use and require SSC physicists to provide the necessary inputs.[23]

The lack of a fully qualified cost-and-schedule control system was a major SSCL problem repeatedly cited by the DOE, Congress, and its principal investigative agency, the General Accounting Office (GAO).[24] This omission could be attributed to several causes, including the failure of Sverdrup Corp. during the first year of the lab's existence, the long delay in hiring a capable project manager, and the failure of top SSC managers to take seriously the need for such a system and use it aggressively (see chapter 4). The lab finally got the system functioning by brute force in late 1991, when Siskin eased Reardon out of the project manager position and into another, working on the SSC detectors.[25] Rees, who had employed similar systems (but on far smaller scales) on SLAC projects, replaced him.[26] Far more comfortable with these programs, and highly regarded in the scientific community, he helped convince SSC Laboratory physicists to use the system and supply the inputs needed to make it perform effectively. Rees collaborated well with Siskin, respecting his engineering expertise and billion-dollar project-management experience. The coordination between high-energy physicists and military-industrial engineers improved markedly under their dual leadership.

Because Rees also had experience working with Parsons-Brinkerhoff, the SSC architectural and engineering firm, his transition into the project manager role went fairly seamlessly. In January 1992 he awarded a contract to begin construction of the Magnet Test Laboratory, adjacent to the Magnet Development Laboratory on the SSC's north campus. This was the second major building to be constructed at the site. Rees also made the official decision to proceed with tunneling for the main collider ring, scheduled to begin that fall, awarding the first tunneling contract in February to Dillingham/Obayashi, a US/Japanese consortium headquartered in San Francisco.[27] And the SSC lab soon awarded Westinghouse a contract to develop

FIGURE 6.2 Two of the early SSC superconducting dipole magnets on the Fermilab test stand. Courtesy of Fermilab.

superconducting dipole magnets for the High-Energy Booster (HEB).[28] That May, construction began on the Linac, the linear accelerator that would feed 600 MeV protons into the first in a series of three booster rings (LEB, MEB, and HEB) for subsequent acceleration to the 2 TeV injection energy.

Prototype superconducting dipoles with 5 cm apertures were beginning to emerge from the Fermilab magnet assembly line, fabricated there by engineers and technicians from General Dynamics and Westinghouse, who were learning the physicists' labor-intensive procedures while industrial facilities to mass-produce these magnets were then under construction in Louisiana and Texas. The first such magnet, assembled at Fermilab by General Dynamics personnel, arrived at the SSC Laboratory in January 1992, while another, built by Westinghouse, awaited testing at Fermilab.[29] More magnets followed that spring and summer.

The immediate goal, one of the critical milestones on the SSC project, was the accelerator systems string test scheduled for October 1992. Commencement of SSC tunneling awaited this all-important test, in which a half-cell containing five collider dipole magnets and a quadrupole magnet, plus the associated cryogenic, electronic, and quench-protection systems, had to operate simultaneously at 4.35 K.[30] The string test was successfully completed two months early, on August 14. Dipole magnets were easily ramped up to an operating current of 6,520 amps, generating magnetic fields of 6.6 T, the official design values.[31] Corks popped. Champagne flowed. But an ominous challenge to the SSC had erupted in Congress that summer; it had to be fully resolved before serious tunneling could begin.

THE HOUSE AND SENATE VOTES OF 1992

With little warning, a political lightning bolt suddenly struck the SSC on June 17, 1992. An amendment to the fiscal 1993 Energy and Water Development Bill, introduced by Ohio Democrat Dennis Eckart to kill the project, succeeded on the House floor 232 to 181, a margin of 51 votes. Instead of the $484 million that the House Appropriations Committee had allocated, DOE was to receive just $34 million to shut the project down. This abrupt reversal came as a big surprise to SSC promoters. A year earlier, the House had quashed a similar amendment by an apparently comfortable 86-vote margin, 251–165 (see chapter 5). But with essentially the same individual members voting in 1992 as in 1991, nearly seventy of them had changed their minds and voted to kill the SSC.[32]

The proximate cause of the reversal was a divisive, rancorous vote on

June 11, in which a balanced-budget amendment narrowly failed to achieve the two-thirds majority needed for passage. Spearheading this effort were two Texans, conservative ("Blue Dog") Democrat Charles W. Stenholm and Republican Joe Barton of the SSC's Ennis district. Another Texas supporter, of a parallel Senate amendment, was Phil Gramm.[33] Leading the House opposition was California Democrat Leon Panetta, chair of the Budget Committee, who argued that Congress should itself shoulder the onerous burden of making difficult budget choices. "Now we've got to roll up our sleeves and get to work on what I think is the effort that really counts," he said after the budget-amendment vote had failed, "so that we truly exercise the discipline that we have to if we're serious about getting the deficit under control."[34] He meant that Congress had to deal with deficit spending on a case-by-case basis, cutting projects and programs of lesser merit—and not by passing a broad, ill-conceived balanced-budget amendment to absolve itself of its Constitutional responsibility.[35]

The SSC was unfortunately the first big project up for a vote after June 11, with a well-organized opposition poised to take advantage of the tightening budget climate. That spring, projections of a nearly $400 billion deficit loomed for 1992, due in part to the continuing recession and to savings-and-loan bailouts.[36] In recognition of these fiscal pressures, the energy and water bill that reached the House floor on June 17 had been pared back in committee to the bare essentials; instead of the $650 million in Bush's budget, the SSC lab was slated for the same $484 million allocation it had received in fiscal 1992.[37] In addition, the high-energy physics base program had been granted $613 million in the bill, down from $628 million in 1992 because funding for Fermilab's Main Injector had been cut in half, from $30 million in the Bush budget to only $15 million—barely enough to keep this project alive.

The federal budget was a pressure cooker that year because of the budget agreement of 1990 and its firewalls against moving money from defense to other areas. A congressional attempt to tear down the firewalls, in view of the Cold War's recent end, had failed decisively earlier that spring, leaving Congress no choice but to pare back on domestic discretionary funding. There would be no peace dividend, especially not with the dawning post–Cold War recognition of the costs of cleaning up the nation's nuclear-weapons complex, estimated to be at least $100 billion.[38] "A company $400 billion in debt cannot pay a dividend," Congressman Stenholm told reporters after that vote.[39]

And a poorly understood, multibillion-dollar basic-science project perceived to be badly managed and experiencing continuing overruns was an

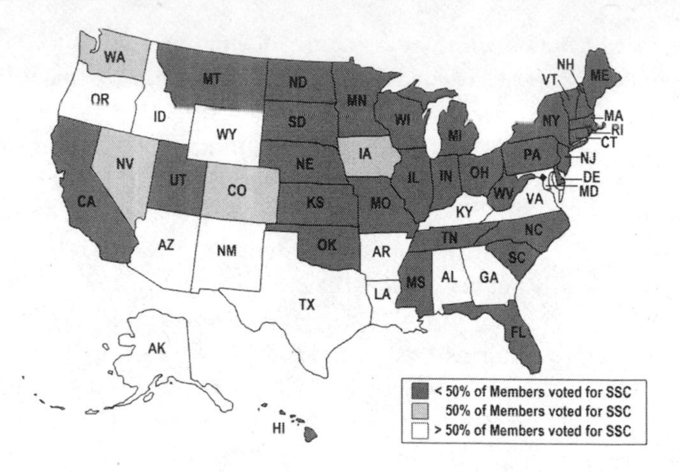

FIGURE 6.3 Map of state-by-state voting patterns on the 1992 House amendment to kill the Super Collider. Courtesy of Shaun Hubbard.

easy political target that would cost its opponents few votes when they ran for reelection in their districts that fall. "They wanted a budget scalp they could take home and wave in front of voters," to show they were taking the budget seriously, said Joe Barton.[40] But his metaphor focused on the wrong part of human anatomy, countered Panetta and Arkansas Democrat Dale Bumpers, SSC opposition leader in the Senate, who argued that "the spines of the Members of Congress still have to be stiffened to make the tough choices."[41]

In this less-than-zero-sum game, it boiled down to a matter of "your project versus my project," at least in the House, where members more often vote for the needs of their district above other considerations because they must face the voters every two years—as they would have to that fall. From 1991 to 1992, the SSC lost 14 votes in California, 7 in Illinois, and 5 in New York, for a combined loss of 26 votes in states with existing high-energy physics labs. Including other states with DOE labs like New Jersey and Tennessee, the total vote erosion came to 32—almost half of what had been involved in this decisive reversal.[42] The vote of the Illinois delegation was particularly telling: 18–2 to kill the SSC, or 90 percent against it, versus 11–9 in 1992.

Representatives in the Northeast-Midwest Congressional Coalition voted very heavily against the Texas project, making it seem like a Rust Belt versus Sun Belt conflict. Nearly half the votes to kill the SSC came from districts in the Northeast and Midwest, which could only muster 57 votes (or 27 percent of all such districts) in support of it. But from 1991 to 1992, the project also lost another 6 votes in Florida, 4 in North Carolina (a losing finalist state), and 3 in Oklahoma—just 100 miles north of Waxahachie.[43] The erosion of support for the SSC was widespread indeed.[44]

Changing one's vote from one year to the next was not easy for a member to do, however, especially on a scientific project. As science generally enjoyed broad bipartisan support, members of Congress usually did not want to be viewed as "flip-flopping" on scientific issues and projects.[45] So there was a certain amount of voting inertia that SSC opponents had to overcome. That June, Boehlert, Eckart, Slattery, Wolpe, and their staffs were armed and ready with arguments against the project. Not only was the SSC mismanaged and its costs continuing to grow out of control, they argued, but there were as yet no firm commitments of foreign support, which DOE had promised and explicitly included in its budget projections for the mid-to-late 1990s. Even though the SSC Authorization Bill of 1990 (H. R. 4380) had not been taken up by the Senate and thus never became law, representatives still regarded it as "the will of the House."[46] That bill capped federal contributions to the SSC at a maximum $5 billion—and it specified that foreign contributions should constitute at least a fifth of the total project cost. By June 1992, absent any concrete Japanese commitment, it was pretty clear that neither goal was going to be met.

So SSC opponents could, and did, assert that the project was going to exceed the bounds established by the House. As Eckart argued on June 17, "Despite the fact that the House has persistently and consistently voted to cap federal government participation at $5 billion, the clock continues to run and the costs continue to soar."[47] The SSC project's violation of the House will therefore justified a member's changing his or her mind and voting against it in 1992. Nearly seventy did. And to many of them, the SSC was an *economic development* project rather than an important basic-science project; after all, this was how it had been promoted during the 1987–88 site competition, as the "Big Pour" that would bring thousands of jobs to the winning state (see chapter 3). Because few members really understood the project's scientific rationale, congressional support for the SSC—based as it was on this rationale—was "a mile wide and an inch deep."[48]

Moreover, it had become increasingly clear that federal funds spent on the SSC would mean less money for other worthwhile science, which in most cases was likely to contribute more rapidly to US competitiveness and economic well-being. "My friends, we are playing a zero-sum game," said Slattery during the House floor debate on the SSC. "Budgets in the next few years are not going to increase; they are going to be cut. Money spent on the Super Collider is money that will not be available for other scientific projects."[49] This troublesome fact had been made painfully clear by the previous autumn's Townes panel proceedings and report, as well as in ensuing HEPAP debates regarding the Main Injector and B Factory. As a protected presidential priority, the SSC would inevitably begin cutting into funding for other DOE labs, including those that might be performing research on nuclear or plasma physics or materials science. One could also argue, as did Slattery and other opponents, that the SSC would have negative impacts on science funding beyond the Energy Department, but such a claim was harder to justify.[50] These arguments helped support some members' changing their votes; they could validly claim that they had been forced to set scientific priorities in difficult budgetary times.

In the final analysis, it was a combination of forces related to the severe 1992 budget pressures that led to the June 17 House vote to terminate the SSC through an amendment to the energy and water bill. But at the focus of the storm was the state of Texas. The project was viewed in the House more as a fat slice of Texas pork than as a major priority for US science. Budget Committee Chair Panetta called the SSC "the poster child for congressional pork."[51] The House action was more an anti-Texas vote than a Rust Belt versus Sun Belt vote—by members envious of all the federal funds and projects Texas had been receiving. And it didn't help that the Texas delegation made a crucial tactical blunder in letting Barton lead the floor debate against the Eckart amendment that day. As a principal sponsor (along with Stenholm) of the recently defeated balanced-budget amendment, he had earned the special ire of House liberals and others like Panetta who thought Congress should maintain its constitutionally mandated control of the federal purse strings. As a House Budget Committee staffer of the early 1990s put it in scatological terms, their reaction to Barton was, "OK [expletive deleted], you want to balance the budget? We're going to start with your project!"[52]

Stunned SSC supporters rallied to revive the project in the Senate. On June 25 a group of forty eminent physicists, including twenty-one Nobel laureates, sent an urgent letter to President Bush and all one hundred senators saying they were "shocked and dismayed by the House rejection of funding for the Superconducting Super Collider."[53] During the following weeks, more than two thousand other scientists added their names to the list of signers. "This sudden rejection stuns and confuses," the letter continued. "To kill an undertaking that is so splendidly fulfilling its expectations and its mission raises fundamental questions about our national commitment and our ability to carry out long-term scientific projects." Bush strongly endorsed the SSC project, meeting publicly with Schwitters and Nobel physics laureates in the White House on July 1. The same day, J. Bennett Johnston convened a public hearing before the Senate Energy and Water Appropriations Subcommittee to tout the SSC, with speakers including Lederman.

More than the House, which has to respond to the desires of four hundred thirty-five individual congressional districts, the Senate takes a broader national perspective, addressing the long-range interests of the United States as a whole.[54] Elected every six years by the entire voting population of a state, a senator is less influenced by the politics of a given district, which is often heavily weighted toward Democratic or Republican voters. Therefore the Senate was the natural forum in which to try to resuscitate the critically wounded SSC project. There it had an eloquent, powerful supporter in Johnston. (And because General Dynamics had decided to build its superconducting magnet factory in Louisiana, he also had parochial economic interests in bringing the ailing project back from the budgetary brink.) Despite the opposition of Senator Bumpers, who also served on the subcommittee, Johnston and his staff skillfully maneuvered the fiscal 1993 energy and water bill through the appropriations process. On July 23 the full Senate Appropriations Committee endorsed the subcommittee's allocation of $550 million for the SSC.[55]

But the project still faced a floor fight in the Senate, where Bumpers promised to introduce another amendment to delete funding for it. So SSC advocates again enlisted the support of Bush, who on July 23 met with five Republican senators and jawboned them on behalf of the project.[56] Facing a difficult (and ultimately losing) reelection bid against Bill Clinton and fellow Texan Ross Perot that fall, he used up some of his increasingly scarce political capital to help ensure its survival (and win the state of Texas). On a flight to California July 30, he stopped off in Dallas to visit the SSC and share a photo-op with Schwitters, Barton, and Bromley in the Magnet Development Lab, inspecting some of the long superconducting dipoles there (see fig. 6.4). "The Super Collider is . . . a big part of our investment in America's future," he told SSC staffers. "And where we once reached for the moon above to explore new frontiers of our universe, soon we'll begin to tunnel below to learn about the fundamental questions of science: how our universe began."[57]

That evening the SSC debate was featured on the PBS McNeil-Lehrer News Hour, with Johnston and Schwitters squaring off against Bumpers and Penn State materials science professor Rustum Roy. A vocal SSC critic,

FIGURE 6.4 SSC Director Roy Schwitters (*right*) leading a tour of the Magnet Development Lab on July 30, 1992. Accompanying him are President George H. W. Bush (*left*) and Deputy Undersecretary of Energy Linda Stuntz. Courtesy of Fermilab Archives, SSC Collection.

Roy argued that the SSC was basically a public-works project funded by a Congress that was "befuddled by hyper language and a lot of exaggerations about benefits."[58] To him, it was clearly a "tremendous welfare program for the state of Texas."[59] Schwitters failed to rebut his fusillades, however, countering only that the core purpose of the project was to do curiosity-driven basic research, not generate commercial or technological spin-offs. But that argument did not sit well with a nationwide audience still reeling from the effects of a nagging recession.

On August 3 the Senate floor debate on the SSC began with Bumpers offering an amendment to delete $516 million from its funding, to be applied only to deficit reduction. "Oh, yes, we need to be on the cutting edge of science," he said, "but . . . this Nation is on the cutting edge of bankruptcy, and nobody can ever seem to find a place to bring this thing under control."[60] As the floor leader of the pro-SSC faction, Johnston called on senators from across the country and both sides of the aisle to speak on its behalf. Unlike Barton, he eloquently addressed the physics goals of the project:

> What we are talking about with the Superconducting Super Collider is breaking the ultimate code of the universe, determining what the ultimate particles and the ultimate forces of the universe are. In breaking that code, . . . scientists believe there will be incredibly useful information which will come forth to the American public. . . .
>
> And finally, . . . we have come to what we believe are the ultimate particles and the ultimate forces, quarks and electrons, and what we are trying to do with the Superconducting Super Collider is to reconcile the whole pattern, the whole code if you will, of the universe.[61]

When the roll-call vote came on the Bumpers amendment after three hours of debate and speechifying, it was no contest. Only 32 Senators voted to kill the SSC, well down from 37 in 1991, while 62 wanted to keep it going another year.[62] It was another stunning victory for Johnston, demonstrating his far-reaching influence in that political forum.

When the conference committee met that September to resolve differences between the House and Senate versions of the Energy and Water Appropriations Bills, the only open question was how much funding the SSC would obtain. This committee was composed largely of members of the appropriations committees in both chambers, who were generally favorable to the project; thus deleting the SSC was out of the question. They compromised at $517 million, splitting the difference between the original SSC allocations in the two bills, still well below the $650 million in Bush's budget.[63] When the compromise bill went to the House floor on September 17, the SSC opponents' only option was to defeat the entire bill, which they tried to do, but it was a losing cause. Members were loath to reject a bill larded with energy and water projects for their districts, especially with the elections approaching. The House voted 245 to 143 to accept the compromise, a comfortable 102-vote margin.[64]

The SSC was granted another year's reprieve, but it had emerged from the 1992 congressional budget battle badly wounded and vulnerable. "Although the House eventually reversed the decision and agreed to continue funding for the project in fiscal 1993, the vote was a shocking blow to project supporters and left the endeavor on precarious political ground," observed the *Congressional Quarterly* in its year-end summary of major legislative actions. "The flip-flop hurts the Energy Department's chances of attracting Japanese or other foreign support for the project, which in turn could erode Congressional support."[65] Indeed, the US/Japan Joint Working Group on the SSC had by then essentially resolved most of the questions about Japan's joining the project as a major partner. But cautious Japanese policymakers decided to take a wait-and-see approach, given all the turmoil in the 1992 electoral process and the likely possibility of significant changes in US leadership after the elections. These officials wanted assurance that the next administration, whichever it might be, firmly supported the project.[66]

STRUGGLING TOWARD INTERNATIONAL PARTNERSHIP

Besides Japan, SSC leaders and DOE officials were still actively pursuing Canada, India, Korea, China, Russia, and Taiwan as potentially significant foreign partners in transforming the SSC into an international laboratory. Though the first three had been approached at least two years earlier, there were as yet no formal nation-to-nation agreements in place by late 1992.[67] Perhaps indicative of the SSC lab's growing desperation here, it was beginning to consider laboratory-to-laboratory agreements with institutes of high-energy and nuclear physics in China and Russia. According to these interlaboratory accords, the Chinese and Russian institutes were to design and manufacture various components for the Super Collider at prices 50 percent or more below what they might cost if produced by US firms. The resultant cost savings would be credited as an in-kind foreign contribution to the SSC by the nation involved.[68] Nearly $200 million worth of SSC components had been identified that could be supplied by Russian institutes, potentially resulting in a total Russian "contribution" of about $100 million.

After the 1991 collapse of the Soviet Union, Russia had a desperate need for such accords to provide sources of external funding for its institutes and foreign currency for the struggling nation. Otherwise, PhD physicists would be forced to find work as taxi drivers or street vendors—or to sell their services to Third World nations interested in bolstering their nuclear capabilities. And with the concomitant collapse of the ruble, many Russian industries could produce SSC components *very* inexpensively, resulting in even larger in-kind contributions. But the approach ran afoul of SSC opponents in Congress, who argued that this was creative accounting at its worst. The DOE was proposing to send US taxpayer dollars overseas to prop up Russian science and industry, they objected, while denying much-needed work to American laborers still suffering from the early-1990s recession.[69]

Despite this criticism, the DOE and the SSC lab proceeded with the plan. On January 3, 1993, Energy Secretary Watkins signed an agreement between the DOE and Russia's Ministry of Atomic Energy endorsing the nation's collaboration on the SSC project.[70] As one measure of how significantly East-West relations had thawed since the Cold War ended, wrote a *New York Times* reporter who visited the SSC lab in March 1993, "Russian laboratories that once designed and built guidance systems for Soviet ballistic missiles aimed at the United States are now using their expertise and equipment, coupled with American computers and advanced computer programs, to build components of the Supercollider's detectors." He quoted one Russian physicist at the SSC as kidding, "Why this is nothing new. For many years we made high-precision components for delivery to the United States!"[71]

The crucial question throughout 1992 and into early 1993 was whether Japan would agree to come on board as a major partner in the SSC Laboratory. Among the thorniest problems to be resolved by the US/Japan Joint Working Group established at the Bush-Miyazawa summit was to ascertain "how this project can be formulated as an international project to enable Japan to participate."[72] How could a project that had been initiated and designed as a US national laboratory be reorganized to allow a foreign nation to participate as an equal partner in managing and operating that laboratory? Another issue to resolve was the exact nature and cost of the Japanese contribution. The principal suggestion made was that Japanese firms contribute the superconducting dipole magnets for one of the two main collider rings, which DOE had estimated at over $700 million dollars if manufactured by US industry. But when the first, rough estimates came in from these companies in January 1992, the total cost of the 4,400 magnets required was projected at more than a factor of two higher—owing in part to the low yen/dollar exchange rate then in effect. SSC Magnet Division leader Tom Bush advised Schwitters, however, that the "Japanese estimates reflect their immature understanding of the magnet development and projected production design and tooling concepts."[73] These kinds of major differences had to be resolved by the Joint Working Group before Japan could ever become a partner in the SSC Laboratory.

A major problem internal to Japan was how to accommodate such a large request within its existing governing structure for science funding.[74] The Japanese ministry Monbusho was the traditional financier of high-energy physics research, but its budget could not accommodate an additional huge project costing some $200 million a year without massive, damaging cutbacks on its other funded research. The Science and Technology Agency was the traditional supporter of big science and technology projects, such as Japanese contributions to the Space Station, but it had no prior experience with or affinity for high-energy physics. The agency was at best a tepid supporter of the SSC. Because of the strong interest in the US project among Japanese industries, the influential Ministry of International Trade and Industry supported Japan's joining the SSC project but was averse to devoting any of its existing funds toward it. Among the other Japanese ministries, Foreign Affairs strongly favored a partnership in the SSC to improve US-Japan relations, while Finance strongly opposed it because of the burdensome cost.[75]

If Japan were to join the project as a partner, at a cost of $1–2 billion, the decision thus had to be made at the very highest levels of the Japanese government, by Prime Minister Miyazawa himself, and additional funds allocated to one or more of these agencies. This was not a matter for the Joint Working Group, which met in Tokyo on April 9–10 and in Washington on July 28–29, 1992, to address the other problems and differences.[76] Despite the higher cost of Japanese companies' (such as Hitachi) fabricating the superconducting dipole magnets, Japan appeared ready to proceed toward partnership by late 1992—as long as the additional funding could be allocated, so that participation in the SSC would not adversely impact its other scientific research.[77] And Miyazawa seemed prepared to make such a commitment in early 1993, despite the weakening Japanese economy, if the new US president were only to request it. "The probability is very high that Japan will contribute if the Clinton Administration assures them [*sic*] of its strong support for the project and if the President personally requests that Japan join in," wrote Burton Richter to John Gibbons, soon to become Clinton's first science advisor, in mid-January 1993 after meeting with Japanese leaders in Tokyo.[78] The hoped-for agreement still had to occur, presumably at the first summit meeting between the two heads of state.

CONSTRUCTION ACCELERATES AMID POLITICAL UPHEAVALS

As the United States lurched into the tumultuous 1992 elections, construction was proceeding apace at the SSC. With a guaranteed budget now exceeding half a billion dollars and its project-management problems largely solved, at least to the satisfaction of DOE officials, the lab could finally concentrate on accomplishing required tasks. Boring machines were being assembled to tunnel the northern arc of the collider ring, beginning with a 2.7-mile segment that started at the edge of the North Campus, where the Magnet Test Lab was nearing completion. The "cut-and-cover" construction of the Linac was also nearly finished at one end of the West campus.

FIGURE 6.5 Aerial view of the SSC North Campus in September 1992, showing the Magnet Development Lab (*left*) and Magnet Test Lab (*center*). The long, narrow building (*at right*) is where Accelerator Systems String Tests occurred; the shaft (*lower right*) leads down to the main ring tunnel, about to begin construction. Courtesy of Fermilab Archives, SSC Collection.

Because of the recession that gripped the nation during the early 1990s, most of these projects were coming in well under budget.[79]

That fall General Dynamics began winding coils at its Louisiana plant for the first, limited production run of superconducting dipole magnets, aimed at resolving manufacturing issues and getting a better grip on long-range costs. The "follower" contractor Westinghouse was close behind at its Round Rock, Texas, factory. It also began fabricating prototypes of the superconducting dipoles for the High Energy Booster. And in a test of one of the last prototype main ring dipoles assembled at Fermilab, the magnet operated at close to 10,000 amps (corresponding to a magnetic field of 10 T) after its liquid helium had been cooled to superfluid temperatures of 1.8 K; this success demonstrated that the new design was indeed robust, given the tremendous forces on the magnet windings at such a high current.[80]

In October the SDC experimental collaboration obtained DOE approval to begin construction of its immense particle detector.[81] Featuring a huge cylindrical superconducting magnet at its core, the 10,000-ton detector was by then expected to cost over $600 million; because DOE had allocated only $550 million for *both* large detectors, however, something over $300 million (or its equivalent in components) still had to be found.[82] Such an amount was an order of magnitude greater than foreign agencies had previously contributed to experiments at Brookhaven, Fermilab, and SLAC. Ever since its early days, the SDC experiment had been organized as a truly international collaboration, including strong European and Japanese research groups. As there were about one hundred Japanese physicists in the SDC collaboration, $100 million might come in if Japan were to join the SSC project as a full partner.[83] But raising another $200 million (or more) was going to be extremely difficult in the budget climate of the mid-1990s.

Tunneling of the initial collider sector began in January 1993, but a major mishap soon occurred. On January 27 a Dillingham/Obayashi worker was crushed and killed by a collapsing segment of the concrete tunnel liner, which was needed because of the unstable composition of the Eagle Ford shale in that sector.[84] The tunneling of that portion was halted pending an investigation, but further boring continued that February in other portions of the north arc passing through the more stable Austin Chalk, which did not require such linings. By late April, four boring machines were in operation on that arc, and four miles (6 km) of tunneling had been completed.[85]

There was great urgency in the SSC construction, noted Malcolm Browne of the *New York Times,* who visited in March and spent almost a week with Schwitters, following him on his rounds. "As scientists, technicians, and workers rush from one task to the next," he observed, "there is a sense that a desperate campaign is under way, to beat the clock and win against tough odds."[86] At the end of a long day, the exhausted SSC director complained volubly about interference from the DOE and Congress. "We should be devoting ourselves to completing this machine as rapidly and cheaply as possible, and getting on with real science," he railed, forgetting that Browne was standing nearby. "Instead, our time and energy are being sapped by bureaucrats and politicians. The SSC is becoming a victim of the revenge of the C students."[87] Published in the *Times,* it was a politically damaging statement,

FIGURE 6.6 View down the main ring tunnel at the base of the vertical shaft shown in fig. 6.5, as seen in early 1993. Courtesy of Fermilab Archives, SSC Collection.

encapsulating high-energy physicists' disdain for the tedious but necessary bureaucratic requirements involved in spending billions of taxpayer dollars.

Tectonic shifts were then going on in the nation's political geology. Not only had Arkansas Governor William J. Clinton defeated President Bush and his fellow Texan Ross Perot that November, ending twelve long years of

Republican White House occupancy.[88] But one hundred twenty-seven new members of Congress—one hundred fourteen in the House and thirteen in the Senate—had also been swept into office, many of them having campaigned on promises to rein in federal spending and bring the budget back under control.[89] Congress had not changed so dramatically since the post-Watergate 1974 election.

With the new administration came a whole new slate of political appointees heading the agencies and departments that interacted with the SSC Laboratory.[90] At the apex was Vice President Al Gore, to whom Bill Clinton delegated most of his responsibilities on science, technology, and the environment.[91] Gore had established his reputation in these areas in the Senate, where he chaired the Subcommittee on Science, Technology and Space. Before that he had served in the House as chairman of the Science Subcommittee on Investigations and Oversight, the same position that Michigan Representative Wolpe held until being redistricted out of office in 1992. Although a friend of science, Gore's personal sympathies lay more along lines of the environment and technological competitiveness, rather than with the basic sciences. He began pushing for the federal government to take a more active role than had the Reagan and Bush administrations in supporting applied research and technology development aimed at enhancing US economic competitiveness.[92]

One of Clinton's first political appointments was Gore's fellow Tennesseean John Gibbons, a nuclear physicist from Oak Ridge National Laboratory with deep Washington political connections, to be his science advisor. Gibbons had been an early advocate of energy conservation, leading a small agency devoted to the subject, before directing the Congressional Office of Technology Assessment from 1979 to 1992. That office issued scores of analytical reports to Capitol Hill evaluating the "economic and political impacts of scientific and technological developments."[93] In the course of these efforts, he had made friends inside and outside of Congress—among them Gore, who recommended Gibbons for the science advisor post.[94] Well liked on both sides of the aisle, he passed his Senate cross-examination with flying colors in late January and was sworn into office on February 1, 1993.

But SSC advocates were worried that Gibbons (like Clinton and Gore) was lukewarm on the project that had been a presidential priority for Reagan and Bush. Asked that February about slowing down the SSC construction, he replied that there was "no reason why we have to find the Higgs boson by the turn of the century."[95] Another key official in the new administration was Leon Panetta, who became the powerful director of the Office of Management and Budget. In a late January 1993 Camp David conclave, he had advocated jettisoning the project, but Clinton's new treasury secretary, Texan Lloyd Bentsen, came to the SSC's defense.[96] Out of that gathering emerged the idea of stretching out the project's construction for three or four years, which would inevitably add more than $1 billion to its cost, pushing the total toward $10 billion. But it would help to keep the project's annual allocations below the $1 billion level. "The Administration is committed to the development of the Superconducting Super Collider as a major contribution to scientific information for the future," read the White House manifesto "A Vision of Change for America," issued after Clinton's State of the Union address on February 17, 1993. But, it noted, the administration believed "that in order to ensure that all the components of this project are technologically effective, the project should be extended."[97]

Questioned a week later by a reporter about the stretch-out, Gibbons replied that "we're trying to recognize that there are certain projects . . . in which the sense of urgency may have been miscalculated compared to other priorities in terms of our national recovery."[98] He was also concerned whether the superconducting magnets for the SSC could indeed be manufactured by the thousands with enough precision and reliability. Giving General Dynamics and Westinghouse more time to address these problems would help resolve them. The stretch-out would also allow foreign countries time to make the contributions required to transform the SSC Laboratory into a truly international partnership.[99] Left unsaid was the fact that without these foreign contributions, the soaring costs of the SSC project would begin to exert intolerable pressures on the budgets of other DOE labs and water projects—and hence lead to political strife in Congress. In a 1996 interview, Gibbons admitted that "the budget profile on the SSC would have just starved everything else to death. My conclusion after looking at it was [that] either this thing gets killed or we stretch it out."[100]

The out-year consequences of both Reagan-initiated gigaprojects—the SSC *and* the Space Station—were what troubled the new administration the most.[101] In the post-1992 budgetary climate emphasizing deficit reduction, they would be easy political targets for lawmakers wanting a quick, high-visibility resolution with little or no electoral consequences back in their districts or states. As Gibbons recalled:

> I remember one meeting in the Roosevelt Room, with the President and a number of other people, [when] we covered both the SSC and the Space Station in the same afternoon, and the President committed to both. But he committed to the Space Station as something that we would internationalize, redesign, scale downward. . . . On the SSC we would try to do a stretch-out on the budget to

> make it less of a near-term, high-cost item, and use that stretch-out to really nail down the problems we were having with the management structure as well as with the magnet designs.[102]

Clinton supported both projects at the outset of his first term, but it was a lukewarm and conditional support. The SSC was no longer a presidential priority; no potential veto awaited an energy and water bill that did not include funding for it.

At the Energy Department, the SSC had lost a strong ally in Admiral Watkins. He was succeeded as secretary by Hazel O'Leary, an attorney who had served in the Federal Energy Administration during the Carter years, then as a vice president at Northern States Power Company in Minnesota. At the outset she demonstrated no special affinity for high-energy physics or interest in the SSC project, which was not as high on her list of priorities as global warming and cleaning up the nuclear-weapons labs.

"I strongly recommend that this important national project be completed on schedule independent of foreign contributions," Watkins had just urged in a letter to House Science Committee Chair George Brown before leaving office in mid-January 1993.[103] He observed that its schedule had already slipped three to six months, due to a $150 million shortfall in fiscal 1993 funding (and a total of $262 million since fiscal 1991) relative to the funding profile set forth in the 1991 Green Book.[104] Cost savings at the SSC Laboratory had helped minimize but not fully eliminate this erosion. To prevent any further schedule slippage and ensure its completion by early 2000, he added, the project would need total funding of over $1.2 billion in fiscal year 1994, including foreign contributions. Only about $50 million was assured, however, without Japanese participation.[105]

In an attempt to help secure a major commitment, O'Leary staffers drafted a letter to Miyazawa requesting a $1.5 billion Japanese SSC contribution for President Clinton's signature and sent it to the White House in mid-February for the needed approvals. But Gibbons had several objections, among them that Clinton had already sent a February 6 letter to Miyazawa addressing economics and trade relations between the two nations. In addition, the SSC schedule was being stretched out, making its estimated costs more uncertain. "For these reasons," he stated in a February 24 memo, "I will not clear on the draft letter."[106] Gibbons, however, agreed to work with Gore, the DOE, and the State Department to craft an appropriate (but nonpresidential) statement of administration support for the SSC.

O'Leary and her staff made another attempt to get the letter cleared in

FIGURE 6.7 The new Energy Secretary Hazel O'Leary (*right*) touring the Magnet Test Lab at the SSC Laboratory on April 8, 1993. Accompanying her are (*l. to r.*) Roy Schwitters, SSC Laboratory Associate Director Ted Kozman, Dallas Congressman Martin Frost (in glasses, behind Kozman), and Texas Senator Bob Krueger. Credit: AIP Emilio Segrè Visual Library, *Physics Today* Collection.

mid-March, sending a memo urging its approval to White House Chief of Staff Mack McLarty.[107] But she was again rebuffed, this time by the National Security Council. "Such a letter could be seen as suggesting that we attach greater importance to Japanese participation in the SSC than we do to Japanese efforts on other fronts, such as aid to Russia," wrote top White House staffers Todd Stern and John Podesta to McLarty on this issue.[108] "NSC believes that we can send the Japanese clear, effective signals of our support for the SSC in other ways." Again, they wrote, the president need not be directly involved.

Besides Watkins, another strong SSC supporter to leave the DOE was OER Director Happer, a Democrat, who at first agreed to remain in the position during the Clinton administration. But he publicly criticized Gore's position about the ozone layer and was asked to leave in May. "This feckless, on-again, off-again behavior of the government is something I neither like nor understand," fumed Happer after returning to Princeton. "As an American, one who isn't a high-energy physicist, I'm confident we can build [the SSC] ourselves, without foreign assistance if we had the will."[109] He was replaced that summer by Martha Krebs, an LBL associate director who had previously served as staff director of a House subcommittee on energy policy.

The Clinton administration was not eager to spend a billion dollars a year rescuing a Reagan and Bush administration gigaproject sited in a solid Republican state. When the president's fiscal 1994 budget was presented on April 8, 1993, there was $640 million in it for the SSC—$10 million less than Bush had requested a year earlier, and $123 million more than what Congress had allocated for 1993.[110] Any more would have to come from Texas and foreign partners. The focus of the Clinton budget was applied science and technology to enhance economic competitiveness. Renewable energy, climate change, and environmental research got big boosts while basic research had to take a back seat. In addition, the administration's goal was to dial back the ratio between military and civilian research to a more balanced 50-50, from the 60-40 level it had gradually fallen to during the Bush years.[111]

Moreover, there were new plums for Democrat-leaning states: funding for construction of the Advanced Neutron Source at Oak Ridge National Laboratory in Gore's home state, the Tokamak Plasma Experiment at Princeton in New Jersey, and a B Factory to be constructed at either Cornell or SLAC.[112] The budget included $36 million for the B Factory, and Fermilab received $25 million for the Main Injector—a $10 million increase, enough to keep that project alive another year.[113] A draft of Bush's suggested 1994 budget had included $858 million for the SSC, but these and other competing projects lowered the figure by more than $200 million and forced the DOE to stretch out the SSC's construction by three years.[114] No longer a

FIGURE 6.8 Japanese Prime Minister Kiichi Miyazawa (*left*) and President William J. Clinton holding a press conference following their April 16, 1993, summit meeting at the White House. To Clinton's left stand (*l. to r.*) Secretary of State Warren Christopher (obscured by the podium), Treasury Secretary Lloyd Bentsen, and US Trade Representative Mickey Kantor. Courtesy of William J. Clinton Presidential Library, Sharon Farmer.

presidential priority, the SSC's budget was being starved by competing, and largely Democratic, political interests.

Because Texas could be counted on for at most $100 million, and only about $50 million might come in from Chinese, Indian, and Russian "contributions," the SSC budget for fiscal 1994 could be at most $800 million—assuming Congress went along with Clinton's $640 million figure, a big if. That would leave SSC construction funding far short of the $1,137 million stipulated in the 1991 Green Book or the $1.2 billion Watkins had suggested was needed to bring it back on schedule. Even a big contribution from Japan would not make up that large a difference.

But SSC advocates were still determined to try to orchestrate the golden handshake when Miyazawa was to visit Clinton in Washington in mid-April for their first summit meeting. On April 14, 1993, a delegation of seven Nobel laureates, including Friedman, Lederman, and Weinberg, plus Schwitters,

Panofsky, and other high-energy physics leaders, met with Gore at the White House after holding a press conference on Capitol Hill.[115] According to Weinberg, the vice president told them that the administration "enthusiastically" supported the SSC. The delegation could not tell, however, whether any of his enthusiasm was communicated to Clinton for the upcoming summit, nor whether it had any impact on the outcome.[116]

At the April 16 meeting, economic concerns were uppermost on the Clinton agenda, particularly obtaining Japanese concessions on imports of US auto parts.[117] "The Cold War partnership between our two countries is outdated," said Clinton in a press conference after the meeting. "We need a new partnership based on a longer-term vision."[118] He observed that the US-Japan economic relationship had become seriously imbalanced and that "rebalancing of our relationship in this new era requires an elevated attention to our economic relations." Clinton also said he was "particularly concerned about Japan's growing global current account and trade surpluses and deeply concerned about the inadequate access for American firms, products and investors in Japan."[119]

Miyazawa and the Japanese officials seemed prepared to agree to partner in the SSC, contingent (expressly or implied) upon the United States softening its demands for import quotas, particularly on auto parts. But US Trade Representative Mickey Kantor was adamantly against backing off on these quotas.[120] Asked at a background briefing whether the subject of Japanese partnership in the SSC had even come up, a "senior administration official" answered that it had not. "The President believes strongly that the economic leg of the stool has to be raised to the forefront and has to be the focus of our attention," the official continued. "Therefore it is important . . . in the nature of our communications to the government of Japan that we make that point clear rather than provide a long . . . menu of things that we care about without indicating priority."[121] Yet again, the golden handshake failed to occur. Japanese participation in the SSC project would thus continue to languish.

In reality, Clinton had only limited political capital at the outset of his first term, and it was already wearing thin. Not only had his health-care initiative met strong opposition, but Senate Republicans were filibustering his economic stimulus plan in mid-April, too. So he could not have devoted much effort to SSC issues.[122]

THE LONG, HOT SUMMER OF 1993

With a quarter of the House seats filled by new representatives, most of them seeking items to delete from the 1994 budget, the SSC had little chance of surviving the annual appropriations battle in that body, especially without firm commitments for major foreign contributions.[123] But under Chairman Tom Bevill, a longtime SSC supporter, the Energy and Water Appropriations Subcommittee was determined to try. Meeting with Energy Secretary O'Leary on April 22 and OER Director Happer on April 26, the subcommittee grilled them about the project's status, especially in light of a February 1993 GAO report claiming the SSC was over budget and behind schedule. Soon to return to academe, Happer disputed these claims, arguing that the tunneling was coming along under budget and the "splendid" superconducting magnets were now being successfully transitioned to industry. If anything, he said (echoing Watkins), project delays owing to funding shortfalls were what was causing the overruns.[124]

When the 1994 energy and water bill came out of committee on June 17, the SSC budget had taken a small hit of $20 million, down to $620 million, while the high-energy physics base program remained at the $628 million level of the original request.[125] This represented more than $1.2 billion in all for high-energy physics, or almost one-third of the OER budget, about as good as could be expected. That same day, Clinton sent a letter to House Appropriations Committee Chairman William Natcher (D-KY) reiterating his support for the SSC.[126] But when the bill reached the House floor on June 24, SSC opponents were ready with their annual amendment to kill the project, again offered by Jim Slattery of Kansas. This time it was no contest, as the project went down to a resounding 130-vote defeat, 280 to 150.[127]

A day earlier the Space Station had survived a similar amendment by a slim one-vote margin, 216 to 215—but only after Clinton and especially Gore got on the phone and jawboned House members on its behalf. These calls doubtlessly made the difference. In this instance, the two had solid political reasons to do so. Not only had Canada, Europe, and Japan already spent about $3 billion on contributions to that project; there were also some 75,000 US jobs at stake, spread over many states. But it is doubtful that such jawboning could have turned the SSC vote around, given the huge margin of defeat. "Maybe the Senate will save it and then we can fight for it in conference," Clinton told reporters afterward. "I always anticipated that if we were going to save the Super Collider, we would fight for it in conference."[128]

But SSC advocates hoping for a sequel to the successful 1992 reversal were dismayed when Congressman John Dingell (D-MI), chairman of the House Energy and Commerce Committee, rewrote the script and called a hearing on the project in his Oversight and Investigations Subcommittee on June 30. A tough, much-feared inquisitor who had represented a blue-collar district near Detroit since 1955, he was an outspoken member of the

Northeast-Midwest Coalition. During the past year he had been calling university scientists and administrators to appear before this subcommittee, berating them for research misconduct and their misuse of taxpayer dollars.[129] More recently, Dingell was investigating the questionable practices of the contractors managing and operating the DOE laboratories. In addition to Secretary O'Leary, he subpoenaed Cipriano, Siskin, Schwitters, and URA President John Toll to appear at the televised hearing. All had to testify under oath.[130]

Dingell was deft at manipulating a largely uncritical press to buttress his attacks. His committee had examined dozens of poorly managed defense contracts, he railed, "but the SSC ranks among the worst projects we have seen in terms of contract mismanagement and failed government oversight."[131] He fired a few frivolous but damaging accusations at URA managers, such as the purchase of potted plants to help convert a drab Waxahachie warehouse into attractive office spaces (which had saved millions of taxpayer dollars). More substantive charges included the long delay in getting the cost-and-schedule control system running and fully validated, and the recent use of parallel sets of books to conceal cost overruns. Toll apologized for the URA lapses, admitting that this dual accounting system was "a dumb idea," and agreed to eliminate it. "The thing that disturbs me most is the perception of our attitude," he pleaded. "There have been errors of judgment and we must correct those."[132]

Another damaging revelation during the nine-hour grilling was the growing likelihood of major cost overruns in manufacturing the superconducting magnets. The February 1993 GAO report had mentioned this possibility earlier that year,[133] but better cost projections were becoming available as General Dynamics began fabricating industrial prototypes at its Louisiana factory. "Mass production of the magnets is much harder than we originally thought," testified its vice president, Walter Robertson. Development costs alone were deemed likely to grow by $50 million—almost 25 percent above the contracted amount. And Dingell revealed that his staff members had found an internal SSC memo estimating that the magnet costs might almost double to nearly $2.7 billion.[134]

Dingell saved some of his fiercest attacks for O'Leary, who withered under fire and blamed the odd DOE management and oversight structure established by Watkins, under which Cipriano and Siskin had direct access to him. This structure "clearly diluted the authority of the Director of Energy Research, who might have exercised closer control over the project." It also meant, she added, that Schwitters no longer had control over the construction costs (and thus he could not be held accountable). But O'Leary had little faith in SSC lab officials, either, claiming that they "lacked the leadership and good judgment necessary for the success of the project."[135] She said the Energy Department was going to review the entire project during the next month and report back with recommendations on how best to address these problems. Among the solutions under consideration was a management shakeup in which SSC construction responsibility might be removed from URA and transferred to an industrial firm.

The lack of a mature, fully qualified SSC cost-and-schedule control system hobbled officials in making the case before Congress that the project was actually on schedule and under budget.[136] Without one, nobody could really tell for sure. In his testimony before the Dingell committee, Siskin claimed that 20 percent of the project was complete while only 4 percent of the $840 million contingency funds had been spent to cover cost overruns.[137] But these figures could easily be challenged, and they were. To try to resolve this issue convincingly, the DOE organized a large team of seventy-five employees and consultants to travel to Waxahachie in mid-July and pore through SSC records and interview laboratory officials. Headed by John Scango of the DOE Field Management Division, they were to prepare a detailed report by mid-August.

Trying to shed a more benign light on the project, Bennett Johnston held off another month before calling a joint hearing on the SSC before the Senate Energy and Natural Resources Committee and the Appropriations Subcommittee on Energy and Water Development, both of which he chaired. Speaking before them on August 4 were Gibbons, O'Leary, Schwitters, Anderson, Weinberg, and Motorola President Robert W. Galvin. "A fundamental question is whether a country which has some budget problems can afford basic science, indeed, whether we ought to be involved in anything other than applied science," said Johnston in opening the hearing.[138] The testimony that morning was largely supportive and uneventful—though Anderson and Weinberg traded barbs. Other senators used the opportunity to pontificate on the sorry condition of the federal budget. "There are a lot of fine desirables that this Senator would love," said Senator Ernest Hollings (D-SC), chairman of the Senate Committee on Commerce, Science and Transportation. "I wish I were allowed loose in the candy shop to pick anything I want, like a Super Collider, but I cannot afford it."[139]

But O'Leary stole the show with her testimony that afternoon, saying she was going to wrest the SSC construction management responsibilities away from URA and transfer them to a company with experience in such matters.[140] By doing so, she promised to deliver on what the DOE had been threatening since Watkins took the helm in 1989. The new contractor, she said, should have "world class experience in managing large construction projects."[141] On the subject of cost and schedule, she said that preliminary findings of Scango's committee confirmed what SSC officials had been claiming all along: that it was indeed on time and on budget.[142] But there were also some major management lapses by URA, as well as "significant cost risks" that "if left unattended, might cause this contract to be much more expensive than we have contemplated."[143] O'Leary did not cite any figures for these potential cost overruns (to use a more loaded phrase), but rumors circulated that it might be as much as another $2 billion over the $8.25 billion figure in the 1991 Green Book. And that was before adding in the unavoidable delay costs resulting from the Clinton stretch-out.

When the Scango Report was released on August 13, SSC advocates and critics alike could put more reliable numbers on these projections. For example, the SSCL had committed about $1.5 billion, or 18 percent, of the allotted $8.25 billion by May 1993 and completed about 20 percent of the assigned tasks.[144] The report cited many "cost risks" (rather than "potential cost overruns"), claiming that these could be addressed by stronger management actions than had occurred to date. And there could be up to $1.7 billion in additional costs, resulting in a total SSC project cost of $9.94 billion, if these cost risks could not be resolved.[145] Add to that the estimated $2 billion cost of a three-year schedule stretch-out, and the SSC had suddenly become a $10 billion—and quite possibly even more expensive—project.[146]

A close reading of the report reveals how serious problems had become. In the area of "superconducting accelerators," for example, which included the collider rings and the High Energy Booster, a $668–861 million cost growth was anticipated. Much of this, $255–320 million, would come in manufacturing the superconducting dipole magnets. And those figures assumed that General Dynamics could indeed lower the cost per magnet by almost $100,000 when it ramped up to full production—or 37 percent below the costs being achieved in the low-production run.[147] The superconducting-magnet development and production was already five to six months behind schedule, and it threatened to slip another three to six months in the coming year. And there were other problems and attendant cost risks omitted from these figures: the need for more spare magnets, whether to include liners inside the beam tubes, and a possible need to test *all* the dipoles (instead of merely 10 percent), etc. These extra measures could add another

$300 million to the cost of the superconducting accelerators and push the cost overrun well above $1 billion in that area alone.[148]

As August faded into September, the mood at the SSC Laboratory was grim but still hopeful. After the disastrous June 24 House vote, Texas decided to hold back $79 million of its 1993 contribution, forcing Schwitters to "jam on the financial brakes" in July. Over three hundred employees were laid off and construction halted in many areas.[149] It didn't help that he and Cipriano were no longer on speaking terms after the *Washington Post* published an acidic Cipriano memo to O'Leary claiming, "Morale is very low, confidence in existing management is practically nonexistent, and cost and schedule trends are worsening at an alarming rate." Firing Schwitters, he added, "may be the only way to keep the Lab from falling apart before the Senate vote."[150]

But tunneling continued apace, and by the end of September nearly fifteen miles had been completed, almost 28 percent of the total; civil construction on the Linac was essentially finished, too. Much effort was now being devoted to "rebaselining" the project, trying to determine a new baseline cost estimate reflecting the likely overruns and the fact that the construction schedule would be stretched out by another three to four years. Initial projections suggested the total cost might balloon to $12 billion.[151]

On September 30 anxious SSC officials and employees clustered around TV sets in Waxahachie, watching the Senate floor debate on C-SPAN. After five tense hours, cheers erupted when senators voted 57–42 to defeat yet another Bumpers amendment to kill the project—thus allocating the full $640 million in the Clinton budget.[152] The Senate margin had narrowed, but it was still substantial. The SSC was granted another reprieve. But its future depended on what was to happen soon in the House/Senate conference committee and on the House floor.

THE OCTOBER CONGRESSIONAL REVOLUTION

Led by Boehlert and Slattery, SSC opponents in the House were bristling for a fight. Based on their experiences of 1992, when the 232-to-181 House vote to kill the project was ignored by the conference committee, they expected a similar maneuver to occur again behind closed doors. So they circulated a letter to House Speaker Tom Foley of Washington state, urging him to include SSC opponents on the committee this time, but it received only 120 signatures.[153] Foley checked House precedents and appointed only Appropriations Committee members, most of whom were solid SSC

FIGURE 6.9 New York Congressman Sherwood L. Boehlert, a leading House opponent of the SSC. Courtesy of Sherwood L. Boehlert.

supporters.[154] A stacked deck, the conference committee acted just as expected and reported out a "compromise" $22 billion energy and water appropriations bill that included the *full* Clinton request of $640 million for the SSC. "The emotional tide of the moment should not direct this committee and its responsibility to science," said Johnston after the October 14 meeting. "To walk away from this project, after $2 billion has been spent, would be something this country cannot do."[155]

"Today's action is an outrage, and it will not stand," declared Boehlert in an angry press release. "The decision of the conferees is an insult to the intelligence and the integrity of the House and to the democratic process."[156] He continued: "More is at stake here than the future of one grossly mismanaged science project. What's at stake is whether the American government can ever cut spending. What's at stake is whether this government will run as the Constitution intended or whether we are going to be governed by a small coterie of Appropriations members. We are going to fight this conference report, and we fully expect to emerge victorious."

To impose the will of the House on Congress and terminate the SSC, however, a majority of its members now had to vote down the *entire* bill, larded as it was with water projects for many of their districts.[157] SSC opponents faced a daunting task. In 1992, just before the elections, such a majority did not exist, and the Senate will had prevailed. But this time the House had one hundred fourteen freshman members determined to make their mark as budget-slashers, and they were not up for reelection. Nor did they have many water projects in their districts to defend. It was a new ball game, about to be contested on a different playing field than the year before.

Another factor working in favor of SSC opponents was the clear fact that the total project cost was now going to exceed $10 billion.[158] And because of the uncertainty bred by its "ever-mounting cost," as underscored by the summer hearings, few felt confident in any particular figure. Some observers worried that the eventual cost might even surpass $15 billion—for an "impossibly esoteric" science project that few lawmakers really understood.[159] At least with the much more expensive Space Station, NASA could show alluring simulations of a sprawling, sun-speckled object serenely orbiting Earth amid the black void of space, while astronauts in space suits drifted weightlessly about.

On October 19, 1993, the conference report on the energy and water bill came to the House floor for a vote. Usually this is a perfunctory affair, with the members rubber-stamping the measure and then moving on to other business. But Boehlert, Slattery, and their allies were primed for battle. They "overcame a maze of parliamentary maneuvers" by SSC supporters intended to blunt any opposition.[160] Slattery then introduced a motion to return the report to the conference committee with explicit instructions to delete funding for the Super Collider. He expressed his outrage that the conference committee had completely ignored the House position after the June 280-to-150 vote to kill it.[161] "Are we going to continue to be big suckers or not?" Slattery asked his fellow representatives.[162] His motion passed by an even more resounding margin, 282 to 143, showing how great a majority, almost two-thirds of the House, shared his feelings.[163] In a broad bipartisan vote, 115 Republican members joined 166 Democrats and an independent to kill the SSC; of the 114 House freshmen, 82 voted against the project.[164] The SSC had suffered yet another stunning—and likely fatal—defeat. "This is a sad day for science," said Johnston after the vote. But he acknowledged that the House's "message on deficit reduction and the SSC was clear and unmistakable."[165]

The conference committee met the next afternoon in a tense confrontation, where "the SSC's most ardent supporters and most vehement critics negotiated the termination of the program."[166] Usually only committee members are present at such meetings, which are generally closed, but this one included Boehlert, Slattery, and a few staffers, plus members of the Texas congressional delegation, lobbyists, and the press. Subcommittee Chairman Bevill, who had been a steady SSC supporter, said he could no longer ignore the will of his fellow representatives after the previous day's decisive vote. Johnston asked Boehlert and Slattery whether they would approve proposed compromise measures, for nothing that lacked their imprimatur was ever going to pass muster in the House. After moving soliloquies and heated debate, the conference committee resolved to keep the $640 million SSC funding in the budget but directed that the money be spent instead to close down the project in an orderly manner. "The SSC has been lynched, and we have to bury the body," Johnston later told reporters.[167]

On October 26 the full House took up the revised conference report. A few minor amendments were debated and voted up or down, but the general sense of the compromise measure prevailed, with the DOE getting $640 million to terminate the SSC. The final House vote on the measure came in overwhelmingly favorable, 332 to 81. The following day the Senate approved the revised report 89 to 11, and the fiscal 1994 Energy and Water Appropriations Bill went to the president's desk for his signature.[168] Although Texas Governor Ann Richards wanted Clinton to veto the bill, he signed it later that month and it became law.[169]

To participants on both sides of the aisle and the two sides of the SSC debate, it was a stunning reversal and a watershed event.[170] What had begun several years earlier as a ragtag militia of dissidents, hailing largely from the Northeastern and Midwestern states, had coalesced into a disciplined army that rebelled successfully against the House leadership, defying the will of the powerful Appropriations Committee.[171] One can suggest over a dozen reasons for this historic turnabout, but at their pivot stood the fact that the SSC project was widely perceived to be out of control at one of those rare moments when Congress was actually trying to reduce federal budget deficits.[172] And it was therefore an easy, high-visibility political scalp for House members to take that would cause few negative consequences in their districts while helping demonstrate that they indeed took deficit reduction seriously.

Leaders of Congress took pains to reassure their constituencies that the SSC termination was not a vote against science. Nor did it portend the death of US high-energy physics, as some worried. The $260 million B Factory at SLAC garnered a healthy $36 million allocation to begin construction, and Fermilab received $25 million to continue with its Main Injector—out of $628 million for the high-energy physics base program.[173] But research at the multibillion-dollar, multi-TeV energy frontier was henceforth going to have to be a truly international affair, as ICFA had hinted more than a decade earlier—before the Wojçicki subpanel, encouraged by George Keyworth, suggested that the nation could go it alone.[174] "SSC is dead because the US Congress, the arm of government that decides what the United States can afford to spend, has now followed other governments in deciding that particle physics is too costly to be a national pursuit," chastised a *Nature* editorial. "The leaders of the US particle physics community must blame themselves for not appreciating that such a point would eventually be reached."[175]

SHUTTING DOWN THE PROJECT

On October 20, Fermilab Director John Peoples was attending a workshop at SLAC on the Next Linear Collider, a futuristic TeV-scale electron-positron collider.[176] That morning he visited SLAC Director Richter to discuss whether Fermilab might collaborate on this effort. As they talked, the phone rang. According to Peoples, it was Senator Johnston's chief of staff Proctor Jones, calling to tell Richter about the House vote the day before to kill the SSC. "Bevill can't carry the water anymore," Jones said ruefully. "It's over."[177] Hearing the SSC's death knell, Peoples quickly decided that he should be the one to shut the project down and told Richter so.

For over a month, Peoples had been considering the possibility of stepping in to take over as SSC director and attempt to save the foundering project. Since June, O'Leary had been threatening to wrest the management of the SSC away from URA and hand it over to an industrial firm.[178] Behind the scenes, DOE officials as well as members of the URA Board of Trustees and SSC Board of Overseers were quietly asking Peoples—then up for review of his five-year contract as Fermilab director—whether he might be willing to replace Schwitters.[179] Aware that the DOE and Congress were unhappy about how the project was faring, he was concerned that the SSC might pull all of URA, including Fermilab, down with it.[180]

When he returned to Illinois, Peoples met with Lederman, Fermilab Deputy Director Ken Stanfield, and Andy Mravca of the DOE site office to discuss shutting down the SSC. Then he began seeking allies, including Brookhaven Director Nick Samios, Sandia Laboratories President Albert Narath, and Frank Sciulli of Columbia—then chair and vice chair of the SSC BOO—who agreed to his plan. Panofsky tried to block the coup, continuing to support Schwitters, whose contract as SSC director was also up for review. But the URA review committee rejected Panofsky's moves and instead accepted the resignation of Schwitters, who recommended that Peoples replace him, later saying he felt "uncomfortable and inexperienced" in the role of "funeral director."[181] Thus the job of closing down the SSC fell to Peoples, who understood his principal task to be following congressional instructions for "the orderly termination of the Superconducting Super Collider Project."[182] On October 28, Schwitters resigned as SSC director in a letter to John H. Marburger, chair of the URA Board of Trustees. "I came to URA and Texas to build the SSC," he stated, "and that is no longer possible."[183]

Meanwhile, O'Leary had been communicating with Governor Richards, the Texas congressional delegation, John Gibbons, and Treasury Secretary Robert Rubin to determine what might be done with the SSC assets and to compensate the state.[184] In an October 27 meeting, Richards told her that "Texas expects full reimbursement for their $500 to $600 million investment in the project"—plus another $270 million for new research projects that might be spawned at the Waxahachie site.[185] That was much more than what Congress had appropriated to shut the project down. White House Chief of Staff Mack McLarty organized a working group to deal with all the SSC termination issues, including the most sensitive one of reimbursing Texas.[186]

The SSC leadership change took a week to become official, because it required O'Leary's approval. At a November 5 meeting in Washington, she rubber-stamped the URA plan to shut down the project. Refusing to speak with URA President Toll, she read from the document that Peoples, Lederman, Stanfield, and Mravca prepared on behalf of URA, making a few small changes and calling Peoples "a knowledgeable director" she trusted.[187]

That same day O'Leary sent a memorandum to Clinton summarizing the delicate Texas political situation as she understood it and asking his guidance.[188] She would be meeting with Richards and members of the Texas delegation during the next week on issues of importance to the Lone Star State, including the requested reimbursement, the legal basis for which was currently under study. Crucial issues to be resolved included a severance package for terminated employees, plus future ownership and use of site assets (e.g., the magnet labs). She recommended "that we develop a plan to reimburse Texas, in whole or in part, for its SSC investment," because of compelling legal and political reasons. In preparation for Clinton's meeting with O'Leary on November 8, a White House adviser wrote McLarty, "I agree with you that we should work with Texas, not because of any political rationale, but rather because it's the right thing to do."[189]

It took another week before Peoples was officially named SSC director, but in reality he was already leading both laboratories after November 5. Upon his return to Washington for a HEPAP meeting on November 8 (the same day Clinton was meeting with O'Leary), he spoke as the Fermilab director. Then he flew to Dallas to begin serving as SSC director and began the task of shutting down the project. Tom Kirk, who had been working as the SDC project manager, agreed to serve as his deputy during the shutdown. Peoples asked Siskin to remain on board to manage the complex SSC reporting system needed to keep congressional staff apprised of progress in terminating the project. The goal was to preserve as much as possible of the DOE investment in the SSC, while reducing the SSC staff of more than eighteen hundred to a minimum while finding jobs for as many employees as possible. They also had to close out about three thousand contracts and deal with the physical SSC remains, such as the magnet prototypes, tunnel sections, portions of the linear accelerator and booster complex, furniture, library, computers, and buildings.[190] Roughly $2 billion had been obligated so far on the SSC: $1.6 billion coming from the US government and the rest from the state of Texas. To support the work of shutting down the SSC, Congress had diverted the fiscal 1994 allocation of $640 million.

Peoples and Kirk divided up the responsibilities. Peoples dealt with contracts, terminating personnel, and anything politically sensitive, while Kirk coordinated with the DOE on distributing equipment and other useful assets. George Robertson, head of the SSC Administrative Services Division began seeking jobs for terminated employees. Aware of O'Leary's hostility toward Schwitters—and of his bitterness about the SSC cancellation—Kirk and Siskin urged Peoples to occupy the director's office before she was to arrive at the site.

Early on Friday, November 12, Peoples met with O'Leary, having reviewed the termination benefits DOE planned to offer SSC employees—including employees of EG&G and the architectural and engineering group Parsons-Brinkerhoff/Morrison-Knudsen. The terms, spelled out in a letter to Cipriano (who by then had left the project), were so harsh that Siskin was worried about O'Leary's personal safety during a large meeting with both federal and roughly one thousand contract employees in which she was to present the contract for termination. But Robertson persuaded her to ameliorate the terms before she met with Peoples.[191]

Peoples described this meeting with her, held shortly before the meeting with employees in the Central Facility, as "not a pleasant interview." As they reviewed the plans for terminating the URA leaders, she reproached him, saying, "I hear you've given all these guys golden parachutes."[192] He told her that Siskin was willing to resign, but it would be useful to keep him on board awhile to help with essential bookkeeping and complex reporting tasks. O'Leary agreed, but she was adamant that Schwitters had to leave.

FIGURE 6.10 Fermilab Director John Peoples at press conference announcing initial evidence for the top quark, April 26, 1994. Courtesy of Fermilab.

At the subsequent meeting with the SSC staff in the converted warehouse, O'Leary presented the revised contract with considerable flair. She brandished the letter containing the harsh terms of the original contract and dramatically ripped it to shreds, promising everyone that they would be taken care of. After presenting the revised contract, she introduced Peoples as the new SSC Laboratory director, calling him "a guy who knows how to run something."[193]

By November 18, the DOE had selected Oak Ridge National Laboratory to manage the project termination process, and O'Leary was leaning toward a 90-day severance package for SSC workers.[194] DOE officials ordered URA "to immediately close out existing cost accounts and begin collection of termination-related costs."[195] Peoples drafted a legally appropriate termination letter, signing the first four hundred of them before heading for Geneva to chair an ICFA meeting at CERN. By the end of the year, termination notices had been sent to over eighteen hundred SSC scientists and other staff. Peoples projected that only seven hundred staffers would still be employed at the SSC by July 1994, and that many of those would be "clerks, contract administrators, and environment, health, and safety workers."[196]

Many members of the high-energy physics community tried to help find jobs for the terminated SSC physicists. Schwitters relocated to the University of Texas at Austin, where he had earlier been named professor of physics. Fermilab hired eighteen of the SSC lab's one hundred and fifty high-energy physicists.[197] Peoples and Kirk recognized that some of the technical staff would be able to find good jobs in Dallas based on their computing skills, and that it would actually be cost-effective to spend some of the authorized close-out funds to provide them with advanced computer training.[198] According to Peoples, this approach saved the DOE approximately $160 million in unemployment compensation.[199] Robertson helped over a thousand employees find new positions.

As O'Leary was eager to close out all the SSC contracts quickly, that complex process began promptly and yielded a formal termination plan by December 1993.[200] Termination notices went out on the six major cost-plus-fixed-fee contracts, valued at $775 million. Peoples sorted out the details with Peter Didisheim, who had been a lead staffer on the House Science Committee but was now working for O'Leary. The detailed work of terminating these contracts was extremely stressful; as Didisheim and Peoples soon realized, they were at considerable risk if mistakes occurred in dealing with the countless legalities involved. For example, official DOE permission was required to terminate a contract; adhering to this requirement took substantial time and effort. But Peoples subsequently found a way around it because he did not need permission to write a letter of his *intent* to terminate a contract. Political deals, however distasteful, helped grease the wheels.

By January 1994 the SSC Central Facility was beginning to look abandoned. "Visitors can stroll a long way into the maze of cubicles at the SSC before they spy a human face," observed a *Dallas Morning News* reporter. "The folks who stay on say it's like working in a morgue."[201]

Kirk meanwhile worked on distributing equipment and other physical assets valued at $10 million to $30 million, whenever possible in ways that would benefit the US high-energy physics program. One idea was to adapt the SSC cryogenics facility to aid US participation in the LHC project at CERN (see chapter 7); another was to use components of the linear accelerator to upgrade Brookhaven facilities.[202] The discussion continued for months with many participants. Over a hundred gathered in late March 1994 to learn more about the SSC facilities and equipment to be dispersed. They "dreamed of inheriting computers, machine tools and testing equipment."[203]

Restoring the land at the SSC site was complicated by costly legal issues such as eminent domain, which took months to resolve. One tricky problem concerned filling the nearly fifteen miles of tunnel already bored. Oak Ridge officials overseeing the lab shutdown suggested an expensive solution, expecting to let a costly conventional construction contract to fill the tunnel. But Peoples stood fast and refused to risk another fatality, as had occurred in January 1993 (see above). Siskin and Parsons-Brinkerhoff officials instead suggested a much cheaper, safer solution: plug only the shaft openings and let the tunnel fill with water; this task was eventually accomplished by filling the openings with drilling spoils on hand. As there was no longer much need to move dirt, Oak Ridge lost interest.[204]

All the while, Texas politicians viewed the SSC shutdown mainly from a financial perspective, trying to recover the state's investment of over $500 million. Governor Richards appointed a group of experts to serve on an advisory committee about what to do with the site and the SSC's other physical assets. This committee, chaired by National Academy of Engineering President Robert White, and including Frank Press, past president of the National Academy of Sciences, initially sought uses that would benefit both US science and the state of Texas.[205] Among its major suggestions were: a superconductivity research center, not only to design and build magnets for future physics projects such as the LHC but also for research on energy storage and conversion; a center for high-performance computing; and a proton cancer-treatment center combined with a radioactive-isotope research facility. These proposals drew on the advanced science and technology capabilities developed at the SSC Laboratory—from the building of superconducting magnets and advanced computer facilities to the linear accelerator, which could be adapted to supply low-energy protons for cancer therapy.[206]

The White panel also suggested building a research and education center near Waxahachie and retaining enough skilled scientists and engineers to staff it. Yet another suggestion, endorsed by the Nature Conservancy of

Texas, was to convert the site into a prairie-restoration project "with grasses several feet high, scores of wildflowers and a small herd of bison."[207] Later in 1994, geologists proposed building a laboratory in the tunnel to study underground processes in the Austin Chalk and the Eagle Ford Shale layers. The White panel recommended that, in addition to the $6 million offered by the DOE, the federal government should allocate another $44 million to support 125 scientists and staffers to develop detailed proposals over the next two years.[208]

Governor Richards and other Texas officials were also intent on recouping whatever they could of the cash and land that the state had given the SSC lab. Texas had sold some $500 million in general-obligation bonds, without guarantees of any cash reimbursement in the original agreement between Texas and the DOE, which said only that if the federal government terminated the project, then the state could claim equipment or facilities to which it had contributed 51 percent or more of the cost.[209]

Arguments about what to do about the debt owed Texas and about the SSC remains grew increasingly strident. Texans were accused of playing politics and attempting to benefit financially from the project's death. In March 1994, during the first congressional hearing on the termination, for example, Boehlert accused Texans of trying to profit from the SSC's demise. Representative Barton angrily rebutted this claim in an opinion piece, trying to refute him with "the facts." He criticized Boehlert's rhetoric as "counterproductive to an honest dialogue" between the two "partners in the SSC"—Texas and the federal government. Barton claimed that Texas has "always done everything we were required to do," pointing out that "approximately one of every four dollars spent on the SSC came from Texans. He also said it was appropriate for the SSC lab to do what it could to "recover a fair settlement for our investment" and to also try "to convince the Energy Department and Congress to put these valuable assets to some productive use."[210]

By this time Peoples was nearing his limits of tolerance. Concerned that the challenges of closing down the SSC project would impair his ability to lead Fermilab effectively, he told Didisheim that he wanted to resign and return to Illinois. And intriguing experimental results on the top quark were beginning to emerge there (see fig. 6.10 and below). But it still took until July for him to transfer the remaining responsibilities of directing the SSC shutdown into Robertson's hands.[211]

July 1994 was also when the SSC termination agreement between the DOE and Texas became public. As announced on July 22, Texas (by then credited with investing $539 million in the SSC) was to receive $210 million in cash and $510 million worth of land and facilities—including buildings and equipment. Of the cash, $145 million would be awarded to Texas in a lump sum, out of which about $4 million would go to repay Dallas, Tarrant, and Ellis counties for the expenses of land acquisition. The remaining $65 million was designated to aid in adapting the linear accelerator for the proposed cancer therapy and radioisotope-production facility. The University of Texas Southwestern Medical Center expressed interest in operating the center and developing additional cancer-therapy programs there. Prospects included developing a high-performance computer center on the site, using the technology in place.[212] Texas received the SSC Magnet Development Laboratory, whose magnet fabrication line and cryogenic plant might be useful in helping the United States remain on the cutting edge of superconducting magnet technology. In return for these concessions, the state agreed to release the federal government from any claims against it for terminating the SSC. On August 1, the Texas National Research Laboratory Commission voted unanimously to accept the settlement.[213] In November 1994 Congress followed suit and approved the terms, which were then included in the 1995 appropriations bill.

Over the next few years, however, these hopeful scientific, technological, and medical uses of the SSC assets foundered on the high costs of implementing them. Funding to create the underground geology laboratory, estimated at $20 million–$55 million, could not be raised.[214] Even though the high-performance computing center could build on the $55 million invested in sophisticated computing capabilities, it also proved too expensive. The cancer treatment and radioisotope-production center, estimated to require about $62 million to complete, was considered financially risky. Although the US government had offered $65 million toward building the center, it was unclear who would use it, as no other medical facilities were close by. Thus it was not clear that the center would be able to generate sufficient revenues to cover its operating costs after completion.[215]

By May 1995, when it became clear that none of the suggested ideas for using the abandoned SSC facilities could be supported, Texas officials began to liquidate the SSC assets, selling all buildings, equipment, and state-owned land, so that the state could use the proceeds (plus the additional $65 million offered by the federal government for building the cancer center) to start paying down the state's bond obligations.[216] Science educators received some benefits in June 1995, when the DOE agreed to give $1.8 million worth of SSC equipment, including spectrum analyzers, mobile laboratories, and office furniture, to the Texas Science Education Collaborative. Each group within the collaborative, consisting of ten school districts and the University of North Texas, agreed to invest $100,000 toward building a computer network and curriculum.

A month later, the TNRLC gave Ellis County the farmland donated for the SSC. County commissioners considered many possible options for the site, including a community college campus and a youth exposition center, but nothing ever came of them.[217] The claims filed by the county and the Ennis and Waxahachie school districts against the DOE and federal government were finally settled in the late summer of 1996. The districts sought payment of the property taxes they had lost because the SSC land, being part of a government project, had been exempt. The county and two school districts received about three hundred acres of property, buildings, and equipment to settle their claims. Texas had to sell the thousands of acres of land set aside for the SSC,[218] but it took until 1997 before the former SSC buildings were actually put on the market.[219] In 1998, the Texas General Land Office sold the SSC's Central Facility for $10 million, bringing the total received for sale of SSC property to $14.7 million. Texas officials hoped that by the end of that summer, all but 4,500 acres of the project's remaining property would be sold.[220]

Many other suggestions for use of the SSC site were proposed and discarded during the 1990s, including converting the tunnel into a wind tunnel, wine cellar, exotic mushroom farm, or a theme park called Six Flags Under Texas. But a good use for the defunct SSC Laboratory was finally achieved in 1999, when the site became part of a picturesque movie set used by Belgian martial artist Jean-Claude Van Damme in production of a sequel to his 1992 science-fiction action adventure *Universal Soldier.* As his location manager Robert Callan remarked, "The thing that's neat about the Super Collider was that we were able to use some buildings for sets, others for our lumber shop, special effects, transportation and set dressing." The *Dallas Morning News* noted that this move gave Ellis County taxpayers "hope that the hungry white elephant they inherited from the feds could be turned into a film studio."[221]

At the turn of the century, after over a decade on the turbulent US political scene, Waxahachie was finally reverting to its languid past as a good movie location, set amid the fallow cotton fields south of Dallas.

The October 1993 congressional termination of the Superconducting Super Collider was a decisive turning point in the evolution of Big Science in America.[222] It marked the official end of the aging Cold War model of doing large-scale physics projects, with the United States taking the lead and other nations climbing aboard as junior partners. Henceforth, such multibillion-dollar enterprises—for example, the Next Linear Collider, which was soon renamed the International Linear Collider—had to be organized from the outset as truly international partnerships.

Had Bush won the 1992 presidential election instead of Clinton, however, the history of the SSC could have been quite different. In that case, a veto pen would likely have been awaiting the energy and water bill in October 1993. And despite its deepening recession, Japan might well have joined the project as a full partner that year, too, making up the difference between what the SSC actually needed for fiscal 1994 and the $800 million or so in the US budget.[223] That involvement could have changed the attitude in the House; it would certainly have dramatically reduced the voting margin. Historians are not supposed to dwell on such counterfactuals, but we do so here to underscore the contingencies of history. The old, Cold War ways of achieving Big Science projects might have worked one more time had a few pivotal events turned out otherwise.

But Clinton and his administration were taking the United States in new directions while trying to reduce federal budget deficits then projected to exceed $300 billion. With the firewalls finally down from the 1990 Budget Enforcement Act, they began moving R&D funding from the military back into civilian purposes, trying to achieve a 50-50 balance and reverse the effects of Reagan's 1980s buildup.[224] Applied science and technology development to enhance national competitiveness were the winners in this new post–Cold War climate, as were biological and medical research, while basic physics—especially an esoteric, nationalistic high-energy physics project—was the big loser.[225]

For a president for whom the SSC had had high priority, it would have been relatively easy to pull a few hundred million dollars a year out of military R&D and devote it to the project in the mid-1990s, when construction costs were projected to peak near a billion dollars per year. Getting Congress to go along with this transfer in a time of acute fiscal austerity was of course another matter entirely. But in the post–Cold War era then beginning, high-energy physics lost the privileged status it had long enjoyed. The once-influential discipline had become just one of many special interests vying for its slice of the federal pie. And high-energy physics could make only a feeble case for its role in enhancing US competitiveness—since Reagan and Keyworth the mantra of the new economic era—based mainly on potential spin-offs of the technologies it might develop or foster.[226]

After 1992 the SSC no longer had the staunch support of the Oval Office that it had enjoyed during the Reagan and Bush years. No longer a presidential priority, the project had to struggle for funding within the DOE laboratory system—and even within high-energy physics itself. In what had become a zero-sum game, the B Factory and Main Injector siphoned off $61 million that *might* have been devoted instead to the SSC. Add to that another $17 million to start the Advanced Neutron Source at Oak Ridge and similar amounts to continue building RHIC at Brookhaven, the Advanced Photon Source at Argonne, and other projects, and one can track down about $100 million that might have been spent instead on the SSC.[227]

Here again we are dwelling in counterfactuals, but they help illustrate the centrifugal political forces that tugged relentlessly on the SSC budget. The annual appropriations process is a give-and-take (mostly take) affair leavened only a little by reason and governed mainly by raw political power.[228] With the Cold War ended and only a weak argument able to be made about the likely SSC impact on economic competitiveness, it was impossibly difficult for Congress to rationalize spending another $10 billion or so to complete a poorly managed Big Science project that few American taxpayers really understood. Given the results of the 1992 elections, both congressional and presidential, the death of the SSC in October 1993 was probably inevitable.

CHAPTER SEVEN

Reactions, Recovery, and Analysis

There comes a time when the cost and effort of the next accelerator are so high that there may be no other way than world cooperation.

—VICTOR WEISSKOPF, May 1984

The immediate reactions of US high-energy physicists to the termination of the SSC by Congress included shock, disbelief, and dismay. Caught largely by surprise at this rejection, many could not believe that the US government, which had generously supported their increasingly expensive activities for decades, could suddenly reverse course and turn its back on their most important project. The killing of the Super Collider was "a tragedy for the field and everyone in it," said Lederman—widely recognized as the project's godfather—who sounded what became a common refrain. "The government decided, in its wisdom, that high-energy physics has no future" in the United States.[1]

At the SSC lab, the reaction was one of stunned disbelief at how far the huge project had fallen from grace since it came to Texas nearly five years earlier. The termination was "an unbelievable waste and tragedy," observed Schwitters ruefully. "You've got people who've left very secure, attractive positions to come here, and the thought of just dumping them on the street is very troublesome."[2] Trying to come to terms with the vote, he observed that "we were held to new standards of explaining ourselves to the public and . . . from now on, all science will have to be justified in stronger terms."[3] But achieving that was not going to be easy. "The whole point of science is to make new discoveries, and if [you] haven't made them yet, how can you explain them?" Schwitters asked. "America is losing its will to do the risky, important endeavors. It's essentially becoming a society that wants to reduce risks at almost any cost."

SSC staffers were "devastated" by the congressional decision, according to Public Affairs Director Russ Wylie. "We're deeply disappointed that so many smart people could do such a dumb thing."[4] Michael Barnett, an LBL physicist on the SDC experiment, lamented that "American science will never be the same." GEM Collaboration spokesman Barry Barish was more reflective. Without the SSC, "the most pressing questions won't be answered," he said. "The field won't be as exciting."[5] Steven Weinberg blamed recent congressional budget-slashing fervor. "Overwhelmingly, this was a year that many members needed a symbolic act of budget cutting," he told *Science*. "The SSC was a large project that many felt their constituents didn't care about."[6]

Other physicists focused on the deteriorating relationship between science and its federal backers.[7] Stanley Wojcicki of Stanford, then chair of HEPAP, as well as the chairman of its 1983 subpanel that recommended pursuing the SSC (see chapter 1), called the abrupt reversal "a very radical change in the partnership between the federal government and the particle physics community that has gone on since World War II and the Manhattan Project."[8] SSC BOO Chair and SLAC Director Emeritus Panofsky later expanded upon this theme in a letter to *Physics Today*. "Major scientific laboratories operated by universities for the government after World War II were not simply contractors but *partners* with the government," he observed. "But this postwar partnership is eroding."[9]

Stanford physicist David Ritson, who had consulted with the SSC lab, doing computer simulations that justified increasing the dipole aperture to 5 cm (see chapter 4), offered a personal analysis of this altered relationship. In a December *Nature* article, he gave several external reasons for the demise, including the inability to attract sufficient foreign SSC contributions.[10] But he focused on the internal strife at the lab between high-energy physicist and military-industrial cultures—epitomized by the "intense animosities" between the DOE officials and physicists, especially Cipriano and Schwitters. "Could this outcome have been avoided?" Ritson asked, referring to the SSC's termination. "When a ship that is undermanned and poorly prepared encounters a violent storm and sinks, one can never say that even if it had been fully manned and well equipped it would not have sunk," he observed, trying to answer his own question. "All one can say is that it would have helped."[11]

Nobody was more prolific in trying to account for the SSC's demise than Lederman, who published a steady stream of articles and letters in the months that followed. "We live in a nation that grew rich by exploring and settling its frontiers," he wrote in November for the *Chicago Tribune*. "Isn't the supercollider a sort of wagon train into the frontier of our comprehension of the universe?"[12] (By implication, then, the expedition must have been ambushed by savages.) Lederman briefly addressed the charges of mismanagement and the impact of opposition to the SSC by other scientists, but he returned to his central, macro-political theme that the nation was unfortunately withdrawing from the risks of scientific exploration. "Thus," he concluded toward the end, "the SSC decision may be viewed in the context of a national mood that favors immediacy at the expense of long-term investment."

Such lofty sentiments ignored the facts that the SSC project *was* poorly managed, and that its inexorably increasing budget was beginning to crowd out other good research, especially at other DOE labs. Few high-energy physicists acknowledged this connection, though it was implicit in some of their comments. Doubling the total project cost "was the greatest cause for killing it," said SLAC Director Richter at the AAAS annual meeting in February 1994, calling the SSC "one of high-energy physics' greatest failures" and claiming it "could have been built in six years" at substantially lower cost.[13]

The American Institute of Physics offered almost a dozen reasons for the demise in an e-mail notice titled "Looking Back: Why the SSC Was Terminated."[14] Topping the list was the steadily escalating SSC cost. Nowhere to be found on this list, curiously, was the vocal opposition of other, primarily condensed-matter physicists, which had so divided the community. Lederman alluded to their impact obliquely in "An Open Letter to Colleagues Who Publicly Opposed the SSC," published in *Physics Today*, while extending them an olive branch, saying he did "not believe that you are dancing on the SSC's grave."[15]

Over the next few months, detailed articles on the SSC's collapse agreed that the escalating cost of the project in a time of heightened austerity was the principal reason for its demise. In addition to a brief *Science* article, veteran SSC-watchers Gary Taubes and Irwin Goodwin weighed in with lengthy analyses in the *New York Times* and *Physics Today*. "Many researchers, including physicists, were privately elated at the death of a project they believed was furiously swallowing research dollars that were in short supply," wrote Taubes.[16] There were contributing factors, too, such as the apparent SSC mismanagement, the lukewarm support of the Clinton administration, and the staunch opposition of the House freshmen,[17] but the seemingly unrelenting cost growth was the real showstopper. "For most members of Congress," Goodwin wrote, "the prospect that the SSC would reach a total cost of at least $11 billion was the main reason for turning it off."[18] And it did not help matters that there were as yet no firm commitments of any major foreign contributions to the project.

But Tom Siegfried, the science editor of the *Dallas Morning News*, who for years had followed the SSC from a Texas perspective, laid the blame largely at the feet of a growing anti-intellectual, antiscientific attitude that he saw emerging in post–Cold War America. "Congress has declared war on basic scientific research, and the Superconducting Super Collider is the first fatality," he complained. "Science is an easy target for an anti-intellectual society obsessed with cutting a budget deficit that science has nothing to do with. . . . Now that the Cold War is over, anti-intellectualism is reasserting itself against science, often dressed in humanistic rhetoric but dedicated to the proposition that ignorance, prejudice and superstition are better guides to action than knowledge."[19]

For SSC foe Boehlert, such high-minded concern about the future of basic scientific research in the United States was overwrought. Congress was not about to renege on its commitments to basic research, despite killing the SSC, he said. Well before construction began in 1990, the House approved a bill (H.R. 4380—see chapter 5) that "capped Federal spending at $5 billion and required foreign contributions to cover 20 percent of the costs," countered Boehlert in replying to a *New York Times* editorial. "The Department of Energy's decision not to take those requirements seriously sealed the project's fate."[20]

Soul-searching was indeed going on at the DOE, but one could not find it in an opinion piece by former Energy Research Director Happer—then back at Princeton. The SSC staff "managed to make good progress on construction while coping with stifling layers of management and human waves of investigators, each determined to send damaging reports back to Washington," he argued.[21] Recalling his rationale for convening the Townes panel (see chapter 6), he went on to criticize the greed of the US high-energy physics community and its insensitivity to the austerity afflicting many sectors of the economy. "While others in America—from construction workers, to military personnel, to materials scientists—were coping with layoffs, it was argued that a substantial buildup of new scientists and engineers was essential for the SSC and the well-deserved growth of existing programs," Happer wrote. "This attitude did not win friends in the rest of the scientific community, and it made it easy for SSC opponents to pose as defenders of less-favored fields of science." In fiscal years 1993 and 1994, in fact, federal outlays for all of high-energy physics exceeded $1 billion—at a time when funding of almost every other science was stagnant or being cut back.

In early 1994 Daniel Lehman, who had replaced Ed Temple as head of the Office of Energy Research's construction management division, was giving talks on the "Lessons from the SSC" that offered a more nuanced interpretation of the interactions between DOE and the SSC lab. He included "increased perception of poor management by *both* DOE and SSCL" as one of four principal reasons for the ebbing public and congressional support.[22] A serious issue was the departure from the "traditional style of building facilities," which allowed customary DOE management oversight practices to be bypassed by Watkins and Cipriano, who had a "*direct reporting line* to [the] Secretary."[23] This cozy relationship meant that the usual DOE checks and balances could be bypassed, too. And it allowed a "considerable DOD style" to infiltrate and influence the decision-making processes at the lab—exacerbating communication problems among headquarters, the DOE site office, and the SSC leadership. Lehman did not explicitly link these problems to the SSC cost growth, although the connection seemed implicit in his remarks.

But in a letter to *Physics Today*, Doug Pewitt, who had served in high-level staff positions in DOE, OMB, and OSTP before joining the project in 1989, put the blame squarely on the shoulders of the physicists managing the new laboratory. Homing in on the "revenge of the C students" comment made by Schwitters, he observed that "this public derogation of the SSC's benefactors, no matter how deeply felt, clearly contributed to the Lab's demise."[24] To him, there was plenty of evidence for the project's mismanagement and of the SSC leaders' reluctance to follow "the bureaucratic procedures that are now part of all large publicly funded projects"—such as a smoothly functioning cost-and-schedule control system. "The SSC did not fail because of lack of motivation or effort," he wrote: "Good intentions, hard work and physics expertise are not enough; success in such an ambitious undertaking demands more. Dealing with Federal sponsors in a manner they find unacceptable and refusing to comply with disagreeable requirements for publicly funded projects is a certain prescription for failure. Success requires much greater attention to the imperatives of large project management."[25]

In December 1994 the Dingell Committee staff finally weighed in with its analysis of events based on its June 30, 1993, hearing and on related reports from the GAO and the DOE Inspector General's Office.[26] As Lehman had done earlier, this damning document blamed *both* URA mismanagement *and* ineffective DOE oversight for many SSC cost overruns. "The performance of both the DOE and URA left much to be desired," its executive summary began.[27] "The project suffered from poor management, inadequate cost controls and cost uncertainties. Management problems plagued the SSC from its inception through its demise." By setting up a DOE oversight structure in which Cipriano reported directly to the Secretary, the report continued, the project was shielded from normal oversight procedures; so "many of the questionable actions the DOE project office and URA undertook went largely unchallenged."[28] As had GAO and others, the report underscored the lack of a fully functional cost-and-schedule control system, which "enabled DOE and URA to present inaccurate information to Congress and prevented adequate oversight until the problems grew so large that they no longer could be hidden."[29] The DOE also exaggerated the anticipated foreign contributions to the SSC detectors, according to the report, which obscured likely cost overruns in those areas, too. In all, it was a devastating commentary that went unchallenged in the press because nobody was left in Waxahachie to rebut it, and few at DOE cared much about the Super Collider anymore. It was history.

Just a year into the Clinton administration, an altered landscape for US science policy had emerged that reflected post–Cold War political and economic realities. Indeed, Clinton had sounded this theme in his February 17, 1993, State of the Union address and the document "A Vision of Change for America" released that day. "Today we must once again find the courage to change," he exhorted. "We must shift our energies from the Cold War priorities of the past to the economic priorities of the future."[30] Following the preferences of Gore, Gibbons, and the Democratic Congress, research funding was about to fall in defense-related areas and begin surging for the environment, medicine, and technology.[31] Being the best in the world in an expensive, esoteric discipline with little direct impact on health, jobs, or industrial competitiveness was not high on their priority list.

Barbara A. Mikulski, the Maryland Democrat who chaired the Senate Appropriations Subcommittee that controlled NASA and NSF budgets, became a prominent spokesperson for the new priority, which she liked to call "strategic research." Instead of striving to fund the best US science and scientists possible, she suggested, perhaps the NSF should "reorganize into a series of institutes on manufacturing, global [climate] change, high-performance computing and other strategic areas."[32] Such a mission-oriented approach resembled what had been accomplished so successfully at Bell Labs in the two decades after World War II, resulting in transistors, lasers, and satellite communications.[33] But it would likely leave more basic sciences like high-energy physics out in the cold.

Mikulski explained her position more fully in an April 1994 *Science* article that included a long commentary on the ill-fated SSC:

> Look what happened to the Super Collider. I voted to keep the Super Collider, and have consistently done that because of its importance in basic physics research. But my colleagues saw another situation. They saw that by one vote on their part, they could cut $8 billion from the budget and not keep one homeless person out of a shelter, not keep one veteran from having his disability benefits, and not keep one school child from having a school lunch program. The Super Collider represented what has happened in science. A wonderful scientific idea, funded with the best intentions, but gone awry. At every turn, there was another cost overrun, a technical complication, or the hubris of the people who ran it who refused to see the situation facing them. No one could adequately articulate how it fit into our national strategy.[34]

In the rest of her article, Mikulski focused on the new world order emerging after the collapse of the Soviet Union and on the appropriate role of US science within it. National security was to be replaced by economic security in the new science policy. "People knew how important it was to win the Cold War," she argued. "We now have to show how important it is to win the economic war." This was a clarion call for scientists to man the barricades of economic competitiveness. "We must focus our science investments more strategically around national goals that are important to economic growth," she explained. "We must train our scientists and engineers, whether they are undergraduates or Ph.D. candidates, so that they are ready to work in strategic areas in the private sector."[35]

Obtaining financial support for basic, or "pure," science has always been difficult in pragmatic, results-oriented America, which has embraced science ever since the nation's Enlightenment origins mainly because of its promise of human betterment. Historian Daniel Kevles voiced such a theme throughout his landmark book *The Physicists*.[36] But during the Cold War, physicists—especially high-energy physicists—managed to circumvent this disinclination and secure generous federal funding for their expensive, purely scientific research, largely because of the close association of their work with national security.[37]

In late 1995 Kevles published a balanced, thoughtful historical analysis of the SSC project and its demise in the preface to the revised edition of his book. Also excerpted in a Caltech magazine, this article returned to his theme of pure and pragmatic physics in America. "In the context of the Cold War," he wrote, "particle physics provided an insurance policy that if something important to national security emerged unexpectedly, the United States would have the knowledge ahead of the Soviet Union."[38]

Physics research benefited greatly from Reagan's determination to fight and win the Cold War. "Although federal funding for all of physics had declined through the seventies," wrote Kevles, "it had been rising dramatically with the Reagan administration's defense buildup . . . and with the national absorption with economic competitiveness."[39] Unfortunately, following urgings of Reagan science advisor Keyworth, the US high-energy physics community elected to hitch its SSC wagon train to this 1980s Reagan buildup.

But with the collapse of the Soviet Union and the end of the Cold War during 1989–91, this rationale had evaporated, almost overnight. High-energy physicists began to cast about for new rationalizations for their expensive project in a nation then struggling with a recession and the growing industrial competition from Europe and Japan. But a good, solid rationale could not be found. "Missing at the national level was what had made physics, including its high-energy branch, so important since World War II—real or imagined service to national security," argued Kevles. "The point was that the SSC had no direct bearing on national security."[40]

"In the end, the collider resolved into a creature of Cold War conservatism at a time when the majority of Congress—both liberals and conservatives—was undergoing a fundamental change to a post–Cold War political order," he observed.[41] The nation was turning inward once again, trying to address serious problems that had been overlooked during decades of East-West confrontation. But America was not about to turn its back on science, Kevles concluded:

> The vote against the SSC was thus not a vote against science or for an end to the longstanding partnership between science and government; rather it signified a redirection of the partnership's aims in line with the felt needs of post–Cold War circumstances. Emphasis would go to what policymakers were calling "strategic" or "targeted" areas of research—fields likely to produce results for practical purposes such as strengthening the nation's economic competitiveness or its ability to deal with global environmental change.[42]

THE FUTURE OF US HIGH-ENERGY PHYSICS

Almost immediately after the SSC's October 1993 termination, US high-energy physicists had begun to conclude that their long-term future would lie, if anywhere, in Europe. For the next decade or so, the collider facilities available or then under construction at Fermilab and SLAC, including the Main Injector and B Factory, could support vigorous, rewarding experimental programs. But the energy frontier of multi-TeV proton interactions had suddenly been closed west of the Atlantic by congressional fiat. If they wanted to probe this frontier, US high-energy physicists now had to make arrangements to work at CERN on its proposed Large Hadron Collider. Many were already doing so.[43]

A day after the fateful House vote, DOE Secretary O'Leary was on the phone to SLAC Deputy Director Sidney Drell, seeking his advice on what to do next.[44] Drell soon advocated a major US move to CERN. "The only good thing to come out of this will be more international collaboration," he told a reporter from *Nature*. "It's important that we move forward on this frontier of research, and the LHC is the one machine that we know how to build."[45] Thus the cancellation of the Super Collider finally forced the US high-energy physics community to embrace seriously what it might have considered doing nearly a decade earlier at the 1984 ICFA workshop in Japan (see chapter 1)—and was also suggested as a less expensive alternative to the SSC by the Congressional Budget Office in October 1988 (see chapter 3, esp. n. 83). But those were times of far more supportive congresses and presidential administrations that encouraged high-energy physics community leaders to believe that the United States could go it alone on a multibillion-dollar basic-physics project that few outside the field really understood or appreciated.

On November 4, 1993, O'Leary wrote HEPAP Chairman Wojcicki a letter asking the panel to address the question of "defining a long-term program to pursue the most important high-energy physics goals now that the SSC has been terminated."[46] She tasked them to consider "the options for a truly international framework for construction, operation and utilization of future high-energy physics research facilities," and specifically requested recommendations on the "practical steps necessary to facilitate enhanced international collaborations for the construction of large high-energy research facilities in the future." Wojcicki subsequently named a subpanel of sixteen members, chaired by Drell, to develop an authoritative report on a "future vision for high-energy physics," to be submitted by May 1994 so the DOE could go to Congress with a new plan that summer.[47]

Later that month, O'Leary and House Science Committee Chair George Brown published a widely read opinion piece in the *Los Angeles Times* about the primary lessons to be learned from the SSC cancellation. "The Superconducting Super Collider as we know it is now dead, yet the quest for a comprehensive understanding of the world around us lives on," they began.[48] "The scientific questions that compelled development of the SSC will not suddenly disappear, nor are they likely to be answered by anything other than a 'big science' endeavor during the next century." But the SSC had suffered mightily from its inability to attract significant foreign contributions.

"The obvious lesson to be learned is that foreign participation must be incorporated into large-scale science and technology pursuits from the very beginning, when prospective partners will have a say in why, where, when and how such projects will be pursued." By unilaterally killing the SSC—as well as withdrawing from other "international" scientific and technological projects—the United States had, however, established its reputation as an unreliable partner in such joint endeavors. "With the help of a blue-ribbon panel on the future of high-energy physics, we are now putting the pieces back together from a project that blew apart after an extraordinary investment of human and national resources." O'Leary and Brown concluded, "Unless we are intent on stopping the pursuit of knowledge that it would have delivered, we must find a way to achieve a truly international framework for large scientific and technological projects."

Europeans were generally favorable to the idea of US participation in the LHC—an idea that CERN's then Director General Herwig Schopper had voiced before Congress as early as 1987 (see chapter 3). "Why build the SSC when the LHC could either demonstrate that the top quark and the Higgs boson exist or suggest . . . how they might be found?" asked *Nature* in late October 1993; its editors advocated that "the United States should pursue the option of a deal with CERN."[49] UK Science Minister William Waldegrave proposed that "CERN should immediately invite the US as a member state," although, he admitted, he had not yet "thought through all the constitutional niceties" of such a move.[50] European physicists were largely in agreement that the United States and Japan should join the LHC project as partners, too. "I personally think CERN should invite the US and Japanese to be more involved in the Large Hadron Collider, and on good terms," said theorist Luciano Maiani, president of Italy's Istituto Nazionale di Fisica Nucleare, who was to become CERN director general in 1999.[51]

Although the projected costs of the LHC, not including the experiments or labor costs for CERN employees, were then estimated at "only" $1.8 billion, this was still a difficult pill to swallow for some European governments.[52] The funds had to come from the annual CERN budget, not additional assessments on its member nations. In particular, German and Spanish delegates to its governing council had expressed reluctance to proceed with the LHC because of economic problems.[53] So inviting the United States and Japan to join CERN, or at least the LHC, could relieve budgetary pressures and allow the project to go forward. But admitting either nation as a *full* member might allow it a dominant role in CERN politics, as each nation's contribution is a fixed percentage of its gross domestic product. Making direct contributions to the LHC project in return for guaranteed access to the facility seemed a better option, but not without its own problems. And getting the US Congress to go along after Europe had largely shunned the SSC might be another possible difficulty. As Oxford theoretical physicist Chris Llewellyn-Smith, then about to replace Carlo Rubbia as CERN director in January 1994, deadpanned, "The first half of next year will be interesting, from a diplomatic point of view."[54]

In the United States, a major concern before the Drell subpanel was how to fund a large US contribution to the LHC without hurting the existing labs in the high-energy physics base program. Something like $60 million to $100 million a year would be needed during LHC construction, and contributions to operating costs after that would be welcome, too.[55] With Fermilab and SLAC proceeding apace with the Main Injector and B Factory, it was difficult to see where such funding might come from without delaying one or both projects—or asking Congress to increase the high-energy physics budget to accommodate an LHC contribution. But physicists were hopeful that the federal government would understand that this was now the *only* way US physicists could ever hope to work at the highest energies. "It's still the last frontier of physics," stated William Happer in reply to a reporter. "As a nation that has prided itself on attacking every frontier, I can't believe we would back out on this."[56]

In late May the Drell panel issued its much-awaited report on the future of the US high-energy physics program, which HEPAP unanimously approved and forwarded to O'Leary.[57] It recommended continued federal funding as planned for the Fermilab Main Injector and the SLAC B Factory, plus smaller upgrades at Brookhaven and Cornell. To support a significant US contribution to the LHC, the subpanel recommended "a modest funding increase" of $150 million over the existing base budget of about $630 million, to be spread over fiscal years 1996–98.[58] As a graph of US high-energy physics funding in the report showed, DOE support for the base program had been declining in real terms during the years of SSC construction (see fig. 7.1); thus additional funding would be needed to reverse this trend and get US physicists involved in LHC construction. If this extra funding did not materialize, deep cuts would have to occur—most likely on the Main Injector or B Factory—to permit LHC participation.

When Drell testified about the report before the House Science Committee a few days later, Boehlert commended his panel "for being realistic," and added that "I haven't always seen this in the high-energy physics community."[59] But there was no guarantee that congressional appropriators would go along with the extra funding—especially after CERN and its previous

HIGH ENERGY PHYSICS FUNDING
(1960-1995)

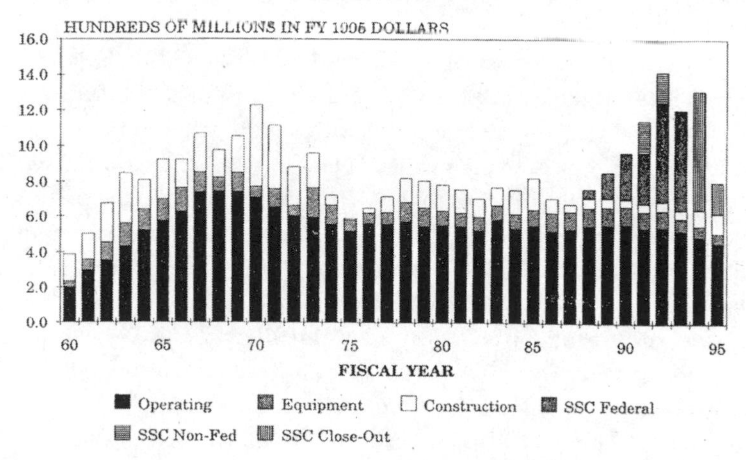

FIGURE 7.1 High-energy physics funding patterns from fiscal 1960 to 1995, in constant 1995 dollars. Source: *Report of the 1994 HEPAP Subpanel on Vision for the Future of High-Energy Physics.*

Director General Rubbia had helped undermine support for the SSC. If this funding bump did not materialize, Drell recommended canceling one of the ongoing domestic projects to support a major LHC contribution. But such an amputation would be painful, remarked Wojcicki: "It's like cutting off your arm or your leg."[60]

It took nearly four years, however, before the United States finally signed a formal agreement to join CERN as a special Observer State making $531 million worth of in-kind contributions to the LHC project,[61] joining nineteen European nations plus Canada, India, Israel, Japan, and Russia. Much had happened in the interim—including the discovery of the top quark at Fermilab (see below)—and political hurdles had to be surmounted on three continents.[62] At first, CERN was seeking up to $600 million from the United States, since five hundred US physicists had joined the two general-purpose LHC experiments, ATLAS and CMS, and more were certain to follow; they outnumbered any other national contingent at CERN. But the DOE initially hewed to the $400 million figure cited by the Drell panel, and different accounting methods employed by the two sides made negotiations difficult.[63] In March 1996 DOE finally blinked, announcing it would allocate "$450 million, give or take $50 million, over the next 8 to

10 years as the US contribution to the LHC and its two large detectors."[64] With an additional $80 million from NSF for the experiments, the US total now came close enough to CERN expectations. Negotiations could proceed to the next level.

"Particle physics is back on track," said a relieved Llewellyn-Smith.[65] The US contribution was to be divided almost equally between construction of the collider and detectors—both areas in which US physicists had special expertise from the SSC efforts. US funds were in fact already being spent for R&D in these areas, mainly at Brookhaven and Fermilab. The added funding would also allow CERN to build the collider in a single step, going directly to its full design energy of 14 TeV rather than following a two-phase "missing-magnet" scheme in which only half the superconducting dipole magnets were to be installed at the outset (see epilogue).[66]

But one hurdle remained to be cleared: the US Congress, which was cool to the idea of sending half a billion dollars overseas—even though most of this was going to be spent on US soil for LHC components. The midterm elections of 1994 had witnessed yet another upheaval in the House, with the Republicans now in the majority and taking over control with Georgia Congressman Newt Gingrich as the new Speaker after his successful "Contract with America" campaign. New spending for physics was not high on his list of priorities. And embittered Texas Congressman Barton threatened to spearhead a House drive to block the DOE agreement with CERN because European countries had snubbed the SSC.[67]

Congressman James Sensenbrenner (R-WI), who had replaced Brown as chairman of the House Science Committee, challenged the agreement DOE had drafted with CERN as too loose. It lacked provisions to limit US responsibility for LHC cost overruns, he said, and had no guarantees of US participation in the LHC decision-making process. Once these issues had been resolved, after a tense Sensenbrenner visit to CERN in July 1997, the revised agreement could proceed. "For the first time, the US government has agreed to contribute significantly to construction of an accelerator outside our borders," observed the new Energy Secretary Federico Pena in a signing ceremony late that year with CERN Council President (and soon to succeed Llewellyn-Smith as director general) Maiani. The agreement, Maiani replied, "represents how scientific progress is made through international efforts."[68] But the arduous four-year negotiation process—navigating the difficult internal politics of many nations—underscored one of the major problems in achieving international scientific collaboration on such multibillion-dollar Big Science projects.[69]

FIGURE 7.2 CERN and DOE officials signing agreement regarding US status as a CERN Observer State. Seated are CERN Director General Chris Llewellyn-Smith (*left*) and DOE Assistant Secretary Martha Krebs. Standing behind them are Peter Rosen (*left*), who replaced Wilmot Hess as head of the DOE Office of High-Energy and Nuclear Physics, and John O'Fallon, director of the Office of High-Energy Physics. Courtesy of CERN.

By the dawn of the new century, the US high-energy physics community was finally recovering from the disastrous termination of the SSC. With the completion of the Main Injector and B Factory projects,[70] Fermilab and SLAC had new vistas to explore at the "intensity frontier"—doing precision measurements and seeking rare subatomic processes in collisions of high-intensity beams of particles with stationary targets or other beams. Even without the Main Injector, physicists had already discovered the long-sought top quark at Fermilab during the mid-1990s. A dozen events turned up in 1994 on the CDF experiment, but cautious physicists in that collaboration waited until the following summer to announce a full-fledged discovery, when members of both CDF and the newer Tevatron experiment DZero concluded that the top quark had finally been found.[71] At a mass-energy of about 173 GeV, top is the heaviest elementary particle encountered thus far. The discovery was the crowning high-energy physics achievement in the

1990s; US physicists could at least exult that all six quarks had turned up west of the Atlantic.

Together with such a massive top quark, precision measurements of the W and Z particles done at CERN, Fermilab, and SLAC were putting more stringent limits on the putative Higgs boson as the decade progressed. Strictly speaking, these indirect arguments applied only to a single Standard Model Higgs boson, while more complicated theoretical scenarios involving multiple bosons were possible. But such a high top-quark mass made Higgs bosons heavier than 200 GeV less likely in almost every possible scenario (see appendix 1.) And the LEP collider at CERN was putting strict lower bounds on the Higgs mass, which by decade's end had to exceed 100 GeV. There was a big flurry of excitement in 2000 when several possible Higgs boson events surfaced at LEP with a mass-energy near 115 GeV. Eager experimenters argued that this electron-positron collider be allowed to keep running for a few months beyond its scheduled shutdown, to determine whether the event cluster might indeed signal a Higgs boson, but LHC advocates won the day when Maiani decided to proceed with construction. After the final LEP shutdown occurred in November 2000, physicists could now conclude that the mass of a Standard Model Higgs boson had to fall between 113 GeV and about 170 GeV—much lighter than most theorists had figured just a decade earlier.[72]

Although no "Drell bump" ever materialized in funding for high-energy physics, US physicists were extensively involved in building the LHC and its detectors. B Factory construction was completed on schedule and on budget in 1998, and the Main Injector came on line in 1999.[73] Thus the construction of these projects was no longer putting pressure on the DOE high-energy physics budget by the end of the century. The funding thereby freed up could be devoted largely to the LHC effort, whose construction schedule had been delayed into the next decade because of the need to complete the high-energy LEP experiments in an orderly manner (see epilogue).

US physicists were then eyeing another highly touted multibillion-dollar project as the best possible future option to allow research at the energy frontier on American soil: the Next Linear Collider, or NLC. Unlike the SSC, the NLC had been planned from the outset as a *truly* international facility open to all interested nations. Accelerator-physics groups at CERN, DESY, KEK, INP Novosibirsk, and SLAC were advocating their differing designs for such a TeV-scale electron-positron collider—expected to stretch at least thirty kilometers—and engaging in collaborative R&D to enable its realization. A HEPAP subpanel chaired by former SSC Research Director Fred

Gilman weighed in with a February 1998 report recommending that SLAC and KEK together pursue a detailed conceptual design for the NLC.[74] But attaining an international consensus on which design to pursue and where to site such a collider remained thorny issues to be resolved at a later date.

WHY DID THE SSC COLLAPSE?

Back in April 1984, in the early days of the SSC, John Deutch had sent a prophetic memo to his fellow members of the SSC Board of Overseers. An MIT professor who had served in the Carter administration as the first director of the Office of Energy Research and undersecretary of Energy (and later as director of the CIA under Clinton), he was well qualified to know the ways of Washington, especially with regard to science-policy matters. "If a significant fraction of the SSC project cost is considered to be an 'add-on' to the HEP programs," he wrote, "it will inevitably compete with demands from other scientific disciplines, of which several are not being shy about proposing big-ticket projects for their own areas."[75] Deutch went on to outline a likely course of events as he saw them:

> A legitimate debate on research priorities and resource allocation among areas of science and technology will occur, perhaps covertly. The outcome of such a process is highly uncertain and it should not be assumed that the high-energy community will be successful. This is especially true when one considers the divisiveness which may emerge when the community fully appreciates the impact of the SSC on the base HEP program, present facilities, and pattern of operation.

The alternative, he suggested, was to pursue the SSC as an international project, citing from the text of his MIT colleague Victor Weisskopf's keynote address to the upcoming ICFA workshop in Japan: "There comes a time when the cost and effort of the next accelerator is so high that there may be no other way but world cooperation."[76] Deutch underscored this insight in his memo, adding that "there comes a point when the magnitude of the project is so large that the chance of success on a strictly national basis is less than on a genuinely collaborative basis despite the difficulties and compromises the latter implies."[77] In a perfunctory reply to this memo, Keyworth assured him that foreign SSC participation was a serious goal.[78]

While the lack of major foreign participation in the SSC is often given as an important reason for its collapse, hardly any of those involved cited this as the principal reason. In oral-history interviews and conversations about the SSC, we frequently asked respondents what they considered the major reasons for the project's collapse. In this way we were able to sample the diverse opinions of the several "communities of interest" involved in the SSC. The answer we heard most often was its steadily growing cost, perceived by many to be careening out of control. This polling process affirmed our early conclusion, also drawn by others, that the primary, proximate cause of the termination was the widespread congressional perception that the project was out of control—that its cost would exceed $10 billion and could easily grow well beyond that figure.[79]

Digging deeper, we find that a major reason for the seemingly relentless cost growth was the absence of strong SSC project management during the first few years in Texas. This problem can be attributed partly to the fact that most of the leaders of the Central Design Group did not join the SSC management team in Texas and returned instead to their academic positions.[80] In addition to the loss of continuity in collider design, their absence led to a management gap that was filled by a succession of weak or temporary project managers until Energy Secretary Watkins inserted his own selected managers into key SSC positions. These individuals hailed not from the physics community but from the nuclear Navy and the US military-industrial complex. While they brought much-needed expertise in large-project management, their interactions with the physicists at the SSC Laboratory became severely strained, making the hiring of additional physicists even more difficult than it already was. Thus a dysfunctional bipolar laboratory culture emerged in Waxahachie that made effective project management harder and hindered compliance with congressional and DOE oversight.[81] It was a serious handicap for such a huge and highly visible project.

Some, but not all, of these problems can be attributed to the difficulty and attendant costs of building a new, multibillion-dollar accelerator laboratory at a greenfield site in an area viewed by the high-energy physics community as an intellectual backwater. Had the SSC been sited instead in Illinois near Fermilab, it would have been far easier to attract top-notch physicists to work on the project. And the existing Fermilab infrastructure, both physical *and* human, could have been devoted to realizing the project, with attendant cost savings. Estimates of the value of this infrastructure have ranged from $495 million to $3.28 billion.[82] If the human infrastructure had been included in DOE life-cycle cost calculations, we believe the true value would have come in toward the middle of this range—perhaps close to $2 billion.[83] For SSC project management and control *could* have been lodged from day one in the hands of people who then had the world's deepest expertise in building superconducting accelerators and colliders—with consequent cost savings. While there certainly would have been increases above the $5.9 billion initially approved by Congress in giving the project a green light, the total cost would probably not have exceeded the $10 billion level that House members eventually came to view as the breaking point.[84] One shortcoming of the SSC site-selection process was that the DOE did not sufficiently account for the value of the existing human infrastructure at Fermilab in estimating overall SSC life-cycle costs at the seven sites, leading in part to the choice of Waxahachie.[85] If the project had instead gone to Illinois, however, it might have been confronted by strong congressional opposition from the Texas delegation, given how high the Waxahachie site had scored in the final DOE site evaluations.

The Department of Energy deserves criticism in other areas, too. While DOE officials correctly identified the project-management gap during the first year of SSC operations, they used a sledgehammer approach in attempting to resolve it. By inserting his own handpicked managers into key positions to try to assume control of the project, Watkins managed to alienate the SSC Laboratory leadership, making its cooperation even more difficult. The DOE also packed its SSC site office with nearly one hundred staffers who had next to zero familiarity with high-energy physics. As the project neared termination, Roy Schwitters was refusing to speak with Joe Cipriano, who was then calling for his ouster. As one observer noted, the Texas settlers had circled their wagons and were firing inwards.[86] And Secretary Watkins established back-channel reporting lines, bypassing the normal DOE checks and balances that could have ameliorated the worsening situation. The poor coordination regarding the SSC project between the Department's Office of Energy Research and its top political appointees and staffers became a difficult, continuing problem after he set up this unorthodox command structure.

Schwitters has been criticized for lacking the strong leadership qualities and management expertise needed to direct such a difficult undertaking. There is truth to these claims, but this criticism raises the question of who *else* could have done better. Nobody in the US high-energy physics community had any experience managing billion-dollar projects; the incoming generation of its leaders epitomized by Schwitters had attempted only projects an order of magnitude or two smaller.[87] This community also had little experience working with large US industrial firms, which were the darlings of the conservative Reagan and Bush administrations. The major components of the Fermilab and SLAC accelerators had all been fabricated on-site; a Brookhaven effort to have Westinghouse manufacture the Isabelle superconducting magnets resulted in failure. To try to address this shortcoming and mollify DOE critics, URA physicists attempted to partner with industrial firms Sverdrup and EG&G. But these partnerships added yet another layer of complexity to an already difficult enterprise. And they were not effectively managed by the SSC leadership—in part because of the lack of a seasoned project manager at the outset. By most accounts, these partnerships largely failed to achieve their stated goals and instead imposed added heavy burdens on the SSC Laboratory.

There is an inevitable tension in building such a Big Science gigaproject. Its leaders have to manage the project responsibly, accounting honestly for the expenditure of billions of taxpayer dollars while trying to establish a laboratory culture within which scientists can thrive and do productive research.[88] Effective project management must be combined with allowing scientists enough freedom to organize their experimental workplace. And the project leaders have to *manage* this difficult interaction. Panofsky and Wilson did this successfully in the process of building SLAC and Fermilab—at a scale of hundreds of millions of dollars. But at the multibillion-dollar scale of the SSC, it was (and will continue to be) a much more difficult task, given the public scrutiny that will always accompany such a large expenditure of public funds.[89] That attention means taking project management seriously from the outset, establishing computerized control systems and setting up the necessary data streams from the various groups involved in the effort, which are needed to keep the project under firm control. Both were sadly deficient until the project was well underway, and costs grew accordingly.

Beyond their troubled interactions with DOE officials, high-energy physicists were on mostly unfamiliar and uncertain terrain. SSC advocates did a poor job, for example, of influencing US public opinion about the project after it moved to Texas. While the Central Design Group had succeeded in such efforts, its task was made easier by the fact that many states wanted to corral the project as a job-generating plum. Once Waxahachie was chosen as the official SSC site, however, public interest and political support withered in other states. The Super Collider now had to be promoted based on its promising scientific research, not on the jobs it might bring to competing states. So it came to be viewed among House members and the general public as just another Texas pork-barrel project—and not as a frontier scientific institution whose output would eventually benefit citizens throughout the country.

The public opposition of other scientists, particularly condensed-matter physicists, was very damaging, too, for it helped undermine what originally was firm congressional SSC support for the project. Unlike NASA, for which public relations is a core part of its mission, DOE officials and high-energy physicists had little expertise in this arena; they were poorly prepared to influence US public opinion. With deep roots in an AEC culture for which public accountability was of little or no concern, they were not equipped to confront the growing criticism.

Although House support was initially wide, even after the Texas site was chosen, this support steadily narrowed as time passed and estimated total costs grew from $5.9 billion in 1989 to $8.25 billion in 1991 to more than $10 billion in 1993. A 1989 House vote on the SSC, for example, came in at over three to one in favor of the project; but by 1993 the margin in that chamber of Congress was almost two to one against it. The power of the Texas delegation had also ebbed during the same period, which was probably part of the reason for the diminished congressional support. And to a certain extent, the project got caught in the increasing crossfire resulting from the deepening regional polarization between Sun Belt and Rust Belt states—especially between Texas and Louisiana, on the one hand, and Illinois and Michigan, on the other. But a principal cause of the deterioration of support for the SSC was that Congress was finally beginning to take budget deficits seriously and trying to limit them during the early 1990s recession.

In addition, the Budget Enforcement Act of 1990 imposed rigid caps on domestic discretionary spending, which meant that the steeply rising annual budget for the SSC was beginning to exert serious pressure on federal funding for other energy and water projects dear to individual members of Congress, which included Brookhaven, Fermilab, and SLAC. A June 1992 House amendment to kill the SSC succeeded, only to be overturned by the Senate that September. But the arrival in early 1993 of 114 new representatives elected in November 1992, many of them determined to establish reputations as budget cutters, spelled doom for the SSC. While a similar scenario played out in Congress that year, with the House voting to kill the project and the Senate voting to restore it, the end game was different. Senate support collapsed in the face of overwhelming, bipartisan House opposition.

Support for the SSC also declined in the White House, which had a new, Democratic occupant in January 1993. While the project had been a presidential priority for adopted Texan George H. W. Bush, it gained only lukewarm support from Bill Clinton.[90] Science-policy priorities for the new administration were shifting dramatically, onto a post–Cold War footing in which applied research and technology development assumed prominence over basic physical sciences such as high-energy physics. Led by Vice President Al Gore and Maryland Senator Barbara Mikulski, the new policy emphasized targeted or strategic research that might lead to more immediate economic, environmental, or health benefits rather than Nobel Prizes for US physicists.[91] In the post-Cold War era, national security—or what science historian John Krige called "civil security"[92]—had markedly weakened as a rationale for pursuing such an abstract, costly science project. The need to demonstrate to developing nations that America still had the world's best physicists essentially evaporated as the Cold War ended.

Had there been billion-dollar foreign participation in the project, however, the SSC outcome might well have been different. That had been the aim from the outset during the Reagan administration, as Keyworth and then Herrington clearly stated, and the Office of Management and Budget established a goal of one-third of the total cost to come from nonfederal sources. In 1990 the House set its own target of 20 percent foreign contributions. While Texas committed to provide almost $1 billion, nearly another billion had to come from foreign sources when the estimated SSC cost was $5.9 billion. But after its estimated cost increased to $8.25 billion in 1991, a total of almost $1.7 billion worth of contributions was needed from other nations to meet this lofty goal. That was going to be a difficult target unless Europe or Japan offered something like a billion dollars.

The Reagan administration had initially promoted the SSC as a *national* project aimed at restoring US leadership in high-energy physics, however. Given that seminal rhetoric, which once established was difficult to reverse, it was challenging to encourage other nations deeply involved in high-energy physics to make large contributions. Especially in Europe, which viewed the proposed Large Hadron Collider at CERN as the long-range future of its high-energy physics research efforts, only modest contributions to the SSC detectors could ever be expected. The bulk of the foreign contributions had to come from elsewhere.

In a belated effort to internationalize the SSC project and obtain a billion-dollar commitment, the Bush administration thus turned to Japan, inviting it to join the United States in a trans-Pacific partnership. But the official invitation had to come from the highest US government level, at a face-to-face summit meeting between the president and the Japanese prime minister, if this partnership were ever going to happen. In two such meetings, one in January 1992 between Bush and Miyazawa and the other in April 1993 between Clinton and Miyazawa, no invitation was forthcoming—most likely because Japan would have insisted upon a US concession to relax demands for auto-parts import quotas in return for a billion-dollar commitment to become a partner in the SSC Laboratory.[93]

At the time of the fateful 1993 congressional votes on the SSC, there were only a few hundred million dollars' worth of (largely in-kind) contributions that nations such as Canada, China, India, Korea, Russia, and Taiwan had promised. It was not nearly enough to satisfy Congress and extend the project another year. With the annual appropriation headed toward $1 billion as the costly project entered its core construction phase and began manufacturing superconducting magnets, the House finally overcame the will of the Senate and voted resoundingly to kill the SSC. What had begun life in the 1980s as a national scientific project had never quite made the hoped-for transition into an international laboratory for high-energy physics.

In retrospect, the Reagan administration launched the SSC project in 1987 based on an aging Cold War model of international scientific collaboration, in which the United States took the lead, expecting other nations to follow. This was the same approach NASA had usually taken on major space projects including the Space Station. But the Cold War was in fact already beginning to wind down during the 1980s, despite Reagan administration efforts to revive and fight it. The Soviet Union was already staggering because of economic reasons linked to the collapsing price of oil. Other US competition—in economic and technological arenas—was coming from Japan and Western Europe.[94] With strong high-energy physics programs of their own to maintain, these nations did not want to be relegated to a role as the junior partners in a project whose goal was to reestablish US hegemony in the discipline.

As Kolb and Hoddeson have written, an effort had been growing for years to establish a truly international laboratory for high-energy physics and build a Very Big Accelerator, or VBA.[95] The nations involved were to be treated as equal partners in this enterprise. But in deciding to embark upon the Super Collider as a national project, the DOE and the US high-energy physics community essentially abandoned the multilateral VBA process in the belief that the United States could build such a machine by itself, supplemented by modest foreign contributions. Despite the Reagan administration's enthusiasm for a multibillion-dollar project—as embodied in Science Advisor Keyworth's 1983 admonition to "think big"—the SSC needed to survive at least another two administrations and several more Congresses before it could ever begin to collide protons.

By contrast, the much costlier Space Station managed to run a similar gauntlet and survive. But it had strong support among a general public that had long endorsed the idea of human presence in space. The project also enjoyed a much larger and politically more powerful industrial base—involving something like seventy-five thousand jobs. When its estimated costs ballooned to the tens of billions in the late 1980s and early 1990s, NASA downsized the project to the grumbling satisfaction of Congress. And the Space Station attracted genuine foreign contributions worth billions of dollars.[96] When the House took up an amendment to kill the Station in June 1993, both Clinton and Gore were on the phone jawboning representatives to save it.[97] The amendment failed by a single vote.

The leaders of the US high-energy physics community seriously erred in neglecting these political dimensions and deciding to proceed with the project unilaterally. And they had been warned as early as 1984, by no less a figure than Deutch. "On the time scale required to complete this project," he told SSC leaders, "the cast of characters is likely to change and these larger issues raised several times."[98] But seduced by Keyworth's siren song, high-energy physicists overestimated their limited supply of what Washington observers recognize as political capital.

In effect, the Super Collider grew too big to succeed. With its price tag then estimated at several billion dollars, the SSC site-selection process *had* to be thrown open to a national competition, if only for political reasons, rather than DOE putting a smaller, less expensive collider by fiat near Brookhaven or Fermilab. As the project moved from computer screens and drafting tables to a greenfield site in Texas, its estimated cost more than doubled—and began to press hard on other competing interests with their own strong political bases. The physicists involved neglected to heed warnings to take the burgeoning federal budget deficits seriously; they did not downsize or suggest building the project in phases in order to reduce its impact on other, competing interests. Instead, they clung resolutely to the original SSC design energy of 40 TeV while pursuing an increasingly conservative and ever more costly design. In retrospect, this choice was based more on politics—the need to exceed the highest possible energy of the European competition by a wide enough margin—than on physics.[99] It proved a costly mistake.

At the multibillion-dollar scale it had attained, crucial decisions about the SSC project had to be made by US political appointees or even heads of state—rather than by career DOE officials who had a far better understanding of the high-energy physics community. Japanese partnership in the SSC, for example, could be achieved only through a hoped-for in-person agreement between the US president and Japanese prime minister, not through the customary agency-to-agency accords. One reason for all the difficulties in securing large foreign SSC contributions was that the relevant science agencies in other nations did not have sufficient funding to accommodate the DOE requests. To be able to participate significantly at the costly scale the SSC had reached, these agencies needed much larger allocations than they already experienced. And sending such money or components overseas or across national borders required approval at the highest government levels.

Given the unprecedented scale of the SSC project and the intense public scrutiny that was to accompany it, the high-energy physics community faced management challenges it had never before encountered and was not equipped to deal with.[100] URA attempted to remedy this deficiency by trying to team with companies and managers from the US military-industrial complex, but it was a difficult, troubled marriage from start to finish. While the URA physicists and DOE officials wrestled with one another in trying to bring the project under control, an impatient Congress decided it had witnessed enough mismanagement, cost overruns, tepid foreign interest, and vocal opposition of other scientists, finally voting to terminate the project in October 1993.

With the benefit of hindsight, we conclude from our research that high-energy physics had reached a point by the mid-1980s that further progress at the energy frontier required multilateral international cooperation as originally envisioned as part of the VBA process. The US high-energy physics community belatedly recognized this watershed fact after Congress terminated the SSC in 1993. As recommended by a panel of high-energy physicists, the United States joined CERN as an Observer State in 1997 and eventually made contributions worth over half a billion dollars to the Large Hadron Collider. US physicists could consequently play important roles in the July 2012 discovery of the Higgs boson at CERN (see epilogue).[101] A similar decision might have been made in the late 1980s, before spending $2 billion in public funds for naught, when the LHC project was in its early design phase and CERN was seeking partners from beyond Europe. But by that time the US high-energy physics community had already been beguiled by the alluring dream of building the SSC as a national project to restore US leadership in this scientific discipline.

In the final analysis, the SSC project collapsed because the Department of Energy and US high-energy physics community remained locked in an obsolete Cold War mind-set and manner of pursuing major projects while the nation and the rest of the world were making a historic transition to a post–Cold War science-policy regime. Other contributing factors include the fact that the project was an order of magnitude larger, in both size and cost, than anything the community had ever before attempted and that its costs appeared out of control. This is one of the principal overarching themes that emerge from our research. Despite the added difficulty of organizing and managing them, pure-science projects at the multibillion-dollar scale should henceforth be attempted only as international enterprises involving interested nations from the outset as essentially equal partners. Nations that attempt to go it alone on such immense projects are probably doomed to failure like the Superconducting Super Collider.

EPILOGUE

The Higgs Boson Discovery

I think we have it! We have a discovery. We have observed a new particle consistent with a Higgs boson.

—ROLF HEUER, July 2012

Beginning in 1984, the planning and preparations for the Large Hadron Collider had been much more deliberate than similar activities were for the SSC.[1] After October 1993 these preliminary activities accelerated markedly. But because all nineteen CERN member nations had to agree to proceed with the new project, complicated negotiations ensued, taking over four years to reach complete agreement among the interested parties, including nonmember nations like the United States that wanted to participate in the project.[2] The talks became especially difficult because of the need of Germany, then facing the tremendous financial burdens of reunification, to reduce its annual contribution to CERN.

European physicists had largely viewed the SSC project as a profligate US attempt to leapfrog CERN's recent success in discovering the W and Z bosons by building a proton collider with well over twice the energy that could be achieved in the LEP tunnel.[3] They argued that the LHC could address the same physics research more cost-effectively, at least in searching for particles with masses up to 1 TeV, by attaining collision rates ten times higher than in the SSC design. By reusing the LEP tunnel plus a chain of proton injectors already in place, LHC designers could save huge sums and hold estimated project costs down to only two to three billion Swiss francs.[4]

Oxford theorist Chris Llewellyn-Smith, who replaced Carlo Rubbia as CERN director general in January 1994, had to navigate a labyrinth of conflicting needs of the CERN member nations. "Attitudes to high-energy physics were hardening in several member states," he recalled, and the CERN Council had granted Germany a temporary reduction in its contribution—the

FIGURE E.1 British physicist Lyndon Evans, who served as LHC project manager from 1993 to completion of its construction in 2008. Courtesy of CERN.

largest of all because of its vigorous economy.[5] The United Kingdom, whose CERN contribution competed directly with other British sciences through the UK Research Councils, was also adamant about cutting the CERN budget. Fortunately, France and Switzerland agreed to increase their contributions slightly to reflect the fact that they benefited more than other member nations from the laboratory facilities being situated on their territory.

Lyndon Evans, the CERN accelerator physicist Llewellyn-Smith named to lead the LHC project after Giorgio Brianti retired in 1993, repeatedly pared the machine's design in attempts to satisfy reluctant council members. Together they finally hatched a daring two-stage "missing-magnet" scheme in which a third of the LHC dipole magnets would not be installed at first, lowering the collision energy to 10 TeV.[6] "It was a totally crazy scenario," recalled Evans. "But one that was swallowed by the German government." Their strategy was to obtain council approval for a two-stage project—which was granted in December 1994—then approach other, nonmember nations like the United States to seek added contributions that would allow CERN to build the LHC all at once.[7]

And it worked! In June 1995, Japan offered in-kind contributions worth $50 million, followed by Canada, India, Israel, Russia, and finally the United

States, which in 1997 agreed to add at least $530 million (see chapter 7). Thus Evans could proceed with the LHC construction in a single stage aimed at reaching the design energy of 14 TeV by 2005. As approved by the council in 1996, the estimated LHC cost was 2.6 billion Swiss francs, plus another 1.05 billion for the detectors—with commissioning to begin in 2006.[8] But to placate still-reluctant Germany and the United Kingdom, which suddenly announced that they would not be able to meet their annual commitments, causing inevitable cash-flow problems, CERN had to borrow the shortfall from the European Investment Bank.[9]

LHC construction progressed gradually during the late 1990s while the LEP experiments continued taking data at ever-higher electron-positron collision energies, eventually exceeding 200 GeV (see chapter 7). Excavation of the experimental hall for the ATLAS detector, for example, proceeded while LEP had two high-energy beams passing through an evacuated tube in the midst of the huge cavern. After the shutdown finally occurred in November 2000 following a six-week extension, LEP magnets and infrastructure were removed and the remaining excavation completed.

As for the SSC, the crucial LHC components were the 1,232 superconducting dipole magnets needed.[10] CERN physicists elected to push the technological limits on these magnets, which used the "two-in-one" yoke design originally pioneered by Brookhaven that the SSC Central Design Group considered but did not adopt (see chapter 2). They also opted for cooling the liquid helium to 1.9 K, enabling strong magnetic fields above 8 T. Magnet design and development took more than a decade before CERN was ready in 1999 to seek manufacturers to produce the many miles of niobium-titanium superconducting cable, wind it into magnet coils, and assemble them within the "cold masses" to surround the twin beam tubes. Of the three magnet-assembly firms selected, two had extensive prior experience in manufacturing superconducting magnets for the HERA collider.[11]

But management problems and attendant cost overruns surfaced in 2001, after CERN Director General Maiani requested a thorough review of the project; it uncovered likely overruns of 480 million Swiss francs (or nearly 19 percent) in the collider alone. In all, another 850 million Swiss francs would be needed. Rather than reveal these problems to council members gradually, as Evans suggested, Maiani elected to drop a bombshell at its September 2001 meeting.[12] Immediate reactions were not pleasant. "How did they get into this without warning us earlier?" asked British council member Ian Halliday. "What CERN has blown over the past few months is its reputation for delivering on time and on budget."[13]

FIGURE E.2 View of superconducting dipole magnets in the LHC tunnel, with the innards of one of them depicted in artist's cutaway drawing. Note the two-in-one cold-mass construction. Courtesy of CERN.

In reality, the original LHC budget approved in 1996 had been cut to the bare bones by Evans and Llewellyn-Smith in an effort to get a pfennig-pinching council to give its required go-ahead.[14] Tender offers had just come in during August from the magnet-assembly companies, resulting in big increases over prior expectations. There was also a major cost increase of 120 million Swiss francs for the advanced computing infrastructure that was needed to channel the terabytes of data the LHC experiments would generate to physics-analysis groups around the world. And there was no contingency included in this budget for unforeseen events, such as the civil-engineering problems encountered in excavating the cavern for the Compact Muon Solenoid (CMS) detector. In retrospect, the big 2001 LHC cost increase was hardly surprising. But at the time it came as a tremendous shock.

In the spate of hand-wringing that followed, Evans came under particular scrutiny for his loose, confederate management style.[15] He favored a hands-on, one-on-one approach emphasizing direct involvement with the myriad technical details of the project, but that could leave important managers out of the loop and lead to confusion.[16] And CERN had a fragmented, largely academic culture in which small groups of experts would assemble spontaneously to address specific problems as they arose. Coordination of this culture to focus on the broader project objectives and set priorities could be difficult. Evans fortunately had at his disposal a web-based project-management control system that told him and other key managers at a glance the status of various LHC systems, subsystems, and components. It was a powerful project-management tool that had been developed at CERN specifically to manage the LHC project using CERN's famous World Wide Web platform.[17]

CERN called in Robert Aymar, a French physicist with deep experience in nuclear-fusion projects to review the LHC project management.[18] Some feared that Evans might be fired. But Aymar recognized his mastery of the technical aspects of the project, as well as his popularity with the staff, and recommended against it.

CERN had the advantage of deep technical expertise in proton colliders and superconducting magnets, most likely the best in the world at that time. Thus it was able to distribute a huge number of procurement contracts all around Europe—as required by political necessity. It then managed quality control on the bewildering variety of components coming back, as well as their integration into one of the most technologically complex systems in existence.[19]

With the superconducting dipole magnets, for example, CERN contracted out the production of NbTi wire and cable, confirmed their quality, then shipped the cable to the three companies—in France, Germany, and Italy—assembling magnet cold masses.[20] They were subsequently inserted into 15-m-long cryostats and the magnets individually tested at CERN to ensure that they operated appropriately at 1.9 K, generating uniform fields of 8.4 T without excessive quenching. Managing such a complex supply chain, which involved shuffling 120,000 tons of material all around Europe, was a technological *tour de force*. It meant assuming the risk that schedules might not be met and costly delays could occur, but CERN's Accelerator Technology Department pulled off this feat with aplomb, realizing substantial cost savings by doing so.[21]

Unforeseen problems cropped up with the magnets and other systems, however, leading to additional delays and cost overruns. The most serious came with the cryogenic system that supplied liquid helium to the magnets; it proved to be manufactured too poorly to be used, and had to be replaced by contracting with another manufacturer. That pushed the completion of LHC construction back by a year and added hundreds of millions to its cost.[22] Mainly because of this problem, the estimated materials costs for the collider had risen to 3.7 billion Swiss francs by mid-2006, a 42 percent increase over the original projection from a decade earlier.[23]

In early September 2008, the LHC construction was complete. Everything finally seemed to be coming together on September 10. Camera crews from around Europe camped out in the control room to catch the moment the huge machine came to life. With the whole world listening and watching on the BBC, CERN accelerator operators guided 450 GeV proton beams through the two rings without any major incidents, needing only about an hour to achieve success in each direction.[24] CERN had a well-known reputation for engineering its accelerators and colliders so thoroughly that they turned on with few hitches. By mid-afternoon that day, celebrations began.

But nine days later, Evans took a panicky phone call from the control room telling him to come quickly. When he arrived, flashing alarms warned that many magnets had quenched; helium gas was filling the tunnel. Later analyses indicated that an electrical splice between two magnets had warmed and gone normal—losing its superconducting properties. After it melted, a powerful spark surged through the magnet vessel, puncturing it and releasing six tons of helium. When CERN workmen entered the tunnel after the gas had cleared, they found dozens of damaged magnets, some ripped from their mounts. A thin film of soot covered the wreckage. Evans later called the disaster "a real kick in the teeth for everyone."[25]

CERN officials initially expected that the magnets could be rebuilt or replaced—and the other damage repaired—in several months, during the normal winter shutdown, so that protons would begin to collide at 14 TeV the following spring.[26] But closer examination of the nearly 10,000 electrical connections between magnets revealed about eighty faulty ones that could not carry the LHC design current of nearly 12,000 amps.[27] Repairing these could delay operations a year or more, and the LHC might have to start up again at a lower energy to avoid another crippling disaster.[28]

The delay in commissioning the LHC gave Fermilab an extra year to search for the Higgs boson without competition. Theoretical and experimental advances, such as the discovery of the top quark at an unexpectedly high mass near 175 GeV (see chapter 7) had suggested that a Standard Model Higgs boson should occur with a mass less than 200 GeV. And theories invoking supersymmetry also required a low-mass Higgs boson.[29] If such particles in fact exist, physicists could possibly discover them in experiments on the

Fermilab Tevatron, even though its protons collided with antiprotons at only 2 TeV, because the interactions between their constituent quarks and gluons would occasionally reach 200 GeV (see appendix 1). Because these highest-energy collision events are rare, and any Higgs bosons that might result would have been extremely difficult to detect amid the overwhelming backgrounds that occur in these messy collisions, the Tevatron had to achieve high luminosity in order to enable such a Higgs discovery. But since it had come back into operation in 2001 after the Main Injector upgrade, Fermilab accelerator physicists had struggled to achieve the required luminosity.[30] So the Higgs search languished.

There were the usual rumors and false alarms that a Higgs discovery might have occurred. A brief flurry of excitement erupted in early 2007, for example, when a group of CDF (Collider Detector at Fermilab) researchers reportedly observed a small cluster of excess events near 160 GeV, perhaps indicative of a supersymmetric Higgs boson.[31] But interest lagged a month later, after the DZero experiment found nothing like it. And as data continued trickling in that year, the CDF experiment's excess withered away.

Accelerator physicists began to resolve the Tevatron's luminosity problems by mid-decade, finally reaching the design goal in 2007. Thus hopes were high that researchers at Fermilab might be able to report evidence for a low-mass Higgs boson (if one existed) before the LHC began operations at a much higher energy. So when the LHC commissioning faltered in late September 2008, and further examinations revealed dozens of faulty magnet interconnections that would take over a year to fix, physicists began to press the DOE and Fermilab leadership to grant the Tevatron a three-year extension beyond its scheduled September 2011 termination.[32]

In October 2010 HEPAP recommended that the DOE approve a Fermilab plan to share the added $150 million cost of operating the collider for another few years, with the lab contributing almost a third of that by cutting back on its operations and delaying other experiments.[33] But US high-energy physics funding had worsened in the aftermath of the financial panic of 2008 and the Great Recession. "Unfortunately, the current budgetary climate is very challenging and additional funding has not been identified," wrote Office of Science Director William Brinkman in denying the HEPAP request.[34] The focus of DOE research during the first Obama administration under Energy Secretary Steven Chu was on applied research in renewable energy, not on basic science. And by then the DOE had made a formal commitment to join the ITER thermonuclear fusion project in Cadarache, France, adding another major burden on its budget. Thus the Tevatron's long, productive twenty-five-year history ended as planned on September 30, 2011, leaving the Higgs search to the refurbished LHC.[35]

When the two Tevatron experiments combined their data that summer to increase the overall sensitivity, small excesses of Higgs-like events turned up at masses between 115 and 155 GeV, but not enough to claim anything significant.[36] Above that mass level a Higgs boson should decay predominantly into a pair of W bosons, making it relatively easy to detect against the backgrounds. Because no such excess occurred, Fermilab could only take satisfaction in excluding a Standard Model Higgs boson having a mass between 156 and 177 GeV. That was not very much to show for a decade's worth of difficult, frustrating research. As CDF researcher John Conway lamented, "We've gone a very long time with no truly new discovery in particle physics, no observation that truly changes the paradigm."[37]

The LHC also features two large, general-purpose detectors. ATLAS is a cylindrical detector 45 m long and 25 m in diameter weighing about 7,000 metric tons. CMS is a more compact cylinder that contains much more iron, so it weighs in at 12,500 tons.[38] Both experiments had completed construction and were ready to begin logging data during the fall of 2008, but they had to wait over a year for the damages to LHC magnets to be repaired. It could have taken even longer had CERN attempted to repair or replace all the defective interconnections, but the laboratory decided to postpone that effort and begin running the LHC at lower energy. "It was time to do some physics," reflected the new director Rolf Heuer in 2012.[39]

The LHC finally collided protons at just 900 GeV on November 23, 2009, with both detectors recording the initial events. Four months later, the machine operators gingerly ramped beam energies to 3.5 TeV, achieving collisions at 7 TeV, over three times what the Tevatron could achieve. Chastened by the 2008 disaster, ATLAS and CMS physicists were resigned to beginning their experiments at half the LHC design energy. Wary of another massive quench, accelerator physicists and engineers led by Steven Myers concentrated on improving the collider's reliability and luminosity.[40] During 2010 they circulated up to 400 proton bunches in each beam, but the collision rate rarely exceeded a thousandth of what the LHC was designed to achieve. Thus, despite its lower energy, the Tevatron experiments remained competitive that year.

The ATLAS and CMS experiments were designed, built, and operated by truly international scientific collaborations, each incorporating around

FIGURE E.3 View inside the ATLAS detector during construction. The eight large tubes arranged around its circumference house superconducting toroidal magnets. Courtesy of CERN.

three thousand physicists—about an order of magnitude larger than for the CDF and DZero experiments. In the case of ATLAS, they hailed from 177 institutions in thirty-eight nations; CMS included 182 institutions from forty nations.[41] The United States has the largest national contingent at CERN, numbering over 1,500 physicists.[42] A participating institution typically contributed a detector component, which was built and tested at home before being shipped to CERN to be integrated into the detector. CERN had agreed to contribute 20 percent of the total detector costs; this share, initially pegged at 210 million Swiss francs in 1996, grew to 260 million in 2001 and reached 310 million in 2008, when the detector construction was complete (a cost growth of 48 percent).[43] Integrating the myriad components from all around the world into these gargantuan detectors, whose tolerances were commonly measured in microns, was a monumental task that CERN staff physicists and engineers successfully managed.

The ATLAS and CMS detectors were optimized to be especially sensitive to heavy subatomic particles such as a Higgs boson and others expected to appear in theories with supersymmetry. Both detectors were designed with excellent energy and momentum resolution (of about 1–2 percent) for electrons, muons, and photons. Although a low-mass Higgs boson was expected to decay predominantly into pairs of W bosons or bottom quarks, which would disintegrate inside the beam tube and strike the detectors as jets of hadrons, these decay modes were likely to be buried under onerous backgrounds billions of times greater—and thus be extremely difficult to identify convincingly. By concentrating on rare decay modes in which electrons, positrons, muons, or photons emerged, ATLAS and CMS physicists had far greater chances of discerning a Higgs signal.[44] It was expected to appear as a narrow peak jutting up from smoother backgrounds in graphs of the data plotted versus possible mass. The position of such a peak would correspond to the Higgs mass (in energy units).

To digest the torrents of electronic signals the experiments produce, up to 300 gigabytes per second, CERN and more than 140 institutions in thirty-five countries involved in the LHC established a worldwide computer network linked by fiber-optic cables and the Internet. The raw data is first reduced at CERN to a manageable subset of the most interesting events, which are then fed over the LHC Grid for more detailed analysis at satellite computers in participating institutions around the globe. Every day, many terabytes of data are handled by distributing and sharing the tremendous computing loads. In this way, researchers can remain at their home laboratories and universities throughout most of the year, while keeping intimately involved in the physics analysis as the data rolls in.[45] The costs of establishing the Grid and related computing infrastructure were not included in the initial 1996 LHC budget, but they were listed as 60 million Swiss francs in 2001 and grew to 90 million by 2008.[46]

Serious data taking began in 2011, as LHC operators nudged the luminosity steadily upwards, and proton collisions began to surge in. From the experiments, there were the usual false alarms and overreactions of reporters and others hovering over the action. A University of Wisconsin group, for example, circulated an excited internal ATLAS report in late April of a potential 115 GeV Higgs boson (reminiscent of the excess events witnessed by the ALEPH experiment at LEP in the fall of 2000) decaying into two high-energy photons.[47] It leaked into the Internet and went viral, forcing researchers to cancel Easter vacations to check it out.[48] This "Easter bump," recalled CMS spokesperson Guido Tonelli of the University of Pisa, helped focus his colleagues' attention on this rare but important decay channel.[49] But the intriguing effect soon proved to be a random fluctuation, for it withered away in early May after more ATLAS data taking.

Rumors abounded that candidate Higgs boson events might be revealed at the summer 2011 conferences, but nothing resembling a discovery was presented. CERN could rule out Higgs masses between 145 GeV and 466 GeV, and found no significant excesses up to 600 GeV, improving dramatically upon prior Fermilab limits. If the Standard Model Higgs boson indeed existed, it now had to be confined to a narrowing range of possible masses between 114 and 145 GeV.[50]

With the Fermilab Tevatron shut down in the fall of 2011, the Higgs search became a race between ATLAS and CMS collaborations. Machine operators steadily increased the LHC luminosity, attaining new records almost every week. By the time the 2011 run ended in November, the machine was producing nearly 20 proton collisions every time two bunches of protons clashed at the points where the beams crossed at the center of each detector, generating hundreds of millions of collisions per second.[51] Although the resulting floods of events were difficult for the detectors to cope with, the expected rarity of Higgs-boson events mandated such extreme event rates.

In the LHC, unlike the Tevatron, the dominant production mechanism was expected to be "gluon fusion," in which two gluons—one inside each colliding proton—merge into a Higgs boson, which disintegrates almost instantaneously. A Higgs boson with a mass below 145 GeV would decay predominantly into a pair of bottom quarks, which immediately transform into two jets of hadrons that strike the detectors. While bottom-quark jets have features that allow physicists to distinguish them from other kinds of jets, they are still buried under tremendous hadron backgrounds that make separation difficult.[52] ATLAS and CMS researchers therefore decided to focus instead on much rarer decay modes involving energetic electrons, muons, and photons. In particular, there are four modes—into a pair of photons, two pairs of muons, two electron-positron pairs, or a pair of muons and an electron-positron pair[53]—in which *all* of the mass of a Higgs boson ends up as readily visible particles in the detectors.[54] By summing up the energy deposited therein by these distinctive particles, researchers can determine the mass of whatever particles might have decayed into them. Although these rare decays occur much less than 1 percent of the time for a Standard Model Higgs boson, they were expected to begin appearing in graphs of the data as narrow peaks jutting up prominently from the remnant backgrounds.

CERN went cautiously public with preliminary results in a two-part seminar on December 13, 2011, viewed via webcast by physicists around the globe. Expectations were high, as discovery rumors had been circulating for a week. Both experiments reported observing dozens of extra two-photon events near 125 GeV—as well as a few excess four-lepton events in that vicinity, too. But the statistical significance in each case was hardly three standard deviations; by high-energy physics standards, this was sufficient to claim "evidence for" a new particle, but not for a "discovery." "We have restricted the most likely mass region for the Higgs boson to 116–130 GeV, and over the last few weeks we have started to see an intriguing excess around 125 GeV," said ATLAS spokeswoman Fabiola Gianotti of CERN. Tonelli agreed, saying CMS physicists "cannot exclude the presence of the Standard Model Higgs between 115 and 127 GeV because of a modest excess of events in this mass region."[55]

The fact that *both* experiments had witnessed excess events near 125 GeV—and in two different decay modes—was intriguing. Having looser standards for claiming a significant result, most scientific disciplines would likely have called this a discovery. But the justification standards in high-energy physics require a five-standard-deviation excess in order to rule out possible random fluctuations with high confidence. That could only come with more LHC data-taking, scheduled to begin in the spring.

Meanwhile, CDF and DZero physicists—many of whom also worked on the ATLAS and CMS experiments—had not yet abandoned the Higgs search. Using sophisticated techniques, they reanalyzed their full data sets, attempting to wring out every possible Higgs boson event. The Tevatron was particularly effective at generating "associated production" events, in which a quark inside a proton merges with an antiquark within a colliding antiproton to yield a W or Z particle plus a Higgs boson. Physicists sought events in which the W or Z decayed into an easily identifiable lepton pair and the Higgs boson into two bottom-quark jets. Indeed, both experiments found telling excesses of such events. Combined into a single result, the CDF and DZero data revealed a broad bump between 115 and 135 GeV—as expected for a 125 GeV Higgs boson decaying into bottom quarks, given the two detectors' low energy resolution for hadron jets.[56]

It seemed to many that the elusive Higgs boson had finally been cornered. While none of these individual results was convincing all by itself, *four* separate experiments had reported intriguing signals near 125 GeV. And the decay rates of whatever it could be accorded well with Standard Model expectations. For some theorists like Gordon Kane, the discovery had already occurred. "When you have four independent signals, they almost never go away," he remarked in a US press conference, claiming that five-standard-deviation ("five-sigma") significance could be obtained by combining the data from all four experiments.[57]

But ever-cautious experimenters, wary of getting too far ahead of their data, suggested everybody wait until after the 2012 LHC runs, which began in early April, with protons colliding at 8 TeV. Operators pulled out the stops, trying to push the luminosity as high as possible, reaching almost 80 percent of the design value by early June. The detectors were flooded with proton collisions, often more than thirty per bunch crossing.[58] Ferreting out the details of interesting events became challenging amid the resulting event "pileup" that occurs at such high collision rates.[59]

This time both ATLAS and CMS did "blind" analyses in which researchers were prevented from looking at the results as events flooded in; this approach was followed to avoid introducing undue experimenter biases. When the LHC was temporarily shut down for maintenance in mid-June, however, both teams took a "look inside the box" to see whether the effects observed in late 2011 near 125 GeV were still present. And they were! But it still took another week of painstaking analysis before the two research teams, working independently, could conclude that each indeed had made a *bona fide* discovery. By combining the results from both the two-photon and four-lepton decay modes, physicists finally achieved the required five-sigma significance, which meant that the chances were less than one in three million that the observed effect was due to random fluctuations in the data. For ATLAS this recognition came on June 25; for CMS it occurred a bit later.[60]

CERN director Heuer got his first glance at the new data on June 22, in a meeting with Gianotti and the new CMS spokesperson, Joseph Incandela of UC Santa Barbara. He realized the evidence was strong enough to make it public. After informing council members, he decided to convene another joint seminar at CERN on the morning of July 4, timed to coincide with the opening of the 36th International Conference on High Energy Physics in Melbourne, Australia, to be broadcast via the Web.[61] On the night before the seminar, hundreds of physicists slept in the hallways outside the CERN auditorium, hoping to get one of the remaining unreserved seats. Evans, Myers, Llewellyn-Smith, Maiani, and Schopper enjoyed front-row seats; even theorists Francois Englert and Peter Higgs were there, having flown in for the event.

Incandela and then Gianotti raced through their slides, covering mainly the new 2012 data but including the 7 TeV 2011 results in combination. As in December, graphs of the two-photon events revealed striking peaks jutting from backgrounds close to 125 GeV, this time containing about two hundred excess events. But now the two experiments revealed over a dozen extra events in which a heavy particle decayed into four leptons near 125 GeV;

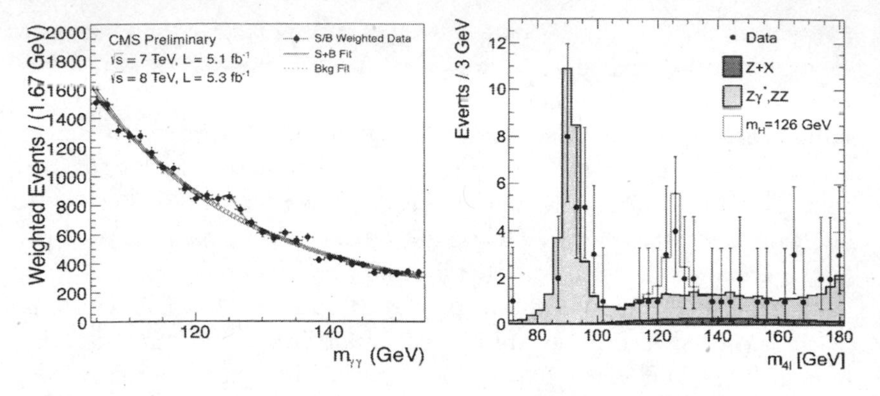

FIGURE E.4 CMS Collaboration proton-proton collision data released on July 4, 2012, revealing enhancements near 125 GeV corresponding to a particle decaying into two photons (*left*) and four leptons (*right*). Courtesy of CERN.

subtle peaks had begun to form in this decay channel, too. After combining this result with the two-photon data, both the ATLAS and CMS experiments independently concluded that they had five-sigma results. In each case, the chances were now less than one in three million that this new apparition was a statistical fluke. "I think we have it," exulted Heuer, wrapping up the seminar. "We have a discovery," he continued, explicitly using the word at last, if guardedly. "We have observed a new particle consistent with a Higgs boson."[62]

It remained to be determined whether this new particle was indeed *the* Higgs boson as predicted by the Standard Model, or something else, but few doubted that a discovery had occurred.[63] To establish that it was indeed a Higgs boson, physicists had to assess its rate of decay into the various possible other particles, including bottom quarks and W bosons. But achieving this goal was much more difficult because of the oppressive backgrounds at the LHC. Here Fermilab had an important insight to add because its CDF and DZero experiments had greater sensitivity to bottom-quark decays. In an article submitted in late July, these physicists presented evidence that a new particle with a mass lying between 120 and 135 GeV was indeed disintegrating into pairs of bottom quarks.[64] Although the statistical significance of this result was not as strong as the ATLAS and CMS results, it provided crucial confirmation that the new boson did indeed decay in this manner, as a 125 GeV Higgs boson must do most of the time.

As theorists pored through the new results, the case for a Higgs discovery got ever stronger. John Ellis and Tevong You of CERN updated a global fit to the data from the LHC and Tevatron, concluding that the strength of the new particle's coupling to other subatomic particles increased in proportion to their masses, as had to be the case for a Higgs boson.[65] Meanwhile, Heuer granted the LHC experiments added running time through the end of 2012, so that they could gather the greatest possible quantities of events before the scheduled shutdown for machine repairs. By the end of October, it was becoming increasingly clear that the new particle also had zero spin as required—the first such elementary particle ever to be encountered.[66]

Still undetermined was whether the new particle is the *one and only* Higgs boson required by the Standard Model, or the lightest member of a series of several Higgs bosons, as occur in theories involving supersymmetry (see appendix 1). And ATLAS was reporting a substantial excess of two-photon decays, well beyond Standard Model expectations, while CMS observed no similar surfeit.[67] If it held up, such an excess might provide indirect evidence for additional heavy particles yet to be discovered. But these kinds of questions had to await the 2015 run, scheduled to occur after the repairs had been completed on the faulty magnet interconnections. After that the LHC could finally be operated near its design energy of 14 TeV.

While the LHC project also experienced trying growth problems and cost overruns, increasing from an estimated 2.81 billion Swiss francs in 1996 to over 4.3 billion Swiss francs in 2009, it managed to survive and discover a low-mass Higgs boson—using only about half its original design energy. When labor costs and contributions from participating nations are included, the total exceeds US$10 billion, a figure often cited in the press.[68] This achievement in the face of problems similar to what the SSC project experienced, if not as difficult, raises the obvious question: why did CERN and its nonmember partners succeed where the United States had failed?

From the SSC's early days, many thought it should have been sited at or near Fermilab, taking advantage of its existing infrastructure (see chapters 3 and 7).[69] CERN had done so for decades, building one machine after another as extensions of its existing facilities and reusing parts of the older machines in new projects, thereby limiting costs. Perhaps as important, CERN had also gathered and developed some of the world's most experienced accelerator physicists and engineers, who worked together smoothly as a team. Fermilab had equally adept machine builders (and substantial physical infrastructure) who could have turned to other productive efforts when inevitable funding shortfalls occurred during the annual congressional appropriations process. And the troublesome clashes that occurred between high-energy physicists and military-industrial engineers during the early years of the SSC project would not have erupted in the already well-integrated Fermilab culture.

These pro-Fermilab arguments, however, ignore the harsh realities of the American political process. A lucrative new project costing over $5 billion and promising more than two thousand jobs cannot be sole-sourced to an existing US laboratory, no matter how good its infrastructure or how powerful its congressional delegation. As politically astute DOE leaders recognized, the SSC project had to be offered up to all states able to provide a suitable site, with the decision based (at least publicly) on objective criteria. A smaller project costing up to $1 billion and billed as a major upgrade of existing facilities *might* have been sole-sourced to Fermilab, given the political climate of the mid-1980s, but not one as prominent and costly as the SSC. It unfortunately *had* to be placed on the US auction block, and Texas made the best bid according to the official DOE judgment criteria.

Unlike the SSC, the LHC project had solid project management throughout by a single physicist, Evans, who had decades of experience with proton colliders.[70] This was undoubtedly an important factor in its success. Despite major problems and cost overruns that eventually exceeded 40 percent,[71] Evans enjoyed the strong support of the CERN management, as well as from a deeply experienced cadre of physicists and engineers who worked together without the cultural clashes that occurred at the SSC lab. And on the LHC project, engineers reported ultimately to physicists, the users of the machine best able to make the required tradeoffs when events did not play out as originally hoped. The LHC project encountered daunting difficulties, serious delays, and major cost overruns, too, but its core management team led by Evans held together and worked through these problems. They also shared a common technological culture—as well as understood and supported the project's principal scientific goals. Similar observations cannot be made regarding the military-industrial engineers who came to dominate SSC construction.

CERN also enjoys an internal structure, governed by its council, that largely (but not entirely) insulates its leaders and scientists from political infighting in and machinations of individual member nations.[72] Unlike in the United States, the lab director or project manager cannot be hauled before a parliamentary investigations subcommittee and required to testify under oath about management problems or cost overruns. Nor did the project face annual appropriations battles and threats of termination, as do major US projects.[73] Serious problems that arose, for example the 2001 cost overrun, had to be addressed in the council, which represents the science ministries of member nations and generally operates by consensus, especially on major new projects like the LHC. This governing structure ultimately helps maintain control of a project within the hands of the scientists involved, instead of allowing politicians or other government officials to intervene.

Because the council must also address the wider interests of national science ministries, CERN leaders have to be sensitive to the pressures its annual budget, new projects, and cost overruns can exert on other European science. The mid-1980s recommendations of the Kendrew panel (see chapter 3, n. 73) had had a chastening effect on the CERN management.[74] In this way, European small science had a valuable voice that was heard within the CERN governing process. The LHC project had to be tailored to address such concerns before the council would grant it final approval.[75] No similar mechanism existed within US science, except for the other, disgruntled scientists to complain openly in prominent guest editorials and before congressional hearings after SSC costs got out of hand in 1989–91. The consequent polarization of the US physics community helped undermine what had originally been fairly broad House support for the project, aiding SSC opponents.[76]

And because of these pressures within its council, CERN had to effectively internationalize the LHC project—obtaining major commitments from nonmember nations such as Canada, China, India, Japan, Russia, and the United States—*before* going full speed ahead. These contributions enabled Evans and his team to proceed with a collider design able to reach the full 14 TeV design energy rather than with the initial phase of a down-scoped, two-phase project that might have been buildable with reduced funding.[77] When the LHC project finally gained council approval in 1996, it was a truly international scientific project with firm financial backing from more than twenty nations worldwide.

And in the final analysis, the LHC was (somewhat fortuitously) much more appropriately sized to its primary scientific goal: the discovery of the Higgs boson. The likelihood that this elusive quarry could turn up at such a low mass as 125 GeV was not well appreciated until the late 1980s, when supersymmetry theories began to suggest that such a light Higgs boson might occur.[78] But by then the SSC die had been cast—in favor of a gargantuan 40 TeV collider that would be able to reveal the causes of spontaneous symmetry breaking, even if such phenomena were to occur at masses up to nearly 2 TeV (see appendix 1).

After that fateful late-1989 decision (see chapter 4), which was endorsed unanimously by a HEPAP subpanel but added billions to the SSC cost, the US high-energy physics community committed itself to a huge project that became increasingly difficult to sustain politically amid the worsening fiscal climate of the early 1990s. With the end of the Cold War and subsequent lack of a hoped-for peace dividend during a stubborn recession, the United States had entered a period of austerity not unlike what has been happening recently in many developed Western countries. In this constrained fiscal environment, a poorly understood basic-science project experiencing large cost overruns and lacking major foreign contributions posed an easy political target for congressional budget-cutters to "sacrifice."

A 20 TeV proton collider—or perhaps just a billion-dollar extension of Fermilab facilities such as the Dedicated Collider proposed in 1983 (see chapter 1)—would likely have survived the budget axe and discovered this light Higgs boson long ago.[79] For another option on the table during the 1983 meetings of the Wojcicki supbanel was to continue building Brookhaven's CBA/Isabelle collider while beginning design work on an intermediate-energy 4–5 TeV Fermilab machine, whose costs were then projected at about $600 million. This more fiscally conservative approach would have maintained the high-energy physics research vitality of these two productive DOE laboratories for at least another decade.[80] And such smaller, cost-effective projects would surely have been more defensible during the economic contractions of the early 1990s, for they accorded better with the US high-energy physics community's diminished political capital in Washington. Their construction would also have proved much easier for physicists to manage and control without having to involve military-industrial engineers.

Instead, the US high-energy physics community elected to "bet the company" on an extremely ambitious 40 TeV collider so large that it would probably have to be sited at a new scientific laboratory in the American Southwest. Such a choice was to abandon the three-laboratory DOE (or AEC) system that had worked so well for nearly two decades and fostered US leadership in the field. Perceived European threats to this hegemony tipped the balance toward making the SSC a national project—and away from it becoming a truly international world laboratory, as others were advocating.

Unlike historians gazing into the past, however, high-energy physicists do not enjoy the benefit of hindsight when planning a new machine. Guided partly by the dominant theoretical paradigm, they work with a cloudy crystal ball through which they can only guess at likely phenomena to occur in the new energy range, and must plan accordingly.[81] And few of them can foresee what may transpire in the economic or political realms that might jeopardize an enormous project requiring about a decade to complete and costing billions of dollars, euros, or Swiss francs—or, relevantly today, a trillion yen. This climate of uncertainty argues for erring on the side of fiscal conservatism and for trying to reduce expenses by building a new machine at an existing laboratory, thus recycling its infrastructure, both physical and human. Such a gradual, incremental approach has been followed successfully at CERN for six decades now, and to a lesser extent at other high-energy physics labs.

But US physicists elected to stray from this well-worn path in the case of the Superconducting Super Collider. It took a giant leap of faith to imagine that they could construct an enormous new collider with over twenty times the energy of any machine that they had previously built, at a green-field site where everything had to be assembled anew from scratch—including its management team—and defend the project before Congress in times of fiscal austerity. A more modest project sited instead at Fermilab (or Brookhaven) would likely have weathered less opposition and still be producing good physics results today. As several leading high-energy physicists acknowledged in hindsight, the SSC was probably "a bridge too far" for this once-powerful scientific community.[82]

Appendix 1
Physics at the TeV Energy Scale

The need for ultramassive fundamental particles—and for huge colliders and detectors to create and search for them—arose from theoretical developments of the 1960s and 1970s that led to the Standard Model, today's paradigm of particle physics. Before that, the heaviest known subatomic particles weighed in below 2 GeV (in equivalent energy units), or about twice the proton mass.[1] But by thirty-five years later, physicists had cornered the top quark at a mass of nearly 175 GeV, almost a hundred times heavier; they were actively contemplating the possible existence of elementary particles beyond the Standard Model with masses an order of magnitude greater, at the TeV scale. To create such massive particles using accessible accelerator technologies required machines like the SSC measuring in the tens of kilometers.

During the late 1950s and early 1960s, groups of quantum field theorists began applying techniques called "spontaneous symmetry breaking" to elementary particle physics. These methods, which had emerged from recent advances in solid-state and condensed-matter physics, offered a means to give fundamental particles distinctive masses while retaining the underlying symmetry of the field equations.[2] This was crucial because attempting to insert particle masses "by hand" into gauge-invariant field theories destroyed this symmetry. A desirable characteristic of these gauge theories, of which quantum electrodynamics or QED is an example, is that the fundamental entities in them carry a conserved charge—e.g., the electric charge in QED. In 1954 Chen-Ning Yang and Robert Mills proposed gauge theories for strongly interacting particles, which worked well as long as the particles had no mass,[3] but the symmetry broke down whenever non-zero masses were introduced. Spontaneous symmetry breaking offered an alternative.

At the time, quantum field theory was in deep decline. It had succeeded marvelously in accounting for electromagnetic interactions of subatomic particles but foundered when it came to describing their weak and strong interactions. The then-dominant theory of elementary particle physics, known as S-matrix theory or the bootstrap model, considered the proliferating menagerie of strongly interacting mesons and baryons as equally elementary, somehow composed of one another.[4] It did not need to appeal to deeper, more fundamental entities to account for this complexity.

Early attempts to deploy spontaneous symmetry breaking in particle theory encountered difficulties. Yoichiro Nambu and Jeffrey Goldstone led the way. Their theories required existence of massless spin-zero bosons, which came to be called Nambu-Goldstone bosons, but such particles did not appear in nature unless one could interpret them as pions.[5] Philip Anderson suggested a possible path forward in a non-relativistic model of the behavior of photons in plasmas, hinting that a fully relativistic theory might resolve the problem.[6] In 1964 six theorists—Robert Brout, François Englert, Peter Higgs, Gerald Guralnik, Richard Hagen, and Thomas Kibble—formulated three independent, equivalent solutions that invoked spontaneous symmetry breaking to confer masses on potential force-carrying gauge bosons without destroying the underlying gauge symmetry.[7] As Higgs recognized most clearly in a follow-up letter, these kinds of theories also required *massive* spin-zero bosons to exist; these have since become known among particle physicists as Higgs bosons.[8]

Beyond the small coterie of quantum field theorists, these papers attracted scant attention until the late 1960s, after Steven Weinberg and Abdus Salam used Yang-Mills theories and this "Higgs mechanism" to generate masses of the W and Z bosons and unify the weak and electromagnetic forces into the "electroweak" force.[9] But it wasn't until the early 1970s, when Gerard 't Hooft and Martinus Veltman demonstrated that Yang-Mills theories could yield calculable results, that they really began to catch on among particle theorists.[10] With the parallel discovery of quarks as fundamental entities making up mesons and baryons, these momentous theoretical advances led to the establishment and widespread acceptance by physicists of the Standard Model during the mid-to-late 1970s.[11]

In the early 1980s, the discovery of the W and Z bosons at CERN with close to the masses expected in the Standard Model, respectively 80 and 91 GeV (or about 86 and 97 times the proton mass), confirmed this theory beyond doubt and opened a new period in the history of particle physics.[12] High-energy colliders able to create particles with masses of 100 GeV or more became the norm; experimenters formed collaborations numbering in the hundreds (and eventually thousands) to design and build the gargantuan detectors required to track down these particles.

There still remained at least two undiscovered fundamental particles in the Standard Model, the top quark and the all-important Higgs boson (or bosons), for which firm mass predictions could not be made. It is helpful to think of this spin-zero particle as the quantum of a ubiquitous scalar field called the Higgs field, which pervades the universe and has exactly the same value everywhere. The Higgs boson is just the physical manifestation of this field, in the same way that photons of every color and stripe are the physical manifestations of vibrations in the electromagnetic field. By virtue of their widely differing interactions (or "couplings") with the Higgs field, the leptons, quarks and bosons of the Standard Model "acquire" their diverse masses.

By the mid-1970s, theorists had already begun to speculate about the possible mass of the Standard Model Higgs boson.[13] In a noteworthy 1976 article, John Ellis, Mary K. Gaillard, and Dimitri Nanopoulos outlined the many ways this boson might become manifest in high-energy experiments, both ongoing and planned.[14] But when it came to predicting its mass, the authors were completely at a loss. "We apologize to experimenters for not having any idea what is the mass of the Higgs boson . . . and for not being sure of its couplings to other particles, except that they are very small," they frankly admitted at the paper's end. "For these reasons, we do not want to encourage big experimental searches for the Higgs boson, but we do feel that people performing experiments vulnerable to the Higgs boson should know how it may turn up."

But experimenters and machine builders paid this admonition little heed. That year initial plans were under discussion at CERN for a huge circular machine able to collide electrons with their antiparticles, positrons, at total energies up to 200 GeV. These discussions soon coalesced into concrete plans for the Large Electron Positron (LEP) collider with a circumference of nearly 27 kilometers, to be tunneled under the French and Swiss countryside immediately adjacent to CERN.[15] In these electron-positron collisions, the full energy of the colliding particles is available to create other particles. At LEP energies, it was envisioned that massive particles such as the W and Z bosons, plus Higgs bosons and other exotica predicted by theories beyond the Standard Model, might be created. For example, an electron and a positron could mutually annihilate each other to produce a Z boson and a Higgs boson—

one of the likely ways the latter might turn up at LEP. That would allow experimenters to search for Higgs bosons with masses upwards of 100 GeV.[16]

As CERN had long been a center for proton accelerators and colliders—and in fact, had pioneered the first proton collider, the Intersecting Storage Rings[17]—the LEP tunnel was intentionally designed wide enough to install magnets for a proton collider once the electron-positron experiments had concluded. Much higher energy protons could be circulated in this tunnel, especially if superconducting magnets were to be used to confine them to their orbits, as projected. With over 1,800 times the mass of electrons, protons lose far less energy per orbit than electrons of similar energy. So they can be accelerated to far higher energies, at the TeV scale in the case of this facility, than electrons and positrons before energy losses become significant.

While much higher proton energies can be attained, however, only a fraction of this energy becomes available to create other subatomic particles during proton-proton (or proton-antiproton) collisions. Because protons are composite particles, their energy is necessarily apportioned among their constituent quarks and gluons. Typically the quark or gluon involved in the interaction carries only a tenth or less of the colliding proton's energy—e.g., 1 TeV in a 10 TeV proton. Thus head-on collisions between two 10 TeV protons are capable of creating subatomic particles with total masses up to about 2 TeV. For example, a quark inside one proton could annihilate with an antiquark in the other to create a Z particle and a Higgs boson H. According to energy conservation, the total energy available in this collision has to be divided among the masses of the Z and the H plus their kinetic energy of motion as they depart the interaction point. That limits the mass of the Higgs bosons that might be created in this way. An even likelier production mode, especially at lower energies (and for high top-quark mass) is the fusion of two gluons—one in each colliding proton—to create a Higgs boson and nothing else.

Another problem physicists encounter is generating sufficient numbers of rare, exotic particles for experimenters to prove that they exist and to enable detailed studies of their physical properties—such as their masses, spins, and decay modes. As the easy discoveries had already been made by the late 1970s, involving particles and processes that occur in abundance, high-energy physicists began to concentrate on searches for rarer, more massive particles like the Higgs boson (and top quark) that by definition had to interact much more weakly with ordinary matter. That meant these particles would be created very infrequently in collisions. Thus, to compensate, physicists' only recourse was to increase the rate of particle collisions, known as the luminosity, in their machines. Brookhaven's Isabelle, for example, was the highest-luminosity proton collider under consideration in the early 1980s, before the SSC came on the scene with a similar design luminosity (of 10^{33} cm^{-2}/s^{-1}).[18]

The luminosities of proton colliders also had to be substantially higher than those of electron-positron colliders like LEP, because proton collisions generate far greater backgrounds of spurious, unwanted events, plus stray particle tracks that have nothing to do with the decays of the massive particles being sought. The main way to cope with these backgrounds is for experimenters to accumulate sufficient numbers of candidate events to be able to distinguish the weak signals from noise. Achieving that goal required either high luminosity, so that enough useful events occurred during a year's running time, or running an experiment for several years (which is hard on grad students and postdocs trying to launch research careers).

Increasing the collision energy was another way to cope with the low rates of production of these rare, massive particles, but that required additional real estate or stronger magnets—or both. And either of these options raises the cost of the collider significantly. One reason for the 40 TeV collision energy selected for the SSC was that production rates for essentially all of the interesting subatomic quarry were expected to increase accordingly, especially for the most massive particles at the TeV scale.[19] That design choice enabled important discoveries to be made even though the SSC luminosity was to be the same as had been planned for Isabelle.

The Large Hadron Collider was however constrained in its collision energy by the radius of the existing LEP tunnel and the maximum magnetic fields that could be attained by the superconducting magnets being designed for it. Initially thought to be 18 TeV, this energy was later lowered to 14 TeV because the original magnetic field estimates proved optimistic.[20] Such a limit naturally restricted the mass range of particles the LHC could generate, but CERN accelerator physicists intended to compensate by boosting its design luminosity by a factor of 10 over the SSC (to 10^{34} $cm^{-2}s^{-1}$).[21] That design choice increased the quantities of rarer, more massive particles it could create, thus extending the LHC's discovery range to become competitive with the SSC's. For comparison, the LHC was originally considered capable of discovering the Standard Model Higgs boson up to a mass of about 1 TeV, while the SSC should have been able to search for it—or what other symmetry-breaking phenomena might be responsible for imbuing particles with mass—up to at least 2 TeV.

The Higgs boson could not possess *any* mass, however. Considerations of unitarity—the fact that the total probability of all the interactions of two particles (in this case W bosons) cannot exceed 100 percent—required that the Higgs mass come in less than about 1 TeV.[22] Above that level, its interaction with the W and Z bosons, as well as with itself, ceased to be weak and instead became strong, like the strong force between hadrons, so that customary means of calculating interaction rates (called "perturbation theory") would no longer be reliable, leading to infinite results. And other new phenomena could then occur above 1 TeV—for example, resonant states involving W and Z bosons bound together.[23] Such a massive Higgs boson or other phenomenon would be much harder for experiments to detect, for it would not appear as a single sharp, readily discernible peak protruding from distributions of proton collision events plotted versus energy or mass. Instead, it would form an extended ledge or plateau easily obscured by backgrounds.

A broad bulge like this would be much more difficult to distinguish at the LHC, which would be able to attain only lower production rates and would experience higher backgrounds due to its greater luminosity. And LHC experiments would have little, if any, sensitivity to symmetry-breaking phenomena if they happened to show up at constituent energies or masses above 1 TeV. One such possibility, for example, was strong WW or ZZ scattering leading to a multiplicity of W and Z bosons, which could occur up to about 1.8 TeV, where another unitarity limit again kicked in.[24] To observe such phenomena above a 1 TeV constituent energy, experiments required more energy than the LHC could deliver. The 40 TeV SSC proton-proton collision energy was thus a conservative design choice that ensured that researchers could discover whatever new phenomena might happen to be responsible for generating elementary particle masses—even if they occurred above constituent collision energies of 1 TeV.[25] SSC advocates often made such arguments to support this expensive alternative.

Other, "dynamical symmetry-breaking" theories that tried to go beyond the Standard Model also suggested the existence of particles with masses above 1 TeV. Theories involving "technicolor," for example, considered the Higgs boson not to be elementary but instead a composite of two massive particles—for example, a top quark plus an antitop quark—bound together by an extremely strong technicolor force modeled after the color force that confines quarks in mesons and baryons.[26] Such an amalgamated Higgs boson could have a mass near 1 TeV. And if technicolor turned out to be the theory that nature obeyed in actual practice, a menagerie of similar composite "techniparticles" was also expected to populate the mass region

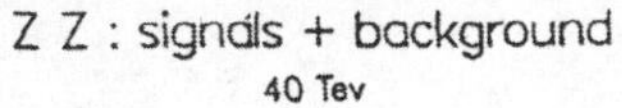

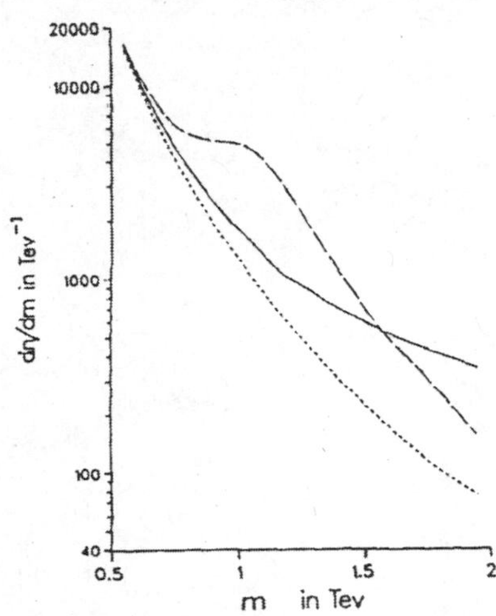

FIGURE A1.1 Computer-simulated production of a 1 TeV Higgs boson at the SSC and its decay into a pair of Z bosons (*dashed curve*). Also shown are backgrounds from other processes, including strong WW scattering (*solid curve*). Courtesy of Michael S. Chanowitz.

up to 2 TeV, providing good, if difficult, experimental targets for SSC physicists to seek.

Another set of theories that attempted to reach beyond the Standard Model involved "supersymmetry," called SUSY for short. They emerged in the 1970s and caught the fancy of many particle theorists in the early 1980s because they could dodge the "hierarchy" or "naturalness" problem of the Standard Model.[27] Briefly put, the mass of a Higgs boson should ordinarily be unstable because of its self-interaction; the resulting quantum corrections should boost its mass upwards by orders of magnitude, which would unfortunately play havoc with the theory. Supersymmetry offered a clever solution, some thought. By positing a plethora of new, as-yet undiscovered particles called "superpartners" or "sparticles"—at least one for each fundamental particle in the Standard Model—SUSY theories could largely cancel out the troublesome effects of these quantum corrections, stabilizing the Higgs boson mass below 1 TeV. And to the delight of avid experimenters, these theories predicted a cornucopia of heavy, weakly interacting particles (the sparticles) that could be sought at masses up to at most a few TeV.[28] In addition, increasing evidence came in during the 1980s and early 1990s

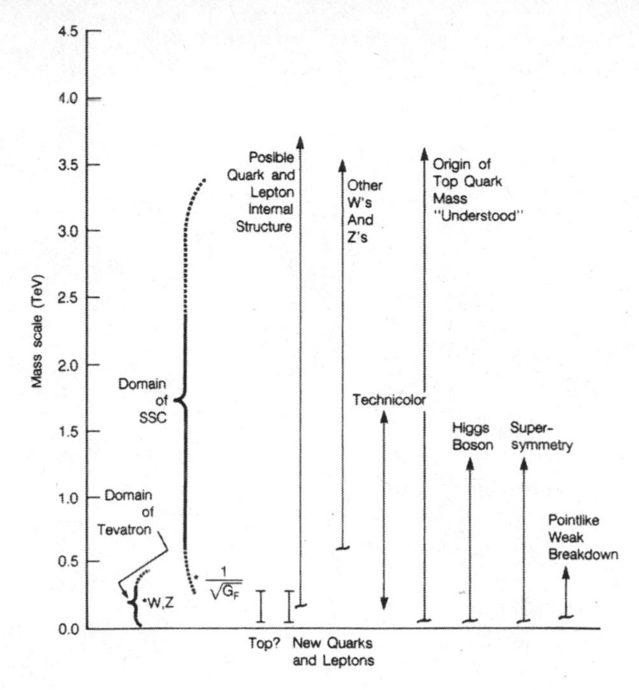

FIGURE A1.2 Phenomena at the TeV mass scale that a 40 TeV SSC could address, from *Report of the Reference Designs Study Group on the Superconducting Super Collider*. Courtesy of Fermilab Archives, SSC Collection.

for the presence of invisible (or "dark") matter gravitating around celestial galaxies and clusters; a leading candidate for this dark matter was a neutral, stable particle called the "lightest SUSY particle."[29]

In a single, bold, beautiful stroke, supersymmetry provided attractive solutions for several intriguing quandaries in physics. Theorists loved it.

SUSY theories also required the existence of *multiple* Higgs bosons. In the simplest versions called "minimal supersymmetric models," there had to be *four additional* Higgs bosons, two of them sporting electrical charges that would make them easy to identify.[30] And indeed, these charged Higgs bosons were high on experimenters' target lists in the 1980s and 1990s, especially when colliders like the Tevatron and LEP pushed into new, higher-energy domains. But the possible masses these machines could address were well below 1 TeV, and no such sparticles ever showed up at either collider.

Of the three neutral Higgs bosons that remained in minimal SUSY models, as theorists gradually realized during the late 1980s, at least one had to be fairly *light*—of the order of the Z boson mass 91 GeV.[31] Due to quantum corrections, this mass could increase to perhaps 50 percent higher than the Z mass, depending on the mass of the (then as-yet undiscovered) top quark.[32] Furthermore, such a light, supersymmetric Higgs boson (should it exist) had to behave very much like the single Higgs boson required by the Standard Model in lieu of supersymmetry. After the top quark showed up at Fermilab in 1995 with a mass close to 175 GeV, these quantum corrections could be calculated much more accurately, with the result that a relatively light Higgs boson with a mass less than 200 GeV became heavily favored. And in minimalist SUSY theories, the mass of the lightest Higgs boson had to come in below about 130 GeV.[33] Higgs boson hunters now had a target particle to seek at energies well below the TeV mass scale. It might even be accessible on lower-energy machines such as LEP and the Tevatron.

But the possibility of a light Higgs boson did not really influence the SSC design energy, which remained fixed at 40 TeV after the 1990 Drell panel supported it. That may have occurred because this terrain was presumably to be explored by LEP and the LHC—where a light Higgs boson eventually showed up in 2012 (see epilogue). For all that high-energy physicists know today, in mid-2015, the 125 GeV Higgs boson discovered at CERN may indeed be the Standard Model Higgs boson, or it may instead be the lightest of three neutral Higgs bosons in a supersymmetric extension of this still-dominant paradigm. As of this writing in mid-2015, we do not know. Only time and further experimentation at the LHC will tell.

Other possible new physics that could be addressed by the SSC and LHC, at least as understood during the 1980s when they were being designed, included new, heavier quarks, leptons, and gauge bosons like the W and Z, plus the possible internal structure of quarks and leptons.[34] Theoretically, there is no limit to the masses of such particles, so the more energy a collider might have to produce them, the better. And the more energy available, the finer any quark and lepton substructure one might be able to examine. These arguments were also used to help justify the 40 TeV SSC collision energy.

In hindsight, 40 TeV was far more than turned out to be required to discover its primary target, the Higgs boson. The SSC had been designed since the mid-1980s to reach the TeV energy scale (up to several TeV) for the collisions of proton constituents, while the more modest LHC design focused on the 100 GeV to 1 TeV energy range that was attainable by installing superconducting magnets in the LEP tunnel. The Higgs boson eventually showed up close to the bottom of this range, as theory increasingly suggested as the 1990s wore on, especially after the 1995 top quark discovery.

Appendix 2
List of Interviews

Interviewee	*Interviewers**	*Date*
Neil Baggett	R. Jacobs	July 31, 1996
Barry Barish	M. Riordan	April 17, 2010
Joseph Barton	S. Weiss	December 3, 1997
Robert Bauer & David Gross	G. Sandiford	July 22, 1997
Edward Bingler	M. Riordan	March 24, 1997
Warren Black	G. Sandiford & S. Weiss	June 10, 1997
Martin Blume	M. Riordan	February 14, 2010
Charles Brown	G. Sandiford	April 10, 1997
George Brown Jr.	M. Riordan	February 18, 1995
Thomas Bush	M. Riordan	April 2, 1998
Alex Chao	L. Hoddeson	December 29, 1999
	M. Riordan	February 3, 2009
James Chapman	M. Riordan	May 3, 2000
Bruce Chrisman	L. Hoddeson & A. Kolb	December 13, 1993
	A. Kolb	December 10, 1997
	L. Hoddeson, A. Kolb, & S. Weiss	January 14, 1998
Mark Crawford	G. Sandiford	May 26, 2010

(*continued*)

Interviewee	*Interviewers**	*Date*
Mary Cullen	G. Sandiford	August 7, 1995
James Decker	S. Weiss	February 11, 1997, & March 4, 1997
	L. Hoddeson & M. Riordan	June 3, 2009
Robert Diebold	M. Riordan	April 20, 1995
Sidney Drell	M. Riordan & S. Weiss	March 23, 1996
Eugene Dretke	M. Riordan	March 24, 1997
Aaron Edmunson	M. Riordan	April 10, 2000
Helen Edwards	A. Kolb & M. Riordan	August 21, 1998
Lyndon Evans	M. Riordan	May 24, 2000
Peter Flawn	A. Kolb	February 10, 2003
Henry Gandy	S. Weiss	December 17, 1997
John H. Gibbons	S. Weiss	August 8, 1996
Paul Gilbert	G. Sandiford	April 3, 2010
Cathy Gillespie	S. Weiss	December 22, 1997
Fred Gilman	S. Weiss	September 3, 1996
David Goldston	S. Weiss	November 21,1997
Edwin L. Goldwasser	L. Hoddeson & A. Kolb	March 1, 1993, & May 8, 1993
Irwin Goodwin	L. Hoddeson, M. Riordan, & G. Sandiford	June 4, 2009
Daniel Greenberg	G. Sandiford	May 25, 2010
David Gross	L. Hoddeson & G. Sandiford	July 22, 1997
Gail Hanson	M. Riordan	December 3, 1999
William Happer	M. Riordan	February 17, 1997
Ezra Heitowit	L. Hoddeson	June 2, 2009
Wilmot Hess	S. Weiss	July 9, 1996
J. David Jackson	S. Weiss	March 25, 1996
	L. Hoddeson & A. Kolb	May 3, 1996
Judith Jackson	G. Sandiford	February 12, 2010

Interviewee	*Interviewers**	*Date*
J. Bennett Johnston & Proctor Jones	M. Riordan	February 14, 2000
Craig Jones	G. Sandiford	February 17, 2010
Lawrence Jones	G. Sandiford	December 15, 1995
Buck Jordan	M. Riordan	March 24, 1997
Drasko Jovanovic	G. Sandiford	Undated
Raphael Kasper	M. Riordan	February 16, 1997
George A. Keyworth	L. Hoddeson	March 12, 2000
Edward Knapp	L. Hoddeson	July 14, 1997
David Kramer	M. Riordan & G. Sandiford	June 2, 2009
John Krige	L. Hoddeson, M. Riordan, A. Kolb, & G. Sandiford	October 25, 2008
Joseph Lach	G. Sandiford	August 8, 1995
Leon Lederman	L. Hoddeson & A. Kolb	November 19, 1999
	G. Sandiford	July 7 & 20, 1995
Daniel Lehman	G. Sandiford & S. Weiss	June 10, 1997
	S. Weiss	June 17, 1997
Peter Limon	L. Hoddeson & A. Kolb	May 5, 1993, & February 5, 2009
Philip Livdahl	L. Hoddeson	April 4, 1989
Chris Llewellyn-Smith	G. Sandiford	December 21, 2009
	M. Riordan	February 18, 2010
Thomas Luce	G. Sandiford	April 19, 2010, & May 12, 2010
Richard Lundy	L. Hoddeson & A. Kolb	February 2, 1989
Paul Mantsch	L. Hoddeson & A. Kolb	April 21, 1993
John H. Marburger III	L. Hoddeson & M. Riordan	June 2, 2009
Hans Mark	M. Riordan	March 31, 1998
Peter McIntyre	M. Riordan	March 20, 1997
John Metzler	M. Riordan	April 14, 1995
James C. Miller III	S. Weiss	January 14, 1997
Frederick Mills	L. Hoddeson	April 4, 1989
W. Henson Moore	S. Weiss	June 25, 1997

(*continued*)

Interviewee	*Interviewers**	*Date*
Richard Nolan	G. Sandiford	March 12, 1998
John O'Fallon	S. Weiss	February 7, 1997
Robert Palmer	M. Riordan & S. Weiss	February 6, 1996
Wolfgang K. H. Panofsky	M. Riordan & S. Weiss	March 29, 1996
Daniel Pearson	S. Weiss	December 14, 1995
John Peoples	L. Hoddeson & A. Kolb	April 5, 1989, April 4, 1990, October 1 & 23, 1993, January 25, 2011, & June 17, 2011
	A. Kolb	July 29, 2011
N. Douglas Pewitt	M. Riordan	May 3, 1998
Jack Pfister	L. Hoddeson & A. Kolb	May 17, 1994
Lee Pondrom	A. Kolb	January 25, 1990
Michael Quear	S. Weiss	November 14, 1995, & February 5, 1996
Chris Quigg	L. Hoddeson & A. Kolb	May 5 & 7, 1993, & May 16, 2011
	L. Hoddeson, A. Kolb, & M. Riordan	August 25, 2011
John Rees	M. Riordan	September 2, 2009, & May 6, 2010
Burton Richter	L. Hoddeson	August 21, 1990
	M. Riordan	September 2, 2009
Robert Roach	S. Weiss	December 13, 1995
Rustum Roy	R. Jacobs	September 30, 1996
Richard Sah	M. Riordan	May 19, 1997
James R. Sanford	G. Sandiford	February 16, 1998
Herwig Schopper	M. Riordan	May 22, 2000
Roy F. Schwitters	L. Hoddeson	March 10, 1988
	M. Riordan	March 22 1997, March 31, 1998, August 30 1999, & January 7, 2000
	A. Kolb	February 13, 2003
Marjorie Shapiro	S. Weiss	March 25, 1996

Interviewee	*Interviewers**	*Date*
Thomas Siegfried	G. Sandiford	Undated
	R. Jacobs	August 1, 1996
Thomas Siegfried & Vigdor Teplitz	A. Kolb & G. Sandiford	June 12, 1996
Edward Siskin	M. Riordan	June 2, 2000
Skip Stiles	M. Riordan	November 1, 1995
L. Edward Temple	G. Sandiford	August 14 & 20, 1997, & February 2, 2000
	A. Kolb	April 9, 2009
Vigdor Teplitz	S. Weiss	March 12, 1996
Dennis Theriot	A. Kolb	November 22, 1993
John Toll	M. Riordan & S. Weiss	February 11, 1996
Timothy Toohig	A. Kolb	January 3, 1994
	L. Hoddeson & A. Kolb	August 24, 1994
Charles Townes	S. Weiss	March 18, 1996
George Trilling	S. Weiss	March 26, 1996
Alvin W. Trivelpiece	S. Weiss	November 6, 1996
	L. Hoddeson	March 29, 2010
Mitchell Waldrop	G. Sandiford	May 26, 2010
William Wallenmeyer	M. Riordan & S. Weiss	October 24, 1995
	G. Sandiford	August 31, 1999
Gregg Ward	M. Riordan & G. Sandiford	Undated
James D. Watkins	M. Riordan & S. Weiss	February 2, 2000
Steven Weinberg	M. Riordan	March 21, 1997
Edward West	G. Sandiford	July 5, 1997
Robert R. Wilson	A. Kolb	December 1990
Bruce Winstein	A. Kolb	September 1990
Stanley Wojcicki	M. Riordan & S. Weiss	October 23, 1996
	A. Kolb & G. Sandiford	May 12, 1997
Howard Wolpe	S. Weiss	February 13, 1996

* Lillian Hoddeson, Robert Jacobs, Adrienne W. Kolb, Michael Riordan, Glenn Sandiford, and Steven Weiss.

Acknowledgments

For a book that has been three decades in the making, there are scores of individuals to acknowledge for their help and encouragement. While we are grateful to them all, we must limit our thanks to those institutions that have given us the greatest support and to a short list of individuals who have been especially generous and helpful in our effort to document, research, and write the history of the Superconducting Super Collider.

Our gratitude is best structured within a skeletal history of this long process. For Adrienne Kolb, it began in 1984, when Frank Cole of the Fermilab library committee suggested she begin collecting SSC documents in the Fermilab archives and assembling a chronology of early events in its history; he said it would be important to track these developments, as the process seemed to be repeating what had occurred during the birth of Fermilab. This work resulted in a lab report published in 1985 and updated in 1989. Fermilab librarian May West assisted Kolb, and Sue Grommes helped by transcribing early interviews, upon which we relied heavily in tracing the history of the Central Design Group.

As this historical research and archival development progressed, Lillian Hoddeson, then Fermilab's historian, joined the effort to preserve SSC documents and tell the story of how the project came into being. Other historians of physics, particularly Catherine Westfall and Peter Galison, became interested in this idea. In 1987 we requested DOE support to research the history of the SSC, but no funding was available. Kolb and Hoddeson continued working on this history, eventually publishing their first scholarly article on the SSC in 1993.

Fermilab Directors Robert R. Wilson, Leon M. Lederman, and John Peoples were especially helpful in supporting the development of the SSC Collection in the archives and also the early stages of our research. Many other physicists and administrators provided detailed information, interviews, and encouragement. In this regard, we thank Bruce L. Chrisman, Jim Cronin, Jim Decker, Edwin L. Goldwasser, Dave Jackson, Joseph Lach, Peter Limon, Robert Matyas, Wolfgang K. H. Panofsky, Chris Quigg, Burton Richter, James R. Sanford, Roy F. Schwitters, Kasuke Takahashi, L. Edward Temple, Maury Tigner, Timothy Toohig, Alvin Trivelpiece, and Stanley Wojcicki.

Meanwhile, Michael Riordan began working independently on the SSC history in 1989, after the Waxahachie site had been selected. He collected documents and recorded conversations about the SSC with his SLAC colleagues, particularly Panofsky, who was soon to become chairman of the SSC Board of Overseers, and Richter. During 1991 Riordan worked at the URA offices in Washington as assistant to President John Toll. While there, he also benefited from frequent discussions about the project's political dimensions with URA Vice President Ezra Heitowit and lobbyist Catherine M. Anderson—as well as with reporters who covered the SSC, especially Mark Crawford, Irwin Goodwin, and David Kramer. Riordan also visited the Texas SSC offices several times that year and spoke with laboratory staff, including Laboratory Director Schwitters, as well as many physicists visiting from other institutions. He recorded numerous events and conversations in his notebooks, and collected hundreds of relevant documents. Riordan is extremely grateful to these individuals and to URA for allowing an invaluable close-in view of the SSC project during that pivotal year.

In 1994, not long after the cancellation of the SSC, Riordan suggested we join forces in an effort to write the definitive history of the SSC. Other scholars were also considering writing about this dramatic failure in modern science—including Steven C. Weiss at George Washington University, who joined us as a postdoctoral researcher in the mid-1990s. In 1995, led by Hoddeson, we secured a major grant supporting our research from the National Science Foundation (SBR 94-11671, April 1995 to April 2000), which allowed us to concentrate on this effort for five years. In addition, in 1995 Hoddeson obtained funds to study the history of the SSC from URA as well as from the University of Illinois Campus Research Board. We are also grateful for additional funding during this period from the Richard Lounsbery Foundation and its President Frederick Seitz. Riordan obtained support from the DOE with the help of Peter Rosen, then associate director for high-energy and nuclear physics. In 1999 he was awarded a John Simon Guggenheim Fellowship to spend the year in Washington doing research and interviews while based at the Smithsonian Institution's National Museum of American History. Riordan also received a small travel grant from the Foundation for High-Energy Accelerator Science in Japan, which allowed him to visit Tokyo and interview Japanese government officials. With all this support, we were able to write and publish three more scholarly articles about various aspects of SSC history, but by 2000 we were still far from having a complete book manuscript.

Weiss left the project shortly thereafter, having contributed many oral-history interviews during the late 1990s, often with political figures we might not otherwise have been able to contact, let alone interview. University of Illinois graduate students Glenn Sandiford and Roberts Jacobs worked with Hoddeson during the same period. Sandiford drafted an article about the SSC site-selection process, and Jacobs researched public perceptions of the project. University of Iowa Professor Joanna Ploeger contributed valuable insights on how the rhetoric of high-energy physics had been deployed to promote the SSC. To varying degrees, ideas derived from their efforts have influenced our thinking during the long process developing the final *Tunnel Visions* manuscript.

During the late 1990s, Riordan did extensive research in archives, starting with the Panofsky Collection at SLAC, where he then worked part time; he is grateful for the guidance of SLAC archivist Jean Deken, who made this collection easily accessible, and for valuable insights from Rene Donaldson on public relations at CDG and the SSC. He also visited the George H. W. Bush Presidential Library in College Station, Texas, and the Ronald Reagan Library in Simi Valley, California, where he obtained documents from the collections of science advisors D. Allan Bromley, William Graham, and George A. Keyworth. Many relevant records of the Clinton administration, including those of John Gibbons, were available online in the digital collections of the William J. Clinton Presidential Library. Riordan was also able to copy many important documents about the SSC from the files of Congressman Sherwood Boehlert and his legislative director David Goldston, including ones they had subpoenaed from the Department of Energy. Other documents on DOE international negotiations he obtained from John Metzler. And Irwin Goodwin of *Physics Today* gave him access to his extensive files about the SSC. Riordan pored through cubic yards of unprocessed documentary remains of the SSC at the NARA records center in Fort Worth, Texas, finding only a few useful records. Fortunately, in 1995, with the help of Peoples and especially Douglas Jones at the SSC Laboratory, Kolb obtained the SSC Collection of crucial historical documents from Waxahachie; West aided her in organizing this collection and adding it to the Fermilab Archives.

Like the SSC project, the history we were writing proved much more complex than anticipated, and we were still far from finished by century's end. After a hiatus to address other responsibilities, including Hoddeson, Kolb, and Westfall's publication of the book *Fermilab* (University of Chicago Press, 2008), we began renewed efforts to complete this history. Led by Riordan, we secured another NSF grant (SES-0823296, September 2008 through August 2011, later supplemented by SES-1012014), which allowed us to fill major gaps in our research; the DOE indirectly contributed about half of these funds. And in parallel, we reorganized the book, rewriting existing chapters and adding extensive new material needed to transform our historical narrative into the present scholarly book. Drawing on his earlier study of SSC siting and doing new research about public perceptions of the project, Sandiford contributed important materials to our treatment of these subjects; but he left the collaboration before we completed the manuscript.

We are deeply grateful to the NSF Science, Technology, and Society Program for critical support of our research. In this regard, we especially thank Program Managers Ronald Overmann, Edward Hackett, and Frederick Kronz. The second NSF grant also supported reviews of an early draft of the book manuscript by Robert Crease, Robert Diebold, David Goldston, John Krige, Thomas Luce, Peter Westwick, and Stanley Wojcicki, who helped us to correct many errors and view the SSC project from their diverse perspectives. Without this generous NSF support, we would never have been able to write *Tunnel Visions*.

Just as Big Science grew, so did our wider "collaboration." We are grateful to over one hundred scientists, politicians, government officials, industrialists, magazine and newspaper reporters, and others (most listed in appendix 2) who sat for oral-history interviews about their interests or involvement in the SSC project. These interviews were transcribed largely by University of Illinois students Kristen Ehrenberger, Nicole Ryavik, Everett Carter, Derek Shouba, Melissa Rohde, and especially Kathryn Dorsh.

We thank the many institutions that supported our efforts, especially Fermilab and the Stanford Linear Accelerator Center, for allowing Kolb and Riordan freedom to work on the SSC history as part of their jobs. Fermilab's Visual Media Services Department, especially Karen Seifrid and Cindy Arnold, provided technical support in digitizing high-resolution images and duplicating drafts of the manuscript, while librarian Kathy Saumell helped on document searches. The manuscript for *Tunnel Visions* was developed in part using funds provided Fermilab by the US Department of Energy under Management and Operating Contract DE-AC02-07CH11359.

The University of Illinois Department of History was also supportive of Hoddeson's SSC research, providing opportunities to discuss this history project with knowledgeable colleagues. We especially thank Thomas Bedwell, business director of the Department of History, and Department Chairs Charles Stewart, James Barrett, Antoinette Burton, Peter Fritzsche, and Diane Koenker, who allowed Hoddeson several leaves from teaching responsibilities to work on this book. At the University of California, Santa Cruz, Abraham Seiden, director of the Santa Cruz Institute for Particle Physics, and Physics Department Chair David Dorfan provided a welcoming home base for Riordan's research efforts, particularly during the latter stages of the research under the second NSF grant. He also thanks administrators Robin Chace and Sharon Collum for their aid in managing this grant. The Department of Energy also contributed to his Santa Cruz research through grant DE-FG00-04ER41286.

We are grateful to the AIP Center for History of Physics for providing travel grants to our graduate students and postdoctoral researchers and for its steady encouragement and support—especially by Spencer Weart, Joan Warnow, Gregory Good, and Joseph Anderson, with whom Riordan worked on establishing the Irwin Goodwin Collection at the Niels Bohr Library and Archives. The APS Forum on the History of Physics sponsored conference sessions at several APS annual meetings in which we discussed aspects of the SSC history, as did the History of Science Society, International History of Science Conference, Illinois History Conference, Columbia History of Science Group, and several Laboratory History Workshops.

We thank our editor Karen Merikangas Darling for guiding us and our manuscript through the process of scholarly book publishing at the University of Chicago Press, during which two anonymous reviewers provided many helpful comments and criticisms that greatly improved the final version. In addition, we are grateful to Louise Kertesz for her expert, detailed copyediting of our manuscript and to Evan White and Yvonne Zipter for managing its production through to the published book.

Finally, we offer heartfelt thanks to our spouses—Peter Garrett, Edward "Rocky" Kolb, and Donna Gerardi Riordan for supporting us in countless ways and for enduring our many anxious moments during the lengthy course of converting *Tunnel Visions* from a hopeful idea into a successful publication.

This has been a wayward book that took on a life of its own during the many years we have been researching and writing it. We are deeply grateful to one and all for their help and forbearance in bringing the SSC history to light.

Notes

PREFACE

1. Dennis Overbye, "Collider Sets Record, and Europe Takes U.S.'s Lead," *New York Times*, December 10, 2009, D1.

2. See, for example, Adrian Cho, "Higgs Boson Makes Its Debut after Decades-Long Search," *Science*, July 13, 2012, 141–43.

3. Robert W. Seidel, "Accelerators and National Security: The Evolution of Science Policy for High-Energy Physics, 1947–1967," *History and Technology* 11 (1994): 361–91. See also Daniel J. Kevles, *The Physicists: The History of a Scientific Community in Modern America*, second ed. (Cambridge, MA: Harvard University Press, 1995), esp. xi–xii and 324–92. For a European perspective on the same period, see John Krige, *American Hegemony and the Postwar Reconstruction of Science in Europe* (Cambridge, MA: MIT Press, 2006).

4. For comparison, the construction costs of the Stanford Linear Accelerator Center and the Fermi National Accelerator Laboratory were about $400 million and $750 million in 1990 dollars. The particle accelerators at these labs stretched about 3 6 kilometers, compared to over 80 for the SSC.

5. J. L. Heilbron and Robert W. Seidel, *Lawrence and His Laboratory: A History of the Lawrence Berkeley Laboratory* (Berkeley, CA: University of California Press, 1989).

6. Robert P. Crease, *Making Physics: A Biography of Brookhaven National Laboratory, 1946–1972* (Chicago: University of Chicago Press, 1999).

7. Lillian Hoddeson, Adrienne W. Kolb, and Catherine Westfall, *Fermilab: Physics, the Frontier, and Megascience* (Chicago: University of Chicago Press, 2008).

8. A. Hermann, J. Krige, U. Mersits, and D. Pestre, *History of CERN, I* (Amsterdam: North Holland Publishing Co., 1987). This is the first of three volumes published thus far on CERN history.

9. Michael Riordan and Lillian Hoddeson, *Crystal Fire: The Birth of the Information Age* (New York: W. W. Norton, 1997).

10. Lillian Hoddeson, Paul W. Henriksen, Roger A. Meade, and Catherine Westfall, *Critical Assembly: A Technical History of Los Alamos during the Oppenheimer Years, 1943–1945* (Cambridge: Cambridge University Press, 1993).

11. Among the best treatments on this topic is Arthur Molella, *Report on Places of Invention: The First Lemelson Institute*, Incline Village, Nevada, August 16–18, 2007 (Washington, DC: Lemelson Center for the Study of Invention and Innovation, 2007); see also Sharon Traweek, *Beamtimes and Lifetimes: The World of High Energy Physicists* (Cambridge, MA: Harvard University Press, 1988).

12. For a discussion of this impact, see the preface to Kevles, *The Physicists*, esp. xxxviii–xlii.

13. A noteworthy exception is Crease's work on the Isabelle project at Brookhaven. See Robert P. Crease, "Quenched! The ISABELLE Saga," parts I and II, *Physics in Perspective* 7, no. 3 (September 2005): 330–76, and 7, no. 4 (December 2005): 404–52.

14. One section of this chapter is based on Adrienne Kolb and Lillian Hoddeson, "The Mirage of the 'World Accelerator for World Peace' and the Origins of the SSC, 1953–1983," *Historical Studies in the Physical Sciences* 24, no. 1 (1993): 101–24.

15. Hoddeson et al., "The Super Collider Affair, 1982–1989," chap. 13 in *Fermilab*, esp. 324–34. An earlier account of this phase is Lillian Hoddeson and Adrienne Kolb, "The Superconducting Super Collider's Frontier Outpost, 1983–1988," *Minerva* 38, no. 3 (2000): 271–310.

16. Michael Riordan, "The Demise of the Superconducting Super Collider," *Physics in Perspective* 2, no. 4 (December 2000): 411–25.

17. An earlier version that focused on the difficult interactions at the SSC lab between high-energy physicists and engineers from the US military-industrial complex is Michael Riordan, "A Tale of Two Cultures: Building the Superconducting Super Collider, 1988–1993," *Historical Studies in the Physical and Biological Sciences* 32, no. 1 (Fall 2001): 125–44.

18. Kevles, "The Death of the Superconducting Super Collider in the Life of American Physics," preface to second ed. of *The Physicists*, ix–xlii.

19. Stanley Wojcicki, "The Supercollider: The Pre-Texas Days" and "The Supercollider: The Texas Days," *Reviews of Accelerator Science and Technology* 1 (2008): 259–302, and 2 (2009): 265–301.

CHAPTER ONE

1. For a comprehensive summary of the Standard Model, see Laurie Brown et al., "The Rise of the Standard Model, 1964–1979," in Lillian Hoddeson et al., eds., *The Rise of the Standard Model: Particle Physics in the 1960s and 1970s* (New York: Cambridge University Press, 1997), 3–35. See also Andrew Pickering, *Constructing Quarks: A Sociological History of Particle Physics* (Edinburg: Edinburgh University Press, 1984; republished by University of Chicago Press, 1999). Historical accounts aimed more at general audiences are Robert P. Crease and Charles C. Mann, *The Second Creation: Makers of the Revolution in Twentieth-Century Physics* (New York: Macmillan and Co., 1986); and Michael Riordan, *The Hunting of the Quark: A True Story of Modern Physics* (New York: Simon & Schuster, 1987).

2. Riordan, *Hunting of the Quark*.

3. Donald Perkins, "Gargamelle and the Discovery of Weak Neutral Currents," in Hoddeson et al., *Rise of the Standard Model*, 428–46. On the cultural differences between American and European high-energy physicists, see Dominique Pestre and John Krige, "Some Thoughts on the Early History of CERN," in Peter Galison and Bruce Hevly, eds., *Big Science: The Growth of Large-Scale Research* (Stanford, CA: Stanford University Press, 1992), 78–99.

4. Riordan, *Hunting of the Quark*, 335–54.

5. Physicists measure the energies of elementary particles in electron volts, the energy acquired by an electron in passing through a potential difference, or voltage, of 1 volt—about the energy it acquires in passing through a flashlight battery. One million electron volts is 1 MeV, and a billion electron volts equals 1 GeV.

6. Sau Lan Wu, "Hadron Jets and the Discovery of the Gluon," in Hoddeson et al., *Rise of the Standard Model*, 600–621.

7. John Krige, "Distrust and Discovery: The Case of the Heavy Bosons at CERN," *Isis* 92, no. 3 (March 2001): 517–40; see also J. Krige, "The ppbar Project. I. The Collider," in John Krige, ed., *History of CERN, III* (Amsterdam: Elsevier/North Holland Publishing Company, 1996), 207–50.

8. On the origins of LEP, see Steven Myers, "The Contribution of John Adams to the Development of LEP," excerpts from the John Adams Memorial Lecture, November 26, 1990, available

at http://sl-div.web.cern.ch/sl-div/history.lep-doc.html. For a detailed description, see Herwig Schopper, *LEP—The Lord of the Collider Rings at CERN, 1980–2000: The Making, Operation and Legacy of the World's Largest Scientific Instrument* (Dordecht, Germany: Springer Verlag, 2009). The possibility of a large electron-positron collider was first proposed by Burton Richter of SLAC while on sabbatical at CERN in 1976. The idea eventually evolved into the Large Electron-Positron Collider LEP. See also n. 145, this chapter.

9. Terrence R. Fehner and Jack M. Holl, *Department of Energy, 1977–1994*, Report no. DOE/HR-0098 (November 1994), 17–31; available at http://www.energy.gov/media/Summary_History.pdf. For historical perspectives on these developments, see Catherine Westfall, "Panel Session: Science Policy and the Social Structure of Large Laboratories," in Hoddeson et al., *Rise of the Standard Model*, 364–83.

10. Comptroller General Report to Congress, US General Accounting Office, *Increasing Costs, Competition May Hinder U.S. Leadership Position in High-Energy Physics*, GAO Report no. EMD-SO-58 (September 16, 1990), esp. 17–19.

11. John M. Logsdon, *Together in Orbit: The Origins of International Participation in the Space Station*, NASA Monographs in Aerospace History 11 (Washington, DC: National Aeronautics and Space Administration, 1998), 1–2.

12. On Congressional Appropriations Committees and processes, see Richard Munson, *The Cardinals of Capitol Hill: The Men and Women Who Control Government Spending* (New York: Grove Press, 1993).

13. John Krige and Dominique Pestre, "The How and the Why of the Birth of CERN," in A. Hermann et al., *History of CERN, I*, 523–44 (see preface, n. 8); J. Krige, "CERN from the Mid-1960s to the Late 1970s," in Krige, *History of CERN, III*, 3–38; and Pestre and Krige, "Some Thoughts on the Early History of CERN." Also J. Krige, private conversations with Hoddeson and Riordan. See also the epilogue, esp. 290–91.

14. Charles R. Morris, *The Trillion-Dollar Meltdown: Easy Money, High Rollers, and the Great Credit Crash*, (New York: Public Affairs, 2008), 9–12.

15. See, for example, John Zysman and Laura Tyson, eds., *American Industry in International Competition: Government Policies and Corporate Strategies* (Ithaca, NY: Cornell University Press, 1983); and William S. Dietrich, *In the Shadow of the Rising Sun: The Political Roots of American Economic Decline* (University Park, PA: Pennsylvania State University Press, 1991).

16. Hoddeson, Kolb, and Westfall, *Fermilab*. Its energy was later raised to 400 GeV.

17. On the late-1970s budgets, see Nicholas P. Samios, "High Energy Physics at Brookhaven National Laboratory," in R. Donaldson, R. Gustafson, and F. Paige, eds., *Proceedings of the 1982 DPF Summer Study on Elementary Particle Physics and Future Facilities*, June 28–July 16, 1982 (Batavia, IL: Fermi National Accelerator Laboratory, 1983), 140–45, esp. graph on 140 for 1968–1982 budgets.

18. Peter Higgs, "Spontaneous Breaking of Symmetry and Gauge Theories," in Hoddeson et al., *Rise of the Standard Model*, 506–10. For scientific details, see F. Englert and R. Brout, *Physical Review Letters* 13 (1964): 321; P. W. Higgs, "Broken Symmetries and the Mass of Gauge Vector Bosons," *Physical Review Letters* 13 (1964): 508; G. S. Guralnick, C. R. Hagen, and T. W. B. Kibble, "Global Conservation Laws and Massless Particles," *Physical Review Letters* 13 (1964): 585–87; and Higgs, "Spontaneous Symmetry Breakdown Without Massless Bosons," *Physical Review* 145 (1966): 1156.

19. For more details, see Kolb and Hoddeson, "The Mirage of the 'World Accelerator for World Peace,'" (see preface, n. 14). Much of this section is based on that publication.

20. Krige and Pestre, "The How and Why"; Pestre and Krige, "Some Thoughts on the Early History."

21. Kolb and Hoddeson, "The Mirage of the 'World Accelerator,'" 103.

22. Robert R. Wilson, *Toward a World Accelerator Laboratory*, Fermilab Report no. TM-811 (August 16, 1978), Fermilab Archives.

23. Robert A. Divine, *Eisenhower and the Cold War* (New York: Oxford University Press, 1981), 143–52.

24. Wilson, *Toward a World Accelerator Laboratory*, 13.

25. Robert R. Wilson, "Ultrahigh-Energy Accelerators," *Science* (May 19, 1961), 1602–07.

26. "Atomic Energy: Scientific and Technical Cooperation in the Field of Peaceful Uses of Atomic Energy," June 21, 1973, Rolland P. Johnson Collection, Fermilab Archives.

27. W. O. Lock, "Origins and Early Years of ICFA," (unpublished, December 1982), Fermilab Archives.

28. Robert R. Wilson, "A World Organization for the Future of High-Energy Physics," *Physics Today* (September 1984), 9, 112.

29. Leon M. Lederman, "New Orleans—A Proposal" (unpublished, 1975), Fermilab Archives.

30. Robert R. Wilson, "A World Laboratory for World Peace," *Physics Today* (November 1975), 120.

31. James D. Bjorken, "Physics Issues and the VBA," (unpublished, May 1976), Fermilab Archives.

32. The name was later modified slightly to "International Committee *for* Future Accelerators."

33. Walter Sullivan, "Physicists Hope to Build a 30-Mile Atom Device to Explore Matter," *New York Times* (October 10, 1976), 30.

34. Leon M. Lederman, "VBA," *IEEE Transactions on Nuclear Science*, NS-24, no. 3 (June 1977): 1903–08. Lederman's map of the New York site was reprinted in *Physics Today* (May 1977), 20.

35. Lock, "Origins and Early Years of ICFA." See also Lee Teng, ed. "Possibilities and Limitations of Accelerators and Detectors," Fermilab, October 15–21, 1978; and Ugo Amaldi, ed., "Possibilities and Limitations of Accelerators and Detectors, Proceedings of the 2nd ICFA Workshop," October 4–10, 1979, Les Diablerets, Switzerland, Fermilab Archives.

36. US Department of Energy, *Report of the 1980 Subpanel on Review and Planning for the U.S. High Energy Physics Program*, Report no. DOE/ER-0066 (June 1980), quote in the transmittal letter from Sidney D. Drell to Edward A. Frieman, July 15, 1980, 1–4. HEPAP is a panel of senior high-energy physicists that meets in public and advises DOE officials on important policy decisions to be addressed in the discipline. To obtain more detailed recommendations, it often appoints "subpanels" of other high-energy physicists—and occasionally including physicists and engineers from other disciplines. These subpanels can meet behind closed doors over longer periods, eliciting detailed input from throughout high-energy physics. The subpanels then report back on the requested issues to HEPAP, which makes the official public recommendations to the DOE Office of Energy Research (or, after 1993, Office of Science). See also n.104, this chapter.

37. US Department of Energy, *Report of the Subpanel on Accelerator Research and Development of the High Energy Physics Advisory Panel*, Report no. DOE/ER-0067 (June 1980), quote in the transmittal letter from Maury Tigner to Sidney Drell, August 26, 1980, iii.

38. See, for example, the discussions in ibid., 1–12, and in *Report of the 1980 Subpanel*, 6–11. A later, much more comprehensive summary and discussion of the status of frontier research in high-energy hadron physics is E. Eichten, I. Hinchliffe, K. Lane, and C. Quigg, "Supercolllider Physics," *Reviews of Modern Physics* 56 (1984): 579–707.

39. Leon Lederman, interview by Hoddeson and Kolb, November 19, 1999. Also George Keyworth, interview by Hoddeson, March 12, 2000.

40. US Department of Energy, *Report of the Subpanel on Long-Range Planning for the U.S. High Energy Physics Program*, Report no. DOE/ER-0128 (January 1982), 46–60.

41. John B. Adams, "Framework of the Construction and Use of an International High-Energy Accelerator Complex," position paper deliberated at ICFA meeting October 21, 1981, Serpukhov, Russia; minutes of the Sixth ICFA Meeting (unpublished draft, November 5, 1981), Fermilab Archives.

42. Robert R. Wilson, "The Tevatron," *Physics Today* (October 1977), 23–30, quote on 23.

43. On the history of the Fermilab superconducting R&D program, see Hoddeson, "The First Large-Scale Application of Superconductivity: The Fermilab Energy Doubler, 1972–1983,"

Historical Studies in the Physical Sciences 18, no. 1 (1987): 25–54. See also Wilson, "The Tevatron," and Helen Edwards, "The Tevatron Energy Doubler: A Superconducting Accelerator," *Annual Reviews of Nuclear and Particle Science* 35 (1985), 605–60. For technical details of the Isabelle project, see J. R. Sanford, "Isabelle, a Proton-Proton Colliding Beam Facility at Brookhaven," *IEEE Transactions on Nuclear Science* NS-24, no. 3 (June 1977): 1845–48. For a scholarly history of this project, see Robert Crease, "Quenched!" (see preface, n. 13).

44. Hoddeson, "The First Large-Scale Application," and Sanford, "Isabelle."

45. Crease, "Quenched!," part I; and Hoddeson, "The First Large-Scale Application."

46. This two-ring design had been pioneered by CERN accelerator physicists during the late 1960s and early 1970s; it would eventually form the basis of the design of the SSC and Large Hadron Collider. For a good summary, see Kjell Johnsen, "The CERN Intersecting Storage Rings: The Leap into the Hadron Collider Era," in Hoddeson et al., *Rise of the Standard Model*, 285–98. For a scholarly history, see Arturo Russo, "The Intersecting Storage Rings: The Construction and Operation of CERN's Second Large Machine and a Survey of Its Experimental Program," in Krige, *History of CERN III*, 97–170.

47. For more details on the Isabelle magnet problems, see William J. Broad, "Magnet Failures Imperil New Accelerator," *Science* (November 21,1980), 875–78; and Gloria B. Lubkin, "Accelerator Superconducting Magnets Give Headaches," *Physics Today* (April 1981), 17–20. The latter is especially good on the technical details of these problems and the efforts to overcome them. See also Crease, "Quenched!," part II.

48. Crease, "Quenched!," part II, 405–13.

49. Ibid., 414–22.

50. Arthur L. Robinson, "Physicists Give ISABELLE a 'Yes, But . . . ,'" *Science* (November 13, 1981), 769–70. See also William J. Broad, "Limping Accelerator May Fall to Budget Ax," *Science* (August 21, 1981), 846–50; and Crease, "Quenched!," part II, 423–34.

51. Hoddeson, "The First Large-Scale Application," 29–33.

52. Ibid., 37–43.

53. Ibid., 36. See also Judy Jackson, "Down to the Wire," *SLAC Beam Line* (Spring 1993), 14–21, for details of the superconducting cable manufacturing program, especially about how Fermilab got industrial firms involved in its successful effort.

54. Crease, "Quenched!," part II, 423–25.

55. Ibid.; Hoddeson, "The First Large-Scale Application," 41.

56. Even after Brookhaven had solved its magnet design problems and built prototypes operating at 5 T, however, Westinghouse engineers said they still could not produce superconducting magnets to Brookhaven specifications. Irwin Goodwin, private communication with Riordan, March 21, 1999.

57. Hoddeson, "The First Large-Scale Application," 48. With saving energy becoming much less a concern during the Reagan years, the accelerator soon became known officially as the Tevatron.

58. Ibid., 48–50.

59. "Lisbon Conference," *CERN Courier* (September 1981), 283–88. In *John Bertram Adams: Engineer Extraordinary* (Amsterdam: Gordon and Breach Science Publishers, 1993), CERN engineer Michael Crowley-Milling claims that Adams was the one who first suggested in the late 1970s that the LEP tunnel should be built large enough to accommodate superconducting magnets for an eventual proton collider.

60. The Snowmass workshop was organized by the American Physical Society's Division of Particles and Fields (DPF), then led by Chair Charles Baltay of Yale University, to give its many members a greater voice in the definition of new high-energy physics facilities. For a good summary of the 1982 workshop and ensuing discussions, see Barbara G. Levi, "A Look at the Future of Particle Physics," *Physics Today* (January 1983), 19–21. The published proceedings are available as

Rene Donaldson et al., eds., *Proceedings of the 1982 DPF Summer Study on Elementary Particles and Future Facilities* (Batavia, IL: Fermi National Accelerator Laboratory, 1983).

61. Hoddeson, Kolb, and Westfall, *Fermilab*, 64–84.

62. On these concerns, see GAO, *Increasing Costs, Competition May Hinder U.S. Position*, n. 10, this chapter.

63. Leon Lederman, "Fermilab and the Future of HEP," in Donaldson et al., *Proceedings of the 1982 DPF Summer Study*, 125–27; quotes on 125.

64. The term "Desertron" does not appear in the *Proceedings*, but Snowmass participants say it was used at the workshop; that fall and winter, it came into common usage in the literature. Lederman referred to a "machine in the desert" and a "desert machine" in his published Snowmass talk. See also Robert Diebold, "The Desertron: Colliding Beams at 20 TeV," *Science* (October 7, 1983), 13–19.

65. See, for example, Rae Steining, "Some Thoughts about a 20 TeV Proton Synchrotron," in Teng, "Possibilities and Limitations."

66. For example, Carlo Rubbia mentioned such a desert in "The Physics Frontier of Elementary Particles and Future Accelerators," *IEEE Transactions on Nuclear Sciences*, NS-28 (June 1981), 3541–48.

67. R. Diebold et al., "'Conventional' 20-TeV, 10-Tesla, $p^{\pm}p$ Colliders," in Donaldson et al., *Proceedings of the 1982 DPF Summer Study*, 307–14, quote on 307.

68. R. R. Wilson, "Superferric Magnets for 20 TeV," in Donaldson et al., *Proceedings of the 1982 DPF Summer Study*, 330–34.

69. On closer examination and with the great benefit of hindsight, the cost estimates for the Desertron were extremely optimistic. For example, in Diebold, "The Desertron," the costs of a 10 TeV on 10 TeV proton collider are projected as only $1.8 billion (in 1983 dollars). Although this estimate included $1 billion for superconducting magnets, it had just $200 million for tunneling costs and $600 million for the remaining laboratory infrastructure (and nothing for experimental detectors, R&D, inflation, or contingencies). These figures assumed the use of 10 T magnets, which eventually proved well beyond reach but would have allowed a collider ring only 30 km in circumference (for 10 TeV beams). These were the kinds of unrealistic, optimistic numbers that high-energy physicists and US government officials were relying upon, however, when they made the 1983 decision to go ahead with the SSC as a predominantly US project.

70. Hoddeson, "The First Large-Scale Application," 53; Hoddeson, Kolb, and Westfall, *Fermilab*, 256–61.

71. On the political and economic forces that brought Reagan to power, see Joel Krieger, *Reagan, Thatcher, and the Politics of Decline* (Cambridge: Polity Press, 1986). For a more general treatment by two conservative Washington reporters, see Rowland Evans and Robert Novak, *The Reagan Revolution* (New York: E. P. Dutton, 1981).

72. On the unemployment rate in mid-1982, see Seth S. King, "Joblessness Remains at 9.5%; Rate for Adult Men Increases," *New York Times* (July 3, 1982), 1. For a breakdown by states, see Edward Cowan, "16 States Forced to get U.S. Loans to Pay the Jobless," *New York Times* (July 19, 1982), 1.

73. Tom Wicker, "Reagan's Fig Leaf," *New York Times* (July 16, 1982), A27.

74. An evaluation of the nation's technological condition in the early 1980s can be found in chapter 6 of Simon Ramo, *The Business of Science: Winning and Losing in the High-Tech Age* (New York: Hill & Wang, 1988), 184–217. See also Ramo, *America's Technology Slip* (New York: Wiley, 1980); and Ian M. Ross, "R&D in the United States: Its Strengths and Challenges," *Science* (July 9, 1982), 130–31.

75. Constance Holden, "Former South Carolina Governor to Head DOE," *Science* (February 6, 1981), 555; Colin Norman, "Commerce to Inherit Energy Research," *Science* (January 8, 1982), 147–48.

76. Eliot Marshall, "An Early Test of Reagan's Economics," *Science* (January 2, 1981), 29–31; Colin Norman, "Reagan Administration Prepares Budget Cuts," *Science* (February 27, 1981), 901–03; Eliot Marshall, "'Black Book' Threatens Synfuels Projects," *Science* (February 27, 1981), 903–06.

77. Congressman George Brown, interview by Riordan, February 18, 1995, and Robert Seidel, private conversation with Riordan. See also Daniel J. Kevles, "The Death of the Superconducting Super Collider in the Life of American Physics," preface to revised edition of *The Physicists*, ix–lxii, esp. x–xii (see preface, n. 18).

78. Colin Norman et al., "Science Budget: Coping with Austerity," *Science* (February 19, 1982), 944–47, quote on 944. Norman noted that military R&D had grown 22.2 percent in real terms from fiscal 1980 to 1982, while civilian R&D had fallen 16.1 percent; meanwhile, funding for basic research in all fields had fallen 5.5 percent in real terms over the same time span.

79. Hans Mark, *The Space Station: A Personal Journey* (Durham, NC: Duke University Press, 1987), 121–25.

80. Colin Norman, "Science Adviser Post Has Nominee in View," *Science* (May 22, 1981), 903–4. Keyworth was initially suggested to Reagan's Chief of Staff Edwin Meese by physicist Thomas Johnson during a meeting at West Point; Teller strongly supported the nomination when Meese subsequently asked him for his advice. Irwin Goodwin and Edward Teller, private conversations with Riordan.

81. Barbara J. Culliton, "Keyworth Gives First Policy Speech," *Science* (July 10, 1981), 183–84; quote on 183.

82. Arthur. L. Robinson, "CERN Sets Intermediate Vector Boson Hunt," *Science* (July 10, 1981), 191–94.

83. N. Douglas Pewitt, interview by Riordan, May 3, 1998.

84. Broad, "Limping Accelerator," 846.

85. Marjorie Sun, "No Boost in Sight for Science Budgets," *Science* (October 23, 1981), 420–21; Barbara J. Culliton, "Frank Press Calls Budget Summit," *Science* (November 6, 1981), 634–35; and Jean Coonan, "Latest Budget Cuts Arouse Concern and Recommendations," *Physics Today* (December 1981), 47–49.

86. A. Robinson, "Physicists Give ISABELLE." See also *Report of the Subpanel on Long-Range Planning*, esp. the cover letter from S. Drell to A. Trivelpiece, February 26, 1982 (see n. 40).

87. Gloria B. Lubkin, "DOE Boosts Particle-Physics Funds," *Physics Today* (April 1982), 20–21.

88. Trivelpiece, quoted in Norman, "Science Budget," 947.

89. Keyworth made these remarks at the June 23, 1982, meeting of the American Association for the Advancement of Science's Colloquium on R&D and Public Policy. His presentation was published as George A. Keyworth II, "The Role of Science in a New Era of Competition," *Science* (August 13, 1982), 606–9, quote on 607.

90. Keyworth, "The Role of Science," 608–9; Colin Norman, "The Making of a Science Adviser," *Science* (November 12, 1982), 658–60, esp. sidebar on "Reagan's Science Policy," 659.

91. Colin Norman et al., "Reagan's Budget Boosts Basic Research," *Science* (February 18, 1983), 747–51.

92. Lederman, interview by Hoddeson and Kolb.

93. Keyworth, interview by Hoddeson.

94. Ibid., also Pewitt, interview by Riordan, and Irwin Goodwin, private conversation with Riordan.

95. David Dickson, *The New Politics of Science* (Chicago: University of Chicago Press, 1988), 16.

96. Ibid., 17.

97. Ibid., viii–ix.

98. Trivelpiece, interview by Steven Weiss, November 6, 1996. These large budget increases were confirmed by Riordan in a brief analysis of OER funding during the 1980s.

99. CERN press release, January 20, 1983. See also Krige, "Distrust and Discovery," 517–18, 533–35.

100. Ibid., and CERN press release, June 1, 1983.

101. Pewitt, interview by Riordan. The "other DOE officials" were permanent staff members of the Office of Energy Research, mainly those responsible for managing the high-energy physics program.

102. Crease, "Quenched!," part II, 441–43.

103. Keyworth, interview by Hoddeson; Pewitt, interview by Riordan.

104. As with many federally supported projects, the organization and funding of high-energy physics is controlled through a system of panels and subpanels. Under the Federal Advisory Committee Act, these panels (e.g., HEPAP) must meet in public session, so subpanels are appointed that can meet and deliberate in private. In high-energy physics, groups of knowledgeable scientists are thus empanelled and charged with responsibilities to make recommendations regarding future research programs and facilities. HEPAP then accepts the recommendations and communicates them to DOE officials, who in turn decide which projects to support. But subpanel recommendations can both drive and reflect a complex negotiation of alliances between the physicists and government representatives who hold the decision-making power. It is here, long before budgets are approved and ground broken, that the "next machine" formally begins. In the seemingly rational world of science policy making, experts are consulted not just to make recommendations but also to provide important "political cover" for decisions that may already have been made by government officials. Pewitt made similar comments in his May 3, 1998, interview with Riordan, claiming that the DOE charge to the Wojcicki panel was deliberately structured to obtain approval for a Desertron-scale project and for terminating Isabelle.

105. US Department of Energy, *Report of the 1983 Subpanel on New Facilities for the U.S. High Energy Physics Program*, Report no. DOE/ER-0169 (July 1983), appendix A.

106. Stanley Wojcicki, "The Supercollider: The Pre-Texas Days," 259–302, esp. 263–64.

107. J. David Jackson, interview by Hoddeson and Kolb, May 3, 1996, emphasis added; see also Wojcicki, memo to subpanel members, March 11, 1983, Riordan SSC files. Keyworth is quoted on "thinking big" in "The Supercollider: The Pre-Texas Days," 264, and in M. Mitchell Waldrop, "Gambling on the Supercollider," *Science* (September 9, 1983), 1038–40.

108. Wojcicki, memo to subpanel members.

109. Maury Tigner, *Report of the 20 TeV Hadron Collider Workshop*, Ithaca, NY, 1983, 1–4, 52–59. In hindsight, this was a more reasonable figure than the ones suggested at Snowmass.

110. Wojcicki, "The Pre-Texas Days," 265–66.

111. "Proposal for a Dedicated Collider at the Fermi National Accelerator Laboratory" (Batavia, IL: May 1983), Fermilab Archives. Much of this proposal was written by theorist James ("BJ") Bjorken, working with Richard Lundy.

112. Jean Coonan, "DOE Boosts Budget for Physics, Especially Materials," *Physics Today* (April 1983), 49–52.

113. On June 1, 1983, CERN issued a press release on the Z discovery that made headlines around the world. See, for example, "Europe 3, U.S. Not Even Z-Zero," *New York Times* (June 6, 1983), A16.

114. Wojcicki, "The Pre-Texas Days," 266.

115. Ibid., 267, which gives the final straw vote as 10–7 against continuing it; Gloria B. Lubkin, "Panel Says: Go for a Multi-TeV Collider and Stop Isabelle," *Physics Today* (September 1983), 17–20.

116. The Energy Saver or Tevatron, as it soon began to be called, achieved a 512 GeV beam on July 3, 1983, a day after the Wojcicki panel had finished its deliberations and written its report; in February 1984, the energy reached 800 GeV. See Hoddeson, "The First Large-Scale Application," and John Peoples, "Introduction," in R. Donaldson and J. G. Morfin, eds., *Proceedings of the 1984 Summer Study on the Design and Utilization of the Superconducting Super Collider*, Snowmass, CO, 23 June–13 July 1984, v.

117. *Report of the 1983 HEPAP Subpanel on New Facilities*, i, vii–viii, 5–6.

118. Ibid.; and letter from Jack Sandweiss to Alvin Trivelpiece, July 12, 1983, i–iii.

119. Lubkin, "Panel Says Go for a Multi-TeV Collider"; Irwin Goodwin, "DOE Answers to Congress as it Officially Kills Brookhaven CBA," *Physics Today* (December 1983), 41–43.

120. For the disadvantages resulting from this cancellation, as well as the complete story of Isabelle's demise from the Brookhaven viewpoint, see Crease, "Quenched!," part II, 449.

121. Lubkin, "Panel Says Go for a Multi-TeV Collider," 19.

122. Waldrop, "Gambling on the Supercollider," 1039.

123. See, for example, Herwig Schopper, "LEP and Future Options," in F. T. Cole and R. Donaldson, eds., *Proceedings of the 12th International Conference on High-Energy Accelerators*, August 11–16, 1983 (Batavia, IL: Fermi National Accelerator Laboratory, 1983), 658–63. With 9 T magnets, CERN physicists thought they could achieve 8 TeV per beam. See S. Myers and W. Schnell, "Preliminary Performance Estimates for a LEP Proton Collider," LEP Note 440, April 11, 1983, which must have been written soon after hearing about the Wojcicki subpanel deliberations, probably from member John Adams. CERN physicists had been thinking along these lines for years. In Robinson, "CERN Sets Intermediate Vector Boson Hunt," Carlo Rubbia is quoted in July 1981 as saying, "Some of us are thinking that we could copy Fermilab's superconducting magnets and put a proton storage ring in the LEP tunnel," 194.

124. Schopper, "LEP and Future Options," 659. At a conversion rate of 1.65 Swiss francs per US dollar, typical of the early 1980s, this figure translates to $550 million, but that included only the costs of external purchases, not CERN salaries, wages, and project management expenses. If those had been included, as is done in US projects, the total LEP project costs would likely have come in close to $1 billion.

125. Ibid., 663. A closer look at this publication suggests that Schopper's material on a proton collider in the LEP tunnel was hastily added in reaction to the Woods Hole recommendation. For example, Figure 6, a cross-section of the LEP tunnel, shows only a rectangular box marked in dashed lines and labeled "S.C. p" above the detailed cross-section of a LEP quadrupole magnet. It had obviously been added as an afterthought, probably just before the Fermilab meeting in August 1983.

126. Burton Richter, memorandum to Ralph DeVries and Wallace Kornack, September 1, 1983, copies in Fermilab Archives and Richter Collection, SLAC Archives. In that memo, he notes that "Europe is now planning on spending about one billion dollars" in the 1980s for new high-energy physics projects, which included the LEP collider at CERN and the smaller HERA collider at DESY.

127. Ibid., 3.

128. For an overview of the international participation in NASA space research projects, see Logsdon, *Together in Orbit*, esp. 1–13.

129. "Summary Conclusions of the Versailles Summit Follow-on Meeting for High Energy Physics," Washington, DC, October 3–4, 1983 (unpublished draft, October 27, 1983, Richter Collection, SLAC Archives).

130. Lederman, interview by Hoddeson and Kolb; Wolfgang K. H. Panofsky, interview by Riordan and Weiss, March 29, 1996.

131. Dieter Trines, "Constructing HERA: Rising to the Challenge," *CERN Courier* (January 21, 2008), and Günther Wolf, "Physics at HERA," *Annals of the New York Academy of Sciences* 461 (December 16, 2006): 699–724.

132. For an overview of TRISTAN construction and physics goals, see Gloria B. Lubkin, "Tristan e^+e^- Collider in Japan Yields 50 GeV Center of Mass," *Physics Today* (January 1987), 21–23.

133. Notebook entry of Burton Richter (who accompanied Keyworth), March 22, 1984, 43; it reads in part, "Object is prelim disc on Japan joining SSC. Get in on beginning & help in planning. Lots beyond HEP involved, inc. Japan's role in basic sci. Better Pacific Ties. We will do it but would rather do it together." Richter Collection, SLAC Archives. See also documents in Keyworth file OA94720 titled "Japan 1984," Ronald Reagan Library Archives.

134. Richter notebook, 49, in which he also notes: "Jay [Keyworth]'s speech gets stronger! We will do it. Join us early to develop it, not just later to pay for it."

135. For details of this program, see Michael Riordan and Kasuke Takahashi, "Cooperation in High Energy Physics Between the United States and Japan," *SLAC Beam Line* (Spring 1992), 1–9.

136. CERN, " Large Hadron Collider in the LEP Tunnel," in *Proceedings of ECFA-CERN Workshop on Hadron Colliders in the LEP Tunnel*, Lausanne and Geneva, March 21–27, 1984, CERN Report no. 84–10, 1, ."ECFA" is the European Committee on Future Accelerators, a European parallel of ICFA; "hadrons" are subatomic particles (which include protons, antiprotons, neutrons, and mesons) that experience the strong nuclear force.

137. Johnsen, "The CERN Intersecting Storage Rings"; John Krige, "Distrust and Discovery." See also Krige, "The ppbar Project. I. The Collider," and Dominique Pestre, "The Difficult Decision, Taken in the 1960s, to Construct a 3–400 GeV Proton Synchrotron in Europe," in Krige, ed., *History of CERN*, III (Amsterdam: Elsevier/North Holland, 1996), 208–50 and 65–96.

138. *Proceedings of ECFA-CERN Workshop*, which states: "A centre-of-mass energy of about 18 TeV could be reached with superconducting magnets of 10 T," 4.

139. Ibid., 75.

140. This account based in part on interview of Burton Richter by Riordan, September 2, 2009, and on Richter's notes of the May 1984 ICFA meeting. Richter Collection, SLAC Archives. See also "Future Accelerators Seminar in Japan," *CERN Courier* (October 1984), 319–22; and Victor Weisskopf, "Keynote Talk," *ICFA Seminar on Future Perspectives in High-energy Physics*, May 14–20, 1984, KEK, Tsukuba, Japan, 9–16.

141. Y. Yamaguchi, "ICFA: Its History and Current Activities," in *Proceedings of the 1985 International Symposium on Lepton and Photon Interactions at High Energies* (Kyoto, 1986), 826–47. Also Goodwin, Llewellyn-Smith, and Richter, private conversations with Riordan.

142. V. L. Telegdi, "Conclusions," *ICFA Seminar on Future Perspectives*, 336–42.

143. Keyworth, quoted in Wojcicki, "The Supercollider: The Pre-Texas Days," p. 266 (emphasis in the original).

144. "Europe 3, U.S. Not Even Z-Zero."

145. "The Supercollider: "The Pre-Texas Days," 265–66. In the final report of the subpanel, Adams and Rubbia dissociated themselves from the statements about restoring US preeminence in high-energy physics. But US high-energy physicists were hardly alone in embracing competition. When CERN began to consider building a 200 GeV electron-positron collider in the late 1970s, Burton Richter suggested at an ICFA meeting that US participation be included on the project, which was essentially the electron VBA under consideration at the time. After European physicists discussed this possibility privately, ICFA Chairman Guy Von Dardel told him, "They don't want to do this." They viewed LEP as the way for Europe to catch and surpass the United States in high-energy physics and didn't want to share the likely prestige. European reluctance to cooperate on LEP may have subtly influenced the United States to pursue the SSC on its own. Richter interview by Riordan.

146. On the decision to proceed with ITER as a truly international project, see Declan Butler, "Japan Consoled with Contracts as France Snares Fusion Project," *Nature* (June 30, 2005), 1142–43; and Daniel Clery et al., "ITER Finds a Home—with a Whopping Mortgage," *Science* (July 1, 2005), 28–29.

147. As the example of ITER amply illustrates, the process of reaching international agreement on how to proceed with and manage a scientific gigaproject is far more difficult and time-consuming than doing it as a national project. But doing it internationally is probably much more sustainable in the long run. See also epilogue, on the difficulties encountered in organizing the LHC project.

148. David Dickson, "A Political Push for Scientific Cooperation," *Science* (June 22, 1984), 1317–19.

149. John Adams, "Some Remarks about New Facilities for High-Energy Particle Physics Research in the USA," April 22, 1983, quoted in Wojciki, "The Pre-Texas Days," 266.

150. Adrian Cho, "More Bad Connections May Limit LHC Energy or Delay Restart," *Science* (July 31, 2009), 522–23.

CHAPTER TWO

1. Frederick Jackson Turner, "The Significance of the Frontier in American History," speech delivered in 1893, reprinted in George Rogers Taylor, ed., *The Turner Thesis: Concerning the Role of the Frontier in American History* (Boston: D. C. Heath and Company, 1956).

2. Vannevar Bush, *Science, the Endless Frontier: A Report to the President on a Program for Postwar Scientific Research* (Washington: National Science Foundation, 1945). The US manned space program had been sold on the same basis, invoking the call of the "final frontier." See, for example, Wernher von Braun, "Crossing the Last Frontier," *Collier's* (March 22, 1952), 24–31, and other articles in that special issue on human space flight, titled "Man Will Conquer Space Soon."

3. Hoddeson and Kolb, "The Superconducting Super Collider's Frontier Outpost, 1983–1988" (see preface, n. 15). See also Hoddeson, Kolb, and Westfall, *Fermilab*, 324–34.

4. Alex Chao, e-mail to L. Hoddeson, December 24, 1999; Chris Quigg, interview by Hoddeson and Kolb, May 5, 1993.

5. *Reference Designs Study for U.S. Department of Energy: Superconducting Super Collider*, draft II (May 8, 1984), Fermilab Archives.

6. Wojcicki, "The Pre-Texas Days," 269–70; Hoddeson and Kolb, "The SSC's Frontier Outpost," 276–78. This study focused on a collider with 20 TeV proton beams.

7. Hermann Grunder et al. to Lab Directors, November 17, 1983, reprinted in appendix A, *Report of the DOE Review Committee on the Reference Designs Study*. A similar funding procedure would later be used in arranging federal support for the Central Design Group.

8. Memo from laboratory directors to Grunder et al., December 14, 1983, Lederman Collection, Fermilab Archives.

9. Mitchell created the Houston Area Research Center (HARC), and McIntyre began the TAC under its aegis. Peter McIntyre, interview by Riordan, March 20, 1997. Mitchell later became famous as the person who pioneered the development of hydraulic fracturing, or "fracking," of shale gas deposits, although much of the early R&D on this approach had been performed under DOE contracts.

10. SSC Reference Design Charter; *SSC Newsletter* published by APS/DPF, (February 15 and March 15, 1984).

11. Tigner, *Report of the 20 TeV Hadron Collider Technical Workshop*, Fermilab Archives. In 1980, Tigner had chaired a DOE HEPAP Subpanel on Accelerator Research and Development, which analyzed and recommended innovative accelerator R&D. *Physics Today* (February, 1980), 92. He also served on the Wojcicki subpanel.

12. The PSSC met on seven occasions between September 30, 1983, and April 30, 1984, to discuss the physics questions related to such design issues as luminosity, whether to employ fixed or internal targets, whether collisions should be proton-proton or proton-antiproton, and whether the energy of beams should be 10 TeV or 20 TeV. They met at Fermilab through the fall of 1983, at Brookhaven, and in Texas (at TAC in The Woodlands) that winter, and at SLAC in the spring of 1984, concluding their work in April back at Fermilab. See *PSSC: Physics at the Superconducting Super Collider Summary Report* (Batavia, IL: Fermilab, June 1984), and PSSC records, in Bruce Winstein Collection, Fermilab Archives.

13. Maury Tigner, ed., *Accelerator Physics Issues for a Superconducting Super Collider*, University of Michigan Report no. UMHE 84-1, 1984; Lawrence W. Jones, letter to A. Kolb, September 16, 2009, Fermilab Archives.

14. See chapters 5 and 6 in Hoddeson, Kolb, and Westfall, *Fermilab*, 95–156.

15. Beginning in January 1984, the Task Coordinators Group of the RDS identified technical problems. It consisted of Fermilab's Don Edwards, responsible for accelerator physics; Peter Limon, accelerator engineering systems, and Lee Teng, injector design. LBL's Tom Elioff headed cost analysis, and Jay Marx was in charge of preparing the report; Jim Sanford of Brookhaven supervised architecture and engineering.

16. The estimates covered the cost of a 40 TeV machine plus equipment and engineering but did not include the costs of research equipment or preconstruction R&D. Costs of site acquisition were to be assumed by the state where it would presumably be located. A machine based on the Brookhaven-Berkeley magnet choice, with a ring of 90 km, would cost $2.724 billion. A machine based on the Fermilab magnet design, requiring a ring of 113 km, came in at $3.05 billion, and a machine based on the Texas superferric magnets, with the much larger ring of 164 km, was estimated at $2.699 billion. The RDS based its estimate of the standard construction costs on a hypothetical "median site" with varying geological and topological conditions devised by an architecture and engineering firm, Parsons Brinckerhoff Quade and Douglas Inc.

17. *SSC Reference Designs Study*, Executive Summary, iii; "Phase 1 Program Milestones," Tigner Files, Central Design Group, Fermilab Archives.

18. *SSC Reference Designs Study*, Section 5.2.1, 11.

19. Gloria B. Lubkin, "SSC Design Goes to DOE; ICFA Discusses CERN Hadron Collider," *Physics Today* (June 1984), 17.

20. *Report of the DOE Review Committee on the Reference Designs Study*, May 18, 1984, iii.

21. M. Mitchell Waldrop, "The Supercollider, 1 Year Later," *Science* (August 3, 1984), 490–91.

22. Ibid.

23. Gloria B. Lubkin, "R&D Funding for the Super Collider," *Physics Today* (October 1984), 21.

24. Paul Mantsch, "Spirit of Snowmass Spreads Across Land," *Ferminews*, (June 14, 1984), 2–3.

25. Gloria B. Lubkin, "SSC Design Goes to DOE"; Irwin Goodwin, "SSC Cost and Size Perplex Congress," *Physics Today* (May 1984), 64; and Goodwin, "DOE Answers to Congress."

26. DOE "program staff" members like Leiss, who oversaw high-energy and nuclear physics, provide long-range program management and direction for scientific research. This kind of management is distinct from project management, which applies to a single construction project like the SSC.

27. Lubkin, "SSC Design Goes to DOE." Thanks to the efforts by CDG physicist Murdock Gilchriese, preliminary Japanese industrial and research interest was gradually won back over the next few years. Waldrop, "The Supercollider, 1 Year Later." At a May 1984 ICFA workshop held in Japan, most European and Japanese physicists appeared to have gotten over their disappointment about the US withdrawal from the VBA process (see chapter 1).

28. D. Hodel, letter to Trivelpiece, August 16, 1984; "R&D Funding for the Supercollider," 21.

29. Norman Hackerman, letter to H. Guyford Stever, July 13, 1983, Lederman/URA Collection, Fermilab Archives.

30. The document trail suggests that this decision took a while to implement in full. The Fermilab BOO was clearly in play by December 1987, according to Fermilab's annual report for that year. Prior to this juncture, Fermilab was overseen by URA's Board of Trustees. URA planned for Fermilab and the SSC to each have its own BOO reporting to the URA President and Trustees. In a memorandum of understanding, URA was named the R&D contractor for the SSC and told to create two distinct boards of overseers "to separate the direct oversight of Fermilab and the SSC Central Design Group." From "Description of Reorganization Plan for URA," undated, ca. late December 1987, URA files, general correspondence, 1986–89, Fermilab Archives.

31. Wojcicki, "The Pre-Texas Days," 270.

32. James Leiss, letter to Guyford Stever, March 1, 1984, URA Collection, Fermilab Archives. Lubkin, "SSC Design Goes to DOE"; Irwin Goodwin, "Tigner Named to Direct R&D Program for SSC," *Physics Today* (August 1984), 69. See also undated three-page document, "Management of

the R&D and Conceptual Design Phase of the Superconducting Super Collider," in Lederman files, Fermilab Archives, Box J3a3, folder marked "URA, March 20, 1984, Washington, DC." Following up on May 27, 1984, URA sent a letter to Trivelpiece "offering the assistance of the Council of Presidents in implementing whatever recommendation the DOE should receive from HEPAP on future accelerator construction."

33. For scientific details of that workshop, see R. Donaldson and J. G. Morfin, eds., *Proceedings of the 1984 Summer Study on the Design and Utilization of the Superconducting Super Collider* (see chap. 1, n. 116).

34. Gloria B. Lubkin, "UA1 at CERN Says It Has Candidates for Sixth Quark, Top," *Physics Today* (August 1984), 17.

35. Mary K. Gaillard, e-mails to Riordan, July 14–31, 2014. See also Gaillard, *A Singularly Unfeminine Profession* (Singapore: World Scientific Publishing Co., 2015).

36. Quigg, e-mail to Hoddeson, July 7, 2014. The article was published as E. Eichten et al., "Supercollider Physics" (see chap. 1, n. 38); see also appendix 1, n. 22.

37. Quigg to Hoddeson, July 7, 2014.

38. Goodwin, "Tigner Named to Direct R&D." Cornell Laboratory of Nuclear Studies staff were notified by Boyce McDaniel in a memo dated June 22, 1984. SSC Collection, Fermilab Archives.

39. Goodwin, "Tigner Named to Direct R&D."

40. Ibid.

41. Wojciki, "The Pre-Texas Days," 271–72. Wojcicki adds, quoting from the minutes of the September BOO meeting, "The Board approved these goals but also added the following statement: 'The CDG will continue to perform the studies . . . which examine tradeoffs between primary physics design goals (energy, luminosity, and number of experimental areas) and total project costs. . . .'"

42. Tigner, e-mail to Hoddeson, July 15, 2014.

43. Quigg, interview by Hoddeson and Kolb, May 5, 1993.

44. Tigner, "Phase 1 Program Milestones," undated copy of a transparency used in 1985. Tigner Files, CDG files, Fermilab Archives.

45. Rene Donaldson, private conversation with Riordan, August 24, 1998.

46. Universities Research Association, "Memorandum of Understanding," appendix E of revised unsolicited "Proposal to Serve as Contractor for the Construction and Operation of the 'Superconducting Super Collider' Laboratory," February 22, 1988, Fermilab Archives.

47. Tigner to Hoddeson, February 10, 1999. DOE frowned on labs holding funds in reserve for incentives or emergencies, because leftover monies were in danger of being reclaimed by Congress.

48. "Knapp Succeeds Stever in URA Presidency," *Ferminews*, August 8, 1985, 1–3. Stever had been the fifth president of URA; Knapp became its sixth. Leon Lederman commented, "We look forward to the regime of Ed Knapp, who will be the first full-time President of URA, and who will face the joint challenge of guiding Fermilab into the era of TEVATRON operations and of bringing SSC into existence." *Fermilab Report*, July–August 1985, 23.

49. Edward Knapp, interview by Hoddeson, July 14, 1997; Ezra Heitowit, interview by Hoddeson, June 2, 2009. Heitowit replaced James Matheson as URA vice president.

50. Lubkin, "R&D Funding for the Super Collider."

51. Mark Crawford, "California Gears Up to Bid for the SSC," *Science* (June 7, 1985), 1181.

52. For example, the group Users of the SSC held a meeting at Berkeley May 20–21, 1985. *Supercollider Newsletter* (March 1985), 1, and (April 1985), 1 (published by CDG), expressed such sentiments.

53. SSC Central Design Group, *Conceptual Design of the Superconducting Super Colllider*, Report no. SSC-SR-2020 (March 1986), Fermilab Archives.

54. Alex Chao, e-mail to Hoddeson, December 29, 1999. Chao cited the reports SSC-3 (1984), SSC-TR-2002 (1984), and SSC-SR-2020 (1986), the SSC Conceptual Design Report. Chao made similar comments in a later interview with Riordan, February 3, 2009.

55. Chao to Hoddeson, December 29, 1999.

56. Ibid. In this regard, Chao cited Tigner to N. Samios, September 10, 1984, and Tigner to B. Richter, October 15, 1984, Chao Collection, LBL Archives (Berkeley, CA).

57. *Magnet Aperture Workshop*, Report no. SSC-TR-2001 (November 1984). Chao e-mail to Hoddeson, December 29, 1999.

58. Chao to Hoddeson, December 29, 1999.

59. See, for example, Hoddeson, Kolb, and Westfall, *Fermilab*, 66, 127, 129–30.

60. Tigner, private communication with Hoddeson, March 5, 2010.

61. Maury Tigner, *Research and Development for the Super Collider, Testimony Before the Subcommittee on Energy Development and Applications, Committee on Science and Technology, United States House of Representatives*, Report no. SSC-53A (October 29, 1985); the Sciulli panel report is included in appendix 3. Lower magnetic-field designs would require larger collider footprints to achieve the same proton energy and thus larger sites available in states like Texas.

62. Bertram Schwarzschild, "Panel Reaffirms High-Field Magnet Choice for Supercollider," *Physics Today* (July 1986), 21–23.

63. William J. Broad, "Supermagnet Design Chosen for a 60-mile Atom Smasher," *New York Times* (September 19, 1985), 1.

64. Ibid.

65. M. Mitchell Waldrop, "Magnets Chosen for Supercollider," *Science* (October 4, 1985), 50.

66. Ibid. One CDG member disagreed with Waldrop's comment about SSC physicists being resigned to a delay, saying the magnet decision "has stimulated a continuing full-scale attack on the remaining major milestones." Moreover, "high energy physicists continue, with unusual unanimity, to support the SSC as the necessary next step" in exploration of the physical universe. Donald Stork, "SSC Design," letter to editor, *Science* (January 10, 1986), 103.

67. M. Mitchell Waldrop, "Congress Questions SSC Cost," *Science* (November 15, 1985), 785.

68. Harry S. Havens, "Gramm-Rudman-Hollings: Origins and Implementation," *Public Budgeting and Finance* (Autumn 1986): 4–24.

69. J. David Jackson, interview by Hoddeson and Kolb, May 3, 1996.

70. A quarter of the projected total cost, or $746 million, was for the superconducting dipole magnets, including $136 million in contingency, or only 22.2 percent of the estimated base cost of $610 million. This was a very low figure for such a high-tech item, for which CDG did not even have a successful full-length prototype—and which had given machine builders fits at Brookhaven and Fermilab. Other, much lower-tech items, such as conventional construction, were assigned similar contingencies of 20–25 percent.

71. SSC Central Design Group, *Conceptual Design of the Superconducting Super Collider*.

72. In July 1985 Brookhaven had successfully tested a 4.5 m demonstration magnet. "Supercollider: Magnet Decision," *CERN Courier* (November 1985), 383–84; Waldrop, "Magnets Chosen for Supercollider." J. D. Jackson, CDG Notebook B4, Fermilab Archives. The Brookhaven magnet team had apparently recovered from the Isabelle fiasco and was again doing excellent superconducting magnet design and fabrication.

73. But see the earlier discussion of the aperture choice and especially n. 70 (above) on the attendant uncertainties in cost, on this question.

74. Along these lines, see L. Edward Temple, "Office of Energy Research Project Performance," February 1986 (unpublished), Fermilab Archives.

75. Quigg, interview by Hoddeson and Kolb, May 5, 1993. Also J. D. Jackson to Kolb and Hoddeson, April 2, 1999, and J. D. Jackson, CDG Notebook B6, 82–89, 104–108, Fermilab Archives.

76. Schwarzschild, "Panel Reaffirms High-Field Magnet Choice."

77. *Report of the HEPAP Subpanel to Review Recent Information on Superferric Magnets*, Report no. DOE/ER-0272 (May 1986).

78. *Report of the DOE Review Committee on the Conceptual Design of the Superconducting Super Collider*, Report no. DOE/ER-0267 (May 1986), i, ii, 7–8 and 9–2. A later Temple review of the project in July 1990 became widely known as "the" Temple report on the SSC.

79. Ibid., 5–3. Having made this observation, the Temple review panel was remiss in not adding substantially to the contingency to reflect the large financial uncertainty involved. The text cites a CDG estimate that the magnet costs might rise by "a maximum of $160 million" if the 5 cm aperture had to be used, but this increase grew to $350 million in late 1989 when the option was about to be exercised by the SSC Laboratory under Roy Schwitters (see chap. 4, esp. n. 75). An additional contingency of $100–200 million, and more, was clearly warranted at this point to reflect the as-yet unresolved issue of the appropriate dipole magnet aperture to be used. By early 1991, in fact, the estimated cost of the SSC superconducting dipoles had nearly *doubled* in constant dollars, to $1.44 billion in 1991 dollars, not including inflation or contingency. See chap. 4, n. 108.

80. Ibid., and Chao to Hoddeson, December 29, 1999.

81. David Ritson, "Demise of the Texas Supercollider," *Nature* (December 16, 1993), 607–10.

82. See, for example, Mark Crawford, "Accelerator Labs Face Austere Year," *Science* (December 5, 1986), 1195; Therese Lloyd, "SSC Faces Uncertain Future," *The Scientist* (February 23, 1987), 1.

83. The Balanced Budget and Emergency Deficit Control Act, known as Gramm-Rudman-Hollings (GRH) after its three principal Senate sponsors, provided for automatic cancellation of budgetary resources (or "sequestration") if the total discretionary appropriations in certain categories exceeded the budget spending thresholds in a fiscal year. Congress and the president imposed it in December 1985 in an attempt to eliminate budget deficits by 1991. These deficits had exceeded $200 billion, owing largely to the Reagan administration's military buildup in the face of sweeping income tax cuts and the lack of major reductions in other budget areas. See Irwin Goodwin, "R&D Budget for Fiscal 1987: Life at the Threshold of Pain," *Physics Today* (May 1986), 55–60. Indeed, by the end of 1989, GRH cuts had been imposed on all federally funded programs, including DOE national laboratories. Correspondence between Tigner and Lederman during 1986 reflected much anxiety over the nation's budget austerity. See Tigner correspondence in the Lederman Collection, Fermilab Archives.

84. Leon Lederman, letter to the editor of the *New York Times*, March 10, 1986 (unpublished), Lederman Collection, Fermilab Archives. A proton-antiproton collider is what Fermilab had advocated to the Wojcicki panel in its Dedicated Collider proposal (see chap. 1, n. 111)

85. See, e.g., Mark Crawford, "Reagan Okays the Supercollider," *Science* (February 6, 1987), 625.

86. Ritson, "Demise of the Texas Supercollider."

87. Leon M. Lederman, letter to Dr. Edward Knapp, October 1, 1985, Lederman Collection, Fermilab Archives. See also Barish communications in the same collection.

88. Schwarzschild, "Panel Reaffirms High-Field Magnet Choice."

89. Crawford, "Accelerator Labs Face Austere Year"; Quigg, interview by Hoddeson and Kolb.

90. Goodwin, "R&D Budget for Fiscal 1987." Although DOE was aware that the delay in construction would raise the overall cost of the project (because of inflation), it did not expect Congress to fund construction until an official site had been selected. Although Tigner's milestones projected site selection for December 1986, this major step was stalled by political considerations.

91. Named as Keyworth's successor in June 1986, Graham was not confirmed until that October. He rose to this position from acting administrator of NASA, where his principal achievement had been troubleshooting in the wake of the Space Shuttle Challenger disaster of January 1986. See Irwin Goodwin, "At Long Last, Graham Becomes Reagan's Science Adviser," *Physics Today* (November 1986), 57–58; Alun Anderson and Joseph Palca, "White House Science Advisor Takes the Reins after Slow Start," *Nature* (October 8, 1987), 476.

92. James C. Miller III, interview by S. Weiss, January 14, 1997. See also Mark Crawford, "Reagan Okays the Supercollider," and Stephen Knott, Russell Riley, and James S. Young, "Interview with James Miller," Miller Center of Public Affairs, University of Virginia, November 4, 2001, 46.

93. Wilmot Hess, interview by Weiss, July 9, 1996. Hess's extensive US government experience had come in agencies that *directly* manage their research laboratories—as distinct from the DOE (or AEC) approach, in which this responsibility was usually farmed out to academic or industrial contractors. His selection to replace Leiss probably reflected a growing DOE desire to bring its national labs under stricter federal control, which Congress began demanding in the late 1980s after serious problems with nuclear wastes had surfaced at several of its weapons labs. The annual DOE costs of cleaning up these wastes were ballooning into the billions of dollars during the mid-1980s.

94. Irwin Goodwin, "Reagan Endorses the SSC, a Colossus Among Colliders," *Physics Today* (March 1987), 47–49. Trivelpiece, interview by Weiss, November 6, 1996.

95. Domestic Policy Council, Memorandum for the President, "Superconducting Super Collider," January 14, 1987. Domestic Policy Council file 459161, folder 2; and various materials dated December 15, 1986, for the DPC meeting on December 17, 1986. Graham file OA-94261-1, Ronald Reagan Library. In this memo, two choices were given: commit to proceed with the SSC, or delay the decision for at least another year. The Departments of Energy, Labor, Defense, and Agriculture, as well as Science Advisor Graham, the US Trade Representative, and the Office of Policy Development were listed as favoring proceeding with the SSC. The Departments of Commerce, Housing and Urban Development, Interior, and Treasury, as well as the Council of Economic Advisors and the Office of Management and Budget favored delaying it. The council was thus almost evenly divided on whether or not to proceed with SSC construction.

96. For high-energy physics, it was also a time of ongoing financial crunches at the national labs. See Crawford, "Accelerator Labs Face Austere Future." The SSC request also came in at a low point in Reagan's presidency, after the 1986 elections in which Democrats regained full control of Congress and as the Iran-Contra "arms for hostages" scandal was being revealed in late 1986. In early January 1987, the Senate established a special committee to investigate the affair, as did the House. See, for example, Bernard Gwertzman, "McFarlane Took Cake and Bible to Tehran, ex-C.I.A. Man Says," *New York Times* (January 11, 1987), and "Calls President Courageous but Weak: Iranian Exhibits Bible Signed by Reagan," *Los Angeles Times* Wire Service (January 28, 1987).

97. The fact that in this same month the Fermilab Tevatron had attained a 1.8 TeV total collision energy undoubtedly boosted Trivelpiece's confidence. Kim A. McDonald, "Reagan Backs Giant $4.4 Billion Particle Accelerator; Scientists Face Major Hurdles in Promoting the Device," *The Chronicle of Higher Education* (February 11, 1987), 7–9; Trivelpiece, interview by Hoddeson, March 29, 2010.

98. Knott et al., interview with Miller, 49. Miller, interview by Weiss, January 14, 1997.

99. On Jack London's "Credo," see the book edited by London's literary executor Irving Shepard, *Jack London's Tales of Adventure* (New York: Doubleday, 1956), vii.

100. Goodwin, "Reagan Endorses the SSC." Reagan's popular association with football derived from his role as George Gipp in the 1940 movie *Knute Rockne, All American*. See also Crawford, "Reagan Okays the Supercollider"; Ben A. Franklin, "Reagan to Press for $6 Billion Atom Smasher," *New York Times* (February 2, 1987), 1; Robert E. Taylor, "President Will Request Funds to Build World's Largest Particle Accelerator," *Wall Street Journal* (February 2, 1987), 1. A copy of the official DPC action memorandum, dated January 27, 1987, but signed January 29, was provided us by Trivelpiece, who obtained it courtesy of the Ronald Reagan Library. It contains the president's initials "RR" next to the option to proceed with the supercollider in fiscal 1988 while seeking maximum cost-sharing from other, nonfederal funding sources. SSC Collection, Fermilab Archives.

101. DOE press release, January 30, 1987, Fermilab Archives. For more detail on the behind-the-scenes maneuvering and discussions of potential foreign SSC contributions, see chapter 3.

102. To avoid a conflict of interest, he resigned from URA's Board of Trustees and joined CDG and URA, each half-time. He agreed to spend half his total time on writing the proposal.

103. "Proposal to Serve as Contractor for the Construction and Operation of the 'Superconducting Super Collider' Laboratory," submitted by Universities Research Association Inc., March 2, 1987, in URA Collection, Fermilab Archives.

104. Goldwasser, interview by Hoddeson and Kolb, March 1, 1993.

105. D. Pewitt's "Red Team Briefing Document," September 30, 1988, Fermilab Archives, suggests that the official was Salgado, but Goldwasser remembers that Herrington asked the question about urgency.

106. Goldwasser, interview by Hoddeson and Kolb, May 8, 1993.

107. John Marburger, interview by Hoddeson and Riordan, June 2, 2009; James Decker, interview by Hoddeson and Riordan, June 3, 2009. They were speaking from perspectives of decades in government service, reflecting on how the relationship between scientists and the Department of Energy had changed since the early 1980s. Marburger served as the Science Advisor to President George W. Bush from 2001 to 2009. In the interview, he said he thought that crossing the billion-dollar line was a crucial step; at that funding level, intense government scrutiny of such a major scientific project became inevitable.

108. Trivelpiece, interview by Hoddeson. He became AAAS executive director on April 27, 1987. See Barbara Culliton, "Alvin Trivelpiece of DOE is Named New Executive Officer of AAAS," *Science* (February 20, 1987), 840. See also Charles M. Susskind, "Alvin W. Trivelpiece: AAAS Executive Officer," *Science* (April 24, 1987), 377.

109. Irwin Goodwin, "Hunter and Nelson Named to DOE Posts," *Physics Today* (August 1987), 45–46; "After a Wait, Hunter Joins DOE, Keyworth, Bernthal to New Jobs," *Physics Today* (November 1988), 52.

110. W. K. H. Panofsky to William Wallenmeyer, January 22, 1987, Panofsky Collection, SLAC Archives.

111. An excellent discussion of the magnet problems can be found in Barbara Goss Levi and Bertram Schwarzschild, "Super Collider Magnet Program Pushes Toward Prototype," *Physics Today* (April 1988), 17–21. See also William J. Broad, "Giant Atom Smasher Encounters Vexing Technical Obstacles," *New York Times* (December 13, 1988), Science Times, 23 and 28.

112. Tigner, e-mail to Hoddeson, February 9, 1999.

113. Ibid. This method had proved successful in the building of Fermilab's Main Ring as well as its superconducting accelerator, the Tevatron. See C. Westfall and L. Hoddeson, "Thinking Small in Big Science," *Technology and Culture* 37, no. 3 (July 1996): 457–92; and Hoddeson, "The First Large-Scale Application." Under Tigner, CDG pursued the same approach to make available "a significantly improved technological capability as cost effectively as possible." But while this R&D program increased the current-carrying ability of the available superconducting cable by 50 percent over what had existed only five years earlier, this achievement was not appreciated beyond the high-energy physics community.

114. There was enormous pressure to solve magnet problems in time to meet goals set for the industrial manufacturers, who had agreed to be ready to begin magnet production once the SSC site was ready.

115. The early Tevatron experience had shown that for good results, testing and developing have to occur hand in hand, at least in the early stages. Hoddeson, "The First Large-Scale Application."

116. Goldwasser, interviews by Hoddeson and Kolb. See also Victor Karpenko, memorandum to M. Tigner (December 7, 1987), Fermilab Archives; and *Super Collider Newsletter*, published by APS/DPF (February 15 and March 15, 1984) for information on this interlaboratory collaboration concept. Also Quigg, interview by Hoddeson and Kolb, May 5–6, 1993.

117. Goldwasser, interviews by Hoddeson and Kolb; Peter Limon, interview by Hoddeson and Kolb, February 5, 2009.

118. L. Hoddeson et al., *Critical Assembly: a History of Los Alamos During the Oppenheimer Years, 1943–1945*. (New York: Cambridge University Press, 1993).

119. Goldwasser, interview by Hoddeson and Kolb.

120. These individuals were Tigner, Alvin Tollestrup, Richard Lundy, and Helen Edwards.

121. John Peoples, interview by Hoddeson and Kolb, October 23, 1993.

122. Ibid. See also Goss Levi and Schwarzschild, "Super Collider Magnet Program," 17–18.

123. Hoddeson, "The First Large-Scale Application."

124. Goldwasser, interview by Hoddeson and Kolb. The magnet development program was also the context in which the vast new capacity of computer modeling first made substantial impact on the SSC. Peoples, interview by Hoddeson and Kolb, October 23, 1993.

125. John Peoples, interview by L. Hoddeson and A. Kolb, June 17, 2011.

126. Goss Levi and Schwarzschild, "Super Collider Magnet Program," 18.

127. John Peoples, "Status of the SSC Superconducting Magnet Program," Report no. SSC-185, submitted to the Applied Superconductivity Conference, San Francisco, CA, September 21–26, 1988, 6–7. SSC Collection, Fermilab Archives.

128. Maury Tigner, letter to editor, *Physics Today* (August 1988), 14. What he neglected to mention was that this performance had been attained by lowering the liquid helium temperature to 3.2–3.3 K, more than a degree below the design SSC operating temperature of 4.35 K.

129. Peoples, "Status of the SSC Superconducting Magnet Program," 7.

130. The Livermore engineer had earlier been project manager of the lab's Magnetic Fusion Test Facility (part of its underground nuclear testing project).

131. Tigner to Hoddeson, February 9, 1999. This observation strongly resembles a prominent concern of Vannevar Bush in setting up the US wartime R&D program under the OSRD and later the NDRC. Despite wartime urgencies, he thought that scientists still had to be free to follow their own research instincts—and not subservient to the short-range tactical whims of military overseers—if they were to be fully effective in contributing to the war effort. See G. Pascal Zachary, *Endless Frontier: Vannevar Bush, Engineer of the American Century* (New York: The Free Press, 1999).

132. Peoples, interview by Hoddeson and Kolb, October 23, 1993.

133. Tigner to Hoddeson, February 9, 1999.

134. On this subject, see also Riordan, "A Tale of Two Cultures."

135. See, especially, F. Russell Huson and Peter M. McIntyre, "Widening the Field of SSC Magnet Competitors," letter to editor, *Physics Today* (August 1988), 14, and M. Tigner reply, 14–15. Also Robert Diebold, private conversation with Riordan, March 20, 2009; and Chris Quigg, interview by Hoddeson, Kolb, and Riordan, August 25, 2011. Quigg recalled that Huson and McIntyre had even contacted Texas Congressman Joe Barton, of the Ennis district in which the candidate Texas site was located, and that Barton or his aides had been questioning Tigner about the magnets.

136. Peoples, "Status of the SSC Magnet Program." Hess was apparently so concerned about the magnet R&D that he visited Berkeley in 1988 and "banged on the table and demanded that four working full-length magnets be completed," according to Wojcicki, e-mail to Riordan, September 9, 2011.

137. According to Quigg, there was an attempt to increase this budget to $35 million for fiscal 1988, in part to support additional magnet R&D. But this increase was eventually denied, and the CDG ended up with just $25 million. See Irwin Goodwin, "Reagan's R&D Budget Looks Great, But Congress Has Some Other Ideas," *Physics Today* (April 1988), 55–61, esp. table, 56, which shows the CDG request of $35 million in fiscal 1988 but $25 million actually allocated.

138. That position was filled in 1988 by physicist John O'Fallon, who would direct the base high-energy physics program throughout the remaining life of the SSC project.

139. Robert Diebold, interview by Riordan, April 20, 1995.

140. Quigg, interview by Hoddeson, Kolb, and Riordan. See also Goss Levi and Schwarzschild, "Super Collider Magnet Program," 18.

141. Quigg, interview by Hoddeson, Kolb, and Riordan. But Goss Levi and Schwarzschild, "Super Collider Magnet Program," 18, indicates that Congress had refused to go along with the reprogramming, at least by April 1988.

142. Quigg, interviews by Hoddeson and Kolb, May 5, 1993, and May 16, 2011.

143. Tigner, private communication to Hoddeson, May 15, 2011, and Wojcicki, e-mail to Riordan, September 9, 2011. Coming from a different federal laboratory culture in which the officials and staff are bona fide government employees, Hess may have considered it normal procedure to issue directives like this that had to be obeyed by subordinates. But such a relationship is unusual in the government-owned, contractor-operated (GOCO) laboratories characteristic of the DOE. Given all the problems with the superconducting magnet R&D in the spring of 1988, and the DOE concerns about the schedule slippage, it seems an ill-considered decision to *order* such a dramatic cutback in CDG funding, which was guaranteed to instill hard feelings among the individuals involved.

144. Revised URA "Proposal to Serve as Contractor for the Construction and Operation of the 'Superconducting Super Collider' Laboratory" (see this chap., n. 103).

145. Quigg, interview by Hoddeson and Kolb, May 5, 1993. Hunter became director of Energy Research at DOE, succeeding Trivelpiece and Decker. His confirmation was held up for over a year while he divested himself of commercial holdings that could be conflicts of interest. He became linked with the new style of "integrated" management, in contrast with the earlier physicist-oriented culture.

146. Quoted in Wojcicki, "The Pre-Texas Days," 292.

147. An announcement of the RFP was made in *Commerce Business Daily*, on August 3, 1988. It was formally issued on August 19, and published in the *Federal Register*.

148. Bruce Chrisman, interview by Hoddeson, Kolb, and Weiss, January 14, 1998. Some physicists felt, as did Richard Lundy of Fermilab, that to comply with the requirements of the RFP would be a Faustian bargain, and that "we're getting in bed with the devil." Quote attributed to Lundy by Chrisman.

149. Request for Proposals no. DE-RP02-88ER40486. URA, "For the Selection of a Management and Operating Contractor for the Establishment, Management, and Initial Operation of the Superconducting Super Collider Laboratory," appendix E, "Key Personnel," J-32.

150. Panofsky, memo to BOO Files, "Discussions with Knapp & McDaniel," August 1, 1988, Panofsky Collection, SLAC Archives.

151. From the DOE viewpoint, physicists did not have sufficient experience to manage a project of this magnitude. In 1997 the former Deputy Secretary of Energy Henson Moore stated: "The sheer size and scope of the collider made it folly for academics to be in charge." And continuing magnet problems buttressed Washington perceptions of indecision and chaos among high-energy physicists. Moore remarked that "the scientists wanted to build this like every other DOE project had been built before, and you just can't do that with something as big as the SSC." Henson Moore, interview by Weiss, June 25, 1997.

152. Memo dated August 26, 1988, in Panofsky Collection, SLAC Archives; "Many Show Interest in SSC Contract, But Few Are Expected to Bid," *Inside Energy*, September 12, 1988. See also Knapp, interview by Hoddeson.

153. Knapp, interview by Hoddeson.

154. Pewitt, interview by Riordan, May 3, 1998, plus Riordan's summary notes of this interview. See also Knapp, interview by Hoddeson.

155. Knapp, interview by Hoddeson.

156. A formal argument for teaming with industrial partners was made in a document dated October 7, 1988, by D. Pewitt and R. Schwitters, "Rationale for Teaming," Fermilab Archives. See also Bruce Chrisman, interview by A. Kolb, December 10, 1997.

157. Panofsky, letter to William Wallenmeyer, January 22, 1987.

158. Draft memorandum, presumably by Pewitt, "Rationale for Teaming," undated, Fermilab Archives. Also Knapp, interview by Hoddeson.

159. Ezra Heitowit, letter to David J. Norton, September 8, 1988, and e-mail to Riordan, November 9, 2011. See also Wojcicki, "The Pre-Texas Days," 293. As of August 26, 1988, the firms that had expressed interest in teaming with URA were: EG&G, IIT, Martin Marietta, Parsons Brinkerhoff, The Ralph K. Parsons Co., SAIC, Stone & Webster Engineering Corp., and Sverdrup Corp., according to a memorandum of that date in Panofsky Collection, SLAC Archives.

160. The agreement with Sverdrup was signed by its Vice President James C. Uselton on September 22, 1988. Knapp recalls that many were resistant to the change, even DOE's Temple, who felt the old way worked well; Knapp, interview by Hoddeson.

161. D. Pewitt, memo to E. Knapp, "SSC O&M Proposal," August 24, 1988, SSC Collection, Fermilab Archives. This memo was circulated to URA in Washington, Wojcicki at CDG, Jim Finks at Fermilab, and the members of the SSC BOO Proposal Oversight Subcommitee (M. Blume, R. Frosch, R. Schwitters, W. Panofsky, and B. McDaniel, ex officio), calling for "input from everyone by COB Wednesday, 31 August," so they could "begin soliciting Red Team reviewers."

162. Ibid.

163. Ibid. Although it wasn't called "systems engineering" in the high-energy physics culture, there were persons responsible for going around and interacting with the various design groups building an accelerator, making sure that all the many parts would fit or work together. In the case of the SSC, however, this kind of work was eventually farmed out to Lockheed, a company with a division formally skilled in systems-engineering practices. John Rees, interview by Riordan, May 6, 2010.

164. Bruce Chrisman, interview by Hoddeson and Kolb, December 13, 1993; Knapp, interview by Hoddeson.

165. Panofsky, "Discussions with Knapp & McDaniel."

166. W. Panofsky, "Talking Paper: Discussions with Trilling and McDaniel," August 8, 1988, Panofsky Collection, SLAC Archives.

167. Panofsky, "Discussions with Knapp & McDaniel."

168. W. Panofsky, memo to BOO Files, "Search for Directorial Nominee by URA," August 26, 1988, Panofsky Collection, SLAC Archives. SLAC attorney Lloyd Sides was asked to investigate whether Trivelpiece might have a conflict of interest in serving as SSC director, even though he had left his position as director of energy research more than a year earlier. Sides concluded that Trivelpiece did indeed have such a conflict because of a requirement of a two-year wait. W. Panofsky, memo to BOO Files, "Information from Lloyd Sides," August 26, 1988, Panofsky Collection, SLAC Archives.

169. The desirable attributes of a director are listed in the URA M&O Proposal, n. 196. According to several sources, an earlier version was circulated in mid-1988 by BOO member George Trilling, but we have been unable to locate a copy of it.

170. Minutes of the Meeting of the SSC Board of Overseers, Chicago, IL, August 19–20, 1988, Panofsky Collection, SLAC Archives.

171. Quoted from Panofsky, "Search for Directorial Nominee," 5. Trivelpiece might have fit this description, as he had extensive business experience at Maxwell Labs and SAIC, but he had been disqualified because of potential conflict of interest. See this chapter, n. 168.

172. On this question, see especially Wojcicki, "The Pre-Texas Days," 294–96. It is telling that Tigner was not among the three top candidates, most likely because he had already been excluded by DOE as a possible SSC director. Ezra Heitowit, private conversations with Riordan, February 16, 1994, and June 25, 2008; Diebold, conversation with Riordan, March 20, 2009.

173. Minutes of the SSC Director Search Committee Meeting, Chicago, IL, August 28, 1988, and Panofsky to BOO Files, September 6, 1988, in Panofsky Collection, SLAC Archives.

174. For details of Schwitters's role in this discovery, see Riordan, *Hunting of the Quark*, 271–85.

175. Wojcicki, "The Pre-Texas Days," 295. But David Jackson, who had served as deputy director for operations until 1986, was contacted that summer at Oxford by Samuel Treiman and asked for his recommendations regarding the SSC director. Jackson e-mail to Kolb, April 13, 2010.

176. Panofsky, conversation with Riordan, June 19, 1998. In this discussion of Tigner's exclusion, Panofsky observed that he "did not suffer fools lightly—and there are a lot of fools in Washington!" Also Diebold, conversation with Riordan, March 20, 2009.

177. Tom Kirk, private conversation with Riordan, June 17, 1999; Diebold, interview by Riordan, April 20, 1995.

178. Quoted by Wojcicki in "The Pre-Texas Days," 295. It also appears in the URA M&O Proposal; see this chap., n. 196.

179. Ezra Heitowit, private conversation with Riordan, February 16, 1994. Other reliable sources have guardedly confirmed this observation; it was and remains an extremely sensitive issue.

180. Martin Blume, interview by Riordan, February 14, 2010.

181. Panofsky, memo to BOO Files, "Directorial Nominee for URA Proposal," September 6, 1988. Also Knapp, interview by Hoddeson.

182. Knapp, interview by Hoddeson.

183. Quigg, interview by Hoddeson and Kolb, May 5, 1993. Marburger had similar recollections of this meeting; Marburger, interview by Hoddeson and Riordan.

184. Schwitters, interview by Riordan, March 22, 1997.

185. The attendees included the following members of URA's staff: Knapp, Catherine Anderson, Ezra Heitowit, and Kenneth Shirley. Others included proposal writers Chrisman, Goldwasser, Wojcicki; and proposal consultants Pewitt, Francis Allhoff of EG&G, and URA attorney Richard Hames. Also at the meeting were the SSC BOO Proposal Oversight Committee members Blume (chair), Panofsky, and Schwitters, along with members of the SSC BOO: McDaniel, John Hulm, and Trilling. Meeting notice of September 14, 1988, Berkeley, Fermilab Archives.

186. Pewitt, interview by Riordan. Tigner has a very different recollection of this encounter.

187. Chrisman, interview by Kolb, Hoddeson, and Weiss.

188. According to Pewitt, CDG did not want to cooperate. Pewitt, interview by Riordan. But according to Tigner, Pewitt never discussed the matter with him. Tigner to Hoddeson, February 9, 1999. Among those who came from CDG and attended for short advisory visits were: Tim Toohig on conventional construction, Tom Kirk on magnets, and Tom Elioff on cost and schedules. Chrisman, interview by Kolb, Hoddeson, and Weiss.

189. Knapp, interview by Hoddeson. J. D. Jackson, CDG Notebook B9, 92–93, Fermilab Archives. The SSC Board of Overseers reviewed the proposal, but the urgent deadline and the complexity of the proposal gave Pewitt and EG&G's Francis Allhoff significant power over its form and content.

190. Schwitters, interview by Riordan, March 22, 1997.

191. Ibid.

192. Wojcicki, interview by Kolb and Sandiford, May 12, 1997; J. D. Jackson, interview by Kolb and Hoddeson, May 3, 1996.

193. Schwitters, interview by Riordan, March 22, 1997.

194. Ibid.

195. Diebold, interview by Riordan.

196. Universities Research Association, "Proposal for the Selection of a Management and Operating Contractor for the Establishment, Management, and Initial Operation of the Superconducting Super Collider Laboratory," Washington DC, November 4, 1988, Fermilab Archives.

197. In a February 9, 1999, e-mail to Hoddeson, Tigner claimed that DOE officials were opposed to the prospect of his serving as SSC project manager; see chap. 4.

198. Knapp, interview by Hoddeson.

199. Those who did go to Texas included Alex Chao, Tom Elioff, Jim Sanford, Tim Toohig, and Roger Coombes. Stan Wojcicki and Dave Jackson later served on the SSC Program Advisory Committee, and in consulting roles at the laboratory.

200. On this question, see especially Wojcicki, "The Supercollider: The Pre-Texas Days," 296; and "The Supercollider: The Texas Days," *Reviews of Accelerator Science and Technology* 2 (2009): 292.

201. Mike Davis, *City of Quartz: Excavating the Future in Los Angeles* (New York: Vintage Books, 1992).

202. Catherine Westfall, "Panel Session: Science Policy and the Social Structure of Large Laboratories," in Hoddeson, Kolb, and Westfall, *Rise of the Standard Model*, 364–83.

203. Pewitt, interview by Riordan.

204. Norman Ramsey, remarks made in a panel discussion at the Third International Conference on the History of Particle Physics, Stanford Linear Accelerator Center, June 1992, summarized in Westfall, "Panel Session: Science Policy," 368.

205. By mid-1989 the estimated total cost had increased even further to $5.9 billion when Congress gave the official go-ahead to begin SSC construction. See chap. 5. And as noted earlier, the $3.0 billion figure did not include sufficient contingency to account for all the existing uncertainties in the superconducting magnet design as well as R&D.

206. Marburger, interview by Hoddeson and Riordan.

207. Pewitt, interview by Riordan.

208. Wojcicki, "The Pre-Texas Days," 293.

CHAPTER THREE

1. For a comparison of the SSC and Human Genome Projects, see Daniel J. Kevles, "Big Science and Big Politics in the United States: Reflections on the Death of the SSC and the Life of the Human Genome Project," *Historical Studies in the Physical and Biological Sciences* 27, no. 2 (1997): 269–97.

2. See, for example, Sheldon L. Glashow and Leon M. Lederman, "The SSC: a Machine for the Nineties," *Physics Today* (March 1985), 28–37. The SSC Collection at the Fermilab Archives has extensive files of press clippings on the project, dating from 1983 to 1994.

3. On the CDG public relations efforts, see Wojcicki, "The Pre-Texas Days," 280–82; also Wojcicki, interview by Riordan and Weiss, October 23, 1996.

4. Donaldson, e-mail to Kolb, February 8, 2011. Among the CDG publications produced by Donaldson were "To the Heart of Matter" (1984), "Supercollider R&D, The First Two Years" (1985), and "The Superconducting Super Collider" (1985). These and other SSC-generated PR documents can be found in the SSC Collection, Fermilab Archives.

5. Wojcicki, interview by Riordan and Weiss. Rene Donaldson, private communications with Riordan.

6. Because of its proximity to Fermilab, the *Chicago Tribune* was an exception in this regard, reporting regularly on the SSC and its scientific goals. In the fall of 1983, it endorsed Fermilab as the best site for the SSC. See "Editorial: Rooting for Fermilab," *Chicago Tribune* (October 22, 1983), 8; Ronald Kotulak, "Fermi's the Favorite in Race for Mammoth Atom Smasher," *Chicago Tribune* (October 23, 1983), E1; Kotulak, "Frontier of Physics," *Chicago Tribune* (December 2, 1984), E1.

7. Kotulak, "Frontier of Physics." See also William Broad, "Physicists Compete for the Biggest Project of All," *New York Times* (September 20, 1983), C1.

8. Goodwin, "DOE Answers to Congress as it Officially Kills Brookhaven CBA" (see chap. 1, n. 119); Gary Taubes, "Onward to the Desertron," *Physics Today* (December 1983), 9; Goodwin, "SSC Cost and Size Perplex Congress" (see chap. 2, n. 25) ; Goodwin, "SSC: Progress on Magnets, Uncertainty on Foreign Collaboration," *Physics Today* (March 1985), 63; Mark Crawford, "House

Committee Questions SSC," *Science* (April 19, 1985), 309; Goodwin, "Congressmen Review SSC with Budget Deficits on Their Minds," *Physics Today* (December 1985), 55.

9. Waldrop, "The Supercollider, 1 Year Later" (see chap. 2, n. 21). In retrospect, it seems curious that URA or CDG did not commission public-opinion polling or hire an agency to raise the SSC's profile in the public's mind, but at the time such efforts did not appear necessary to the physics-centered effort. Individual states interested in the project, such as Illinois and Texas, did commission such polls, at least within their own borders. See below, and n. 47, this chap.

10. See, for example, Kotulak, "Frontier of Physics."

11. For an early example of the use of such rhetoric, see Diebold, "The Desertron: Colliding Beams at 20 TeV" (see chap.1, n. 64). In terms of prominence in the article, cost came a distant second to technical and scientific issues, even though the SSC's estimated price was given as only $1.7–$2.2 billion—still "an unprecedented sum for a single research facility," Diebold admitted.

12. Kotulak, "Frontier of Physics."

13. Goodwin, "DOE Answers to Congress."

14. "Keyworth Decries Scientists' Negative Reaction to ABM," *Physics Today* (June 1983), 45.

15. Riordan, "A Tale of Two Cultures" (see preface, n. 17). Well into the 1980s, for example, the DOE still had strict rules about using color in the reports and other documents issued by the Office of Energy Research and the DOE national laboratories. To get around these injunctions, CDG used university funds from URA in printing its color booklets about the SSC. Donaldson, private communications with Riordan.

16. See, for example, Hoddeson, Kolb, and Westfall, *Fermilab*, 54-61, 56–59, and 88–90 on high-energy physicists' efforts to promote what became the Fermilab project before the Joint Committee on Atomic Energy. Their reports and publications were aimed at Congress and its staff, not the public at large.

17. Waldrop, "The Supercollider, 1 Year Later." In his words, "everyone flinches at the price tag." See also Waldrop, "Gambling on the Supercollider" (see chap. 1, n. 107); and Goodwin, "SSC Cost and Size Perplex Congress."

18. Broad, "Physicists Compete for the Biggest Project."

19. Kotulak, "Frontier of Physics."

20. Goodwin, "DOE Offers SSC Site Document, But Sidesteps Its Endorsement," *Physics Today* (September 1985), 53–55. See also Broad, "Supermagnet Design Chosen for a 60-mile Atom Smasher" (see chap. 2, n. 63).

21. The Site Parameters Report described the SSC collider ring, marked every 2–4 miles with a cluster of buildings, as "some vast Stonehenge on the open landscape."

22. Taubes, "Collision over the Super Collider," *Discover* (July 1985), 62.

23. Ibid., 69.

24. These talks and interviews provided material that Leon Lederman would use later in writing (with Dick Teresi) *The God Particle: If the Universe Is the Answer, What Is the Question?* See chap. 5.

25. Kotulak, "Frontier of Physics."

26. This meeting is attended by scores of journalists and science writers from around the country and world.

27. Jon Van, "Scientists Hear Plea for New Accelerator," *Chicago Tribune* (May 29, 1985), 5B. See also Broad, "Supermagnet Design Chosen."

28. See, for example, Chris Quigg, "Elementary Particles and Forces," *Scientific American* (April 1985); J. David Jackson, Maury Tigner, and Stanley Wojcicki, "The Superconducting Supercollider," *Scientific American* (March 1986); James Cronin, "The Case for the Supercollider," *Bulletin of the Atomic Scientists* (May 1986), 8–11; Chris Quigg and Roy Schwitters, "Elementary Particle Physics and the SSC," *Science* (March 1986), 1522–27; and Roy Schwitters, "Super Collider," *American Politics* (July 1986), 5–7.

29. This connection cited in Wojcicki, "The Pre-Texas Days," 281.

30. Glashow and Lederman, "The SSC: A Machine for the Nineties."

31. Goodwin, "Amazing Race: the SSC Contest Generates Disorder and Discord," *Physics Today* (May 1988), 69–74.

32. Charles J. Hailey, Gordon R. Freedman, Pedro M. Echenique, and Sheldon L. Glashow, "Superconducting Super Collider," *Physics Today* (December, 1986), 11–15. See also Freeman Dyson, "Alternatives to the Superconducting Super Collider," *Physics Today* (February, 1988), 77.

33. A noteworthy exception was Oak Ridge National Laboratory Director Alvin Weinberg, who wrote a series of influential articles about the impact of Big Science beginning in the 1960s. See especially Alvin M. Weinberg, "Criteria for Scientific Choice," *Minerva* (Winter 1963), 159–71. These articles were eventually published as a book, *Reflections on Big Science* (Cambridge, MA: MIT Press, 1967).

34. J. G. Bednorz and K. A. Muller, *Zeitschrift für Physik* B64 (1986): 189. See also Walter Sullivan, "Team Reports Breakthrough in Conductivity of Electricity," *New York Times* (February 16, 1987), 1.

35. James M. Krumhansl, letter to John S. Herrington, February 19, 1987, emphasis in the original, Goodwin Collection, Niels Bohr Library and Archives. See also Goodwin, "Will High-T_c Superconductivity Affect the SSC's Design?" *Physics Today* (August 1987), 50–52.

36. Wil Lepkowski, "Recent Discoveries Stir Debate over Superconducting Super Collider," *Chemical and Engineering News* (May 11, 1987), 14. Anderson and Krumhansl's comments were subsequently excerpted in "Dear Colleague" letters circulated in the House, identifying the two physicists as opposing construction of the SSC. See also Wojcicki, "The Pre-Texas Days," 290–91. These overly optimistic statements by prominent theorists reflected little practical comprehension of how long it could take to commercialize such a discovery. Over twenty-five years later, researchers were just beginning to manufacture suitable wire and cable from these essentially ceramic materials. See, for example, David Larbalestier, presentation at the APS annual meeting, Anaheim, CA, May 1, 2011, briefly described by George Zimmerman in "History of Superconductivity Sessions," *History of Physics Newsletter* XI, no. 4 (Spring 2011): 4–7, on 5.

37. Wojcicki, "The Pre-Texas Days," p. 291.

38. An example of the debate was the view expressed in the *New York Times* editorial, "Super Hasty on the Supercollider," *New York Times* (April 28, 1988).

39. Goodwin, "Amazing Race," 69.

40. Ibid.

41. Eliot Marshall, "Big Versus Little Science in the Federal Budget," *Science* (April 17, 1987), 249; William J. Broad, "Atom-Smashing Now and in the Future: A New Era Begins," *New York Times* (February 3, 1987), C1.

42. The meeting was sponsored by the Division of Particles and Fields of the American Physical Society and hosted by the state of Colorado and a group of Colorado universities and industries. Its proceedings, *SSC Status Report to the Nation: A National Symposium on the Superconducting Super Collider*, ed. by George W. Morgenthaler and Uriel Nauenberg (University of Colorado, October 24, 1988), reprinted talks by physicists, politicians, and administrators expressing their views on the wide-ranging benefits of building the SSC.

43. Goodwin, "After a Wait, Hunter Joins DOE, Keyworth, Bernthal to New Jobs."

44. In the same Denver speech, Keyworth recognized and encouraged the growing trend toward interdisciplinary research and partnerships between universities and industry.

45. Kotulak, "Frontier of Physics"; Mark Crawford, "Supercollider Faces Budget Barrier," *Science* (April 17, 1987), 246–48.

46. Kim McDonald, "States Spend Millions in Stiff Competition to Provide Site for Proposed Supercollider," *Chronicle of Higher Education* (April 8, 1987), 4–6.

47. "An Overview of Citizen Reactions to the Proposed Superconducting Super Collider," Northern Illinois University Center for Governmental Studies Public Opinion Laboratory, April 1987, Box N4b7, Lederman Collection, Fermilab Archives.

48. Craig Jones, interview by Glenn Sandiford, February 17, 2010.

49. Goodwin, "Amazing Race."

50. Gayle Golden, "Elections Could Boost Odds for Texas Super Collider Bid," *Dallas Morning News* (July 31, 1988); staff and wire reports, "Convention May Boost Martin's Campaign," *Greensboro News & Record* (August 19, 1988).

51. Hoddeson, Kolb, and Westfall, *Fermilab*, 54–55.

52. Leon Lederman and Chris Quigg, *Appraising the Ring: Statements in Support of the Superconducting Super Collider* (Washington: Universities Research Association, 1988).

53. Ibid. But Fitch was incorrect about Japanese spending for basic research. See chap. 5.

54. "Reagan Calls Super Collider 'Doorway to New World,'" *Los Angeles Times*, March 31, 1988. Text of his speech in "Remarks by the President in Meeting with Supporters of the Superconducting Super Collider Program," White House press release, March 30, 1988, SSC Collection, Fermilab Archives.

55. Chris Quigg, interview by Hoddeson and Kolb, May 5, 1993. The SuperCollider T-shirts were produced by CDG under Donaldson's direction.

56. Eliot Marshall characterized the SSC and the Space Station as "budget bombs" in "Big Versus Little Science." The Reagan administration had also been substantially weakened by the Democrats retaking control of the Senate in November 1986 and by the 1987 Iran-Contra scandal; see chap. 2, n. 96.

57. Quigg, interview by Hoddeson and Kolb, May 5, 1993.

58. Goodwin, "SSC Cost and Size Perplex Congress"; Crawford, "House Committee Questions SSC"; Waldrop, "Congress Questions SSC Cost"; Goodwin, "Congressmen Review SSC with Budget Deficits on Their Minds"; Crawford, "The SSC's Price Tag Troubles Congress," *Science* (February 20, 1987), 837–38; Crawford, "Supercollider Faces Budget Barrier"; Goodwin, "Race for the Ring: DOE Reacts to Congress's Anxieties on SSC," *Physics Today* (August 1987), 47–50; Goodwin, "Amazing Race"; Crawford, "Budget Crunch Stalls Super Collider," *Science* (April 1, 1988), 17–18.

59. Crawford, "CBO Cautions Congress on SSC," *Science* (October 14, 1988), 186; Robert Gillette, "Supercollider's Cost Could Rise by 50%," *Los Angeles Times* (November 8, 1988), 4. Much of this cost growth, however, reflected inclusion of items that were not part of the original CDG estimate, such as further R&D, commissioning, experiments, and inflation.

60. Memorandum, "Superconducting Super Collider (SSC)" from Presidential Science Advisor William R. Graham to Chief of Staff Donald T. Regan, November 19, 1986: "Japan could be persuaded to provide as much as $0.5B in superconducting magnets . . . Canada may share in the cost—depending on where it is sited—by perhaps as much as $100M. Western Europe may provide detectors and staff—worth perhaps as much as $100M to $200M." William R. Graham file A04261-1 (CFoA 990), Ronald Reagan Library.

61. Memorandum, "Superconducting Super Collider (SSC) Project," from Secretary of Energy John S. Herrington to the Domestic Policy Council, December 15, 1986. Graham file A04261–1 (CFoA). In the text of this memo, Herrington gives a figure of $4.5 billion in 1988 dollars, but the attached table gives $4.375 billion. These figures included a 10 percent contingency added to the estimated core cost of $4 billion. Trivelpiece to Hoddeson, January 21, 2010.

62. One draft of language on the SSC for Reagan's 1987 State of the Union address that January states, "The U.S. would provide 70% of the funding for the SSC but would not begin the project until there was a commitment by other countries to cost share in this project to the level of 30%." Domestic Policy Council file 459161 [series FG010–03], folder 2. This language was not used in the address, but its use is indicative of the fact that by mid-January 1987, the Reagan administration had already begun to discuss 30 percent nonfederal contributions to the SSC.

63. Domestic Policy Council, "Memorandum for the President on the Superconducting Super Collider," January 14, 1987. Domestic Policy Council file 459161, folder 2. Ronald Reagan Library.

64. Ibid. This was an optimistic assessment based more on hope than reality. By early 1987, European governments had already decided that a proton collider in the LEP tunnel had higher priority than contributions to the SSC.

65. Edwin W. Meese III, "Memorandum for the Domestic Policy Council on the Superconducting Super Collider (SSC)," January 30, 1987, Domestic Policy Council file 459161 [series FG010-03], folder 2. Ronald Reagan Library.

66. US Department of Energy press release, "DOE Requests Funding to Build the Superconducting Super Collider," January 30, 1987. Also cited in transcript of the press conference by Ace-Federal Reporters Inc., Irwin Goodwin Collection, Niels Bohr Library and Archives, American Institute of Physics, College Park, MD. During the ensuing question-and-answer period, reporter Chris George of the British science magazine *New Scientist* asked, "What indication do you have from foreign countries that they are even interested in participating monetarily at all?" Herrington replied, "We have indications from foreign governments. . . . There is tremendous interest outside of the borders of the United States in this project," thus dodging the question on specific countries or commitments. Transcript, 10–11.

67. Transcript of Herrington press conference, 12.

68. James F. Decker, Statement before House Subcommittee on International Scientific Cooperation, March 16, 1988, SSC files of NY Congressman Sherwood Boehlert.

69. Ibid., 3.

70. Memorandum from Todd Schultz, Republican technical consultant, to Republican members of the House International Scientific Cooperation Subcommittee, March 16, 1988, Boehlert files.

71. Ibid., 1.

72. There were frequent rumors and suggestions that Italy, which had a long history of strong support for high-energy physics, might be able to come up with a contribution of as much as a billion dollars. In a telephone interview and conversation with Hoddeson, Trivelpiece recalled that he had such a verbal agreement with Italian Foreign Minister Guilio Andreotti, who later became prime minister. The Italian firm Ansaldo was manufacturing superconducting magnets for the HERA project and was thus well positioned to supply them for the SSC. But nothing concrete ever came of this interest. Trivelpiece interview by Hoddeson, 2010. See also Wojcicki, "The Supercollider: The Pre-Texas Days," 283.

73. In fact, the United Kingdom had suggested it might have to reduce its CERN contribution because of the resulting pressures on the rest of the nation's scientific program, which was funded from the same pot of money. See William Sweet, "Abragam and Rubbia Reports Chart Future for CERN," *Physics Today* (September 1987), 71–75. Also Chris Llewellyn-Smith, interview by Riordan, February 18, 2010; in the mid-1980s he had served as a member of Great Britain's Kendrew committee, which recommended cutting the CERN budget by 25 percent to reduce these pressures.

74. Along with the proposal of this cross-border site came the suggestion that Canada could supply cheap hydroelectric power from extensive resources in Quebec, which would reduce the long-term operating costs of the SSC. The DOE summarily ruled out this site because it did not conform to the prescription that a site had to be entirely within US borders (see n. 111, this chap.). Notes of conversation among Boehlert, Riordan, and Weiss, October 30, 1997; David Goldston, Riordan notes of presentation at APS meeting, April 29, 2001, plus e-mail to Riordan, May 20, 2010. Also Boehlert, letter to Allan Gottlieb, Canadian ambassador to the United States, October 29, 1987, and Boehlert press release, "Boehlert Says Briefing on SSC Strengthens His Resolve to Require International Funding of Project," February 2, 1988, SSC files of Boehlert/Goldston.

75. A good summary of this process, especially as regards the SSC, can be found in a memorandum by Panofsky, "Remarks on International Participation in SSC," October 24, 1988, Panofsky Collection, SLAC Archives.

76. This R&D work was being done under the auspices of the Implementing Agreement on U.S./Japan Cooperation in High-Energy Physics, signed initially in 1979 and thereafter renewed

every five years. See Riordan and Takahashi, "Cooperation in High-Energy Physics" (see chap. 1, n. 135), 7–8.

77. Lubkin, "Tristan e^+e^- Collider in Japan" (see chap. 1, n. 132), 21.

78. For comparison, NASA vigilantly tried to avoid possible transfer of any sensitive technologies as part of its international collaboration efforts. That is a principal reason why its foreign collaborators were generally limited to contributing scientific payloads while NASA built the satellite launchers all by itself; there are strong parallels between launchers and accelerators (or colliders), on the one hand, and between payloads and particle detectors, on the other. John Krige, private communications. See also John Logsdon, *Together in Orbit*.

79. Panofsky, interview by Riordan and Weiss.

80. Panofsky, "International Collaboration for the SSC," memorandum to the URA Board of Overseers, February 13, 1985, Panofsky Collection, SLAC Archives. In discussions with Riordan, he subsequently estimated that the actual cost avoidance factors came in somewhere between 50 and 80 percent, depending on the nation and its contribution. See also the observations of Volker Soergel on HERA contributions, cited in Wojcicki, "The Pre-Texas Days," 284.

81. The SSC M&O Contract between URA and DOE was signed in January 1989. See chap. 4.

82. *Superconducting Super Collider: Hearings Before the House Committee on Science, Space, and Technology*, April 7–9, 1987, 288–93.

83. Congressional Budget Office, *Risks and Benefits of Building the Superconducting Super Collider*, October 1988, xiii–xix, 63–79. The CBO report was subsequently criticized for lending too much credibility to projects such as the LHC and a TeV-scale electron-positron linear collider, for which designs were not as mature as that of the SSC in 1988. See also Sweet, "Abragam and Rubbia Reports," 73–75; Crawford, "CBO Cautions Congress on SSC"; and Wojcicki, "The Pre-Texas Days," 282–84. Robert Diebold mentioned this report in his April 20, 1995, interview with Riordan.

84. According to LHC Project Manager Lyndon Evans, the final construction cost of the collider, not including the experimental detectors and computers, nor the salaries, wages, and benefits of CERN employees working on the project, was 3.245 billion Swiss francs in 2008 (then $3.15 billion), about twice what had been projected in the late 1980s. Evans e-mail to C. Llewellyn-Smith, March 10, 2010.

85. Schultz, memo to Republican members, 1.

86. Wojcicki, "The Pre-Texas Days," 292.

87. This one-third figure for nonfederal contributions appears to have originated within the Office of Management and Budget, which was skeptical of the project throughout its history, but we have not been able to obtain documentary evidence for this statement. It also corresponds roughly to what the DOE was claiming it could expect from state and foreign sources in the late 1980s. James C. Miller, interview by S. Weiss, January 14, 1997.

88. Congressional Budget Office, *Risks and Benefits of Building the Superconducting Super Collider*, 53.

89. Hoddeson, Kolb, and Westfall, *Fermilab*, 70–84; Catherine L. Westfall, "The Site Contest for Fermilab," *Physics Today* (January 1989), 44–52; Theodore J. Lowi, *Poliscide: Big Government, Big Science, Big Politics* (New York: Macmillan, 1976), 61–80.

90. William Wallenmeyer, interview by Riordan and Weiss, October 24, 1995; L. Edward Temple, interview by G. Sandiford, August 14, 1997. See also April 1987 letter from Trivelpiece, "Prospective Proposers," explaining this position, SSC Collection, Fermilab Archives; and Trivelpiece, interview by Weiss, November 6, 1996.

91. Herwig Schopper, presentation at APS Meeting, Washington, DC, April 29, 2001; Llewellyn-Smith, interview by Riordan. See also Dominique Pestre, "The Difficult Decision, Taken in the Late 1960s, to Construct a 3–400 GeV Proton Synchrotron in Europe" (see chap. 1, n. 137).

92. Leon Lederman, "Fermilab and the Future of HEP" (see chap. 1, n. 63).

93. Lederman also stated later that the best laboratories mix youth and experience, with senior researchers refining and otherwise mentoring the creative energies of their junior colleagues. Lederman, interviews by G. Sandiford, July 7 and 20, 1995.

94. James Sanford, interview by G. Sandiford, February 16, 1998. A particle physicist known to work well with engineers, Sanford had been a key manager in the early days of NAL and later served as project manager on the Isabelle project at Brookhaven.

95. Ibid.

96. Ibid. See also *SSC Reference Designs Study* (March 1984), Section 5.2.1; SSC Central Design Group, *Conceptual Design of the Superconducting Super Collider* (see chap. 2, n. 53), 3; SSC Monthly Reports, August and September 1985, SSC Collection, Fermilab Archives.

97. Trivelpiece, interview by Weiss.

98. Trivelpiece, interview by Hoddeson.

99. Temple, interview by Sandiford.

100. William Wallenmeyer, interview by Sandiford, August 31, 1999. He described the likelihood of "political reasons" justifying evaluation of the site to best ensure the future outcome of the scientific research to be performed at the SSC.

101. Temple, interviews by Sandiford, August 14 and 20, 1997; Wilmot Hess, interview by Weiss, July 9, 1996. The SSC was characterized as the "crown jewel of particle physics" by Goodwin, "DOE Submits 36 SSC Site Bids While House Seeks to Micro-manage Project," *Physics Today* (November 1987), 45–46.

102. Peter T. Flawn, interview by Kolb, February 10, 2003. See also Peter T. Flawn, *The Story of the Texas National Research Laboratory Commission: How Texas Won . . . and Lost* (Austin, TX: Bureau of Economic Geology, 2003), 47–50. Texas A&M physicist Peter McIntyre made a similar observation in an interview with Riordan, March 20, 1997.

103. McDonald, "States Spend Millions in Stiff Competition to Provide Site for Proposed Supercollider." Five had spent over $2 million; they were Arizona, California, Colorado, Illinois, and Ohio.

104. Central Design Group, *Superconducting Super Collider: Siting Parameters Document*, Report no. SSC-SR-2040 (February 1987); according to Flawn, *The Story of the TNRLC*, 8, an earlier draft of this report was issued in June 1985; Sanford, interview by Sandiford.

105. Warren Black, interview by Sandiford and Weiss, June 10, 1997; Sanford interview by Sandiford.

106. US Department of Energy, *Invitation for Site Proposals for the Superconducting Super Collider*, Report no. DOE/ER-0315 (April 1987). The actual SSC site announcement occurred on November 10, 1988 (see below).

107. Ibid., Section 3.2.

108. Ibid., Section 3.3.

109. David L. Gross, interview by Hoddeson and Sandiford, July 22, 1997; Temple, interviews by Sandiford.

110. Richard Slansky, *SSC Site Atlas*, 2nd ed., Los Alamos Report no. LA-UR-84-3893 (1984), SSC Collection, Fermilab Archives; McDonald, "States Spend Millions"; Goodwin, "DOE Submits 36 SSC Site Bids"; Daniel Lehman, interview by Sandiford and Weiss, June 10, 1997; Robert Bazell, "Quark Barrel Politics," *New Republic* (June 22, 1987), 9–10.

111. James Decker, interview by Hoddeson and Riordan, June 3, 2009. The cross-border site also did not meet Herrington's rhetorical position, stated publicly earlier that year, that the SSC would be "an American project." (See above, this chap.)

112. The summary DOE rejection of the cross-border site convinced New York Congressman Sherwood Boehlert and his staff that its leaders were not really serious about seeking foreign participation in the SSC project, for such a truly international site, like CERN, would have been much more attractive to foreign governments considering this prospect. They acknowledged that the site may not have had very good geology, and that the closest major airport was in Montreal (about

fifty miles away over less than the best roads). But they felt strongly that the site should have been allowed to proceed to the next level of review and not thrown out at this early stage. Boehlert, private conversation with Riordan and Weiss. In a presentation to the American Physical Society meeting in Washington, DC, April 29, 2001, Boehlert aide David Goldston, then staff director of the House Science Committee, said that the DOE rejection of this site was "the first signal that the SSC was not going to be an international project." At a time of intensive discussions of the North American Free Trade Agreement with Canada, he joked, it was entirely appropriate that "protons might cross the border, too." David Goldston, April 29, 2001, in Riordan notes.

113. "Super Collider Review Trims 7, Leaving 36 Sites in Running," *Science and Government Report* (October 1, 1987), 1–4. David Goldston, interview by Weiss, November 21, 1997.

114. Temple, interviews by Sandiford. DOE later added another category to deal with factors related to life-cycle costs of the finalist sites (see below).

115. Raphael Kasper, interview by Riordan, February 16, 1997; Temple, interviews by Sandiford. See also Super Collider Site Evaluation Committee (NAS and NAE), *Siting the Superconducting Super Collider* (Washington: National Academy Press, 1988), appendix A. The quoted text of the work statement is on 39–40.

116. US Department of Energy, *Best Qualified Sites for the Superconducting Super Collider*, Report of the United States Department of Energy Superconducting Super Collider Site Task Force (January 1988).

117. US General Accounting Office, *Federal Research: Determination of the Best Qualified Sites for DOE's Super Collider*, Report no. GAO/RCED-89-18 (January 1989). States have sites listed in parentheses here because some of them submitted other sites that did not make the Best Qualified List.

118. Wojcicki, deputy director of the CDG, who served on the NAS/NAE Committee, recalled the top four finalists as Texas, Illinois, North Carolina, and Tennessee, "roughly in that order." Wojcicki, interview by Kolb and Sandiford, May 12, 1997. See also Kasper, interview by Riordan.

119. GAO, *Determination of the Best Qualified Sites for DOE's Super Collider*; see also John N. MacLean, "Illinois Gains in Bid for Supercollider," *Chicago Tribune* (December 30, 1987), 1; DOE, *Best Qualified Sites for the SSC*, 7; Kasper, interview by Riordan.

120. Gross, interview by Hoddeson and Sandiford; William Hart, "Evaluation Leaves State Super Sore," *Dallas Morning News* (February 14, 1988). The state referred to here is New Mexico, not Texas.

121. Mark Crawford, "SSC Sites: Then There Were Eight," *Science* (January 8, 1988), 133–34.

122. Gross, interview by Hoddeson and Sandiford.

123. Dave Barry, "Superconductivitexpialidocious," *Washington Post* (October 18, 1987), Style, 1.

124. Danny Lineberry, "No-funds Collider May Enrich Plan," *Durham Morning Herald* (April 6, 1988); Max Gates, "Politics a Key to Super Collider Site Selection," *Ann Arbor News* (February 21, 1988); Gates, "State Must Try Harder to Win SSC, Expert Says," *Ann Arbor News* (March 13, 1988).

125. Mike Magner, "Collider Finalists Trying to Boost Support for Project," *Kalamazoo Gazette* (March 24, 1988). Lobbyist quote excerpted from Sean Griffin, "Teamwork Aids Arizona Supercollider Effort," *Phoenix Gazette* (June 20, 1988), A1.

126. John N. Maclean, "New York Withdraws Collider Bid," *Chicago Tribune* (January 15, 1988), 1. Cuomo and NY Senator Daniel Moynihan asked the DOE to substitute the international site on the US-Canadian border, but to no avail.

127. Representative Don Ritter, "Quark Barrel Politics (Should We Spend $5 Billion for the Higgs Boson?)," *Policy Review* (Spring 1988): 70–72. See also Bazell, "Quark Barrel Politics"; and "Quark Barrel Politics," *Economist* (March 26, 1988), p. 24.

128. NAS/NAE Committee, *Siting the Superconducting Super Collider*, 26–37.

129. "SSC at Fermilab Means $3.28 Billion Savings," *SSC for Fermilab Newsletter*, Spring 1988, 1, 9.

130. NAS/NAE Committee, *Siting the Superconducting Super Collider*, 36.

131. Don Hayner, "Fermilab Collider Boosted," *Chicago Sun-Times* (February 16, 1988), 5. Released in February 1988, the study was done by SSC for Illinois (renamed "SSC for Fermilab"), a public-private partnership formed in 1985 by business consultants interested in landing the SSC at Fermilab. Even the *Los Angeles Times*—the leading newspaper in President Reagan's home state—gave Fermilab a strong vote of support, saying: "Other than the federal pork barrel, there is no apparent reason to build the supercollider anywhere but Fermilab, which is surrounded by flat and lightly populated farmland with solid limestone beneath it." Editorial, *Los Angeles Times*, February 3, 1987, B4.

132. Sanford, interview by Sandiford.

133. Temple, interviews by Sandiford.

134. DOE, *Invitation for Site Proposals for the Superconducting Super Collider*, 30.

135. Flawn, interview by Kolb; also Flawn, *The Story of the TNRLC*.

136. Crawford, "States Want More Time to Prepare SSC Bids," *Science* (June 12, 1987), 1422; see also Goodwin, "Race for the Ring."

137. Lehman, interview by Sandiford and Weiss; Richard Nolan, interview by Sandiford, March 12, 1998.

138. Lehman, interview by Sandiford and Weiss; Sanford, interview by Sandiford; Nolan, interview by Sandiford; NAS/NAE Committee, *Siting the Superconducting Super Collider*, 24–25.

139. NAS/NAE Committee, "Siting the SSC," n. on 30. According to one view, it was unfair that Illinois be given any credit for a laboratory built with federal dollars originating from taxes on all the states. See, for example, Judith Crown, "Illinois' Fermi Pushing Hard for Super Collider," *Houston Chronicle* (October 30, 1988).

140. Robert Siemann et al., "Analysis of the Fermilab Tevatron as an Injector for the SSC," RTK—A Joint Venture, October 1987 (unpublished), Fermilab Archives.

141. Another $22 million was allowed for physical infrastructure such as dining areas, libraries, and roads. That figure, a paltry sum in light of the extensive campus development at Fermilab, reflected a DOE stipulation in 1987 that Illinois could use Fermilab as a host site only if the SSC was positioned, constructed, and operated so as to minimize interference with the Tevatron research program. The agency, saying it planned to continue research with the Tevatron as long as it remained scientifically productive, required that Illinois's site design allow for the dual use of the Tevatron both as a stand-alone facility and as an injector for the SSC. Lawrence Davenport to Governor James Thompson, June 23, 1987, cited in Siemann, "Analysis of the Fermilab Tevatron," appendix C, 53.

142. US Department of Energy, *SSC Site Evaluations: A Report by the SSC Task Force*, Report no. DOE/ER-0392, 98–99, also called the "DOE Green Book"; Transcript of DOE's debriefing of Illinois, December 7, 1988, 30–33, in SSC files, Library of the Illinois State Geological Survey.

143. Joseph Salgado, "Superconducting Super Collider (SSC): Proposer Assistance," March 26, 1987, in file "Misc. SSC Politics," Box H3b3, Leon M. Lederman Collection, Fermilab Archives.

144. Lehman, interview by Sandiford. See also DOE, *Best Qualified Sites for the SSC*, 7; and DOE, *SSC Site Evaluations*, 23.

145. NAS/NAE Committee, "Siting the Superconducting Super Collider," v, emphasis added.

146. Temple, interviews by Sandiford. Others have suggested that this human infrastructure should have been evaluated at about a billion dollars, above the value of the Fermilab physical infrastructure (see chap. 4, n. 212).

147. Temple, interviews by Sandiford.

148. Ibid.

149. Kathy O'Malley and Hanke Gratteau, "Inc.," *Chicago Tribune* (May 23, 1988), 24.

150. During the summer of 1988, Texas TV viewers witnessed commercials featuring buxom models speaking enthusiastically about the big scientific instrument soon to be built in their state. Hoddeson's personal memories of a trip driving across Texas in June 1988.

151. Stevenson Swanson and Katherine Siegenthaler, "U.S. Scientist Team to Visit Collider Site," *Chicago Tribune* (May 17, 1988), 1–2. Hess, interview by Weiss.

152. Craig Jones, interview by Sandiford. For Jones, leader of CATCH-Illinois, SSC opposition boiled down to "trust." He felt the secrecy exhibited by state officials, both before and after their announcement of the proposed SSC footprint around Fermilab, undermined public trust in them and in the governmental process. According to the map of the proposed Illinois site, however, the ring placement was *not* set at the time the proposal was submitted, and included the caption: "It is not the actual and final placement of the SSC. If and when Illinois is selected as the host state, the final ring location will be determined." *Site Proposal for the Superconducting Super Collider in Illinois*, fig. 3-1a, SSC Collection, Fermilab Archives. See also Stevenson Swanson and Andrew Bagnato, "State Shows Where Collider Would Go," *Chicago Tribune* (January 13, 1988), 1.

153. Stevenson Swanson, "Collider Foes Speak Out, and Governor Answers," *Chicago Tribune* (February 19, 1988), 1.

154. Various sources give widely different numbers. The DOE Green Book, *SSC Site Evaluations*, says that over 800 parcels would have to be acquired. See also Goodwin, "Amazing Race."

155. Gayle Golden, "Image Woes Plague North Carolina Collider Bid," *Dallas Morning News* (July 14, 1988). North Carolina, Illinois, and Tennessee proposed sites in regions undergoing rapid suburban development. This made documentation of possible relocation difficult, which may have contributed to the communication and trust problems.

156. Lawrence Jones, interview by Sandiford, December 15, 199.

157. Flawn, interview by Kolb; Flawn, *The Story of the TNRLC*, 7, 40, 47–50; Peter McIntyre, interview by Riordan, March 20, 1997.

158. Kasper, interview by Riordan; Lehman interview by Sandiford and Weiss; Flawn, *The Story of the TNRLC*, 13–50; Thomas Luce, interview by Sandiford, April 19, 2010.

159. Hoddeson, Kolb, and Westfall, *Fermilab*, 4. See also R. R. Wilson, "Starting Fermilab: Some Personal Viewpoints of a Laboratory Director (1967–1978)," *Fermilab Annual Report*, 1987.

160. John N. MacLean, "Early Decision Urged on Super Collider," *Chicago Tribune* (April 10, 1987), 14.

161. Gross, interview by Hoddeson and Sandiford. According to Gross, who headed the Illinois state environmental assessment team, members of its site team were constantly reminded by Galen Reiser in the state's Washington office that "the hardest work would begin if we won it, because then there would be the political obligation of Illinois helping DOE get many billions of dollars in annual appropriations."

162. James W. Cronin, letter to John H. Marburger, August 19, 1988, Cronin SSC Correspondence file, Fermilab Archives.

163. James W. Cronin, "Dear Colleague" letter, August 22, 1988, Cronin SSC Correspondence file, Fermilab Archives.

164. Nolan, interview by Sandiford; Lehman, interview by Sandiford and Weiss; Black, interview by Sandiford and Weiss.

165. DOE, *SSC Site Evaluations*. Also private communication by Robert Diebold with Riordan.

166. Tennessee did well, thanks primarily to its superb limestone geology, but its chances were hurt by the Murfreesboro site's setting, which the task force considered deficient in such areas as schools and labor force, along with inadequate coordination with local governments. Robert Diebold, then head of the DOE Office of the SSC, thought it still came in second to Texas; private communication with Riordan. Colorado had also put forth a strong proposal, but the geology & tunneling subcommittee concluded that its tunnel would require a liner along its entire length, while the regional resources team felt the site was too remote from the nearest metropolitan area, Denver—at least a seventy-minute drive away. Nolan, interview by Sandiford.

167. The Illinois Department of Business and Economic Development had handled the acquisition of 6,800 acres in 1967–72 for the site that became Fermilab. The Illinois Department of Transportation might have more easily facilitated the acquisition, according to Gross, interview

by Hoddeson and Sandiford. Had the Illinois site been selected for the SSC, the state would have had to acquire many parcels among the sprawling housing developments along the Fox River Valley, according to one proposed placement of the collider ring.

168. Kathy Seigenthaler, "Opposing Forces Clash Again at Hearings on Collider," *Chicago Tribune* (October 8, 1988).

169. Lehman, interview by Sandiford and Weiss; Temple, interviews by Sandiford; Diebold, interview by Riordan; DOE, *SSC Site Evaluations*; DOE debriefing of Illinois, 25.

170. Black, interview by Sandiford and Weiss.

171. Ibid.; DOE, *SSC Site Evaluatons*, 11–12.

172. DOE Facts, "The Environmental Impact Statement on The Superconducting Super Collider," September 1988; transcript of DOE debriefing of Illinois; State of Illinois, "Review and Comment on the Site Selection Evaluations by the US Department of Energy for the Superconducting Super Collider" (March 1989), SSC Collection, Fermilab Archives.

173. W. Hess, transmittal memo to R. Hunter, November 7, 1988, in DOE, *SSC Site Evaluations*, i. Temple, interviews by Sandiford; Lehman, interview by Weiss, June 10, 1997; Black, interview by Sandiford and Weiss. Some task force members say they never openly ranked the sites at the meeting with the advisory board, nor recommended a particular site. Such claims amused task-force manager Richard Nolan, who said with regard to the proposals that "by definition, they're all going to be relative after they have an independent grade." In other words, Secretary Herrington would have had little problem determining that Texas was the strongest site. Nolan, interview by Sandiford.

174. Goodwin, "DOE Picks Texas for 'Gippertron,'" *Physics Today* (February 1989), 95–98.

175. US General Accounting Office, *Federal Research: Information on Site Selection Process for DOE's Super Collider*, Report no. GAO/RCED-90-33BR (October 1989); GAO, *Final Site Selection Process for DOE's Super Collider*, Report no. GAO/RCED-89-129BR (June 1989). Five years after the announcement, *Physics Today*'s veteran reporter Irwin Goodwin, who covered the SSC story from beginning to end, remained skeptical that the selection was not a political choice. Goodwin, "After Agonizing Death in the Family, Particle Physics Faces Grim Future," *Physics Today*, (February 1994), 89–94.

176. Decisions of this magnitude, involving the commitment of billions of dollars in federal funding, are not made in a political vacuum, by the Energy Secretary acting alone. He would have made sure that he had the approval of the White House in so doing.

177. "It's Here!" *Waxahachie Daily Light* (November 10, 1988), 1. Peter T. Flawn, "Night-and-Day Work Pays Off for Collider Commission," *Austin American Statesman* (November 11, 1988), A17+. John Williams, "Super Collider Funding to Be 'Hardest' Part," *Houston Chronicle* (November 11, 1988), p. A1+.

178. Clay Robinson, "Super Collider Decision is Big Hit in Texas," *Houston Chronicle* (November 11, 1988), A25.

179. "Superheroes—Thanks for the Extraordinary Sales Job" (editorial), *Dallas Morning News* (November 14, 1988), 12a.

180. Gregory Jaynes, "Super Deal," *Life* (February 1989), 100–104. At a February 6, 1989, meeting of the Waxahachie City Council, however, Mayor James Self took umbrage at how *Life* had portrayed his city and urged citizens to write the magazine and complain if "their feelings were the same as his." Minutes of the City Council meeting, Waxahachie, TX, February 6, 1989.

181. Paul Weingarten, "Supercollider a Big Win When Texas Needed One Most," *Chicago Tribune* (November 13, 1988), A4.

182. Richard Lundy, interview by Hoddeson and Kolb, February 2, 1989. These predictions were recorded in a notebook in October 1988 by Fermilab physicist Drasko Jovanovic.

183. Hoddeson, Kolb, and Westfall, *Fermilab*, 335–38; see also "Hats Off to Waxahachie," *FermiNews*, November 18, 1988, 1, 3, 6–7; Lederman, interview by Hoddeson and Kolb, November 19, 1999; and video recording of the event, Fermilab Archives.

184. Lederman, private communication to Hoddeson, November 10, 1988.

185. Wojcicki, "The Pre-Texas Days," 292.

186. In fact, Philip Anderson had written a scientific paper, based on insights about mass from plasma and condensed-matter physics, which was a precursor of the 1964 articles by Peter Higgs and others that predicted the existence of a scalar boson responsible for generating particle masses. See P. W. Anderson, "Plasmons, Gauge Invariance and Mass," *Physical Review* 130 (1963): pp. 439–42 (see appendix 1).

187. Quoted from Karen A. Frenkel, "Nobelist Steven Weinberg Calls for Bigger Science, More Taxes," *Science Now* (June 6, 2011), http://news.sciencemag.org/2011/06/nobelist-steven-weinberg-calls-bigger-science-more-taxes.

188. On the cultural interactions in high-energy physics, see Peter Galison, *Image and Logic: A Material Culture of Microphysics* (Chicago: University of Chicago Press, 1997), esp. chap. 9: "Coordinating Action and Belief," 781–844.

189. James McAuley, "The City with a Death Wish in Its Eye," *New York Times* (November 16, 2013), available at http://ww.nytimes.com/2013/11/17/opinion/sunday/dallass-role-in-kennedys-murder.html.

CHAPTER FOUR

1. Dennis Theriot, interview by Kolb, November 22, 1993. A native of Louisiana, Theriot was left "very, very disappointed" by his visit to the Waxahachie area, with its Confederate memorial and racially divided sections of town. "I really thought Waxahachie was not a good place to start an international laboratory," he said. Similar concerns about Dallas contributed to his decision not to relocate to Texas. For more about physicists' perceptions of life in the Dallas and Waxahachie areas, see Bruce Chrisman, interview by Hoddeson and Kolb, December 13, 1993.

2. Theriot, interview by Kolb. In fact, 300 acres of the designated SSC site had been used for decades between 1880 and 1960 as a work farm for a penal colony.

3. Chrisman, interview by Hoddeson and Kolb.

4. Bingler was executive director of the Texas National Research Laboratory Commission (TNRLC), a position he would continue to hold throughout the life of the SSC Laboratory.

5. Theriot, intcview by Kolb.

6. Hoddeson, Kolb, and Westfall, *Fermilab*, 106–112.

7. For a detailed analysis of this culture clash, see Riordan, "A Tale of Two Cultures" (see preface, n. 17). See also Wojcicki, "The Supercollider: The Texas Days" (see preface, n. 19).

8. For a discussion of the origins of the AEC/DOE national laboratory system, see Peter J. Westwick, *The National Labs: Science in an American System, 1947–1974* (Cambridge, MA: Harvard University Press, 2003). The GOCO approach to lab management and operations is discussed on 49–55.

9. W. K. H. Panofsky, Memo to BOO Files, December 20, 1988, Panofsky Collection, SLAC Archives; Edward A. Knapp, letter to James R. Bieschke, December 29, 1988, Douglas Pewitt file "SSC Contract," Fermlab Archives; Roy Schwitters, private conversation with Riordan, January 14, 1989, recorded in Riordan notes. See also Goodwin, "DOE Picks Texas for 'Gippertron'"; John H. Marburger III, *Science Policy Up Close* (Cambridge, MA: Harvard University Press, 2015), esp. chap. 2.

10. Panofsky, private conversation with Riordan, January 19, 1989; telegram from Schwitters to W. Panofsky, January 19, 1989, Riordan SSC files.

11. Panofksy, converstation with Riordan, January 19, 1989. See also Panofsky, letter to the editor, *Physics Today* (March 1994), 13–15.

12. Goodwin, "DOE Picks Texas." See also DOE press releases, "Texas is Site for Super Collider" and "DOE Selects Super Collider Management and Operating Contractor," January 18 and 19, 1989.

13. Former DOE Assistant Secretary for Energy Research Alvin Trivelpiece, who had served under Energy Secretary Herrington in the Reagan administration, however, suggested that naming the SSC after such a popular president would have made it difficult for Republican members of Congress—for example, New York Congressman Boehlert—to oppose the project as openly and vigorously as they later did. Trivelpiece, interview by Weiss.

14. Wojcicki, "The Texas Days," 267.

15. Schwitters, interview by Riordan, March 22, 1997.

16. Ibid. SLAC Director Emeritus Burton Richter agreed that a laboratory director should never grant such sweeping authority to a project manager. Private communications with Riordan, April 27, 2011.

17. For example, the three CDG deputy directors, David Jackson, Chris Quigg, and Stan Wojcicki returned to research at UC Berkeley, Fermilab, and Stanford, respectively. They subsequently interacted with the SSC Laboratory as occasional consultants, at best. In February or March 1989, Schwitters invited Jackson to serve as the SSC deputy director, but Jackson declined, according to an e-mail from Jackson to Kolb, April 13, 2010.

18. For example, Schwitters said, "One can only assume that they were planning to manage the project out of DOE." Interview by Riordan, March 22, 1997.

19. M. Tigner e-mail to Hoddeson, February 9, 1999.

20. In fact, setting up such a cost-and-schedule control program was an explicit requirement of the SSC M&O contract between the DOE and URA.

21. Pewitt, interview by Riordan, May 3, 1988; Chrisman, interview by Hoddeson and Kolb.

22. J. Mervis, "Supercollider Suffering Birth Pangs," *The Scientist*, October 2, 1989, 1; Chrisman, interview by Hoddeson and Kolb.

23. R. Hunter, private conversation with Riordan, May 5, 1998. Briggs had been the MIT undergraduate roommate of President Bush's first chief of staff, former New Hampshire Governor John Sununu, which could have been an important consideration in his selection as deputy director.

24. J. Rees, private conversation with Riordan, February 3, 2009. His principal distinction between the two was that projects eventually come to an end and are (usually) completed, while programs continue.

25. Hunter, conversation with Riordan; B. Richter, private conversation with Riordan, April 27, 2011.

26. Schwitters, interview by Riordan, March 22, 1997, and T. Toohig, interview by Kolb, January 3, 1994. In fact, Schwitters stated that the relationship with Sverdrup was "a dismal failure." A high-level DOE official (who requested anonymity) claimed that the Navy had recently (in the late 1980s) given Sverdrup a "B-minus" rating as a construction contractor. See also Wojcicki, "The Texas Days," 267–68 and n. 9. By all accounts, the SSC lab's experience with Sverdrup was an unmitigated disaster.

27. Hunter, conversation with Riordan. Throughout this hour-long conversation, Hunter repeatedly used the word "manage" to describe the DOE relationship with the SSC lab. His decision to set up the OSSC was confirmed by Diebold, interview by Riordan.

28. Schwitters, interview by Riordan, March 22, 1997. See also "SSC Management Implementation Plan" and "DOE SSC Management Plan," (unpublished drafts, May 1989); and US Department of Energy, "Superconducting Super Collider Management and Implementation Plans," August 30, 1989, SSC Collection, Fermilab Archives.

29. For more details on this subject, see Riordan, "A Tale of Two Cultures," and Robert P. Crease, *Making Physics: A Biography of Brookhaven National Laboratory* (Chicago: University of Chicago Press, 1999), 69–92.

30. The most ardent and eloquent advocate of this viewpoint was Panofsky, the founding director of SLAC, who stepped in as chair of the SSC Board of Overseers in 1990. See his letter to the editor of *Physics Today* (see this chap., n. 11).

31. On the construction of Fermilab, see Hoddeson, Kolb, and Westfall, *Fermilab*, 95–156.

32. A detailed analysis of these projects, with comparisons to cost control on other AEC, DOE, and other agency projects as well as nuclear power plants, is given by L. E. Temple, in an internal DOE document, "Office of Energy Research: Project Performance" (unpublished, February 1986), Fermilab Archives. See especially tables 2 and 6 comparing the actual vs. projected costs for DOE accelerator projects with other DOE and other large-scale construction projects.

33. On the Isabelle cost overruns, see Crease, "Quenched!" (see preface, n. 13). On construction of the Fermilab Tevatron, see Hoddeson, Kolb, and Westfall, *Fermilab*, 196–224.

34. "Workshop on Master Builders: The Reconstruction of American Science after World War II," April 16–17, 1999, Johns Hopkins University (unpublished). Other such "master builders" include Lloyd Berkner of Brookhaven and James E. Webb of NASA. The concept is similar to the "system builders" discussed by historian of technology Thomas P. Hughes.

35. To be fair, there had been few opportunities for younger physicists to manage large projects during the mid-to-late 1970s, largely because of the severe decline in high-energy physics funding during that period. As CDF cospokesman, Schwitters probably had among the best management experience of his generation of high-energy physicists. But it did not come close to that of older physicists like Panofsky, Wilson, Lederman, and Richter. Schwitters admitted that there was "a rather shallow pool of management experience in our community, especially at that time." Interview by Riordan, March 22, 1997. The fact that the top CDG managers did not go to Texas severely exacerbated this problem. Schwitters then had to turn more and more to industry to fill the many available positions at the SSC Laboratory.

36. On Watkins at DOE, see Mark Crawford, "Watkins Named Energy Secretary," *Science* (January 20, 1989), 309; and "Watkins Takes Helm at DOE," *Science* (March 3, 1989), 1136. An article about Watkins aimed at general audiences is Gregg Easterbrook, "Radio Free Watkins and the Crisis at Energy," *Washington Post Magazine* (February 18, 1990), 16–40.

37. Schwitters, interview by Riordan, March 22, 1997.

38. Ibid.

39. James D. Watkins, Memorandum for Robert O. Hunter Jr., "SSC Management Plan," September 7, 1989. See also Mervis, "Supercollider Suffering Birth Pangs," 25.

40. Goodwin, "Hunter Departs DOE after Riling Key Lawmakers and Top Texans," *Physics Today* (January 1990), 49. See also "Minutes of the Meeting of SSC Board of Overseers," October 4–5, 1989, Panofsky Collection, SLAC Archives.

41. G. Ross (*sic*) and T. Kirk, "Report of the SSC Collider Dipole Review Panel," SSCL Report no. SSC-SR-1040, appendix A, 25. Perhaps indicative of the haste with which this document was prepared, the name of one of its editors, Gustav Voss, was misspelled and the error not caught or corrected.

42. Ibid., vii. See also Mark Crawford, "Will Magnet Problems Delay the SSC?" *Science* (May 17, 1989), 1425–26, for a science writer's prescient coverage of the unresolved problems with the SSC dipole magnets. Crawford also discusses the questions then being raised about the 4 cm magnet aperture.

43. These persistent currents had first been encountered at HERA, which added correction coils to its superconducting dipole magnets to help cancel them out. In a note added in proof to the SSC magnet report, HERA leader Voss suggested that higher injection energies "would greatly alleviate the problem."

44. Ross and Kirk, "Report of the SSC Collider Dipole Review Panel," 1.

45. Ibid., 5. Emphasis added.

46. Alex Chao, interview by Riordan, February 3, 2009; Tigner, private conversation with Riordan, September 30, 1998; Chao and Tigner, "Requirements for Dipole Field Uniformity and Beam Tube Correction Windings" Report no. SSC-N-183 (May 27, 1986), Fermilab Archives.

47. Ross and Kirk, "Report of the SSC Collider Dipole Review Panel," 21–24; John Rees, private communication to Riordan, November 3, 2009.

48. J. R. Sanford and D. M. Mathews, eds., "Site-Specific Conceptual Design of the Superconducting Super Collider," Report no. SSCL-SR-1056 (July 1990).

49. Chao, interview by Riordan. Boyce McDaniel, "Trip Report, Visit to Dallas, 6/21–22," July 26, 1989, Panofsky files, SLAC Archives. McDaniel served as chair of the SSC Board of Overseers until the fall of 1990, when Panofsky replaced him; he visited the SSC Laboratory in late June 1989 and reported back to the board on progress being made there. Speaking about accelerator physicist Rae Stiening, who was doing computer simulations of SSC performance with David Ritson, McDaniel noted: "Rae has the philosophy of 'greater margin' which implies throwing out some decisions reached by Maury and his group. Helen's philosophy is much the same as Rae's."

50. McDaniel trip report, which noted, "Results of aperture calculations for 100 thousand turns were presented. Beam loss was still continuing at that stage."

51. Ibid., 2. On the early 1989 discussions of redesigning the SSC dipole magnets or raising the injection energy, see Crawford, "Will Magnet Problems Delay the SSC?"; and "Lab Report Puts SSC Magnets in Limbo," *Science* (August 25, 1989), 809–10.

52. In this regard, it is noteworthy that SSC Associate Director for Administration Bruce Chrisman had already returned to Chicago in May 1989, after having served less than six months—well before suitable business-management and project-management systems had been established at the SSC Laboratory. Chrisman, interview by Hoddeson and Kolb.

53. McDaniel trip report, 1.

54. Ibid., 3. Emphasis added.

55. "Workshop on Physics and Experiments for the Superconducting Super Collider," Dallas, TX, October 1–4, 1989 (unpublished notes), Goodwin Collection, Niels Bohr Library.

56. Chao, interview by Riordan. Chao claimed that he did not agree with the decision, having played a central role in making the parameter choices for the CDG design. But Edwards had decided to pursue a more conservative, less risky design, and he accordingly supported her decision.

57. H. Edwards, "SSC Site Specific Design," workshop presentation transparencies, October 3, 1989, Irwin Goodwin Collection, Niels Bohr Library and Archives.

58. Schwitters, interview by Riordan, March 31, 1998.

59. Chao, interview by Riordan.

60. Panofsky, "Meeting of the SSC Board of Overseers Executive Committee," November 21, 1989, 1, Panofsky Collection, SLAC Archives. The FY 1990 funding for the SSC was later reduced to less than $220 million by across-the-board budget cuts.

61. Goodwin, "Trying Times: Cost of Remodeling SSC Causes Texans to Circle Their Wagons," *Physics Today* (January 1990), 45–47; B. Schwarzschild, "SSC Design Revisions Call for Thinner Beams and Fatter Magnets," *Physics Today* (January 1990), 47–49.

62. R. Jeffrey Smith, "Supercollider Could Face Cutbacks," *Washington Post* (November 19, 1989), A1; brackets and ellipsis in the original. Henson Moore, interview by Weiss.

63. Goodwin, "Trying Times," 46; Goodwin, conversation with Riordan, March 19, 2009.

64. In this meeting with him, Schwitters recalled, "Watkins went berserk and started screaming at me and swearing, yelling about all this stuff." Schwitters, interview by Riordan, March 31, 1998.

65. Copies of transparencies, notes, e-mails, and reports from the files of Lyndon Evans, project manager of the Large Hadron Collider, labeled "Documentation on SSC, Meeting #1 (4–5 December 1989)."

66. Hoddeson, Kolb, and Westfall, *Fermilab*, 148–52. These problems could be largely attributed to Wilson's minimalist design philosophy and its implementation in manufacturing the magnets. As the first SSC business manager, Bruce Chrisman later observed, "Maury was going with the old Wilson 'shave it, cut it, we'll make it work somehow' approach." Chrisman, interview by Hoddeson and Kolb. Also Richter, private communication to Riordan, April 2011.

67. Draft "Report of the SSC Machine Advisory Committee, December 4–5, 1989," 3, Riordan SSC files; R. Diebold, private communication to Riordan, July 2010.

68. In addition, use of a 5 cm magnet aperture would enable easier insertion of a "liner" just inside the beam tube, then beginning to be considered by SSC designers to enable the proton beam intensities to be increased in future years. Peter Limon, interview by Hoddeson and Kolb, February 5, 2009. Such a liner would help protect the tube from X-rays emitted by the circulating protons, which would increase to unacceptable levels as their number increased beyond the design intensity. Doing so was considered very difficult to impossible with a 4 cm aperture.

69. Schwarzschild, "SSC Design Revisions," 47. In 1998 Tigner claimed that the CDG design with 4 cm magnet apertures involved a "significantly conservative approach, with an acceptable design risk." Private conversation with Riordan, September 30, 1998. Had he come to Texas and brought the CDG team with him, it seems clear that they would have proceeded with something closer to the original design, accepting its greater risks—and achieving significantly lower costs. One can only speculate what might have happened during the SSC commissioning process. Here it's worth noting that the Large Hadron Collider experienced serious commissioning problems, with the result that it began experimental runs in 2010 at only half the design energy, or 7 TeV, but physicists still managed to discover the Higgs boson using this collider during the next two years.

70. Panofsky, "Meeting of the SSC Board of Overseers Executive Committee." On 4, he states that "at this point it does not seem to be necessary to increase the magnet aperture beyond the 4.0 centimeter value."

71. Panofsky, "Meeting of the SSC Board," 2. It is ambiguous whether this figure includes redesign and testing of the magnets for 5 cm apertures; probably not. Of the $1.1 billion increase, $350 million was attributed to costs that the DOE had omitted in its FY1989 presentations to Congress.

72. Ibid., 3. At about the same time, the Space Station was experiencing even greater cost overruns, and NASA responded with similar "descoping" efforts. See, for example. Eliot Marshall, "Space Station Science: Up in the Air," *Science* (December 1, 1989), 1010–12. CERN's Large Hadron Collider uses 1.9 K superfluid helium to obtain the maximum possible magnetic fields and therefore the highest proton energies; it also aims for luminosities an order of magnitude larger than the SSC design luminosity to compensate for much lower collision energies.

73. Panofsky, "Meeting of the SSC Board," 3. Panofsky may, however, have overstated the opinion of board members here, as he himself held these opinions strongly at the time, and he was well known for his ability to influence a committee that he chaired or served on. And he wrote the memo, too.

74. Panofsky, "A Personal View on SSC Project Cost Growth and Its Control," December 13, 1989, Panofsky Collection, SLAC Archives. There must have been an earlier draft because there are earlier memos from B. McDaniel and B. Richter referring to his memo, but not in the SLAC archives.

75. According to Panofsky's memo, the added cost of the aperture increase was only $350 million, but other contemporaneous documents indicate that this figure did not include the costs of development and testing of the redesigned magnets, nor the impact of the schedule stretch-out required to do this additional work. B. McDaniel, letter to J. Toll, December 11, 1989, with the attached "SSC Project Cost Estimates," telefax dated December 11, 1989, Panofsky Collection, SLAC Archives.

76. Panofsky, "A Personal View," 3. He included the needs of the wider scientific community here.

77. Ibid., 5. Panofsky noted that (except for the W and Z boson discoveries at CERN), theorists had been notoriously unable to predict the energy scales for new phenomena with accuracy. Indeed, the Higgs boson (or at least the first one to be found) turned up on the LHC at a mass far less than the 1 TeV being discussed during the late 1980s as the upper limit for this particle's possible mass. See also this chapter, n. 79, and appendix 1.

78. Ibid., 5.

79. "Report of the Ad Hoc Committee on SSC Physics," December 11, 1989, SSCL Report no. SSC-250.

80. The LHC game plan was to compensate for its far-lower energy by attaining higher luminosity and thus more proton collisions. It is instructive to note that this machine started experimentation in 2010 at a total energy of only 7 TeV, just half of its design energy and less than *one fifth* that of the SSC, but physicists began to observe evidence for the Higgs boson in the fall of 2011. After raising the collision energy to 8 TeV the following spring, physicists announced the discovery of the Higgs boson at a mass close to 125 GeV on July 4, 2012. See epilogue for a detailed discussion.

81. "Report of the Ad Hoc Committee on SSC Physics," 17.

82. Ezra D. Heitowit, "Minutes of the Joint Informational Meeting of the URA Board of Trustees and the SSC Board of Trustees (*sic*)," December 19, 1989, Panofsky Collection, SLAC Archives.

83. Formerly president of the University of Maryland and chancellor of the University System of Maryland, Toll had just stepped in on December 1, 1989, to replace Knapp, who returned to research at Los Alamos National Lab. Toll would have a baptism of fire. See Goodwin, "Toll Heads URA; Moore Leaves NSF and Sanchez Joins; Foster at DOD," *Physics Today* (February 1990), 72.

84. Heitowit, "Minutes of the Joint Informational Meeting," 5. This figure probably included only $350 million for the aperture increase, which did not include the development and testing costs for the new magnets, nor the costs of schedule delays. On the same page, for example, Schwitters states, "The cost impact of the 5 cm aperture is estimated to result in an increase of between $200 million and $400 million, due essentially to the requirement for additional materials for magnet production." A better total cost figure here was at least $7.6 billion, if these development and testing costs were included in the total.

85. Ibid., 6. He went on to note that the 17 TeV option would reduce the overall costs to $6.7 billion, an increase of less than $1 billion that could allow DOE to rationalize the cost growth as due in part to its earlier omissions in submitting the $5.9 billion figure to Congress.

86. Panofsky, interview by Riordan and Weiss, March 29, 1996.

87. Heitowit, "Minutes of the Joint Informational Meeting," 9.

88. Panofsky, interview by Riordan and Weiss; Marshall, "Space Station Science."

89. The likely impacts of the SSC cost overruns on other national labs and scientific research sponsored by the DOE Office of Energy Research (now the Office of Science) were particularly worrisome. Though the Reagan and Bush administrations claimed that the SSC would be funded by "new money," those familiar with the congressional appropriations process realized that this was a convenient fiction.

90. D. Kramer, "Watkins to Weigh Changes in Cost of SSC," *Inside Energy*, December 18, 1989, 1.

91. J. M. Decker, letter to F. E. Low, December 21, 1989, appendix B in *Report of the 1990 HEPAP Subpanel on SSC Physics*, DOE Report no. DOE/ER-434 (Washington, January 1990). Emphasis added.

92. Goodwin, "HEPAP and Its Subpanel Approve Redesign and Higher Cost of SSC," *Physics Today* (February 1990), 67–68. The five Nobel laureates were James Cronin of Chicago, T. D. Lee of Columbia, Leon Lederman of Fermilab, Norman Ramsey of Harvard (and the founding president of URA), and Steven Weinberg of the University of Texas. Douglas Pewitt, who had served in the Office of Management and Budget, the DOE, and the Office of Science and Technology Policy, observed that there are essentially two different kinds of scientific advisory panels: those that provide advice that an agency subsequently follows, and others that provide political cover for decisions already made. The Drell Subpanel fell more into the latter category. Pewitt, interview by Riordan.

93. F. E. Low, letter to J. M. Decker, January 12, 1990, in *Report of the 1990 HEPAP Subpanel*.

94. This question had been explicitly raised over a year earlier by the Congressional Budget Office in a detailed report, *Risks and Benefits of the Superconducting Super Collider* (Washington, DC, October 1988); see esp. 63–70. See also chap. 3.

95. J. F. Decker, "Remarks to High Energy Physics Advisory Panel," January 12, 1990. Irwin Goodwin Collection, AIP Niels Bohr Library and Archives.

96. Admiral James D. Watkins, interview by Riordan and Weiss, February 2, 2000.

97. Sanford and Mathews, "Site-Specific Conceptual Design."

98. David P. Hamilton, "The SSC Takes On a Life of Its Own," *Science* (August 17, 1990), 731–32.

99. US Department of Energy, Office of Energy Research, *Report of the 1990 HEPAP Subpanel on SSC Cost Estimate Oversight*, Report no. DOE/ER-0464P (Washington, DC, July 1990).

100. Ibid., i–ii, quote on i.

101. Mark Crawford, "SSC Cost Estimates Climb," *Science* (August 3, 1990), 473.

102. David Kramer, "Study Puts SSC Cost at Nearly $12 billion," *Inside Energy* (August 6, 1990), 1; D. Kramer, private conversation with Riordan, March 19, 2009.

103. Independent Cost Estimating Staff, "Independent Cost Estimate for the Superconducting Super Collider" (Washington, DC, September 1990), Goodwin Collection, Niels Bohr Library and Archives.

104. Ibid.

105. Ibid. Emphasis in the original.

106. David Kramer, "OMB Said to Challenge SSC Cost Projections," *Inside Energy* (October 8, 1990), 1.

107. The SSC cost-estimating process is also discussed in Wojcicki, "The Texas Days," 271.

108. US Department of Energy, *Report on the SSC Cost and Schedule Baseline*, Report no. DOE/ER-0468P (Washington, DC, January 1991). See also Irwin Goodwin, "As SSC Project Accelerates, Its Cost Exceeds $8.2 Billion," *Physics Today* 44:3 (1991), 51–53; and David P. Hamilton, "The SSC Gets Its (Official) Price Tag: $8.3 billion," *Science* (February 15, 1991), 741. The cost of the superconducting dipole magnets, which both the Townsend Panel and the ICE panel considered among the riskiest items, came in at $1,444 million, exclusive of contingencies and inflation, compared with $610 million in the original SSC Conceptual Design Report of 1986. See chap. 2, n. 79. This was nearly a *doubling* of the cost of these magnets in real terms, far greater than the 22 percent contingency that the DOE had allowed.

109. Sanford and Mathews, "Site Specific Conceptual Design." A more readable, digestible executive summary is M. G. C. Gilchriese and K. Metropolis, eds., *Site Specific Design of the Superconducting Super Collider*, Report no. SSCL-SR-1055 (July 1990).

110. According to Pewitt, "Dick Briggs is a good friend of mine, but he's no project manager.... He just wasn't interested in it. He didn't understand it. The project manager has to lead the project." Pewitt, interview by Riordan. John Rees, who served as SSC project manager from 1992 to 1993, echoed these sentiments. Private conversation with Riordan, February 3, 2009.

111. Schwitters, "Project Management Changes," SSCL memorandum, March 12, 1990, Riordan SSC files.

112. Pewitt, interview by Riordan, and Rees, interview by Riordan (September 2, 2009), were helpful in understanding these CS^2 systems.

113. Richard Sah, who had worked in the DOE Office of High Energy Physics in the early 1980s and then in the CDG project management group, provided important insights into the reluctant adoption of these systems in high-energy physics, including at CDG. He claimed that CDG physicists began to embrace these systems only grudgingly. Interview by Riordan, May 19, 1997, and e-mails of March 20 and 31, 1997, Riordan SSC files.

114. Pewitt, interview by Riordan; Schwitters, interview by Riordan, March 31,1998; Robert Diebold, interview by Riordan; Chrisman, interview by Hoddeson and Kolb.

115. Thomas Elioff, private conversations with Riordan, October 1, 1998, and September 1, 1999. He recalled a 1990 meeting with Schwitters and other top SSC managers at which he advocated using his cost-and-schedule control system but was rebuffed by Helen Edwards, who bluntly told him, "This is not the way I build an accelerator." Cornell and Fermilab physicists seemed particularly averse to using these systems.

116. Timothy Toohig, private conversation with Riordan, December 22, 1999.

117. Pewitt, interview by Riordan. Also D. L. Pells, "Meeting Minutes: PMS Workshop Group Meeting of April 11, 1989," and attached "SSC Project Management System," April 18, 1989, Pewitt files, SSC Collection, Fermilab Archives.

118. For a project manager's views about the SSC lab's lack of a cost-and-schedule control system, see E. Payson Willard, "The Demise of the Superconducting Super Collider: Strong Politics or Weak Management?" in David I. Cleland et al., eds., *The Project Management Casebook* (Newtown Square, PA: Project Management Institute, 1994), 493–502, esp. 497.

119. Wojcicki, private conversation with Riordan, May 4, 2009. See also Wojcicki, "The Texas Days," 267–68, esp. n. 9, where he states, "Everyone I talked to at the SSC during its early days agreed that the Sverdrup partnership was a disaster." Tim Toohig had similar unfavorable comments about the firm, interview with Kolb and conversation with Riordan.

120. Schwitters, interview by Riordan, March 22, 1997. Pewitt, "Sverdrup Corporation Performance," memorandum, October 11, 1989, Pewitt files, Fermilab Archives.

121. David Ritson, "Demise of the Texas Supercollider" (see chap. 2, n. 81), quote on 608. He also observed that, compared with existing high-energy physics laboratories, "SSC staffing was very light on experienced machine personnel and top heavy on administrative personnel."

122. It is difficult to put a number on these costs, but it was certainly at least tens of millions of dollars. For other observations on this question, see Wojcicki, "The Texas Days," 292.

123. Pewitt, interview by Riordan.

124. Casani had served as project manager on NASA's Voyager and Galileo space probes. Forsen was then a senior vice president and manager of technology at Bechtel.

125. Hoddeson, Kolb, and Westfall, *Fermilab*, 201–2.

126. Panofsky, "Changes in Top Management Structure of the SSC," memorandum, October 8, 1990, Panofsky Collection, SLAC Archives. According to the attached résumé for Siskin, he had served as a "Field Office Manager reporting directly to Admiral H. G. Rickover, the Director, Division of Nuclear Reactors." Stone and Webster had a long history of construction management, stretching back before the Manhattan Project, in which it managed construction at Oak Ridge.

127. Ibid., 2.

128. Ibid.

129. Watkins, interview by Riordan and Weiss. Joseph Cipriano, in a private conversation with Riordan, January 28, 2000, said much the same thing.

130. Admiral Rickover had a similar management strategy of installing his own lieutenants at all levels of his organization to obtain reliable information about what was happening, said Panofsky. A good leader needs this kind of information to manage properly. The problem occurs, he continued, when these people are used to bypass the regular organizational structure in the process of making or executing decisions. That often leads to chaos. Private conversation with Riordan, June 6, 1997.

131. Cipriano, conversation with Riordan.

132. Watkins, interview by Riordan and Weiss.

133. Cipriano, conversation with Riordan.

134. DOE press release, "Watkins Names Cipriano Project Manager of the Superconducting Super Collider," April 19, 1990. The choice of Cipriano's title is revealing, for it seemed to make him the very person for whom URA was then searching. Later he was renamed "Project Director," causing even more confusion, for it apparently put him on an essentially equal footing with SSC Laboratory Director Roy Schwitters (as Watkins had admittedly intended).

135. Watkins, interview by Riordan and Weiss.

136. Redacted copies of these memoranda were obtained from the SSC files of Congressman Boehlert and his legislative aide David Goldston. They had been obtained under subpoena by the House Science Committee's Investigations and Oversight Subcommittee, on which Boehlert served as ranking Minority member in the early 1990s.

137. Joseph R. Cipriano, "Memorandum for the Secretary of Energy," September 14, 1990, Boehlert/Goldston files.

138. David Kramer, "Parsons Brinckerhoff–Morrison Knudsen Win $1 Billion SSC Contract," *Inside Energy* (February 26, 1990), 3–4.

139. Students of military history will immediately recognize "NUTS." On December 22, 1944, US Army troops under General Anthony McAuliffe were surrounded by German armored forces in the Belgian town of Bastogne during the Battle of the Bulge. When presented with a message from the German commander suggesting that they surrender to avoid inevitable annihilation, McAuliffe wrote "Nuts!" on the typewritten document, rejecting the offer, and had it carried back to the German lines.

140. Watkins, interview by Riordan and Weiss. In the 1990 Superconducting Super Collider Authorization Act (H.R. 4380), the House had specifically authorized the Office of the SSC to report directly to the Secretary of Energy (but the bill was never taken up by the Senate). In a sense, Watkins was acting appropriately within executive powers granted him, at least by the House.

141. US Department of Energy Office of Inspector General, *Summary Audit Report on Lessons Learned from the Superconducting Super Collider Project*, Report no. DOE/IG-0389, April 1996. Cipriano was also given Head of Contracting Authority status at the SSC Laboratory, which would ordinarily have been the purview of the DOE Chicago Operations Office; it made him even more powerful.

142. Other high-energy physicists such as Robert Diebold and John Rees had come from the national laboratories to Washington to work in the DOE high-energy physics program office, bringing with them their understanding of this culture. So did the "detailees" from the labs, who worked at AEC or DOE for a few years before returning to their lab positions. Rees, in fact, became an AEC employee in the late 1960s and worked in the HEP program office during the construction of Fermilab. Then he returned to SLAC and served as project manager on its PEP and SLC electron-positron colliders, bringing with him a thorough understanding of the bureaucratic needs of his Washington counterparts. These kinds of interactions and cross-fertilizations helped foster better communication between the two disparate cultures. Diebold, interview by Riordan, April 20, 1995; Rees interview by Riordan, September 2, 2009.

143. Thomas Bush, interview by Riordan, April 2, 1998; see also T. Bush testimony, *Status of the Superconducting Super Collider Program: Hearing Before the Investigations and Oversight Subcommittee of the House Science, Space and Technology Committee*, May 9, 1991, 102–3.

144. Schwitters, interview by Riordan, March 31, 1998.

145. For example, Fermilab had built the original magnets for the Main Ring almost entirely in-house, as it had done on the superconducting magnets for the Tevatron; see Hoddeson, Kolb, and Westfall, *Fermilab*, 142–52 and 246–58. Brookhaven had tried and failed to transfer its (flawed) superconducting dipole magnet technology to Westinghouse for manufacturing; see Crease, "Quenched!," part II, 404–22.

146. Harvey Lynch, private conversation with Riordan, October 22, 1996; T. Bush, interview by Riordan.

147. It was customary to "black box" military development and construction projects, in part because of the requirements of secrecy, said historian of technology Thomas P. Hughes, in a private conversation with Riordan, February 17, 1999. The development of the various parts of a military system, such as the Atlas missile, would be walled off from each other, with only the performance requirements of a given component specified to the engineers working on it. Systems engineers would then "manage the interfaces" among all the various parts, making sure they would work together effectively. For more details, see Thomas P. Hughes, *Rescuing Prometheus* (New York: Pantheon Books, 1998), esp. chap. III, "Managing a Military-Industrial Complex: Atlas," 69–139.

148. Joseph Cipriano, memorandum to James D. Watkins, January 12, 1991.

149. For comparison, the largest projects managed to date by US high-energy physicists had been the construction of SLAC and Fermilab, by Panofsky and Wilson, which would have been

about $400-million and $750-million projects if evaluated in 1990 dollars. Other, more recent high-energy physics projects, such as PEP and SLC projects at SLAC and Fermilab's Tevatron, fell in the $100–$200-million class. Brookhaven's ill-fated Isabelle project would have fallen somewhere in between.

150. Schwitters, interview by Riordan, March 22, 1997.

151. The moniker "Texas collider cartel" was coined by Gayle Hudgens, "Muons and Megabucks: Super Collider—Super Bust?" *The Nation* (March 19, 1990), 365ff. On the TNRLC, see Peter T. Flawn, *The Story of the TNRLC.*

152. Clements, who stated soon after taking office that his top priority would be bringing the SSC to Texas, remained closely involved in the project until its termination. Hans Mark, "Site Selection Criteria for Site Selection Process," *Proc. Philosophical Society of Texas* 56 (1993), 37; Flawn, *Story of the TNRLC*, 12–14; Tom Luce, interview by Sandiford.

153. Flawn, *Story of the TNRLC*, 21; Edward Bingler, interview by Riordan, March 24, 1997; Schwitters, interview by Riordan, March 22, 1997; Chrisman, interview by Hoddeson and Kolb.

154. Flawn, *Story of the TNRLC* 86–88, 127–46.

155. Barbara Deckard, "State Party Delegations in the U.S. House of Representatives—a Comparative Study of Group Cohesion," *Journal of Politics* (February 1972): 202–22, and "State Party Delegations in the U.S. House of Representatives—an Analysis of Group Action," *Polity* (Spring 1973): 312–34; Tom Luce, interview by Sandiford; Hans Mark, interview by Riordan, March 31, 1998. Also Cathy Gillespie, interview by Weiss, December 22, 1997.

156. Luce, interview by Sandiford. Initially, some observers expected the Texas victory to boost the SSC's momentum on Capitol Hill, because of the state's political clout. But that expectation naively assumed the losing states would continue to support the project. Crawford, "Texas Lands the SSC," *Science* (November 18, 1988), 1004; Goodwin, "DOE Picks Texas for 'Gippertron'"; William D. Marbach, "When Protons—and Politics—Collide," *Newsweek* (July 6, 1987), 44; Gillespie, interview by Weiss.

157. Luce, interview by Sandiford.

158. Congressional Budget Office, *Risks and Benefits of Building the Superconducting Super Collider*, 44–50. Though disputed, this prediction turned out to be low.

159. "The Unaffordable Atom-Smasher" (editorial), *New York Times* (November 16, 1988), http://www.nytimes.com/1988/11/16/opinion/the-unaffordable-atom-smasher.html.

160. In the Texas political system, exceptional executive power is vested in its commissions rather than the governor's office. This practice can be traced to Reconstruction after the Civil War, when it became a way to keep power in the hands of Texans and out of the grasp of the "Carpetbaggers" then flooding in from Union states. See, for instance, James R. Novell, "The Railroad Commission of Texas: Its Origin and History," *Southwestern Historical Quarterly* 68, no. 4 (April 1965), 465–80.

161. Crawford, "CBO Lists Options for Cutting R&D," *Science* (February 24, 1989), 1001.

162. Luce, interview by Sandiford.

163. Thomas W. Luce III, letter to J. Bennett Johnston, May 30, 1989, Riordan SSC files.

164. Ibid.

165. H. L. Lynch, "Detector R&D," in Gerald F. Dugan and James R. Sanford, eds., *Superconducting Super Collider: A Retrospective Summary, 1989–1993*, Report no. SSCL-SR-1235, 227–29. Over $25 million was spent in all for generic detector R&D in fiscal 1990–91, before official DOE commitments were made to the two large experiments. Most of this funding came from the Texas contributions via the TNRLC; see table 19-1 on 227.

166. Luce, interview by Sandiford, May 12, 2010. At the time, the TNRLC envisioned paying these subcontractors directly, but this did not occur. Texas funds were transferred to the DOE, which handled disbursements to the SSC Laboratory and its subcontractors.

167. Goodwin, "At Last, Congress Agrees to Build SSC, after Texas-Type Wheeling and Dealing," *Physics Today* (October 1989), 51–52; Luce, interview by Sandiford, May 12, 2010.

168. Goodwin, "At last, Congress Agrees"; Atin Basuchoudhary, Paul Pecorino, & William F. Shughart II, "Reversal of Fortune: the Politics and Economics of the Superconducting Supercollider," *Public Choice* 100 (September 1999): 189. The $225 million allocation was later reduced by across-the-board DOE cuts.

169. Flawn, *Story of the TNRLC*, 326; Goodwin, "Making News by Calling It Quits, Bucy Leaves a Message for the SSC," *Physics Today* (August 1991), 52–53; Bingler, interview by Riordan.

170. "Super Collider Plans," *Washington Post* (March 17, 1990), A18.

171. John Racine, "Texas Commission to Price $250 Million of Lease Bonds for Super Collider Project," *Bond Buyer* (December 10, 1991), E114.

172. Mary Jacoby, "Superconductor Already Colliding with Some Texans," *Washington Post* (July 6, 1990), A8.

173. Goodwin, "Trying Times."

174. Riordan, "A Tale of Two Cultures," 136; Texas State Library and Archives Commission, http://www.tsl.state.tx.us/arc/appraisal/tnrlc.html; Wojcicki, "The Texas Days."

175. Goodwin, "As SSC Project Accelerates."

176. Dugan and Sanford, *Superconducting Super Collider*, graph of staffing history on 302.

177. Ibid.; "The Supercollider: It's Crunch Time," *Business Week* (May 6, 1991), 131; Riordan, "A Tale of Two Cultures," 16.

178. Wojcicki, "The Texas Days," 266.

179. As former CEO of Texas Instruments, Bucy had good foreign contacts and sources, especially in Japan, mainly through his involvement in the semiconductor industry.

180. Flawn, *Story of the TNRLC*, 326; "Fund-Raiser for Particle Accelerator Resigns," *New York Times* (July 9, 1991), A14; Goodwin, "Making News by Calling It Quits."

181. Bingler, interview by Riordan; Luce, interview by Sandiford.

182. Goodwin, "As SSC Cost Accelerates," 53.

183. Robert Diebold, memorandum, "SSC Scientific Policy Committee Meeting," January 5, 1990, which stated that "the main recommendation discussed was the Committee's strong feeling that it is important to maintain the 20 TeV SSC beam energy; some physics will need the full energy." Riordan SSC files.

184. Like the 1990 Drell subpanel, the SPC did not consider the financial and political impacts of its policy recommendations. Nor was it asked to do so.

185. SSC Laboratory internal memorandum, "Summary of the Meeting and Recommendations of the Superconducting Super Collider Program Advisory Committee" (Snowmass, CO, July 1990).

186. Office of Energy Research, *Report on the SSC Cost and Schedule Baseline*," table 7, on 55, which gives $760 million (in 1990 dollars) as the estimated cost of experimental systems. Adding inflation at an estimated rate of about 20 percent over the life of the project brought this figure up to $910 million.

187. Robert P. Crease, "Choosing Detectors for the SSC," *Science* (December 21, 1990), 1648–50.

188. Schwitters, "Decision Memorandum—on Aspects of the Initial Scientific Program for the SSC," January 4, 1991, Riordan SSC files.

189. Hoddeson, Kolb, and Westfall, *Fermilab*, 288–97. Riordan, *Hunting of the Quark*, 255–57, 271–72, 274–75, 278–85, 287, photo on 256. Trilling had been a spokesman for this SLAC-LBL collaboration, whose discovery of the psi particle in 1974 led to the 1976 Nobel Prize for Burton Richter of SLAC. Trilling and Schwitters were old friends and fellow colleagues who knew each other well. Also George Trilling, interview by Weiss, March 26, 1996.

190. Crease, "Choosing Detectors." Ting had led the MIT-BNL collaboration that discovered the J particle at Brookhaven just before the psi particle showed up at SLAC; he shared the 1976 Nobel Prize in physics with Richter. See Riordan, *Hunting of the Quark*, 262–92.

191. Ting had led the Mark-J collaboration at DESY's PETRA and the L3 collaboration at CERN's LEP collider. The designs of these general-purpose detectors emphasized calorimetry over

magnetic fields and particle tracking. The L3 collaboration had experienced major cost overruns that imposed a burden on DOE high-energy physics funding in the mid-1980s, leading to concerns it might happen again at the SSC. On the L* proposal and decision, see also Wojcicki, "The Texas Days," 274–75.

192. Schwitters, "Report to the PAC," May 3, 1991, Riordan SSC files, plus personal notes of this meeting. In this memo, Schwitters noted that the anticipated cost of the L* detector might exceed $764 million, which came close to what DOE had allocated for the entire SSC experimental program.

193. David P. Hamilton, "Ad Hoc Team Revives SSC Competition," *Science* (June 21, 1991), 252; Robert Pool, "SSC Gets New Detector," *Nature* (June 17, 1991), 681. Also personal notes of Riordan (who worked for URA at the time) and Barry Barish, interview by Riordan, April 17, 2010.

194. Barry Barish and William Willis, spokespersons, "GEM Letter of Intent," November 30, 1991, SSC Collection, Fermilab Archives; also Barish, interview by Riordan.

195. As one measure of his popularity and influence in the high-energy physics community, Gilman was named to replace Wojcicki as HEPAP chair in 1996.

196. Schwitters, interviews by Riordan, March 22, 1997, and March 31, 1998.

197. Edward Siskin, interview by Riordan, June 2, 2000; Toohig, conversation with Riordan; Rees, interview by Riordan, May 6, 2010. Rees said that this was not an excessively large number of employees for a multibillion-dollar project, considering the need to input accurate data into the system and interpret the results.

198. Cipriano, conversation with Riordan.

199. Cipriano, "Memorandum for the Secretary of Energy," January 12, 1991, Boehlert/Goldson SSC files.

200. Siskin, interview by Riordan.

201. Rees, interview by Riordan, May 6, 2010. He also noted that SLAC had always used such project-management control systems on its major projects, whether on paper at its inception or on computers in later years. In a May 2000 meeting, LHC Project Manager Lyndon Evans showed Riordan the computerized web-based project-management system they were using on this multibillion-dollar project, which allowed him (and other CERN managers) to determine the status of its components at a moment's glance. Lyndon Evans, private communication with Riordan, May 24, 2000.

202. Siskin, interview by Riordan.

203. Kim A. McDonald, "Supercollider Scientists Are Embroiled in Dispute; Outcome Could Raise Project's $8-Billion Cost," *Chronicle of Higher Education* (April 10, 1991), A19. The need for such a liner was only gradually becoming apparent; one finds hardly any evidence in SSC documents that it was discussed very seriously before 1990, when the dipole aperture was officially increased.

204. "Report of the SSC Machine Advisory Committee," November 9–10, 1991. See also Ritson, "Demise of the Supercollider," 609, and Wojcicki, "The Texas Days," 277–78.

205. Siskin, interview by Riordan. In a heavily redacted copy of a March 8, 1991, "Memorandum to the Secretary of Energy," Cipriano wrote, "The technical people at the Lab feel strongly that we should increase [redacted items]. They cannot say that the machine will not meet spec if we don't do it. They do say it gives the design more margin and would certainly make upgrades easier. The change along with a beam tube liner that would go along with it would cost by their estimate [more redacted items] nor I believe it is necessary; however, there is strong support among the physicists at the Lab and within DOE for doing it as a risk reduction measure." In the margin is the handwritten decision of Watkins on this matter: "NO to item d. unless you convince me!" The redacted items obviously refer to the quadrupole aperture and its cost, and that Siskin opposed the change, too. Boehlert/Goldston SSC files.

206. Actually, Schwitters probably did not have sufficient authority to do so by then, given what had recently been happening at DOE headquarters. This matter of the quadrupole aperture

change was apparently *still* unresolved when the SSC project was terminated by Congress in October 1993. In the "SSC Project Monthly Progress Report" of September 1993, there is an entry in "Figure 6.3, Listing of Laboratory Approved CCB/CCP Actions," entitled "Quadrupole Aperture and Associated 3B Arc Spec. Changes (40 to 50 mm)" that states: "Awaiting DOE Approval of Contingency." The estimated cost of this change was given there as $70.29 million. SSC Collection, Fermilab Archives. When asked about the status of the quadrupole aperture problem at termination, Rees could not recall it but said he thought it had been resolved in favor of an increase to 5 cm. Rees, interview by Riordan, May 6, 2010.

207. DOE had formally agreed to this oversight capacity in the URA M&O contract, but there were still substantial ambiguities in the construction-management area. Panofsky, conversations with Riordan.

208. Cipriano's March 8, 1991, memorandum to Watkins states, "The change control process is in full swing and decisions that have been hanging for months are getting made." The primary point here is that Cipriano—not Siskin or Schwitters—retained control of the contingency funds, applying them only *after* CCB approval. Panofsky and Wilson had controlled earlier contingency funds in building SLAC and Fermilab.

209. Similar conflicts between scientists and engineers occurred in deciding to construct the Intersecting Storage Rings at CERN, where the desires of accelerator physicists and engineers to build a new kind of atom smasher took precedence over the scientific goals of high-energy physicists; see Dominque Pestre, "The Second Generation of Accelerators for CERN, 1956–1965: The Decision-Making Process," in Armin Hermann et al., eds., *History of CERN*, vol. 2 (Amsterdam: Elsevier Publishing Co., 1990), 679–780. On similar issues, see also Peter Galison, "Time Projection Chambers: An Image Falling Through Space," in Peter Galison, *Image and Logic*, 553–688, esp. section 7.7, "Physicists' Dreams, Engineers' Reality," 614–40; and Robert W. Smith, *The Space Telescope* (New York: Oxford University Press, 1989), 187–220.

210. One noteworthy exception was Ted Kozman, a mechanical engineer who became associate director of the Accelerator Systems Division. Before that he had worked at the Lawrence Livermore Laboratory and then LBL, participating in the Central Design Group. Because of such a cross-cultural background, he worked well and cooperatively with the physicists he interacted with.

211. Here it is worth noting that DOE efforts to impose stronger control over the SSC project were part of a larger campaign going on at all the DOE national labs during the Bush administration, in part because of the 1980s revelations about nuclear wastes at the weapons labs. Congress was insisting that the DOE get the labs under better control, but this ran counter to the government-owned, contractor-operated (GOCO) approach that dated back to the late 1940s. Laboratory scientists objected that the nonweapons labs should be treated differently, but they were all lumped together as one.

212. For more on the problem of establishing a laboratory at a greenfield site, see Riordan, "A Tale of Two Cultures," esp. 141–43, and Wojcicki, "The Texas Days," 291–92. Chris Llewellyn-Smith, the CERN director general at the 1994 outset of the LHC project, also commented on this problem in "International Scientific Collaboration," presentation at the American Physical Society meeting in Denver, May 2, 2009. He estimated that had the SSC been sited at Fermilab, the project could have saved infrastructure costs of up to $2 billion (lecture notes of Riordan). Similar topics were discussed by Llewellyn-Smith, interview by Riordan. Perhaps half of this figure was the added cost of building up the needed human infrastructure.

213. Such a system was used during the mid-1990s on the $250 million SLAC B Factory project, which came in on schedule and on budget.

214. On Evans and the building of the LHC, see Geoff Brumfiel, "The Machine Maker," *Nature* (December 18–25, 2008), 862–68. According to Evans, the costs of building the LHC, not including its detectors, computing, and internal CERN labor costs, experienced an overrun of 48 percent, which CERN was able to accommodate. Evans, private conversation with Riordan, April 30, 2011.

A member of the SSC Machine Advisory Committee, he also observed that the engineers who came to the SSC project from the US military-industrial complex "didn't know how to do things inexpensively." On the LHC cost figures, see A. Unnervik, "The Construction of the LHC: Lessons in Big Science Management and Contracting," in L. Evans, ed., *The Large Hadron Collider: A Marvel of Technology* (Lausanne: EPFL Press, 2009), 38–55, esp. table 1, 40.

215. Llewellyn-Smith, interview by Riordan.

CHAPTER FIVE

1. For a presidential perspective on these pivotal years in world history, 1989 to 1991, see George Bush and Brent Scowcroft, *A World Transformed* (New York: Alfred A. Knopf, 1998), esp. 369–71.

2. But Wright was charged (by Newt Gingrich) with ethics violations in early 1989 and resigned as Speaker that May, to be replaced by Washington Democrat Tom Foley, a big loss for the Texas delegation. Had Wright remained in power, House Democrats would have feared opposing the Speaker's pet project.

3. James Chapman, interview by Riordan, May 3, 2000.

4. Goodwin, "Numbers Game: Bush's 1990 R&D Budget Uses Reagan's Figures in Making Deals," *Physics Today* (May 1989), 43–49, esp. the table of DOE funding requests on 44.

5. Ibid. The Space Station request was more than double its FY89 allocation of $900 million, and SDI would grow by 46 percent over its $4 billion allotment, if Congress appropriated this amount.

6. It is worth noting that two of these four congressmen hailed from states that had been selected as finalists in the SSC site-selection process but did not win the contest, New York and Michigan. And they all came from states in the US Northeast and Midwest, which had been hardest hit by the deep recessions and consequent job losses of the late 1970s and early 1980s.

7. Tom Bevill, *Congressional Record—House* (June 28, 1989), H-3308.

8. James D. Watkins, letter to Tom Bevill, June 27, 1989, Riordan SSC files.

9. Goodwin, "At Last, Congress Agrees to Build SSC." The *Congressional Record* put the margin at 330 to 93, with 9 abstentions, but a member may have subsequently changed his or her vote, as often happens. The Illinois delegation voted largely in favor of the supercollider (i.e., against the amendment), partly because the Texas delegation had agreed to support Fermilab's Main Injector project (see below and chap. 6, n. 9).

10. Ibid., 51.

11. Ibid., 52.

12. Ibid., 51. The $225 million was later decreased by Gramm-Rudman-Hollings sequestrations and other reductions to $218.6 million, including $129 million for construction. Goodwin, "Future Shocks: Bush's 1991 Budget Boosts R&D but Deficit Threatens It," *Physics Today* (June 1990), 51–58, esp. table of DOE figures, 52.

13. On regional economic conflicts and their relationship to US foreign policy, see Peter Trubowitz, *Defining the National Interest: Conflict and Change in American Foreign Policy* (Chicago: University of Chicago Press, 1998), esp. chap. 4: "The Rise of the Sunbelt: America Resurgent in the 1980s." The origin and early years of the Northeast-Midwest Coalition are discussed on 221–24.

14. Ibid., 225ff.

15. David E. Rosenbaum, "Southwest to Get Economic Benefits in Savings Bailout," *New York Times* (June 25, 1990), D5.

16. David V. Gibson and Everett M. Rogers, *R&D Collaboration on Trial: The Microelectronics and Computer Technology Corporation* (Boston: Harvard Business Review Press, 1994). The role of the Texas congressional delegation in securing SEMATECH for Austin is discussed on 490–94.

17. R. Jeffrey Smith, "Supercollider Could Face Cutbacks" (see chap. 4, n. 62).

18. Serge Schmemann, "Clamor in the East: East Germany Opens Frontier to the West for Migration or Travel; Thousands Cross," *New York Times* (November 10, 1989). This happened the exact same day that Roy Schwitters visited Washington to tell DOE officials about the SSC redesign and likely cost overruns.

19. In *The Age of Extremes: A History of the World, 1914–1991* (New York: Vintage Books, 1994), Marxist historian Eric Hobsbawm uses this period as one of the two bookends of a singular era. As he stated in his introduction, 5, "There can be no serious doubt that in the late 1980s and early 1990s an era in world history ended and a new one began." In *A World Transformed*, Bush and Scowcroft provide a personal account of the discussions and negotiations going on behind the scenes during the transition from a Cold War footing to the "new world order" they hopefully envisioned.

20. Goodwin, "Future Shocks," 52–53. SDI funding was $3.95 billion in FY1989 and $3.82 billion in FY1990; the FY1991 request was for $4.66 billion, but it ultimately received "only" $3.0 billion. The Space Station swelled from $900 million in FY1989 to $1.75 billion in FY1990; the request for FY1991 was $2.45 billion. In general, the burgeoning military spending experienced during the Reagan years did not continue into the Bush administration. For example, in the president's FY 1991 budget, 61 percent of R&D funding was slated for military R&D, versus 69 percent five years earlier.

21. Ibid., 53. Costs of nuclear-weapons lab remediation grew from $1.66 billion in FY 1989 to $2.22 billion in FY1990; the FY1991 Bush administration request was for $2.79 billion.

22. Mark Crawford, "Super Collider Advocates Tangle with Cost Cutters," *Science* (January 12, 1990), 152–53. Crawford noted, "An aide to Representative Robert Roe (D-NJ), Chairman of the House Committee on Science, Space and Technology, predicts there will be little or no opposition to an increase on the order of $1 billion to $1.5 billion."

23. Goodwin, "HEPAP and Its Subpanel Approve Redesign," 68.

24. Whip Advisory sheet from William H. Gray III, Majority Whip, on "The Superconducting Super Collider Authorization Act of 1990 (H.R. 4380)," April 20, 1990, Boehlert/Goldston files.

25. Ibid., emphasis added. This wording dovetailed with what the DOE was trying to achieve at the time with regard to the SSC management structure, perhaps because DOE officials were working closely with congressional staffers to have this wording inserted in the bill (see chap. 4, n. 140).

26. The Texas contribution to SSC construction was reduced to $875 million because it committed $125 million to things like generic detector R&D. This figure assumes a $1 billion Japanese contribution.

27. Democratic Study Group Legislative Report, "Schedule for the Week of April 23, 1990," 19, Boehlert/Goldston files.

28. David Goldston, e-mail to Riordan (May 20, 2010). In entering the amendment, he said, Boehlert was trying to hold the DOE to its public position.

29. Whip Advisory sheet from William H. Gray, 1.

30. Sherwood Boehlert, *Congressional Record*, May 2, 1990, H-1958. H.R. 4380 passed easily, 309–109.

31. Goldston, private communications with Riordan.

32. Wojcicki, "The Supercollider: The Texas Days," 278. A group led by Rubbia was just then putting the finishing touches on a conceptual design study for the LHC project. See The LHC Study Group, *Design Study of the Large Hadron Collider (LHC)*, Geneva: CERN Report no. 91-03, May 1991.

33. Italy, for example, had long been viewed as a good possibility for making contributions worth up to a billion dollars, thanks mainly to the efforts of physicist Antonino Zichichi of the University of Bologna and his strong connections in the Christian Democratic Party then in power. A. Trivelpiece e-mail to L. Hoddeson, March 29, 2010. As late as January 1990, some SSC advocates still thought that as much as $200 million might be possible, according to US physicist

Lawrence Sulak of Harvard. Robert Diebold memorandum, "Meeting of Sciulli HEPAP Subpanel at SSCL," January 23, 1990. But a closely held DOE internal report of October 1990 stated: "Italy is fully committed to the continuation of CERN and the LHC program and currently claims not to have the means within its present science budget to participate in the construction of the SSC. Italy is strongly behind the LHC program because the director general, Carlo Rubbia, is an Italian national and a Nobel laureate, and because of their large investment in CERN." US Department of Energy, "Superconducting Super Collider Status Report," October 1990 (unpublished), Riordan SSC files, copied from SSC files of Boehlert/Goldston.

34. "Superconducting Super Collider: Framework for International Participation," (unclassified) draft, April 1990, 1, Riordan SSC files. It specifies the goal of one-third nonfederal contributions: "The President has set as an objective for the Department of Energy (DOE) to construct the SSC on schedule and within established estimates of cost with one third of the total project cost from non-Federal sources, such as the host State, the private sector, and foreign countries," 2.

35. Ibid., 3.

36. Ibid., 8.

37. "International Collaboration and the SSC," undated position paper attached to memo, February 27, 1990, Panofsky Collection, SLAC Archives. At the multibillion-dollar scale of the SSC and LHC, however, there were increasing discussions that this policy might be inadequate. Some kind of additional *quid pro quo* contributions to construction of the facility might have to be required, that is, before a nation's scientists would be allowed to join experiments using it.

38. Riordan and Takahashi, "Cooperation in High-Energy Physics Research Between the United States and Japan," 5–6. Discussions with Kasuke Takahashi, who served as director of the Washington Office of the Japan Society for the Promotion of Science during the early 1990s, and earlier as KEK deputy director, proved invaluable in understanding the Japanese scientific culture and system of science funding.

39. Ibid., 7.

40. Ibid., 1. Also Ichita Itabashi, private conversation with Riordan, November 1, 2000.

41. Henson Moore, interview by Weiss, June 25, 1997.

42. Ibid.; John Metzler, interview by Riordan, April 14, 1995.

43. Letter from President Bush to Toshiki Kaifu, May 25, 1990, from SSC files of John Metzler (copy in Riordan SSC files), who claimed he drafted the letter earlier that year. Metzler, interview by Riordan.

44. "Equipment That Might Be Provided by Japan," undated list, files of John Metzler, international program manager at the DOE Office of the Superconducting Super Collider (copy in Riordan SSC files). Included in the list were "one half of all the superconducting magnets" at $950 million, "components of one of the large SSC detectors" at $200 million, and "grants of money" at $300 million.

45. Jacob M. Schlesinger, "U.S. Asks Tokyo to Fund Basic Science," *Wall Street Journal* (May 31, 1990). See also Goodwin, "Conversation with D. Allan Bromley on Major Issues in Scientific Research," *Physics Today* (July 1990), 49–55, wherein President Bush's science advisor stated, "It's not a question of inviting another country to participate in the design and use of our accelerator. It's a question of inviting them to take an equity position in this facility." (Quote on 52.)

46. Unclassified DOE cable, June 5, 1990, which states, "The Japanese asked a number of questions concerning the nature of the 'partnership' being proposed by the United States." Metzler SSC files, copy in Riordan SSC files.

47. Ibid., 3.

48. Ibid., 4. See also Louis C. Ianniello memorandum, "Trip to Japan and Korea," June 7, 1990, which states: "It will take some time for the Japanese to arrive at an internal consensus and respond to our invitation." Metzler SSC files, copy in Riordan SSC files.

49. Ibid. Also, "Korean Participation in S.S.C. Project (To be discussed on June 5, 1990)," list in Metzler SSC files, copy in Riordan SSC files.

50. Bill Loveless and David Kramer, "Moore Finds Korea Bullish over SSC; Japan's Reaction is Restrained," *Inside Energy* (July 11, 1990).

51. Akhito Arima, in a private conversation with Riordan, October 31, 2000, mentioned Kaifu's leadership of Monbusho and familiarity with high-energy physics.

52. Benjamin Weakley, memo to Harold Jaffe, "Meeting with Toichi Sakata on SSC," July 13, 1990, Metzler SSC files, copy in Riordan SSC files.

53. To his September 14, 1990, memo to Watkins about SSC project management (see chap. 4, n. 137), Cipriano added a handwritten note: "Fred Bucy just called and told me that his sources in Japan (Kaifu's secretary) say Japan will provide no funding for the SSC because of Iraq contribution. He said the only way to get the decision turned around would be for President Bush to talk to Kaifu directly."

54. Metzler, interview by Riordan.

55. Ibid.

56. "Superconducting Super Collider Status Report," October 1990, Riordan SSC files. This unpublished report summarizes the status of negotiations with potential foreign partners as they stood in the fall of 1990. Its executive summary states: "Total foreign contributions for the SSC are targeted at $1.7 billion, an amount that is highly unlikely without a significant investment by Japan and the Soviet Union; other countries . . . will most likely not make significant contributions."

57. Ibid.

58. Ibid., 8.

59. Memorandum to URA Washington staffers from John Toll, "SSCL–INP General Collaboration Agreement," December 28, 1990, Riordan SSC files.

60. Mark Crawford, "SSC's Forlorn Quest for Foreign Partners," *Science* (April 5, 1991), 25.

61. Charles Vacek, private communication with Riordan, April 3, 1991, Riordan notes. Vacek, the administrative assistant to John Easton, mentioned that Bromley was suggesting a figure of $80 million for the US contribution to KAON, then considered a $400 million project.

62. "Superconducting Super Collider Status Report," 8.

63. Colin Norman, "Science Budget: Growth Amid Red Ink," *Science* (February 8, 1991), 616–18; Goodwin, "Bush, His Faith in R&D, Raises Hope for Higher 1992 Budget in Hard Times," *Physics Today* (April 1991), 79–86, esp. table on 80.

64. Crawford, private communication to Riordan, Riordan notes, February 10, 1991. He said he'd heard that Texas had threatened to withdraw its $149 million FY1991 contribution.

65. The previous November, Bush had signed the Omnibus Budget Enforcement Act of 1990, which superseded the Gramm-Rudman-Hollings process, with its onerous sequestrations, and established a pay-as-you go procedure that required any real, noninflationary budget increases to be supported by new taxes. Thus real increases in discretionary spending, such as high-energy physics, had to come from other nondefense budgets. For more details on the consequences of this Act, see below and n. 79.

66. Goodwin, "Bush, His Faith in R&D," table on 80.

67. Riordan, notes of IASSC meeting February 5, 1991, transcribed and recorded February 10, 1991, Riordan notes.

68. Ibid.

69. US Department of Energy, *Report on the Superconducting Super Collider Cost and Schedule Baseline*, Report no. DOE/ER-0468P (Washington, DC: US Department of Energy, January 1991); hereafter, the "1991 Green Book."

70. Riordan, notes of February 5, 1991, meeting.

71. David P. Hamilton, "The SSC Gets Its (Official) Price Tag"; Riordan, notes of February 5, 1991, meeting. The 1991 Green Book gives a figure of $2.6 billion for the total estimated nonfederal contributions, table 9, 62. Subtracting the $875 million Texas commitment leaves $1.725 billion in needed foreign contributions.

72. Riordan, notes of February 5, 1991, meeting.

73. "IASSC Member Companies," February 1991, Irwin Goodwin Collection.

74. "Superconducting Super Collider: A New Frontier for Science and Technology," with attached business card of Henry Gandy and US map titled "National SSC Impact." Americans for the SSC, undated. Goodwin Collection, AIP Niels Bohr Archives. This sheet noted that the DOE had already awarded over $100 million to ninety institutions in over thirty states.

75. Ibid. The sheet noted that almost $2 million had already been awarded to twenty universities in 1990, and another $2 million in SSCL fellowships that year had been supported by TNRLC funding. In 1991, it said, Texas expected to award another $3 million for twenty grants and twenty-four fellowships.

76. Alissa J. Rubin, "Supercollider Surges Ahead Despite Concerns Over Cost," *Congressional Quarterly* (April 6, 1991), 856–57, quote on 857.

77. Ibid.

78. Ibid., 857. Panetta added: "There's no question we have a set of members, particularly from the Southwest, whose vote for the budget will largely depend on the supercollider project."

79. Colin Norman, "Science Increases Will Test New Regime," *Science* (February 8, 1991), 617. See also C. Norman, "Science Budget," and this chap., n. 65.

80. D. Hamilton, "The SSC Gets Its (Official) Price Tag." See esp. chart of SSC annual budgets, reproduced in this book as fig. 5.3.

81. Catherine M. Anderson, private conversation with Riordan, May 15, 1991.

82. Ibid.

83. Boehlert, "Floor Statement on Energy and Water Appropriations," May 29, 1991, Boehlert /Goldston files.

84. It could have been closer. The Illinois delegation split its vote 11–9 in favor of the amendment (i.e., to kill the SSC), but only after $10 million had been pried out of the DOE budget for accelerator R&D and earmarked for the Main Injector construction start in a separate floor amendment. See "House Vote Revives Fermilab Injector," *Science* (June 7, 1991), 1373.

85. According to a "Vote Analysis" by the SSC lab's Washington watcher Coby Chase, who worked closely with the TNRLC, 105 of the 165 votes (or almost 64 percent) of the anti-SSC "aye" votes came from states in the US Northeast or Midwest, especially the latter, which voted 55–43 in favor of the amendment. In contrast, representatives from Texas and contiguous states like Louisiana voted 46–2 against the measure. California, a key Sun Belt state, also voted strongly against it, 34–11, while Best Qualified List finalist Michigan voted 14–3 in favor of it, to kill the SSC. "Vote Analysis: The Slattery Amendment to Strike SSC Funding from H.R. 2427," Riordan SSC files.

86. Goodwin, "Making News by Calling It Quits," 52.

87. Ibid.

88. In addition, Japan had recently signed a formal agreement to partner with the United States, Canada, and European nations in the Space Station, which would entail a contribution of over a billion dollars and further drain Japanese sources of R&D funding. See Logsdon, *Together in Orbit*, 36–39, 42.

89. "Superconducting Super Collider Status Report," 2.

90. "News Conference of President Bush and Prime Minister Toshiki Kaifu of Japan," Kennebunkport, Maine, July 11, 1991, The American Presidency Project, UC Santa Barbara; text of the press conference available at http://www.presidency.ucsb.edu/ws/?pid=19775. The "explanation" that Kaifu mentions is probably the May 25, 1990, Bush letter and supporting documentation. See this chap., n. 43.

91. Ibid.

92. Ibid.

93. Maureen Dowd, "Kaifu Visits Bush and Brings Word of War Payment," *New York Times* (July 12, 1991), A1. See also Bucy observation, this chap., n. 53.

94. Letter from George H. W. Bush to Toshiki Kaifu, October 11, 1991, Metzler SSC files, copy in Riordan SSC files.

95. Yukihiro Hirano, "Public and Private Support of Basic Research in Japan," *Science* (October 23, 1992), 582–83.

96. Arima conversation with Riordan.

97. D. Allan Bromley, *The President's Scientists: Reminiscences of a White House Advisor* (New Haven and London: Yale University Press, 1994), 213, emphasis in original. Bromley and Moore carried with them a glossy booklet titled "Superconducting Super Collider: An International Partnership in Basic Research," Metzler SSC files, copy in Riordan SSC files. For a different viewpoint from a reporter who was familiar with Japanese culture, see David E. Sanger, "Japan Wary as U.S. Science Comes Begging," *New York Times* (October 27, 1991), E13.

98. Bromley, *The President's Scientists*, 213. Other knowledgeable Japan-watchers disagree with his too-optimistic assessment, saying the Japanese consensual decision-making process is much more complicated than this, and that Bromley had a strong tendency to overstate his political influence.

99. See Wojcicki, "The Texas Days," 279, on the Happer delegation to Japan.

100. Letter from James D. Watkins to James A. Baker III, October 25, 1991, Metzler SSC files, copy in Riordan SSC files. Riordan witnessed these events from a close-in perspective while at URA, as both Watkins and Baker wished to bring with them copies of the Japanese edition of his book *The Hunting of the Quark* as gifts to Japanese ministers. Watkins insisted he be the one to do so, and the Department of State deferred to the DOE.

101. "Secretary Watkins, Trip to Japan, December 3–6, 1991." In "Briefing Book: Global Partnership for the Superconducting Super Collider and Energy Cooperation," November 29, 1991, Metzler SSC files, copy in Riordan SSC files.

102. Frederick Shaw Myers, "SSC: The Japan That Can Say No," *Science* (December 13, 1991), 1579. Myers notes that the Japanese "government has already had to cut budgets to pay Japan's $13 billion contribution to the Gulf War" and that "several of Watkins' arguments for Japanese contributions . . . are wearing thin in Tokyo."

103. Bromley, *The President's Scientists*, 213. See also Jeffrey Mervis, "Bromley Memoirs Take Off the Gloves," *Science* (September 2, 1994), 1357. Educated as a nuclear engineer at MIT, Sununu was much more inclined than Skinner to favor science in general and the SSC in particular. Schwitters recalled that this was a crucial turning point in the Bush administration attitudes toward obtaining a Japanese commitment, for Skinner was much more focused on trade quotas rather than scientific cooperation. Private conversation with Riordan, December 19, 2014.

104. Bromley, *The President's Scientists*, 213. It is telling indeed that Bromley did not accompany the president on this trip while eighteen top business executives did, according to Glen S. Fukushima, "Bush-Miyazawa Meeting Is 'Staged' for Success," *Los Angeles Times* (January 2, 1992). Riordan recalls that in early December 1991, while working at URA offices in Washington, he heard that the SSC rose as high as *sixth* on Bush's list of talking points for the summit meeting. But it probably did not remain at that level after auto industry executives became involved.

105. Michael Wines, "Bush in Japan: Bush Reaches Pact with Japan, But Automakers Denounce It," *New York Times* (January 10, 1992), 1. Trade issues are cited throughout this lengthy article, but the SSC is never once mentioned. Nor was it ever mentioned in the January 9, 1992, press conference by Bush and Miyazawa, available at http://www.presidency.ucsb.edu/ws/?pid=20437.

106. Kasuke Takahashi and Ichita Itabashi, private conversations with Riordan. To embark on such a process and then say they were uninterested would have entailed serious "loss of face" for Japanese officials. But few in the US government understood these subtleties and viewed the US/Japan working group as merely a delaying tactic. See also B. Richter memo to John Gibbons, January 14, 1993, "Japan and the SSC," Richter Collection, SLAC Archives, in which he states, "The [Japanese] politicians imply that the creation of the International Working Group on the SSC . . . was a clear sign that Japan was prepared to proceed. Indeed, [LDP faction leader Ichiro] Ozawa said as much directly."

107. Rene Donaldson, private communications with Riordan. They had both been willing to move to Texas but were not courted very enthusiastically. Since the 1982 birth of the idea of the Desertron at Snowmass, Donaldson had devoted her extensive talents and countless hours to publicizing the SSC and explaining its goals in print. Nobody else had anything close to her experience in this area.

108. P. W. Anderson, "Outline Testimony on the SSC," February 24, 1989, Goodwin Collection, AIP Niels Bohr Library and Archives.

109. William J. Broad, "Vast Sums for New Discoveries Pose a Threat to Basic Science," *New York Times* (May 27, 1990), 1.

110. Howard Wolpe, in "Status of the Superconducting Super Collider Program," Subcommittee on Investigations and Oversight, House Committee on Science, Space and Technology (May 9, 1991), 1.

111. Boehlert, in ibid., 7. Newspaper reports of this hearing included Warren E. Leary, "Supercollider Outlook: Higher Cost, Less Income," *New York Times* (May 10, 1991), A20; and Thomas W. Lippman, "Super Collider's Cost, Technology Criticized," *Washington Post* (May 10, 1991). On the impact of the 1990 budget resolution, see David Rogers, "Deficit Law Turns the Budget Debate Into Process of Cutting and Spending," *Wall Street Journal* (February 7, 1991), A6.

112. Goldston, private conversation with Riordan and Weiss, October 30, 1997. Goldston later became staff director of the House Science Committee when Boehlert stepped up as chair.

113. See, for example, Rogers, "Deficit Law Turns Budget Debate"; Thomas W. Lippmann, "Energy's 'Mountain Building Up,'" *Washington Post* (February 12, 1991); Daniel S. Greenberg, "Are We Really Shortchanging Science?" *Washington* Post (February 12, 1991); Rogers, "House Panels Vote to Kill 1992 Funds for Space Station, Trim Collider Request," *New York Times* (May 16, 1991); "Big Science" (editorial), *Dallas Morning News* (May 29, 1991); Edwin Chen, "Big Science Faces Big Troubles," *Los Angeles Times*, (June 5, 1991), A1; Leon Jaroff, "Crisis in the Labs," *Time* (August 26, 1991), 45–51, esp. sidebar, "Big Ventures that Swallow Dollars by the Billions," 48; Gregg Easterbrook, "Big Science on Easy Street," *New York Times* (November 29, 1991); "Between an Atom and a Hard Place" (editorial), *Newsweek* (December 2, 1991).

114. An excellent account of the 1991 congressional efforts to kill the Space Station is Richard Munson, *The Cardinals of Capitol Hill: The Men and Women Who Control Government Spending* (see chap. 1, n. 12). This book also provides insight into the congressional appropriations process and how it is controlled by powerful Appropriation Subcommittee chairs—the "Cardinals" of the book's title.

115. Tom Kenworthy and Curt Suplee, "House Panels Vote to Kill Space Station, Trim Super Collider," *Washington Post* (May 16, 1991); see also Rogers, "House Panels Vote to Kill 1992 Funds."

116. Jim Sasser, "The Budget Follies Have to Stop," *New York Times* (July 29, 1991). Sasser was using the ICE estimate of SSC costs; at the time, estimates of Space Station costs ranged as high as $40 billion.

117. Steven Weinberg, *Dreams of a Final Theory: The Search for the Ultimate Laws of Nature* (New York: Pantheon Books, 1992), 273.

118. Goodwin, interview by Hoddeson, Riordan, and Sandiford, June 4, 2009.

119. "The American Physical Society Council Statement on Physics Funding and the SSC," (College Park, MD: American Physical Society, January 20, 1991).

120. Robert L. Park, "What's New," e-mail newsletter of the APS Office of Public Affairs, April 19, 1991.

121. Ibid. High-energy physicists *could* legitimately take credit for driving high-volume production of superconducting NbTi cable and thus its price down to the point where MRI became cost effective. See Judy Jackson, "Down to the Wire," *SLAC Beam Line* (Spring 1993), 14–21.

122. Christopher Myers, "Congressional Actions Suggest Lawmakers Are Skeptical About Continued Support for Big-Science Projects," *Chronicle of Higher Education* (May 29, 1991), A15ff.

Park is also quoted therein as stating, "Members of Congress are realizing now that with big projects they make trouble for themselves next year."

123. For example, the October 18, 1991, issue opens with the headline "It Seems That the Japanese Have No Yen for the Supercollider!"

124. SSCL Technology Transfer Office, "Not for Scientists Only: Technology Spin-offs from High Energy Physics and the Super Collider," SSCL Report no. SSCL-Pub-0001.

125. Jackie Koszczuk, "Collider's Backers Start Spinoff Talk," *Fort Worth Star Telegram* (December 10, 1991), A18. It is, however, ironic that during that very same month, SLAC physicists were adapting new Internet communications software from CERN called the World Wide Web and establishing the first US website as well as the first graphical user interface, or web browser, called MIDAS.

126. For example, Texas Democrat Bill Sarpalius enthused: "I do not look at the Super Collider as a cost factor, but an investment—an investment in the future. Imagine what it could do by compressing energy. Our kids will be able to see a battery about this size that will have enough energy to operate an automobile. One a little larger will provide enough energy to heat and cool our homes. . . . The Japanese have developed a Dick Tracy watch, a watch that has a telephone in it, but they cannot put it on the market because they do not have a battery small enough to develop that watch." See William Sarpalius, *Congressional Record—House* (May 29, 1991), H-3673.

127. Dennis Eckart, *Congressional Record—House* (May 29, 1991), H-3682.

128. A detailed analysis of the debate in the *Congressional Record—House* (May 29, 1991), H-3663 to H-3685, reveals twelve instances of concerns being expressed about negative impacts of the SSC on other sciences and scientific projects, including other high-energy physics labs. By comparison, there were twenty instances of concerns about the SSC costs and nine each about the lack of foreign contributions and technical problems or uncertainties with the project, principally its superconducting magnets.

129. "Testimony of Dr. Leon M. Lederman," *Congressional Record* (May 29, 1991), H-3678 to H-3680.

130. George Brown, *Congressional Record—House* (May 29, 1991), H-3670.

131. In fact, Riordan came to Washington from SLAC in 1991 to work with URA as assistant to the president and try to rectify this public communications problem regarding the SSC's physics goals. He wrote an article on SSC physics, "Why Are We Building the SSC?" *Beam Line* (Summer 1991), 1–8, and attempted to get versions of it published more widely in general science magazines, but to no avail. The common reaction of editors was that their publications had already covered the topic and didn't need to do so again, at least not until the SSC began producing scientific results. As explained by a few science reporters, the SSC science was not "news," while its teething problems were. The *Beam Line* article was, however, circulated to Congress by Joe Barton.

132. Michael D. Lemonick, "The Ultimate Quest," *Time* (April 16, 1990), 50–56. The contributing reporters included Madeleine Nash, spouse of Fermilab physicist Tom Nash. The magazine cover shouted "Smash! Colossal Colliders Are Unlocking the Secrets of the Universe." Recalling this article, David Goldston called it a "puff piece" on high-energy physics. Goldston, interview by Weiss.

133. Lemonick, "The Ultimate Quest," quote on 54.

134. Faye Flam, "The SSC: Radical Therapy for Physics," *Science* (October 11, 1991), 194–96, quote on 194. Even *Science,* however, did not flinch from covering the global chase. This article included a sidebar by David P. Hamilton, "CERN's Horserace with the SSC," 194, on Rubbia's fond hopes of building the LHC inside the LEP tunnel and beating the hapless SSC to some of its prized discoveries. Experienced science writers like Hamilton somehow could not perceive the tremendous roadblocks in Rubbia's way and continued to publish inadequately researched articles like this suggesting that an LHC before 2000 was indeed possible. Experiments on it actually began over a decade later, in 2010. See epilogue.

135. Ibid., "The SSC: Radical Therapy," quote on 195.

136. An early account of this merger activity was Riordan and David N. Schramm, *The Shadows of Creation: Dark Matter and the Structure of the Universe* (New York: W. H. Freeman, 1991); see esp. 254–55 on the possible appearance of dark-matter particles at the SSC.

137. Weinberg, *Dreams of a Final Theory*. The manuscript for this book was written and delivered to its publisher in 1991, just after congressional efforts to kill the SSC began to get serious.

138. Ibid., 274.

139. Leon Lederman with Dick Teresi, *The God Particle: If the Universe Is the Answer, What Is the Question?* As noted in chap. 3 and n. 24, that chap., this book was a distillation of Lederman's countless popular lectures and writings about particle physics stretching back over a decade or more.

140. Ibid., 22. That something was the Higgs field, or something like it, which should manifest itself as a particle or particles at a sufficiently high-energy collider like the SSC. "This invisible barrier that keeps us from knowing the truth is called the Higgs field," Lederman continued. "Its icy tentacles reach into every corner of the Universe, and its scientific and philosophical implications raise large goose bumps on the skin of a physicist."

141. See, for example, R. Harré, *The Philosophies of Science* (New York: Oxford University Press, 1985), 141–67.

142. An early—and perhaps the best—exposition of this alternative philosophy is P. W. Anderson, "More Is Different: Broken Symmetry and the Nature of the Hierarchical Structure of Science," *Science* (August 4, 1972), 393–96. Anderson was also author of a key 1963 article in a series of theoretical papers that led to the conception of the Higgs boson and Higgs field: "Plasmons, Gauge Invariance, and Mass." In fact, the theoretical insights of the early 1960s that led to the 2012 discovery of the Higgs boson at CERN were derived from concepts of spontaneous symmetry breaking that originated in condensed-matter physics. See Frank Close, *The Infinity Puzzle: Quantum Field Theory and the Hunt for an Orderly Universe* (New York: Basic Books, 2011), esp. chap. 8, "Broken Symmetries," 127–50.

143. For another perspective on this debate, see Joseph D. Martin, "Fundamental Disputations: The Philosophical Debates that Governed American Physics, 1939–1993," *Historical Studies in the Natural Sciences*, vol. 45, no. 5 (2015).

144. Alvin M. Weinberg, "Impact of Large-Scale Science in the United States," *Science* (July 21, 1961), 161–64. Weinberg subsequently published this essay in *Reflections on Big Science*.

145. Johnston's state stood to benefit from the manufacture of most of the superconducting dipoles by General Dynamics at its Louisiana plant and would therefore have been the state most impacted by sending some of this work overseas. Sending part of the high-tech manufacturing overseas would also have violated another of the stated SSC goals—to enhance the competitiveness of US industry.

146. Congressional Budget Office, "Risks and Benefits."

147. Future CERN Director Chris Llewellyn-Smith had had to confront this very question of the impact of high-energy physics on other sciences when he served on the British Kendrew Commission in 1984. This experience helped guide and temper his thinking in the early 1990s as he worked to get the LHC approved by the CERN Council. See epilogue. Llewellyn-Smith, interview by Riordan.

148. Logsdon, *Together in Orbit*.

149. Weinberg, *Dreams of a Final Theory*, 5.

CHAPTER SIX

1. This account is based largely on notes of Riordan, who attended this meeting as a URA observer. See also David P. Hamilton, "Allocating the Pain in Energy Science," *Science* (September 27,

1991), 1482; David Kramer, "Watkins Sees More Global R&D Partners, *Inside Energy* (September 23, 1991), 1–2, and "Budget Ax Hangs over Major Projects, *Inside Energy* (September 30, 1991), 1.

2. Kramer, "Watkins Sees More Global," 2.

3. See, for example, Crease, *Making Physics*, 257–59, on Brookhaven Director Leland Haworth's efforts to ensure his laboratory's continuing research vitality.

4. On the Trivelpiece plan, see Catherine Westfall, "Retooling for the Future: Launching the Advanced Light Source at Lawrence's Laboratory, 1980–1986," *Historical Studies in the Natural Sciences* 38, no. 4 (2008), 569–609, esp. 599–601.

5. Trivelpiece, interview by Weiss.

6. Hamilton, "Allocating the Pain."

7. According to the Hamilton article, the Main Injector was a $181 million construction project, and the B Factory was then estimated at $200 million. Other reporters put the Main Injector at $178 million. Hess's budget also included major amounts for laboratory operations and university groups doing high-energy physics research at these facilities, as well as at CERN and other overseas laboratories.

8. Riordan notes of Townes panel sessions. On the second day, it seemed as if decisions had already been reached in a meeting behind closed doors or perhaps at dinner. The public sessions appeared to be deliberately choreographed to provide political cover for a DOE decision to defer the Main Injector project until the overall funding climate improved.

9. "Proton-Proton Collider Upgrade (Main Injector, New Tevatron)," Fermilab Project no. 90-CH-400, Technical Components and Civil Construction (May 1988), Fermilab Archives.

10. As of 1991, based on fruitless searches at CERN and Fermilab, the top-quark mass had to be larger than 89 GeV, else it would already have been discovered. The top quark showed up at Fermilab in 1994–1995 at a mass close to 175 GeV (see chapter 7), discovered *without* any benefit from the increased collision rate eventually provided by the Main Injector, which began construction in 1993 and came online in 1999.

11. Department of Energy Office of Energy Research, *Report of the HEPAP Subpanel on US High Energy Physics Program for the 1990s*, Report no. DOE/ER-0453P (April 1990).

12. Riordan, conversation with Mark Crawford, September 20, 1991. Other reporters made similar comments. Hess's projected budget for DOE's high-energy physics base program topped $800 million in 1995–1996, the very same years in which the SSC budget was scheduled to be peaking.

13. Kramer, "Budget Ax," 9, which notes: "At the same time, Happer held out hope for proposals advanced by the Stanford Linear Accelerator Center and Cornell University to build so-called B-factories. . . . He said the High Energy Physics Advisory Panel may want to re-prioritize B-factory proposals against the Fermilab upgrade project. . . ." An atomic and molecular physicist who specialized in laser research, Happer seemed to have little patience for the proton-smashing high-energy physics community, whose insatiable demands were putting such enormous pressures on his OER budgets.

14. Letter from Illinois Congressional Delegation to James D. Watkins, October 17, 1991, Boehlert/Goldson files.

15. A moderate Republican first elected in 1987, Hastert later became Speaker of the House in 1999.

16. Robert Michel and J. Dennis Hastert, letter to James D. Watkins, October 22, 1991, Riordan SSC files. Also quoted in David Kramer, "New Bid to Cut Fermilab Draws Protest from Illinois Delegation," *Inside Energy* (October 28, 1991), 9.

17. For example, Michel, Hastert, and Sidney Yates had all voted against the May 1991 House amendment to kill the SSC, while Rostenkowski had voted for it, i.e., to terminate the SSC. As a whole that year, the Illinois delegation voted 11–9 to terminate the SSC project, after supporting it solidly in 1989–1990.

18. David Kramer, "Watkins to Ask for More Funds for Research," *Inside Energy* (December 2, 1991), 1–2, quote on 2.

19. Ibid., 2.

20. Goodwin, "Cheers for Bush's 1993 R&D Budget Cut Short by Problems and Pessimism," *Physics Today* (June 1992), 55–61, esp. table of DOE funding on 56–57.

21. Ibid.

22. Like other established physicists, Rees joined the project initially on a consulting basis, intending to work in Texas for only two years or so. He remained officially on the SLAC payroll, with the SSCL reimbursing his salary and benefits, and paying him an additional stipend to cover an apartment rental in Dallas. The "warm" machines included the linear accelerator (or Linac), Low-Energy Booster (LEB), and Medium-Energy Booster (MEB), which did not require superconducting magnets, an area in which Rees had little experience. John Rees, interview by Riordan, September 2, 2009.

23. Rees, interview by Riordan, September 2, 2009. He also recalled that Reardon frequently missed important meetings and that crucial decisions were consequently not being made in a timely fashion. Siskin had said much the same thing in an earlier interview, although he placed the blame for the lack of a cost-and-schedule control system more broadly on the high-energy physicists, saying, "I won't tell you how many different people told me, 'Do you want me to do my job, or do you want me to crank out all this paper?'" Siskin interview by Riordan.

24. DOE Office of Inspector General, *Follow-up Audit, Department of Energy's Superconducting Super Collider Program*, Report no. DOE/IG-0305, March 13, 1992; US General Accounting Office, *Federal Research: Implementation of the Super Collider's Cost and Schedule Control System*, Report no. GAO/RCED-92-242, July 21, 1992; House Committee on Science, Space and Technology, *Status of the Superconducting Super Collider Program: Hearing Before the Subcommittee on Investigations and Oversight*, May 9, 1991 (Washington, DC: US Government Printing Office, 1991).

25. Siskin, interview by Riordan; Rees, interview by Riordan, September 2, 2009.

26. Rees, interviews by Riordan. Rees had also worked in the AEC high-energy physics program office during the building of Fermilab, and thus understood the bureaucratic needs for reliable information about progress (or the lack of it) on a major construction project. When he stepped in as SSC project manager, he became a regular laboratory employee, rather than continuing to serve as a consultant on leave from SLAC.

27. Dillingham Construction Corp. and Obayashi America Corp. both had major Japanese ownership.

28. Rees, interview by Riordan. See also "SSC Project Monthly Progress Report," January 1992, SSC Collection, Fermilab Archives. Westinghouse was the "follower" contractor to General Dynamics in the "leader/follower" contract to develop and manufacture the collider dipole magnets, which it planned to do at a new factory under construction in Round Rock, Texas, near Austin.

29. "SSC Project Monthly Progress Report," January 1992, 3. See also R. W. Baldi et al., "An Overview of the Collider Dipole Magnet Program," in *Supercollider 5*, ed. Phyllis Hale (Dordrecht, Netherlands: Plenum Press, 1994), 23–26.

30. A full cell thus consisted of ten dipole and two quadrupole magnets; there were forty-three cells in each of ten SSC sectors, or 4,300 dipole and 860 quadrupole magnets in each of two rings, plus other, more specialized, magnets close to the interaction regions.

31. Goodwin, with Bertram Schwarzschild, "Good News for the SSC as Senate Approves Funds and Magnets Work," *Physics Today* (September 1992), 54–56. See also "SSC News," (an e-mail newsletter about the SSC, available at *www.hep.net/ssc/new/history/sscnews*), August 1992.

32. David P. Hamilton, "Lightning Strikes the SSC," *Science* (June 26, 1992), 1752–53. For a deeper analysis of congressional dynamics, see also Holly Idelson, "House Denies Atom Smasher Its 1993 Expense Account," *Congressional Quarterly* (June 20, 1992), 1773–85; and George Hager,

"Surprise Vote Sounds Alarm for 1993 Spending Bills," *Congressional Quarterly* (June 20, 1992) pp. 1773–77.

33. It is ironic that Republicans were leading this budget battle, given that the legacies of the Reagan administration—its defense buildup, SDI, Space Station, etc.—were a major cause of the deficit problem.

34. John R. Cranford, "Defeat of Budget Amendment Fans Anti-Deficit Flames" and box on Panetta's role and reactions, "A Winner at Last," *Congressional Quarterly* (June 13, 1992), 1683–89; quote on 1688. Panetta later became OMB director and then White House chief of staff under Clinton (see below).

35. Leon Panetta, telephone conversation with Riordan, June 25, 2002. In his weekly column, APS spokesman Robert Park observed, "Democrats were furious with the Texas delegation for leading the fight for the [balanced-budget amendment] and were delighted to have an opportunity to retaliate." *What's New*, June 19, 1992, downloaded from http://bobpark.physics.umd.edu/WN92/wn061992.html.

36. George Hager, "Is the Deficit Now Too Big for Congress to Tame?" *Congressional Quarterly* (May 2, 1992), 1140–47. At that point the Congressional Budget Office was projecting a 1992 deficit of $368 billion, of which $99 billion was attributed to the recession and $65 billion to the bailouts; see sidebar, "How Big Is the Deficit?" on 1141. The eventual deficit figure came in at "only" $292 billion.

37. Holly Idelson, "Less Energy Spending Spreads the Pain Around," *Congressional Quarterly* (June 13, 1992), 1692–93. It's worth noting, however, that high-energy physics had still been allocated a total of nearly $1.1 billion in the House Energy and Water Bill, as reported out of committee.

38. Idelson, "Nuclear Weapons Complex Braces for Overhaul," *Congressional Quarterly* (April 25, 1992), 1066–73, esp. sidebar, "The Costs of Cleaning Up," 1069.

39. George Hager, "Rejection of Walls Bill Spells Spending Squeeze at Home," *Congressional Quarterly* (April 4, 1992), 866–67. Stenholm quote on 867. "This is the year when we're all going to the dentist without Novocain," quipped Congressman Joseph McDade (R-PA), ranking minority member of the House Appropriations Committee. "All the major items will be scrubbed hard."

40. Goodwin, "Victim of House Budget Balancing War, SSC Now Faces Uncertain Fate in Senate," *Physics Today* (July 1992), 53–54; quote on 53.

41. Ibid., 53.

42. Idelson, "House Denies Atom Smasher," chart on 1783. Also "Net Change in State SSC Votes Between 1991–1992," vote analysis from the files of House Science Committee staffer Robert Palmer (unpublished, 1992) and in a similar URA vote analysis (unpublished). Republicans split evenly on the vote, 79–79, while Democrats voted for the Eckart amendment (against the SSC), 152–102.

43. Ibid., especially Idelson article and Palmer analysis.

44. Mark Crawford, editor of *Energy Daily* and *New Technology Week* in the early 1990s, pointed out that as the annual SSC budget inevitably approached $1 billion, it would press not only on other energy projects but also on the water projects that were grouped in the same energy and water bill. A quick and easy way for members to relieve this pressure on their own projects was to kill the SSC before it could do any such damage in the mid-1990s. Crawford, conversation with Riordan, April 18, 2000.

45. Daniel Pearson, private conversation with Riordan and Weiss, October 25, 1995, Riordan notes.

46. Ibid. Also private conversations of David Goldston with Riordan.

47. Dennis Eckart, *Congressional Record* (June 17, 1992), H-4811.

48. Sherwood Boehlert, conversation with Riordan and Weiss; David Goldston, interview by Weiss. The very same comment was echoed by others we talked with, including DOE Deputy

Secretary Moore and SSC General Manager Siskin. Goldston later noted that few congressmen understand the science of such projects, but that the main question here was "whether they thought this general type of research was important enough, given everything wrong with the project." Goldston, e-mail to Riordan, May 22, 2011.

49. Jim Slattery, *Congressional Record* (June 17, 1992), H-4819.

50. To maintain such a position, one would have to argue that the burgeoning SSC budget was cutting into the allocations for other Appropriations Subcommittees involved with science, such as those that oversee the budgets for NASA, NSF, the National Institutes of Health, and the US Geological Survey. Although the SSC budget was approaching $1 billion, such impacts were unlikely to occur. But SSC opponents could argue convincingly that "the SSC exerted a kind of general pressure on the entire discretionary budget, esp. DOE." Goldston, e-mail to Riordan, May 22, 2011.

51. Panetta, conversation with Riordan, Riordan notes. He initially supported the SSC, voting for it in 1991, but changed his mind in 1992 and voted with the majority to terminate the project.

52. Michael Telson, private conversation with Riordan, March 15, 2000. On March 20, 2000, Telson extended his remarks about the anti-Texas and anti-Barton sentiments driving the anti-SSC vote.

53. Goodwin, "What's Gone Wrong with the SSC? It's Political, Not Technological," and "2100 Physicists Use a Democratic Process for the SSC," *Physics Today* (August 1992), 58–60; quotes on 59.

54. Goodwin, "Senator Bennett Johnston Talks About Physicists in Politics," *Physics Today* (January 1996), 51–55. On the special character of the Senate, see also David S. Broder, "Sen. Robert Byrd's Vanished Ethic," *Washington Post* (July 1, 2010), A15.

55. Here he had the able help of the subcommittee clerk, Proctor Jones, one of the most powerful staff members in the entire Senate, who oversaw the process of dividing up the subcommittee's "603(b) allocation" among all the parties vying for a slice of the annual energy-and-water pie. J. Bennett Johnston and Proctor Jones, interview by Riordan, February 14, 2000. For an excellent analysis of the congressional appropriations process in the early 1990s, see Munson, *The Cardinals of Capitol Hill*, which focuses on the budget and appropriations activities of the 102nd Congress from 1991 to 1992.

56. Goodwin, "What's Gone Wrong with the SSC?," 58–60.

57. Quoted from Goodwin, with Schwarzschild, "Good News for the SSC," (September 1992), 55.

58. Ibid. For more about the small science versus Big Science debate on the SSC, see chap. 3 and chap. 5.

59. As quoted by Dale Bumpers, *Congressional Record*, August 3, 1992, S-11146.

60. Ibid., S-11144.

61. J. Bennett Johnston, *Congressional Record*, August 3, 1992, S-11150.

62. Technically, the vote was taken on a Johnston motion to table the Bumpers amendment, effectively killing it. For details of the vote, see *Congressional Record*, August 3, 1992, S-11211.

63. Idelson, "Conferees Save Supercollider But Can't Resolve Test Ban," *Congressional Quarterly* (September 19, 1992), 2806–08.

64. Ibid. In a May 22, 2011, e-mail to Riordan, Goldston pointed out that it is "extraordinarily rare and procedurally difficult" to defeat such a conference report on an appropriations bill.

65. "10. Superconducting Super Collider," *Congressional Quarterly* (December 19, 1992), 3867.

66. Burton Richter, "Japan and the SSC," memorandum to John Gibbons, January 14, 1993, Richter Collection, SLAC Archives; Kasuke Takahashi and John Toll, conversation with Riordan; Akhito Arima, conversation with Riordan, November 20, 2000.

67. US General Accounting Office, *Federal Research: Foreign Contributions to the Superconducting Super Collider*, Report no. GAO/RCED-93/75 (December 1992), gives the best summary of these activities as of late 1992.

68. See, for example, Paul Reardon, letter to Joseph R. Cipriano, October 15, 1991, Reardon files, SSC Collection, Fermilab Archives: "The total amount of payments will be significantly less than the baseline cost estimate for the same items, thereby constituting an in-kind donation by the Union of Soviet States (USS)." (This letter was written during the late-1991 transition from the USSR to the successor Commonwealth of Independent States.)

69. See, for example, Mark Crawford, "SSC's Forlorn Quest for Foreign Partners," and Dennis Eckart, *Congressional Record* (June 17, 1992), H4810–11.

70. *SSC News* (February 1993), 1.

71. Malcolm W. Browne, "Building a Behemoth against Great Odds," *New York Times* (March 23, 1993), C1.

72. "The Tokyo Declaration on the U.S.-Japan Global Partnership," official summit document, in file "Bush Originals—1992," George Bush Presidential Library, College Statión, TX, 10.

73. T. Bush, memorandum to R. Schwitters, "Inflated Japanese Cost Estimates," January 31, 1992, SSC Collection, Fermilab Archives.

74. An excellent resource on the state of science and science funding in Japan at that time is the special issue of *Science* magazine devoted to the topic, vol. 258 (October 23, 1992), 561–91. See esp. Alun Anderson, "Japanese Academics Bemoan the Cost of Years of Neglect," 564–69; Yukihiro Hirano, "Public and Private Support of Basic Research in Japan," 582–83; and Akhito Arima, "Underfunding of Basic Science in Japan," 590–91.

75. Private conversations of Riordan with Kasuke Takahashi, Akhito Arima, Ichita Itabashi, and Yoshikazu Hasegawa (November 1, 2000).

76. In addition, preliminary meetings occurred in Tokyo in February and in Washington in March 1992. Subgroups also met at the SSC lab in April, May, and October, as well as in Tokyo in June and Hawaii in July, mainly to address the question of superconducting magnet costs. Another subgroup to address "international aspects" met in Tokyo in October and November. It was a very busy year.

77. The opinions of knowledgeable observers on this question are mixed but generally favor Japanese partnership in the SSC, as long as an accommodation could be reached between the Japanese premier and US president. Jiro Kondo, the president of the Science Council of Japan in 1991–92 (and thus principal science advisor to Miyazawa) agreed that Japan was ready to join the project by late 1992, needing only a formal request from the US president; Jiro Kondo, private conversation with Riordan, October 31, 2000, Riordan notes.

78. Richter memo to Gibbons, "Japan and the SSC." He ended it with the observation: "It is my impression that if the President says the SSC is a priority in this Administration, the Japanese will join. If the President is ambiguous, the Japanese will delay. If the President kills the SSC, the Japanese will be relieved."

79. *SSC News*, October, November, and December 1992.

80. *SSC News*, October 1992. Operation at this current level, over 50 percent above the design value of 6,550 amps, indicated that the circumference of the collider ring *could* have been reduced in the major 1989–90 redesign, as Panofsky had argued, and 40 TeV proton collision energies eventually achieved by lowering the liquid-helium temperature to 1.9 K, as is now routinely accomplished in the CERN Large Hadron Collider.

81. *SSC News*, November 1992.

82. Gary Taubes, "SSC Detectors Desperately Seeking Donors," *Science* (February 5, 1993), 757.

83. Ibid. See also Wojcicki, "The Texas Days," 275.

84. *SSC News*, March 1993; Bertram Schwarzschild, "Tunnel Boring Begins at Superconducting Super Collider," *Physics Today* (March 1993), 19. The Eagle Ford Shale was also apparently going to be a problem in digging the nearby experimental halls at the SSC West Campus. See Allan Freedman, "Second Thoughts on the Super Collider," *D Magazine* (July 1991); also Raymond Stefanski, private communications with Kolb.

85. *SSC News*, May 1993.

86. Browne, "Building a Behemoth," C1.

87. Ibid., C3. SLAC physicist Karl Brown, who was present, told Riordan that this offhand remark was made at a cocktail party after work, when Schwitters thought he was speaking off the record. Karl Brown, private conversation with Riordan, July 9, 1997.

88. It is ironic that two Texans split Republican and independent voters, allowing a Democrat with no special allegiance to the SSC to win the presidency with only 43 percent of the total popular vote.

89. Goodwin, "Despite Retirements and Defeats, Congress Retains Friends of Science," *Physics Today* (February 1993), 71–72. Three staunch SSC opponents were gone, however: Dennis Eckart, Don Ritter, and Howard Wolpe.

90. This political convulsion would also inevitably affect relations with potential foreign partners like Japan. As Henry Kissinger eloquently wrote in *On China* (New York, Penguin Press, 2011), 377: "One of the obstacles to continuity in America's foreign policy is the sweeping nature of its periodic changes of government. As a result of term limits, every presidential appointment down to the level of Deputy Assistant Secretary is replaced at least every eight years—a change of personnel involving as many as five thousand key positions. . . . For countries relying on American policy, the perpetual psychodrama of democratic transitions is a constant invitation to hedge their bets."

91. Goodwin, "President Clinton Picks John Gibbons as Science Adviser to Reinvent Policy," *Physics Today* (March 1993), 73–74.

92. On this issue, see President William J. Clinton and Vice President Albert J. Gore Jr., "Technology for America's Economic Growth: A New Direction to Build Economic Strength," February 22, 1993, delivered by Clinton in a press conference held that day at Silicon Graphics, Mountain View, CA. Republicans were and have continued to be far less interested than Democrats in pursuing such "technology policy."

93. Goodwin, "President Clinton Picks John Gibbons," quote on 73.

94. John Gibbons, interview by Weiss, August 8, 1996.

95. Goodwin, "Clinton's Hands-On Economic Plan: Technology Gains, Big Science Loses," *Physics Today* (April 1993), 43–46; quote on 45. It was finally discovered in 2012, nineteen years later, at CERN. See epilogue.

96. Ibid. See also Sharon Begley, "Sparing Those Sacred Cows," *Newsweek* (March 1, 1993), 17. Panetta confirmed his Camp David opposition to the SSC project via Panetta Institute archivist Ellen Wilson, e-mail to Riordan, August 7, 2013. His opposition continued well beyond this Camp David meeting to at least May 1993. In his autobiography (written with Jim Newton) *Worthy Fights: A Memoir of Leadership in War and Peace* (New York: Penguin Press, 2014), Panetta discusses the 1993 budget process in detail, 103–21. The SSC is never once mentioned.

97. Office of Management and Budget, "A Vision of Change for America," February 17, 1993, 87.

98. John Gibbons, White House press briefing, February 23, 1993, Clinton Digital Library Archives, available at http://clinton6.nara.gov/1993/02/1993-02-23-press-briefing-by-john-gibbons-bowman-cutter.html.

99. Ibid.

100. Gibbons, interview by Weiss.

101. In addition, Reagan's Strategic Defense Initiative was then consuming \$4–\$5 billion per year.

102. Gibbons, interview by Weiss.

103. James D. Watkins, letter to George E. Brown Jr., January 14, 1993, Riordan SSC files.

104. US Department of Energy, *Report on the Superconducting Super Collider Cost and Schedule Baseline*, 62.

105. Watkins, letter to Brown. These "contributions" entailed SSCL purchase of accelerator and detector components at substantially less than US prices, then crediting "donor" countries with the difference (see, e.g., this chap., n. 68).

106. John H. Gibbons, memo for John D. Podesta, "O'Leary Draft Letter from President Clinton to Kiichi Miyazawa," February 24, 1993. FOIA request 2010-450-F, Clinton Library Digital Archives. The Clinton letter to Miyazawa, mentioned in this memo, was dated at February 6, 1993, using a withdrawal/redaction sheet listing that letter as "National Security Classified Information."

107. This letter is mentioned in another document, Todd Stern and John Podesta, memo for Mack McLarty, "Hazel O'Leary/Superconducting Supercollider," March 18, 1993, Clinton Digital Archives. We could not locate a copy in these archives, possibly because it had been withdrawn and not yet released.

108. Ibid.

109. Goodwin, "Happer Leaves DOE under Ozone Cloud for Violating Political Correctness," *Physics Today* (June 1993), 89–91; quotes on 90.

110. Goodwin, "Clinton's Budget Boosts Technology, Making Research Scientists Jittery," *Physics Today* (June 1993), 83–89, esp. table on 84. Adjustments had cut the fiscal 1993 SSC allocation to $517 million.

111. Ibid., 84. See also Eliot Marshall and Christopher Anderson, "Clinton's Mixed Broth for R&D," *Science* (April 16, 1993), 284–85; and Gibbons, interview by Weiss.

112. Kim A. McDonald, "3 Projects Pushed by White House, Reversing Energy Dept.'s Position," *Chronicle of Higher Education* (May 5, 1993), A31–34. SLAC's Burton Richter and Stanford University lobbyists had appealed through the California congressional delegation and new OMB Director Panetta to get the B Factory added to the budget, as a reward for the state's decisive role in Clinton's election, but New York Senator Daniel P. Moynihan insisted that Cornell be allowed to compete for this project, too. The B Factory eventually was awarded to SLAC later that year, after DOE officials decided it had made the better proposal. Personal notes and recollections of Riordan, who worked in the SLAC director's office at the time and witnessed events at close hand.

113. Goodwin, "Clinton's Budget," table on 84.

114. McDonald, "3 Projects Pushed." "Such delays are plainly poor management," Watkins had observed in his January 14 letter to Congressman Brown. "It is better that a project be cleanly terminated than left to die a slow, lingering and expensive death."

115. Catherine M. Anderson, "Nobel laureates and HEP leaders visited Washington for the SSC," *URA Members Bulletin* (April 30, 1993), p. 7.

116. Steven Weinberg, private conversation with Riordan, May 2, 2001. Riordan could find no evidence in the Clinton archives that Gore's enthusiasm had in fact been communicated to Clinton at a meeting the following day to prepare for the Miyazawa summit. Photographs of this meeting show US Trade Representative Mickey Kantor seated facing Clinton, indicating the prominence being given trade issues.

117. This conclusion was reached by consulting a variety of sources, including Wojcicki, "The Texas Days," 280; Will Happer, interview by Riordan, February 17, 1997; John Toll, interview by Riordan, February 11, 1996; and Karl Erb, private conversation with Riordan, June 12, 2000.

118. William J. Clinton, "Remarks by the President and Prime Minister Miyazawa in Joint Press Availability," April 16, 1993, Clinton Library Digital Archives, at http://clinton6.nara.gov/1993/04/1993-04-16-clinton-p-m-miyazawa-joint-press-availability.html.

119. Ibid. There is no mention whatsoever of the SSC in this press briefing, but see n. 121 below.

120. Wojcicki, "The Texas Days," 280; Toll, interview by Riordan. "You have a lot of things you want to ask the Japanese [for], ranging from the SSC to auto parts," said Happer, interview by Riordan. "And I would guess that for Clinton things like auto parts rate more heavily, because they have many more jobs [associated with them] than the SSC." As Richard Barth, a PhD chemist who served on the National Security Council during the Bush administration and continued into the Clinton administration for a few months, observed in 2011, "You're talking about something that a few thousand scientists had a strong interest in versus something that a few hundred thousand

autoworkers were interested in." Barth, telephone conversation with Riordan, August 22, 2011, Riordan notes.

121. Background briefing by senior administration official (unidentified), White House Office of the Press Secretary, April 16, 1993, Clinton Library Digital Archives, http://clinton6.nara.gov/1993/04/1993-04-16-baclkground-briefing-on-japan.webarchive.

122. Marshall and Anderson, "Clinton's Mixed Broth," 285. Also Gibbons, interview by Weiss.

123. Christopher Anderson, "SSC Deathwatch Starts Again," *Science* (June 4, 1993), 1421. See also Yukiya Awano, "US-Japan SSC Collaboration, *Science* (July 9, 1993), 146.

124. Catherine M. Anderson, "DOE/Sec. O'Leary Testified at House Energy & Water Appropriations," and "Happer: DOE General Science Testimony at House Energy & Water Apps," *URA Members Bulletin* (April 30, 1993), 9–12, quote on 11; US General Accounting Office, Federal Research: *Super Collider Is Over Budget and Behind Schedule*, Report no. GAO-93-87 (February 1993). These shortfalls totaled $310 million in fiscal 1992 and 1993. A telling GAO criticism that does not seem to have been discussed at these sessions was that over $400 million in nonfederal contributions would be needed to build the big SDC and GEM detectors as planned, a figure that was not even on the horizon of possibility at the time; see discussion on 37–39, especially table 5.1, 38.

125. Anne Marie Kilday and Beth Silver, "House Panel Gives a $620 Million Vote of Approval," *Dallas Morning News* (June 18, 1993), A4. Catherine M. Anderson, "Text of Committee Report on FY94 Energy and Water Development Appropriations," *URA Members Bulletin* (June 17, 1993), 1–4. The Main Injector and B Factory stayed in the budget at the request levels of $25 million and $36 million, respectively.

126. Eliot Marshall, "Clinton Backs SSC, Space Station," *Science* (June 25, 1993). A close reading of this letter, however, reveals only lukewarm support; he was not about to fall on his sword over the SSC. Copy of Clinton letter in Riordan SSC files.

127. Colin McIlwain, "SSC Falls Victim to Congressional Austerity," *Nature* (July 1, 1993), 6; Goodwin, "Collision Course: Besieged by Congress, SSC Awaits Key Decisions on Its Fate," *Physics Today* (August 1993), 43–45. Slattery's amendment allowed $220 million to close down the SSC.

128. Goodwin, "Collision Course," quote on 43–44. Clinton seemed unwilling to use his ebbing political capital and step out into the line of fire on the SSC battle. A DOE chronology prepared for the Clinton Administration History Project in 2000 notes that Secretary O'Leary and Vice President Gore "made telephone calls to uncommitted House members . . ." but that "the Administration's appeals were to no avail." Clinton phone logs for June 23–24 reveal that he made twenty-eight calls to senators but only one to a congressman, probably about the Space Station, for which he did indeed jawbone. Political insiders claimed that he had struck a deal with Democratic Texas Governor Ann Richards to defend the Space Station vigorously but let the SSC fend for itself in 1993. Hans Mark, interview by Riordan. But see Kilday and Silver, "House Panel," and Jane Hickie, letter to Professor Hans Mark, September 14, 1993, Riordan SSC files.

129. Two leading Dingell targets were David Baltimore and Donald Kennedy, the presidents of Caltech and of Stanford, respectively.

130. Goodwin, "Collision Course." See also Christopher Anderson, "University Consortium Faulted on Management, Accounting," *Science* (July 9, 1993), 157; Colin McIlwain, "Barrage of Mud Fails to Stick to Super Collider," *Nature* (July 8, 1993), 92; Mike Mills, "Oversight Questions Nag Efforts to Save Super Collider Again," *Congressional Quarterly* (July 31, 1993), 2031–34.

131. Goodwin, "Collision Course," 44. Also quoted in Anderson, "University Consortium Faulted." In this hearing, Dingell relied heavily on investigations by a new staff member, Robert Roach, who had formerly worked for Howard Wolpe but joined his staff after the 1992 elections, in which Wolpe chose not to run because of redistricting. Robert Roach, interview by Weiss, December 13, 1995.

132. Anderson, "University Consortium Faulted"; Goodwin, "Collision Course," 45.

133. US GAO, *Federal Research: Super Collider Is Over Budget and Behind Schedule*, 34–35.

134. Goodwin, "Collision Course," 44–45, General Dynamics quote on 45. In June 2000, Siskin disputed these estimates, saying it was normal to experience such overruns during the startup of production, but unit costs were expected to come down as General Dynamics moved along the learning curve and began manufacturing thousands of magnets. Siskin, interview by Riordan.

135. Goodwin, "Collision Course," 45.

136. Although the cost-and-schedule control system was up and running by late 1991, there were still difficult problems in implementation, particularly in getting regular, reliable inputs from the physicists at the SSC lab. Siskin, interview by Riordan.

137. Goodwin, "Collision Course." Siskin's claim was based on the May 1993 SSC progress report; he projected that only 12 percent of the contingency would eventually be needed.

138. Senator J. Bennett Johnston, *Joint Hearing Before the Committee on Energy and Natural Resources and the Subcommittee on Energy and Water Development of the Committee on Appropriations, United States Senate*, August 4, 1993, Senate Hearing 103–185, US Government Printing Office, 3. On this hearing, see also Goodwin, "As the SSC Faces 'Live or Die' Vote, DOE's O'Leary Shakes up Management," *Physics Today* (September 1993), 52–53; and Mike Mills, "Collider 'Problems' Lead O'Leary to Replace Its Main Contractor," *Congressional Quarterly* (August 7, 1993) 2156.

139. Senator Ernest F. Hollings, in Johnston, *Joint Hearing*, 4.

140. This pending management shake-up is mentioned in Hazel O'Leary, memorandum to Mack McLarty, "Weekly Report," August 5, 1993, Clinton Library Digital Archives: "The new contractual arrangement will divide the responsibilities into two categories: Design/Operate for design and scientific work and Execute/Integrate for conventional construction work. . . . Our goal in this effort is to establish a new management team comprised of two major contractors that have complementary strengths necessary for this project to proceed in a manner that demonstrates the Administration's commitment to spend the taxpayer's money more efficiently."

141. Hazel R. O'Leary, *Joint Hearing*, 90; also quoted in Goodwin, "As the SSC Faces," 53. See also Christopher Anderson, "DOE Pulls Plug on SSC Contractor," *Science* (August 13, 1993), 822–23. For O'Leary's opening remarks and prepared statement, see *Joint Hearing*, 81–92.

142. O'Leary, *Joint Hearing*, 82–83; Goodwin, "As the SSC Faces"; Anderson, "DOE Pulls Plug."

143. O'Leary, letter to John H. Dingell, September 1, 1993, quote on 3, Riordan SSC files. Similar comments can be found in *Joint Hearing*, 84.

144. US Department of Energy Office of Field Management, *Report of the DOE Review Committee on the Baseline Validation of the Superconducting Super Collider*, Report no. DOE/ER-0594P, August 1993 ("Scango Report"). The report writers admitted that evaluating the percentage completion was a subjective affair: "The Committee determined that the project progress is reasonably reported in the May 1993 Monthly Progress Report at approximately 20 percent, but is very difficult to track using the current project and business management systems." Executive Summary, i.

145. Ibid., table on iv. This table puts the total potential "cost risk" at $1.475 billion, but that figure ignores a $219 million "1991 contingency adjustment" on the foreign contributions to the detectors—typical of the accounting sleight of hand in this report. The Executive Summary, viii, says that the total project cost "could increase by $1.6 billion if all potential increases were to occur" and projects a one-year delay in completion of construction. These figures do not include the additional financial impacts of a three-year schedule stretch-out, which were then roughly estimated at nearly $2 billion.

146. In "DOE Pulls Plug," Anderson called the SSC an "$11 billion accelerator" in mid-August 1993. In September, Goodwin dubbed it a "$10 billion proton-proton collider" in "As SSC Faces." Mills called it "the $11 billion project" in "Collider 'Problems' Lead O'Leary." Another GAO report put the total cost at $11 billion. By late summer 1993, the dominant Washington perception was that the SSC price tag would exceed $10 billion. See Sharon LaFraniere, "Super Collider Quest Mired in Murky Cost Equations," *Washington Post* (September 6, 1993), A1.

147. Scango Report, 59–65. At the time, General Dynamics was projecting the cost per dipole at $270,000 for the initial, low-rate production run. The 1991 Green Book estimate had been

$146,400, while the Scango Report figured that unit costs could be driven down to $171,000 in full production. Siskin and SSC Magnet Division head Tom Bush made similar comments in interviews by Riordan.

148. According to the Scango Report, the added spares would cost $45 million, the liners $70 million, and the extra magnet testing $97 million. There were other, less costly items, too.

149. "Politics Takes Toll on SSC," *Science* (July 30, 1993), 539.

150. Sharon LaFraniere, "Energy Dept. Official Urges Firing Super Collider Chief," *Washington Post* (August 2, 1993), A3. A copy of this Cipriano memo is printed in *Joint Hearing*, 41–42, along with a letter from eleven top SSC managers rebutting it and endorsing Schwitters's leadership, 43.

151. John Rees, private conversation with Riordan, February 3, 2009. This was before excess costs had been "scrubbed" from the budget, he said, but that would have lowered it only to $11.5 billion.

152. Goodwin, "Senate Rescues SSC, But Final Act Awaits Conference," *Physics Today* (October 1993), 110–11; Christopher Anderson, "Senate Vote Lifts Prospects for SSC," *Science* (October 8, 1993), 171. See also Mike Mills, "Super Collider, Advanced Reactor Withstand Assaults in Senate," *Congressional Quarterly* (October 2, 1993), 2628–30; he now cited a $13 billion SSC price tag.

153. Goodwin, "Senate Vote Rescues SSC," and Anderson, "Senate Vote Lifts Prospects."

154. Mike Mills, "Negotiators Preserve Collider, Setting Up House Showdown," *Congressional Quarterly* (October 16, 1993), 2800–01. On the maneuvering regarding conference committee members, see especially sidebar "Power of Tradition," 2801.

155. Quoted in ibid., 2800. It was surprising that the conference committee completely ignored the resounding House vote of the previous June, making no effort at all to compromise on this figure. This decision could only have enhanced the resolve of House opponents to reject the entire bill.

156. "Boehlert Is 'Outraged' about Conference Action on SSC," press release, office of Congressman Sherwood Boehlert, October 14, 1993, Boehlert/Goldston SSC files.

157. Mills, "Negotiators Preserve Collider."

158. One of the House staffers who had worked tirelessly behind the scenes to defeat the project said two years later that he doubted SSC opponents could have killed the SSC if its estimated total cost had been convincingly kept below $10 billion. According to him, this was the effective watershed level for the House, not the $5 billion cap for US funding that had been spelled out in the 1990 House authorization bill. Daniel Pearson, interview by Riordan and Weiss.

159. Quotes from LaFraniere, "Super Collider Quest." This detailed article, which appeared on the *Washington Post* front page in early September, fanned the growing uncertainty over SSC costs. Later that month it was reprinted in the *Washington Post Weekly*, its national edition. In Washington, where perception counts as much as reality, this article severely damaged the project's tottering public image. In a letter to all House members, O'Leary claimed the DOE was bringing the project under control and that "these management actions will allow us to complete the SSC in the year 2002 for less than $11 billion." O'Leary letter to Neil Abercrombie, October 19, 1993, SSC files of Kim McDonald.

160. Eric Pianin, "House Deals a Big Defeat to Atom Smasher in Texas," *Washington Post* (October 20, 1993), A1.

161. Catherine M. Anderson, "Today's Bump in the Road (Slattery Vote on SSC)," *URA Members Bulletin* (October 19, 1993); Jeffrey Mervis, "The Endgame," *Science* (October 29, 1993), 646.

162. Eric Pianin and Sharon LaFraniere, "Proponents of Collider Give Up," *Washington Post* (October 21, 1993), A1, quote on A12.

163. Pianin, "House Deals a Big Defeat," and Anderson, "Today's Bump in the Road." Apparently the Clinton administration was unaware of the seriousness of the looming battle on the House floor. There is absolutely no mention of the SSC in Clinton's or O'Leary's daily schedules for October 19, 1993. Their attention that day was focused on a Rose Garden ceremony where Clinton, Gore, and O'Leary introduced "a broad package of initiatives to curb the threat of global

warming." See Gary Lee, "Clinton Offers Package to 'Halt Global Warming,'" *Washington Post* (October 20, 1993), A4.

164. Pianin and LaFraniere, "Proponents of Collider Give Up."

165. Pianin, "House Deals a Big Defeat."

166. Quoted from Catherine M. Anderson, "Dramatic Conference Committee on SSC's Orderly Termination," *URA Members Bulletin* (October 21, 1993). See also Pianin and LaFraniere, "Proponents of Collider Give Up."

167. Eric Pianin and Tom Kenworthy, "Super Collider Takes Last Breath" *Washington Post* (October 21, 1993), A11.

168. Catherine M. Anderson, "Today's House Action on the Energy and Water Conference Report," and "Today's Senate Action on the Energy and Water Conference Report," *URA Members Bulletin* (October 26 and 27, 1993).

169. T. J. Glauthier, Keith Mason, and Sylvia Mathews, memorandum for the President and Mack McLarty, "Summary of Richards/O'Leary Meeting," February 27, 1993, FOIA Request 2010-450-F, Clinton Library Digital Archives.

170. According to David Goldston, who was present at the historic meeting on October 20, 1993, this was a "precedent-shattering event"—the first time that the House had ever overturned such an Appropriations Conference Committee report. David Goldston, private conversation with Riordan and Weiss, October 30, 1997, Riordan notes.

171. Besides the Appropriations Committee, major House leaders who backed the SSC included Speaker Tom Foley, Majority Leader Dick Gephardt (D-MO), and Science Committee Chairman George Brown. But Energy and Commerce Committee Chairman John Dingell opposed the project.

172. Subcommittee on Oversight and Investigation, Committee on Energy and Commerce, US House of Representatives, *Out of Control: Lessons Learned from the Superconducting Super Collider* (Washington: US Government Printing Office, December 1994).

173. Faye Flam, "Is There Life After the SSC?" *Science* (October 29, 1993), 644–47.

174. See, for example, Hazel R. O'Leary and George E. Brown Jr., "Resuming the Pursuit of Knowledge," *Los Angeles Times* (November 21, 1993).

175. "Disappointing End for the SSC" (editorial), *Nature* (October 28, 1993), 771.

176. *Proceedings of the Fifth International Workshop on Next-Generation Linear Colliders*, Stanford Linear Accelerator Center, October 13–21, 1993, Report no. SLAC-436.

177. John Peoples, interview by Hoddeson and Kolb, January 25, 2011. This account is largely based on his recollections. Burton Richter has notebook records of a telephone conversation with Johnston that day, but not of a meeting with Peoples. Richter e-mail to Riordan, September 29, 2011.

178. Hazel O'Leary, letter to John H. Marburger III, June 24, 1993, Box J3a6, URA Collection, Fermilab Archives.

179. Peoples, interview by Hoddeson and Kolb, July 25, 2011.

180. Peoples, interview by Kolb, July 29, 2011.

181. Goodwin, "After Agonizing Death in the Family, Particle Physics Faces Grim Future," 90, see chap. 3, n. 175.

182. *SSCL Termination Project Monthly Report*, SSC Laboratory, December, 1993, 1. See also Minutes, Fermilab Board of Overseers meeting, November 10, 1993, Fermilab Archives.

183. Schwitters, letter to John H. Marburger III, October 28, 1993, URA 1993 Correspondence, Fermilab Archives.

184. Hazel R. O'Leary, memorandum for the President, "Status of the Superconducting Super Collider," October 21, 1993. FOIA Request 2010-450-F, Clinton Library Digital Archives.

185. O'Leary, memorandum to Mack McLarty, "Weekly Report," November 5, 1993, 2, Clinton Library Digital Archives. According to President Clinton's daily schedule, accessed online at the Clinton Library website, Richards also met with him and Vice President Gore on that same day.

186. Bill Burton, memorandum to President Clinton, "Energy Secretary's Memo re SSC," November 6, 1993, Clinton Library Digital Archives.

187. Peoples, interview by Hoddeson and Kolb, January 25, 2011.

188. O'Leary, memorandum for the President, "Successful Closeout of the Superconducting Super Collider," November 5, 1993, FOIA Request 2010-450-F, Clinton Library Digital Archives.

189. Bill Burton, memorandum to Mack McLarty, "President's Discussion with Secretary O'Leary re SSC," November 8, 1993, FOIA Request 2010-450-F, Clinton Library Digital Archives.

190. Gary Taubes, "Fight Heats Up Over SSC's Remains," *Science* (December 3, 1993), 1506–07.

191. According to O'Leary, "I had originally planned to announce a severance package to the workers, but 500 employees signed a letter to me asking that a committee be elected to negotiate a package with the Department which addresses employees' needs. I agreed to this request and staffs have already met on the issue." Hazel R. O'Leary, memorandum to Mack McLarty, "Weekly Report," November 12, 1993, 2, Clinton Library Digital Archives.

192. Peoples, interview by Hoddeson and Kolb, January 25, 2011.

193. Ibid.

194. O'Leary, memorandum to Mack McLarty, "Weekly Report," November 18, 1993, 2, Clinton Library Digital Archives.

195. *SSCL Termination Project Monthly Report*, 1.

196. Ibid.

197. Goodwin, "After Agonizing Death," 89–91.

198. In fact, a few SSC physicists made their way to Wall Street, finding good work as "quants" and applying their computer skills in fashioning new investments then becoming known as "derivatives." See Gary Taubes, "Young Physicists Hear Wall Street Calling," *Science* (April 1, 1994), 22.

199. Peoples, interview by Hoddeson and Kolb, January 25, 2011.

200. *SSCL Termination Project Monthly Report*, 2.

201. Jennifer Nagorka, "Super Collider Dying a Quiet Death," *Dallas Morning News* (January 6, 1994), A1.

202. Gary Taubes, "Fight Heats Up Over SSC's Remains."

203. Nagorka, "More than 100 Scout out SSC Equipment; Agency Awaits Ideas on Use of Project Assets," *Dallas Morning News* (March 23, 1994), A29. But it would take months before these requests were processed.

204. Peoples, interview by Hoddeson and Kolb, January 25, 2011.

205. Nagorka, "Technical Experts Discuss Uses for Collider Assets," *Dallas Morning News* (January 6, 1994), A25.

206. Laura Beil, "Proposal Outlines Use for Collider in Radiation Therapy for Cancer," *Dallas Morning News* (June 4, 1994), A33.

207. Associated Press, "State Officials Tout Plan to Turn Collider Site into Prairie," *Dallas Morning News* (April 19, 1994), A21. This suggestion reflected what had been done at Fermilab.

208. Nagorka, "3 Options Chosen for SSC Assets; State Panel Focusing on Technological Uses," *Dallas Morning News* (February 22, 1994), A21.

209. Nagorka, "Texas Comes out Ahead in SSC Deals, Officials Say; Congress, Clinton Still Must Approve Settlement," *Dallas Morning News* (July 23, 1994), A1.

210. Joe Barton and Shelton Smith, "Texas Isn't Playing Politics with Corpse of Collider," *Dallas Morning News* (March 20, 1994), A6.

211. Peoples, interview by Hoddeson and Kolb, January 25, 2011.

212. "Super Collider: Settlement Greatly Improves Site's Options" (editorial), *Dallas Morning News* (July 23, 1994), A28.

213. Nagorka, "State Approves Super Collider Settlement; Texas Would Get $210 Million in Cash, $510 Million Worth of Land, Equipment," *Dallas Morning News* (August 2, 1994), A17.

214. Matt Crenson, "Tunnel Vision; Geologists Lobby for Underground Lab at SSC Site," *Dallas Morning News* (December 26, 1994), D6.

215. Goodwin, "After Agonizing Death"; Nagorka, "SSC Site May Be Reborn as Medical Research Center," *Dallas Morning News* (November 27, 1994), A33.

216. "The Epilogue: Super Collider's Assets Will Not Be Reused" (editorial) *Dallas Morning News* (May 17, 1995), A20. See also Jonathan Eig, "Legislators Reject Plan for SSC Site; Some Criticize Decision to Scrap Cancer Center, Sell Land, Assets," *Dallas Morning News* (May 16, 1995), A17.

217. Alexei Barrionuevo, "State Officials Vote to Return Land to Ellis County," *Dallas Morning News* (July 19, 1995), A22.

218. Chris Kelley, "Officials Settle SSC Claims with U.S. Government," *Dallas Morning News* (August 27, 1996), A18. Ennis schools received portable office buildings, tools, equipment, and computers, while Waxahachie schools received computers, equipment, portable buildings, fiber optic equipment, and 150 acres of vacant land.

219. Mike Jackson, "Remains of Collider Project up for Sale," *Dallas Morning News* (October 5, 1997), 9.

220. Jackson, "State Reaches Deals on 2 Super Collider Buildings," *Dallas Morning News* (February 12, 1998), A35. Associated Press, "Super Collider Lab Sold for $10 million," *Dallas Morning News* (May 16, 1998), A31.

221. Jane Sumner, "Hollywood Pirates on the Prairie; Abandoned Super Collider Site Finds New Life as Film Set," *Dallas Morning News* (February 13, 1999), C1.

222. On this point, see also Kevles, "Big Science and Big Politics in the United States" (see chap. 3, n. 1).

223. The draft Bush budget for fiscal 1994, which was passed along to the Clinton administration, had $858 million for the SSC. Had he been reelected, Bush would likely have met with Premier Miyazawa and asked Japan to join the project as a partner. By most reports, Miyazawa was ready to agree to make such a commitment, in return for an easing of the US stance on economic and trade issues.

224. Gibbons, interview by Weiss. But none of this peace dividend found its way to the SSC budget.

225. Goodwin, "Congress Cancels SSC and Allocates High Budgets for Technology in 1994," *Physics Today* (November 1993), 77–80.

226. It bears noting, however, that one of those spinoffs, the World Wide Web, has been transforming the world economy and modern culture, contributing trillions of dollars per year to the gross international product. Spawned at CERN, the web made its way into the United States via Fermilab and especially SLAC, which established the first US website in 1991 and developed the first web browser, MIDAS, shortly thereafter. The $10–$15 billion that might have been spent on SSC construction is negligible compared to this contribution, whose size and importance were unfortunately poorly understood when Congress terminated the project in October 1993. Such is the unpredictable nature of spin-offs.

227. See Goodwin, "Congress Cancels SSC," esp. the table on 78, for a quick summary of the amounts allocated for these projects in fiscal 1994.

228. For a revealing study of this fascinating process, focused on the fiscal 1992 budget with emphasis on Space Station funding, see Munson, *The Cardinals of Capitol Hill*.

CHAPTER SEVEN

1. Goodwin, "After Agonizing Death in the Family," 87.

2. Pianin and LaFraniere, "Proponents of Collider Give Up" (see chap. 6, n. 162).

3. Anne Reifenberg, "SSC's Fall Linked to Its Complexity," *Dallas Morning News* (October 24, 1993), A1–A15.

4. Russell Wylie, as quoted by Peter Rodgers in "How the Super Collider Ground to a Halt," *Physics World* (November 1993), 6–7, quote on 6.

5. Faye Flam, "Is There Life After the SSC?," quote on 644 (see chap. 6, n. 173).

6. Christopher Anderson, "The Anatomy of a Defeat," *Science* (October 29, 1993), 645.

7. Colin McIlwain, "SSC Decision Ends Post-war Era of Science-Government Partnership," *Nature* (October 28, 1993), 773–74.

8. Flam, "Is There Life?" 647.

9. Panofsky, "The SSC's End: What Happened? And What Now?" letter to the editor, *Physics Today* (March 1994), 13–15, 88, quote on 88, emphasis in the original. Panofsky originally submitted this letter in December 1993 as an article for publication in the *SLAC Beam Line*, but its editorial board quashed that possibility, after which Panofsky's text was forwarded to *Physics Today*, which published other letters about the SSC demise on 88–91.

10. David Ritson, "Demise of the Texas Supercollider," quote on 610.

11. Ibid., 610.

12. Leon M. Lederman, "Scientific Retreat: Demise of the SSC is a National Loss," *Chicago Tribune* (November 18, 1993), 31. Another version appeared in *The Scientist* (November 29, 1993), 12.

13. David Kramer, "SLAC's Richter, in SSC Postmortem, Urges Shorter Construction Times," *Inside Energy* (March 7, 1994), 7–8. See also Kim A. McDonald, "Who Killed the SSC?" *Chronicle of Higher Education* (March 2, 1994), A10.

14. Richard Jones, "Looking Backward: Why the SSC Was Terminated," FYI no. 142 (Washington, DC: American Institute of Physics Public Information Division, October 27, 1993).

15. Leon M. Lederman, "An Open Letter to Colleagues Who Publicly Opposed the SSC," *Physics Today* (March 1994), 9–11. Congressman Boehlert and aides David Goldston and Dan Pearson agreed that the opposition of other physicists to the SSC helped them to build a case for House members to vote against the project, for they would not be perceived as voting "against science"—only as establishing difficult priorities within science. Sherwood Boehlert, David Goldston, and Daniel Pearson, private conversations with Riordan and Weiss, October 30, 1997.

16. Gary Taubes, "The Supercollider: How Big Science Lost Favor and Fell," *New York Times* (October 26, 1993), C1–3, quote on C1.

17. Clifford Krauss, "Knocked Out by the Freshmen," *New York Times* (October 26, 1993), C3.

18. Goodwin, "After Agonizing Death," 87.

19. Tom Siegfried, "Super Collider's Death Is Sign of Bad Times for Basic Science," *Dallas Morning News* (November 1, 1993), D9.

20. Boehlert, "Collider Project Doomed Itself by Excess," letter to the editor, *New York Times* (November 4, 1993). In 2001 Boehlert became chair of the Science Committee and served in that capacity for most of the second Bush administration, until the Democrats won the majority of seats in 2006; David Goldston served as his chief of staff. Throughout his tenure as chair, Boehlert was a strong advocate of basic research, despite a Republican administration that paid little attention to scientific evidence and facts.

21. William Happer, "Diversity Needed in Federal Support of Basic Science after the SSC," *APS News* (March 1994), 12.

22. Daniel Lehman, "Lessons from the SSC: An ER-65 Critique," PowerPoint presentation (February 1994), emphasis added, Fermilab Archives. See also Lehman, interview by Weiss.

23. Lehman, "Lessons from the SSC," emphasis in the original.

24. Douglas Pewitt, letter to the editor, *Physics Today* (March 1994), 88–89.

25. Ibid.

26. Subcommittee on Oversight and Investigations, House Committee on Energy and Commerce, *Out of Control: Lessons Learned from the Superconducting Super Collider*. Among the staffers who wrote this report was Robert Roach, who had served on Congressman Howard Wolpe's staff before joining Dingell's staff in 1993. See also Roach, interview by Weiss.

27. Ibid., 2.

28. Ibid., 3.

29. Ibid., 5.

30. Office of Management and Budget, "A Vision of Change for America," 1.

31. On the new social contract for science, see Radford Byerly Jr. and Robert A. Pielke Jr., "The Changing Ecology of United States Science," *Science* (September 15, 1995), 1531–32.

32. Jeffrey Mervis, "A Strategic Message from Mikulski," *Science* (February 4, 1994), 604. Remarks quoted from Mikulski talk at a government-sponsored forum, "Science in the National Interest," January 31–February 1, 1994, at the National Academy of Sciences (NAS). See also Goodwin, "White House Forum on Science Produces only a Reassuring Hug," *Physics Today* (March 1994), 41–42.

33. On the postwar reorganization of research at Bell Labs along mission-oriented lines, see Riordan and Hoddeson, *Crystal Fire: The Birth of the Information Age* (New York: W. W. Norton, 1997), esp. 115–17. See also Riordan and Hoddeson, "Transistor's Father Knew How to Tie Basic Industrial Research to Development," *Research and Technology Management* 41, no. 1 (January 1998), 9–11, and Riordan, "No Monopoly on Innovation," *Harvard Business Review* 83, no. 12 (December 2005), 18–20.

34. Barbara A. Mikulski, "Science in the National Interest," *Science* (April 8, 1994), 221–22.

35. Mikulski's speech and article sent shudders through the US scientific establishment. Donna Gerardi Riordan, then working on the NAS staff, recalled how its leaders reacted, saying that they had to reveal the connections between basic research and human betterment. This concern led to the establishment of the NAS Office of Public Understanding of Science, with her serving as its first director, and the initiation of a series of booklets, *Beyond Discovery: The Path from Research to Human Benefit*, which she developed and edited. Private conversation with Riordan, October 25, 2011.

36. Kevles, *The Physicists*. See esp. chap. IV: "Pure Science and Practical Politics," 45–59.

37. Ibid., esp. chap. XXIII: "The Physicists Established," 367–92. See also Robert W. Seidel, "Accelerating Science: The Postwar Transformation of the Lawrence Radiation Laboratory," *Historical Studies in the Physical Sciences* 13 (1983), 375–400.

38. Kevles, "The Death of the Superconducting Super Collider in the Life of American Physics," in *The Physicists*, ix–xlii, quote on xi. See also Kevles, "Good-bye to the SSC: On the Life and Death of the Superconducting Super Collider," *Caltech Engineering & Science* 58, no. 2 (Winter 1995), 16–25; and Kevles, "Big Science and Big Politics."

39. Kevles, "Good-bye to the SSC," 18.

40. Ibid., 22. Contrast this sentence with Robert R. Wilson's statement before the Joint Committee on Atomic Energy that the proposed National Accelerator Laboratory had "nothing to do directly with defending our country except to help make it worth defending." Hoddeson et al., *Fermilab*, 14.

41. Kevles, *The Physicists*, xxxviii.

42. Ibid., xli.

43. For example, University of Wisconsin physicist Sau Lan Wu, who began doing research at CERN on the LEP collider during the 1980s, joined the LHC collaboration ATLAS in 1993.

44. Peoples, interview by Hoddeson and Kolb, January 25, 2011. But Drell did not recall this call when asked to confirm it.

45. McIlwain, "SSC Decision Ends," 773. Sidney Drell, interview by Riordan and Weiss, March 23, 1996.

46. Hazel O'Leary, letter to Stanley Wojcicki, November 4, 1993, asking HEPAP to organize a "Subpanel on Vision for the Future of High-Energy Physics," in Report no. DOE/ER-0614P, appendix B.

47. Goodwin, "After Agonizing Death," 91.

48. O'Leary and George E. Brown Jr., "Resuming the Pursuit of Knowledge." This essay was initially crafted by Brown and his staff. Science Committee Deputy Director Peter Didisheim, who transferred to DOE to assist O'Leary in closing down the SSC, played an important role. On the

need to internationalize large science projects, see also George Brown Jr., interview by Riordan, February 18, 1995.

49. "Disappointing End for the SSC" (editorial), *Nature* (October 28, 1993), 771.

50. Philip Campbell and Peter Rodgers, "Europe Beckons US High-Energy Physicists as SSC Is Cancelled," *Physics World* (November 1993), 5–6, quote on 5.

51. Ibid.

52. Peter Aldous, "CERN: Alone on the Frontier," *Science* (December 17, 1993), 1808–10. The official estimate was 2.63 billion Swiss francs, which was then about $1.8 billion in US dollars. If labor and the cost of detectors were included, the estimates increased to $3–$4 billion—about a third of the $11 billion estimated SSC cost at termination. Its design energy was also about a third of the SSC's 40 TeV design energy.

53. Ibid., 1809. Normally wealthy Germany was then encountering "economic hardship" due to the tremendous costs of reunifying the country after the fall of the Berlin Wall and the end of the Cold War. See epilogue, 275–76.

54. Ibid., 1808.

55. William Sweet, "1994 Expected to Be Year of Decision for European Super Collider," *Physics Today* (February 1994), 93–96. See also Faye Flam, "Physicists Struggle for Consensus About the Future," and sidebar "The Message From CERN: Help Wanted," *Science* (February 11, 1994), 749. The total US contribution most often discussed at this point was $400 million.

56. Flam, "Physicists Struggle for Consensus."

57. US Department of Energy, *HEPAP Subpanel on Vision for the Future of High-Energy Physics*, Report no. DOE/ER-0614P (May 1994).

58. Faye Flam, "Panel Presents Vision for Physics after the Supercollider," *Science* (June 3, 1994), 1397. This proposed three-year increase became known as the "Drell bump" on the base budget.

59. Ibid.

60. Ibid.

61. Goodwin, "Four Years after SSC's Demise, US Reaches Agreement on 'Unprecedented' Collaboration in CERN's LHC," *Physics Today* (January 1998), 43–44. The DOE was to spend a total of $450 million for collider construction and on the detectors; NSF would contribute another $81 million toward the LHC detectors.

62. Chris Llewellyn-Smith, "How the LHC Came to Be," *Nature* (July 19, 2007), 281–84; see also Llewellyn-Smith, interview by Riordan.

63. Goodwin, "CERN Leaders Troubled by Mixed Reactions in Washington on Visit to Raise Funds for LHC," *Physics Today* (April 1996), pp. 47–48.

64. Irwin Goodwin, "LHC 'Back On Track' as DOE Proposes to Ante Up $400 million to $500 million," *Physics Today* (May 1996), 61.

65. Ibid.

66. Goodwin, "LHC 'Back On Track,'" 61; Llewellyn-Smith, "How the LHC Came to Be"; Llewellyn-Smith, interview by Riordan. The LHC collision energy had been lowered from 16 TeV to 14 TeV, because of more realistic limits on the dipole magnet current. And because of unforeseen problems with these magnets during the 2009 LHC commissioning process, its start-up energy fell to 7 TeV, or 3.5 TeV per beam; in 2012, it operated at 8 TeV. And in 2015, the LHC was restarted after two years of repairs at 13 TeV. See epilogue for more details.

67. Goodwin, "Four Years after the SSC's Demise," 43.

68. Ibid., quotes on 43 and 44.

69. Llewellyn-Smith, "How the LHC Came to Be."

70. Colin McIlwain, "Stanford Accelerator Takes Lead in Race to Quantify CP Violation," *Nature* (February 10, 2000), 586–87. Mike Perricone, "A Banner Day for the Main Injector," *Fermi News* (June 18, 1999), 2–5.

71. Faye Flam, "Taking a Gamble on the Top Quark," *Science* (April 29, 1994), 658–69. Antonio Regalado, "With Quark Discovery, Truth Comes out on Top—Twice," *Science* (March 10, 1995), 1423. For a scholarly history, see Kent W. Staley, *The Evidence for the Top Quark: Objectivity and Bias in Collaborative Experimentation* (Cambridge: Cambridge University Press, 2004). See also Hoddeson et al., *Fermilab*, 349.

72. For a brief summary of the Higgs boson status after LEP, see Riordan, P. C. Rowson, and Sau Lan Wu, "The Search for the Higgs Boson," *Science* (January 12, 2001), 259–60. A major factor in this conclusion was the 1995 discovery of such a high-mass top quark at Fermilab, which, when included in precision fits to the existing Standard Model parameters, favored a low-mass Higgs boson.

73. McIlwain, "Stanford Accelerator Takes Lead."

74. US Department of Energy, Office of Energy Research, *HEPAP Subpanel Report on Planning for the Future of US High-Energy Physics*, Report no. DOE/ER-0718 (February 1998), 8, available at http://science.energy.gov/~/media/hep/files/pdfs/hepap_report.pdf. In 2004, NLC proponents teamed with the other groups in a joint effort called the International Linear Collider, or ILC. As this book goes to press, the best hope for the ILC appears to be a machine sited in Japan using superconducting accelerating cavities.

75. John M. Deutch, memorandum to members of the Board of Overseers of the SSC, April 19, 1984. Deutch mentioned this memo to Riordan in an encounter at the Cosmos Club in the mid-1990s, urging him to find and cite it. Riordan located a copy of the memo in the files of George A. Keyworth, "Superconducting Super Collider," file OA94676-1 (Box 1), Ronald Reagan Library. See also John M. Deutch, interview by Weiss, March 7, 1997.

76. Deutch, memorandum, 2. An eminent theoretical physicist who had been involved in the development of quantum mechanics, Weisskopf served as director of CERN from 1961 to 1968.

77. Ibid.

78. Keyworth, letter to Dr. John M. Deutch, June 11, 1984, Keyworth file OA94676-1 (Box 1).

79. See, for example, Taubes, "The Supercollider"; Goodwin, "After Agonizing Death"; and Riordan, "The Demise of the SSC." This point was made especially forcefully in Subcommittee on Oversight and Investigations, *Out of Control*.

80. Hoddeson and Kolb, "The SSC's Frontier Outpost."

81. Ritson, "Demise of the Texas Supercollider"; Riordan, "A Tale of Two Cultures."

82. US Department of Energy, *SSC Site Evaluations*, 98–99, gives figures ranging from $495 million to $1,033 million; on the $3.28 billion figure from the state of Illinois, see "SSC at Fermilab Means $3.28 Billion Savings" and Hayner, "Fermilab Collider Boosted" (see chap. 3, n. 131).

83. In a lecture at the annual APS Spring meeting in May 2009, for example, former CERN director Chris Llewellyn-Smith estimated this value at $2 billion. But in a subsequent interview, he could not provide any basis for this figure when asked. Llewellyn-Smith, interview by Riordan.

84. Given all the problems with and limitations of the 4 cm aperture magnets, the decision to go to a 5 cm aperture was the better choice, but it meant an unavoidable increase in the SSC total cost of many hundreds of millions of dollars, at least. Tigner himself might have made this decision, had he gone to Texas, when confronted by all the problems with the 4 cm aperture magnets.

85. Wojcicki makes a similar point in "The Texas Days," 291–92.

86. Ritson, "Demise of the Texas Supercollider."

87. Unlike Panofsky and Wilson, none of the new generation of high-energy physicists had any experience stretching back to the Manhattan Project. Although Schwitters was a member of the JASON group that advised the Defense Department on military matters, he did not enjoy the same confidence that they had earned in Washington's *real* corridors of power, including in the Department of Defense. Alvin Trivelpiece, who had deep industry experience, political savvy, *and* a good ability to communicate with high-energy physicists, probably would have made a good SSC director. In January 1989, the same month that Schwitters became SSC director, Trivelpiece became director of Oak Ridge National Laboratory (and by most accounts did well in this position),

despite not having been out of political office in the DOE for the required two years. See chapter 2, n. 168.

88. For more detail on this question, see Riordan, "A Tale of Two Cultures."

89. Marburger, interview by Hoddeson and Riordan. See also his *Science Policy Up Close* and "The Superconducting Super Collider and US Science Policy," *Physics in Perspective* 16, no. 2 (June 2014), 218–49.

90. Clinton delegated almost all his authority for science policy to Vice President Al Gore, so the SSC project no longer had a staunch advocate in the Oval Office, as it had under Reagan and Bush. Although Gore may have jawboned Congress on the project's behalf, which is difficult to substantiate, this did not carry anywhere near the political weight as the US president doing so.

91. Kevles makes a similar point in "The Death of the Superconducting Super Collider."

92. Krige makes an important distinction between national security and "civil security"—the attempt to demonstrate a nation's scientific and technological superiority to the emerging Third World. See Krige, "La science et la sécurité civile de l'Occident," in A. Dahan and D. Pestre, *Les sciences pour la guerre, 1940–1960* (Paris: EHESS, 2004), 369–97.

93. Other trade areas were under consideration at the April 1993 summit, for example, in the semiconductor industry, but quotas for auto parts were the most frequently cited in press accounts.

94. China was establishing itself as the leader of the Third World of developing nations unaligned with either the United States or Soviet Union. See Henry Kissinger, *On China*. Although it was not perceived very well during the 1980s, the bipolar world of the Cold War era was already beginning to fragment into today's multipolar world.

95. Kolb and Hoddeson, "The Mirage of the World Accelerator." See also chap. 1.

96. Logsdon, *Together in Orbit*. The ability to modularize the Space Station and assign specific units to be designed and built by Canada, Japan, and Europe helped in this regard. So did the fact that the project was located in space rather than on Texas soil.

97. According to the official Clinton Administration History Project and Clinton phone logs, only Gore and Hazel O'Leary were calling members of the House to support the SSC in late June 1993.

98. Deutch, memorandum to SSC Board of Overseers, 3.

99. In July 2012, physicists doing research on the Large Hadron Collider announced the discovery of the Higgs boson at 125 GeV despite the fact that it was operating at only 8 GeV. A smaller US proton collider might have discovered this long-sought particle much earlier. See Riordan, "A Vision Unfulfilled: The Hopeful Birth and Painful Death of the Superconducting Super Collider," in Oliver Brüning and Steven Meyers, *Challenges and Goals for Accelerators in the 21st Century* (Singapore: World Scientific Publishing Company, 2015). See also the epilogue.

100. Marburger, *Science Policy Up Close*, chap. 2, and "The Superconducting Super Collider and US Science Policy." Also Marburger, interview by Hoddeson and Riordan. In that interview, with the perspective of having served as a presidential science advisor, he said that a billion dollars was probably a threshold figure for the community, beyond which Congress would not trust physicists to manage a project on their own, without bureaucratic management controls that had become standard in industry and most large-scale government contracting.

101. For major contributions of US physicists to the Higgs boson discovery, see esp. Dennis Overbye, "Chasing the Higgs Boson," *New York Times* (March 5, 2013), D1, available at http://www.nytimes.com/2013/03/05/science/chasing-the-higgs-boson.html. See also Riordan, Guido Tonelli, and Sau Lan Wu, "The Higgs at Last?," *Scientific American* (October 2012), 66–73. One of the two huge scientific collaborations making the Higgs discovery announcement on July 4, 2012, was headed by Professor Joseph Incandela of the University of California, Santa Barbara. Another important figure in making this discovery, as highlighted by Overbye in his article, was Professor Sau Lan Wu of the University of Wisconsin, who worked on the ATLAS experiment headed by Fabiola Gianotti of CERN.

EPILOGUE

1. CERN, "Large Hadron Collider in the LEP Tunnel," March 1984 (see also chap. 1, n. 136).

2. By far the best reference on this political process is Llewellyn-Smith, "How the LHC Came to Be" (see chap. 7, n. 62); available at http://dx.doi.org/10.1038/nature06076.

3. For example, Llewellyn-Smith wrote in ibid., "I doubted that the SSC—which seemed profligate to me, despite its enormous potential—would ever be funded."

4. At typical early-1990s exchange rates of 1.3 to 1.5 Swiss francs per US dollar, this amounted to only $1.5–2.3 billion, but European accounting is different from the US approach. Only the external costs of procurements are included in the total, not the costs of CERN labor, management, and overheads. To make a rough comparison with US accounting standards for high-energy physics projects, which include these costs, multiply the figures for total project costs by a factor of 2.

5. Llewellyn-Smith, "How the LHC Came to Be," 282. A member state's annual CERN contribution is generally a fixed, uniform percentage of its gross domestic product, but exceptions like Germany do occasionally occur.

6. Llewellyn-Smith, "How the LHC Came to Be"; Brumfiel, "The Machine Maker" (see chap. 4, n. 214), 865–66, quote on 866.

7. "CERN Council Gives Go-ahead for Large Hadron Collider," CERN press release, December 16, 1994.

8. "LHC Cost Review to Completion," October 16, 2001, http://user.web.cern.ch/user/LHCCost/2001-10-16/LHCCostReviewToCompletion, (downloaded October 19, 2009). These figures are for materials only and do not include labor or overheads, and no contingency was added. CERN was to pay 20 percent of the detector costs, with the remainder coming from members of the experimental collaborations.

9. Llewellyn-Smith, "How the LHC Came to Be," 283; Brumfiel, "The Machine Maker," 866.

10. Lucio Rossi and Ezio Todesco, "Superconducting Magnets," chap. 4.1 in Evans, *The Large Hadron Collider*, 68–85. See also Oliver Brüning and Paul Collier, "Building a Behemoth," *Nature* (July 19, 2007), 285–89.

11. The three firms were the Italian company AS-G (previously Ansaldo Superconduttori Genova), the French consortium Alstom MSA-Jeumont Areva, and the German company Babcock Noell Gmbh. Ansaldo and Babcock Noell had prior experience manufacturing HERA superconducting magnets.

12. Brumfiel, "The Machine Maker," 868. This meeting occurred hardly over a week after the September 11 attacks on the Pentagon and World Trade Center in New York. For details of the cost overruns, see "LHC Cost Review to Completion." The overruns would have been much larger had Maiani granted LEP another few months of experimental running time during early 2001.

13. Toni Feder, "CERN Grapples with LHC Cost Hike," *Physics Today* 54:12 (2001), 21–22, quote on 21.

14. Evans, interview by Riordan, May 24, 2000; Llewellyn-Smith, interview by Riordan.

15. Brumfiel, "The Machine Maker," 867–68.

16. Ibid.; Steven Myers, private conversation with Riordan, June 27, 2012.

17. Evans demonstrated this impressive project-management control system to Riordan in May 2000.

18. Brumfiel, "The Machine Maker," 868.

19. Anders Unnervik, "The Construction of the LHC: Lessons in Big Science Management and Contracting," chap. 3.1 in Evans, *The Large Hadron Collider*, 38–55.

20. Lucio Rossi, "The Longest Journey: The LHC Dipoles Arrive on Time," *CERN Courier* (October 5, 2006), http://cerncourier.com/article/cern/2973 (downloaded December 17, 2013); Rossi and Tedesco, "Superconducting Magnets."

21. Ibid.; Unnervik, "The Construction of the LHC."

22. Brumfiel, "The Machine Maker," 868.

23. Table, "Budget for LHC—May 16, 2006," given Riordan by James Gillies. In "LHC: The Guide," a brochure published by CERN in 2008, the total materials cost for the "LHC machine and areas" is given as 3.68 billion Swiss francs, of which 430 million came from in-kind contributions.

24. A series of articles by Jon Cartwright in the online edition of *Physics World* chronicles the events of that day: "LHC Milestone Day Gets Off to a Fast Start"; "Home Run Complete, LHC Set to Repeat It Backwards"; "Mission Complete for LHC Team," *Physics World* online, September 10, 2010, available at: http://physicsworld.com/cws/article/news/2008/sep/10/lhc-milestone-day-gets-off-to-fast-start; http://physicsworld.com/cws/article/news/2008/sep/10/home-run-complete-lhc-set-to-repeat-it-backwards; http://physicsworld.com/cws/article/news/2008/sep/10/mission-complete-for-lhc-team.

25. Jon Cartwright, "LHC Report Confirms Electrical Fault," *Physics World* online, October 17, 2008; http://physicsworld.com/cws/article/news/2008/oct17/lhc-report-confirms-electrical-fault. See also Ian Sample, *Massive: The Missing Particle That Sparked the Greatest Hunt in Science* (New York: Basic Books, 2010), Evans quote on 207.

26. See, for example, Cartwright, "LHC Report Confirms Electrical Fault."

27. Adrian Cho, "More Bad Connections" (see chap. 1, n. 150).

28. The estimated costs of these repairs eventually came to approximately 100 million Swiss francs, including labor costs. Steven Myers, private conversation with Riordan.

29. J. Ellis, G. Ridolfi, and F. Zwirner, "Radiative Corrections to the Masses of Supersymmetric Higgs Bosons," *Physics Letters* 257 (1990), 83–88; H. Haber and R. Hempfling, "Can the Mass of the Higgs Boson of the Minimal Supersymmetric Model Be Larger than m_z?" *Physical Review Letters* 66 (1991): 1815–18. See appendix 1 for further discussion.

30. Gordon Kane, private conversations with Riordan, April 2012. Having served on the Fermilab Program Advisory Committee during the early 2000s, Kane was well aware of the luminosity problems. See also John Conway, "The End of the Tevatron," Cosmic Variance blog, *Discover* magazine, http://blogs.discovermagazine.com/cosmicvariance/2011/01/10/the-end-of-the-tevatron/. This account shows a useful graph of the Tevatron luminosity from its 1986 startup through early 2010.

31. An excellent, if journalistic, account of this false alarm is found in Sample, *Massive*, 192–98.

32. Adrian Cho, "Higgs or Bust? Fermilab Weighs Adding 3 Years to Tevatron Run," *Science* (September 10, 2010), 1266–67.

33. Adrian Cho, "Fermilab to End Its Quest for Higgs Particle This Year," *Science* (January 14, 2011), 131.

34. Ibid. Brinkman had served as research director at Bell Telephone Laboratories. Chu had worked at Bell Labs under him, doing research on laser cooling that earned him the 1997 Nobel physics prize.

35. Conway, "The End of the Tevatron."

36. TEVNPH Working Group for the CDF and D-Zero Collaborations, "Combined CDF and D-Zero Upper Limits on Standard Model Higgs Boson Production with up to 8.6 fb^{-1} of Data," May 13, 2013, updated September 20, 2011, http://arXiv:1107.5518v21.

37. Conway, "The End of the Tevatron."

38. For more details on these detectors, see the collaboration websites http://www.atlas.ch/detector.html and http://cms.web.cern.ch/news/cms-detector-design.

39. Rolf Heuer, private conversation with Riordan, June 27, 2012.

40. Myers, conversation with Riordan.

41. Figures from http://atlas.ch/what_is_atlas-html#2a and http://cms.web.cern.ch/content/people-statistics, as viewed on January 6, 2014. These numbers will inevitably change with time.

42. According to the website "US Participation in the Higgs Discovery," 23 percent of the ATLAS physicists and 33 percent of CMS physicists came from US institutions. See http://uslhc.us/higgs for up-to-date numbers.

43. "LHC Cost Review to Completion"; "LHC: The Guide." These figures are for materials only, not labor, and include CERN contributions to other detectors, principally ALICE and LHCb. If we take them at face value and multiply by 5 (reflecting the stated 20 percent CERN contribution), the total cost of materials for the LHC detectors comes to 1.55 billion Swiss francs. At an average of US$0.75 per Swiss franc during the construction period, that's about $1.2 billion for detector materials costs, not including labor at either CERN or the contributing institutions. As these costs vary widely with nations of origin, this number should be taken only as a ballpark figure for LHC detector costs. From "US Participation in the Higgs Discovery," the DOE and NSF contributed $164 million to construction of ATLAS and $167 million to CMS; given US accounting standards, these figures probably include salaries in addition to materials costs.

44. Michael Riordan, Guido Tonelli, and Sau Lan Wu, "The Higgs at Last?," *Scientific American* (October 2012), 66–73, esp. sidebar on 70–71.

45. For details on the Grid, see http://home.web.cern.ch/about/computing/worldwide-lhc-computing-grid.

46. "LHC Cost Review to Completion"; "LHC: The Guide." Again, these figures are for the materials only and represent the CERN 20 percent share of total costs. "LHC: The Guide" gives an additional 90 million Swiss francs for CERN "personnel" costs; assuming a similar proportion of personnel costs at participating institutions, the total costs for computing infrastructure come to approximately 900 million Swiss francs in all, or about $675 million, using an average US$0.75 per Swiss franc.

47. Y. Fang et al., "Observation of a $\gamma\gamma$ Resonance at a Mass in the Vicinity of 115 GeV/c^2 at ATLAS and its Higgs Interpretation," CERN: ATLAS Internal Note no. ATL-COM-PHYS-2011-415, first revealed by Peter Woit on his blog http://www.math.columbia.edu/~woit/wordpress/?p=3643. See also Tushna Commissariat, "Things That Go Bump in the Night," *Physics World* online (April 26, 2011). Sau Lan Wu, private conversations with Riordan, June 2012.

48. Overbye, "Chasing the Higgs," *New York Times* (March 5, 2013), available at http://www.nytimes.com/2013/03/05/science/chasing-the-higgs-boson-how-2-teams-of-rivals-at-CERN-searched-for-physics-most-elusive-particle.html.

49. Guido Tonelli, conversation with Riordan and S. L. Wu, June 22, 2012.

50. Matthew Chalmers, "Hunt for Higgs Enters Endgame," *Physics World* online (August 22, 2011).

51. See, for example, http://twiki.cern.ch/twiki/bin/view/CMSPublic/LumiPublicResults for the luminosity measured by CMS in 2010–12.

52. The LHCb detector at CERN was specifically optimized to study events involving bottom-quark production, but it was not involved in the Higgs boson search.

53. The last three of these are called "four-lepton" events because electrons, positrons, and muons are known collectively as leptons. In general, they are grouped together in the ATLAS and CMS analyses.

54. Riordan, Tonelli, and Wu, "The Higgs at Last?," esp. sidebar on 70–71.

55. "ATLAS and CMS Present Higgs Search Status," CERN press release, December 13, 2011.

56. Adrian Cho, "Last Hurrah: Final Tevatron Data Show Hints of Higgs Boson," *Science* (March 9, 2012), 1159. It probably helped that Fermilab researchers now knew approximately where to look.

57. Gordon Kane, press conference at APS annual meeting, Atlanta, GA, April 8, 2012. But Kane was hardly objective on the question of a 125 GeV Higgs boson, having coauthored a paper the previous December, before the CERN seminar, predicting a 122–129 GeV Higgs boson based on considerations of supersymmetry and string theory.

58. See http://twiki.cern.ch/twiki/bin/view/CMSPublic/LumiPublicResults for graphs of the CMS luminosity.

59. Tonelli and Wu, private conversations with Riordan.

60. An account of this moment of discovery can be found in Riordan, Tonelli, and Wu, "The Higgs at Last?," esp. 72. See also Overbye, "Chasing the Higgs."

61. Riordan, Tonelli, and Wu, "The Higgs at Last?," 72. This early scheduling unfortunately meant that physicists at Fermilab and other US institutions would have to watch the webcast in the wee morning hours of Independence Day, an unintended irony.

62. Ibid., quotes on 72–73. Note especially his use of the indefinite article "a."

63. One such doubter was Mieczyslaw Witold Krasny of the CMS Experiment, e-mail communications with Riordan, May 2013 and June 2014.

64. T. Aaltonen et al., "Evidence for a Particle Produced in Association with Weak Bosons and Decaying to a Bottom-Antibottom Quark Pair in Higgs Boson Searches at the Tevatron," *Physical Review Letters* 109, 071804 (2012); http://link.aps.org/doi/10.1103/PhysRevLett.109.071804. The statistical significance of this result was just over three standard deviations, corresponding to a probability of only about 0.1 percent that it might be just a random fluctuation.

65. John Ellis and Tevong You, "Global Analysis of the Higgs Candidate with mass ~125 GeV," *Journal of High Energy Physics* 2012:123 (September 2012), also at http://arXiv/1207.1693v2 (August 17, 2012).

66. ATLAS Collaboration, "Evidence for the Spin-Zero Nature of the Higgs Boson Using ATLAS Data," *Physics Letters B* 726 (2013), 120–44; S. Chatrchyan et al. (CMS Collaboration), "Study of the Mass and Spin-Parity of the Higgs Boson Candidate via its Decays to Z Boson Pairs," *Physical Review Letters* 110 (2013) 081803: 1–15.

67. ATLAS Collaboration, "Measurements of Higgs Boson Production and Couplings in Diboson Final States with the ATLAS Detector at the LHC," *Physics Letters* B726 (2013), 88–119.

68. Of the 2.81 billion figure, 2.60 billion was estimated for the collider and experimental areas but did not include the detectors or computers, which were less well understood in 1996; it grew by nearly 45 percent to 3.756 billion CHF in 2009, including inflation—and probably should include another 2 to 4 percent to repair damages that occurred on September 19, 2008. Much greater cost growth occurred on the detectors and computers, from 210 million to 576 million Swiss francs (CERN materials costs only), a 274 percent cost growth. These figures do not include CERN personnel costs, which raise the total cost of the LHC to about 6.5 billion Swiss francs. Including outside contributions, the total LHC project cost exceeded 10 billion dollars. These figures do not include the value of the LEP tunnel or other CERN infrastructure reused in constructing the LHC. Nor do they include the costs of repairing and replacing the destroyed magnets in 2009–10. James Gillies, e-mail communications with Riordan, October 2009; Evans and Llewellyn-Smith, e-mail communications with Riordan. See also Unnervik, "The Construction of the LHC," esp. table 1, 40, which puts the total CERN cost of the LHC project, as evaluated in 2009, at 6.51 billion Swiss francs, not including repair costs.

69. See, for example, Wojcicki, "The Texas Days," 291–92.

70. Brumfiel, "The Machine Maker."

71. But the original estimated costs for the LHC project contained *no* contingency. Perhaps a 40 percent contingency should be allowed in similar high-technology projects, not the mere 10–20 percent that had been typically used in high-energy physics projects.

72. There were certainly severe political problems to be addressed, such as when Germany and Great Britain initially refused to support the proposed project in 1994. Another pressing political problem was the troublesome claim that the LHC might create a black hole that could swallow and destroy the earth. Heuer, conversation with Riordan.

73. Exceptions to this rule include major military projects such as aircraft carriers, which can secure guaranteed funding for the life of the project, typically five years. The Department of Energy had attempted to secure such funding for major projects like the SSC but never succeeded. Robert Hanfling, telephone conversation with Riordan, July 2, 2014.

74. See, for example, R. Herman, "High-energy Physics Splits the Scientists," *New Scientist* (June 20, 1985), 10; Llewellyn-Smith, interview by Riordan. Another panel that took up the question of CERN's impact on other European science during the mid-1980s was headed by French physicist Anatole Abragam. Discussions with Llewellyn-Smith and John Krige were helpful in

understanding the council's crucial role in representing the interests of other European scientists within the CERN governing structure.

75. Llewellyn-Smith, "How the LHC Came to Be."

76. Had such a voice been present in DOE deliberations, project advocates might have taken more seriously Panofsky's proposal to reduce the SSC project's scope when it faced cost increases of $2 billion due to design changes in late 1989 and early 1990.

77. Llewellyn-Smith, "How the LHC Came to Be"; Brumfiel, "The Machine Maker." In effect, the LHC took such a two-phase approach in reaching its full design energy. On June 3, 2015, it finally began experiments at 13 TeV, or 1 TeV shy of its 14 TeV design energy.

78. See, for example, the review by Michael S. Chanowitz, "Electroweak Symmetry Breaking: Unitarity, Dynamics and Experimental Prospects," *Annual Reviews of Nuclear and Particle Science* 38 (1988), 323–420, esp. 364–68. See also Ellis, "Radiative Corrections to the Masses"; Haber and Hempfling, "Can the Mass of the Lightest Higgs Boson." By the end of the 1990s, the Fermilab discovery of the top quark with a mass near 175 GeV, when employed in fits to precision measurements of other Standard Model parameters, suggested that a light Higgs boson should occur with a mass less than 200 GeV. See Riordan, Rowson, and Wu, "The Search for the Higgs Boson."

79. While the proposed machine, a 4–5 TeV proton-antiproton collider, might not have achieved sufficient luminosity to discover the 125 GeV Higgs boson, a subsequent proton-proton Dedicated Collider suggested by Lee Teng would have had enough, given how close the 2 TeV Tevatron came to a discovery. This design, using superconducting dipole magnets developed for the SSC, could have generated a collision energy of 6 TeV at a luminosity equal to that of the SSC. See L. Teng, "A 3 TeV on 3 TeV Proton-Proton Dedicated Collider for Fermilab," *Fermilab Technical Report No. TM-1516* (March 30, 1988). We mention this hypothetical machine mainly to illustrate the wisdom of pursuing a conservative, gradualist approach to accelerator design in a constrained fiscal environment. In this regard, see also "Proposal for a Dedicated Collider" (see chap. 1, n. 111).

80. In hindsight, CBA would not have produced any major discoveries, for the top quark mass turned out to be much too high, well beyond its experimental reach. But in the mid-1980s, theorists expected this mass to come in much lower. See, for example, Eichten et al., "Supercollider Physics," which often used top quark masses of 30 and 70 TeV in its simulations of likely event rates. And CBA would have given US physicists an excellent proving ground to achieve and use its high design luminosity.

81. As a savvy CERN director emeritus (who will remain anonymous) once told Riordan in this regard, "In designing a new machine, do not trust theorists too much, only a little!"

82. One such physicist was Wolfgang Panofsky, in a conversation with Riordan, paraphrasing James Cronin of the University of Chicago (see chap. 3, nn. 162–63).

APPENDIX ONE

1. For example, see Riordan, *Hunting of the Quark*, 1959 chart of subatomic particles on 70.

2. The best, though fairly technical, reference on these theoretical developments is Frank Close, *The Infinity Puzzle: Quantum Field Theory and the Hunt for an Orderly Universe*, esp. chap. 8, "Broken Symmetries," 127–50. Another good resource is Laurie Brown et al., "Panel Session; Spontaneous Breaking of Symmetry," chap. 28 of Hoddeson, Brown, Riordan, and Dresden, eds., *Rise of the Standard Model*, 478–522; it includes individual contributions from three leading theorists who were deeply involved in these breakthroughs: Robert Brout, Peter Higgs, and Yoichiro Nambu.

3. Close, *The Infinity Puzzle*, 80–91. C. N. Yang and R. Mills, "Isotopic Spin and a Generalized Gauge Invariance," *Physical Review* 95 (1954): 631; "Conservation of Isotopic Spin and Isotopic Gauge Invariance," *Physical Review* 96 (1954): 191–95.

4. Riordan, *Hunting of the Quark*, 73–86.

5. Y. Nambu and G. Jona-Lasinio, "Dynamical Model of Elementary Particles Based on an

Analogy with Superconductivity," *Physical Review* 122 (1961): 345–58; J. Goldstone, "Field Theories with 'Superconductor' Solutions," *Nuovo Cimento* 19 (1961): 154–64.

6. Close, *The Infinity Puzzle*, 142–50; P. W. Anderson, "Plasmons, Gauge Invariance and Mass" (see chap. 3, n. 186).

7. Close, *The Infinity Puzzle*, 151–81. F. Englert and R. Brout, "Broken Symmetry and the Mass of Gauge Vector Bosons"; P. W. Higgs, "Broken Symmetries, Massless Particles and Gauge Fields," *Physics Letters* 12 (1964): 132–33; Guralnik, Hagen, and Kibble, "Global Conservation Laws and Massless Particles" (see chap. 1, n. 18).

8. Higgs, "Broken Symmetries and the Mass of Gauge Vector Bosons" (see chap. 1, n. 18). Higgs recounts this observation in Brown et al., "Panel Session," 508–09.

9. S. Weinberg, "A Model of Leptons," *Physical Review Letters* 19 (1967): 1264–66; A. Salam, "Weak and Electromagnetic Interactions," in N. Svartholm, ed., *Elementary Particle Theory* (New York: John Wiley & Sons, 1968), 367–77. Others, especially in Continental Europe, have called this mechanism the "Brout-Englert-Higgs mechanism" or the "BEH mechanism."

10. Gerard 't Hooft, "Renormalization of Gauge Theories," 179–98, and Martinus Veltman, "The Path to Renormalizeability," 145–78, in Hoddeson et al., *Rise of the Standard Model*.

11. Pickering, *Constructing Quarks*; Riordan, *Hunting of the Quark*; L. Brown et al., "The Rise of the Standard Model, 1964–1979," in Hoddeson et al., *Rise of the Standard Model*, 3–35.

12. The best scholarly historical account of these discoveries is Krige, "Distrust and Discovery" (see chap. 1, n. 7). The mass of the Z boson was determined much more accurately as 91 GeV in electron-positron collisions at LEP and SLAC; it was originally thought to be closer to 93 GeV.

13. Henceforth we use the phrase "the Standard Model Higgs boson" to denote a unique spin-zero particle that arises in the simplest version of the Standard Model, that published by Weinberg in 1967. More complicated particle theories, especially those that invoke supersymmetry (see below) to cancel out troublesome infinities, usually require multiple Higgs fields and corresponding Higgs bosons.

14. J. Ellis, M. K. Gaillard, and D. V. Nanopoulos, "A Phenomenological Profile of the Higgs Boson," *Nuclear Physics* B106 (1976): 292–340, quote on 334. See also their retrospective article, "A Historical Profile of the Higgs Boson," *http://arXiv.org/1201.6045v1.pdf*, which appeared in January 2012, six months before the Higgs boson was officially discovered (see epilogue).

15. Herwig Schopper, *LEP—The Lord of the Collider Rings*.

16. Ellis, Gaillard, and Nanopoulos, "A Phenomenological Profile," 322–24. Originally, physicists thought that experiments at LEP could discover a Higgs boson only up to about 70 GeV, but this limit was later extended, and they eventually were able to exclude all possible Higgs masses up to 113 GeV (see chap. 7).

17. Kjell Johnsen, "The CERN Intersecting Storage Rings," in Hoddeson et al., *Rise of the Standard Model*, 285–98.

18. US Department of Energy, *Report of the 1983 HEPAP Subpanel on New Facilities*.

19. The most comprehensive review of the physics to be addressed by hadron colliders at multi-TeV energies, at least as viewed in the mid-1980s, is Eichten et al., "Supercollider Physics."

20. The 18 TeV figure was initially given in the report "Large Hadron Collider in the LEP Tunnel," in *Proceedings of the ECFA-CERN Workshop*, Lausanne and Geneva, March 21–27, 1984, vol. 1. In *Design Study of the Large Hadron Collider (LHC)*, by the LHC Study Group, CERN Report no. 91-03 (May 1991), the maximum collision energy given is 15.4 TeV, assuming 10 T magnets could be fabricated with NbTi cooled to superfluid helium temperatures below 1.8 K. After beginning LHC operations in 2010 at 7 TeV and increasing to 8 TeV in 2012, CERN physicists reached a collision energy of 13 TeV in the 2015 run, (see epilogue, n. 37).

21. *Design Study of the LHC* gave $2 \times 10^{34} cm^{-2} s^{-1}$ in 1991; the goal has since been lowered to 10^{34}.

22. Benjamin W. Lee, C. Quigg, and H. B. Thacker, "Strength of Weak Interactions at Very High Energies and the Higgs Boson Mass," *Physical Review Letters* 38, no. 16 (April 18, 1977): 883–85.

Prior theoretical work along similar lines had been published by D. A. Dicus, V. S. Mathur, and M. Veltman.

23. Ibid., 885.

24. Michael S. Chanowitz and Mary K. Gaillard, "The TeV Physics of Strongly Interacting W's and Z's," *Nuclear Physics* B261 (1985): 379–431. For a more accessible explanation of this limit, see Michael S. Chanowitz, "Electroweak Symmetry Breaking: Unitarity, Dynamics, and Experimental Prospects," (see epilogue, n. 78), 323–420, esp. 327–28. E-mails with Mary K. Gaillard, July 14–31, 2014, were very helpful in understanding these questions.

25. SSC Central Design Group, *Conceptual Design of the Superconducting Super Collider*, esp. 26–39. In his article "The Supercollider—The Pre-Texas Days," Wojcicki states that the 40 TeV design choice was made during the summer of 1984 and approved by the SSC Board of Overseers at its September 1984 meeting.

26. Eichten et al., "Supercollider Physics," 650–65; Chanowitz, "Electroweak," 349–61.

27. Chanowitz, "Electroweak," has a good discussion of this problem on 337–39, and its resolution by supersymmetry on 363.

28. Sally Dawson, E. Eichten, and C. Quigg, "Search for Supersymmetric Particles in Hadron-Hadron Collisions," *Physical Review* D31 (1985): 1581–1640. This mid-1980s paper focused on SUSY particles with possible masses in the tens or hundreds of GeV that might be discoverable at existing or planned hadron colliders like the Tevatron. Eichten et al., "Supercollider Physics," observed that "the low-energy artifacts of supersymmetry, including the superpartners, should occur on a scale of ~1 TeV or less." Quote on 666. Later works, too numerous to cite here, extended these arguments to the TeV mass scale, often a few TeV, which could only be addressed at the SSC or LHC.

29. Riordan and Schramm, *The Shadows of Creation*, 196–97.

30. Dawson, Eichten, and Quigg, "Search for Supersymmetric Particles," 1582–84.

31. John F. Gunion et al., *The Higgs Hunter's Guide* (Menlo Park, CA: Addison-Wesley Publishing Co., 1990), 233–40. This intriguing possibility, which might have been discoverable at LEP, was curiously not mentioned in Eichten et al., "Supercollider Physics," or Dawson, Eichten, and Quigg, "Search for Supersymmetric Particles," suggesting it had not yet appeared prominently by 1985 on high-energy physicists' target lists. Nor was any mention made of it in SSC Central Design Group, *Conceptual Design of the Superconducting Super Collider*, which was focused on the TeV scale. Chanowitz, "Electroweak," briefly mentions the possibility of a light Higgs boson on 364, 367–68, but that review appeared in 1988, and he included no further discussion of its possible detection.

32. Ellis, Ridolfi, and Zwirner, "Radiative Corrections to the Masses"; Haber and Hempfling, "Can the Mass of the Lightest Higgs Boson."

33. Haber and Hempfling, "Can the Mass of the Lightest Higgs Boson."

34. Eichten et al., "Supercollider Physics," 642–50 and 684–95.

Bibliography

Aaltonen, T., et al. "Evidence for a Particle Produced in Association with Weak Bosons and Decaying to a Bottom-Antibottom Quark Pair in Higgs Boson Searches at the Tevatron." *Physical Review Letters* 109 (2012): 071804.

Adams, John B. "Framework of the Construction and Use of an International High-Energy Accelerator Complex." Paper presented at ICFA meeting, Serpukhov, Russia, October 21, 1981. Fermilab Archives.

Aldous, Peter. "CERN: Alone on the Frontier." *Science,* N.S. 262:5141 (1993), 1808–10.

Amaldi, Ugo, ed. "Possibilities and Limitations of Accelerators and Detectors." In *ICFA Workshop Proceedings*. Les Diablerets, Switzerland, October 4–10, 1976. Fermilab Archives.

Amato, Ivan. "House Vote Revives Fermilab Injector." *Science,* N.S. 252:5011 (1991), 1373.

"American Physical Society Council Statement on Physics Funding and the SSC." College Park, MD: American Physical Society, January 20, 1991.

Anderson, Alun. "Japanese Academics Bemoan the Cost of Years of Neglect." *Science,* N.S. 258:5082 (1992), 564–69.

Anderson, Alun, and Joseph Palca. "White House Science Advisor Takes the Reins after Slow Start." *Nature* 329:6139 (1987), 476.

Anderson, Catherine M. "DOE/Sec. O'Leary Testified at House Energy & Water Appropriations." *URA Members Bulletin,* April 30, 1993, 9–12.

———. "Dramatic Conference Committee on SSC's Orderly Termination." *URA Members Bulletin,* October 21, 1993, 1–2.

———. "Happer: DOE General Science Testimony at House Energy & Water Apps." *URA Members Bulletin,* April 30, 1993, 9–12.

———. "Nobel Laureates and HEP Leaders Visited Washington for the SSC." *URA Members Bulletin,* April 30, 1993, 7.

———. "Status of Conference Proceedings." *URA Members Bulletin,* October 7, 1993, 1.

———. "Text of Committee Report on FY94 Energy and Water Development Appropriations." *URA Members Bulletin,* June 17, 1993, 1–4.

———. "Today's Bump in the Road (Slattery Vote on SSC)." *URA Members Bulletin,* October 19, 1993, 1.

———. "Today's House Action on the Energy and Water Conference Report." *URA Members Bulletin,* October 26, 1993, 1–2.

———. "Today's Senate Action on the Energy and Water Conference Report." *URA Members Bulletin*, October 27, 1993, 1–2.

Anderson, Christopher. "The Anatomy of a Defeat." *Science*, N.S. 262:5134 (1993), 645.

———. "DOE Pulls Plug on SSC Contractor." *Science*, N.S. 261:5123 (1993), 822–23.

———. "Senate Vote Lifts Prospects for SSC." *Science*, N.S. 262:5131 (1993), 171.

———. "SSC Deathwatch Starts Again." *Science*, N.S. 260:5113 (1993), 1421.

———. "University Consortium Faulted on Management, Accounting." *Science*, N.S. 261:5118 (1993), 157.

Anderson, Philip W. "Plasmons, Gauge Invariance and Mass." *Physical Review* 130 (1963), 439–42.

———. "More Is Different: Broken Symmetry and the Nature of the Hierarchical Structure of Science." *Science*, N.S. 177:4047 (1972), 393–96.

———. "Outline Testimony on the SSC." February 24, 1989. Irwin Goodwin Collection, AIP Niels Bohr Library & Archives.

Arima, Akhito. "Underfunding of Basic Science in Japan." *Science*, N.S. 258:5082 (1992), 590–91.

Associated Press. "State Officials Tout Plan to Turn Collider Site into Prairie." *Dallas Morning News*, April 19, 1994, A21.

———. "Super Collider Lab Sold for $10 million." *Dallas Morning News*, May 16, 1998, A31.

"ATLAS and CMS Present Higgs Search Status." CERN press release, December 13, 2011.

ATLAS Collaboration. "Evidence for the Spin-Zero Nature of the Higgs Boson Using ATLAS Data," *Physics Letters B* 726 (2013): 120–44.

———. "Measurements of Higgs Boson Production and Couplings in Diboson Final States with the ATLAS Detector at the LHC." *Physics Letters B* 726, 2013: 88–119.

"Atomic Energy: Scientific and Technical Cooperation in the Field of Peaceful Uses of Atomic Energy." June 21, 1973. Rolland P. Johnson Collection, Fermilab Archives.

Awano, Yukiya. "US-Japan SSC Collaboration." *Science*, N.S. 261:5118 (1993), 146.

Baldi, R. W. et al., "An Overview of the Collider Dipole Magnet Program." In *Supercollider 5*. Edited by Phyllis Hale. Dordrecht, Netherlands: Plenum Press, 1994.

Barish, Barry, and William Willis. Spokespersons. "GEM Letter of Intent." November 30, 1991. SSC Collection, Fermilab Archives.

Barrionuevo, Alexei. "State Officials Vote to Return Land to Ellis County." *Dallas Morning News*, July 19, 1995, A22.

Barry, Dave. "Superconductivitexpialidocious." *Washington Post*, Style, October 18, 1987, 1.

Barth, Richard. Telephone conversation with Riordan. August 22, 2011. Riordan notes.

Barton, Joe, and Shelton Smith. "Texas Isn't Playing Politics with Corpse of Collider." *Dallas Morning News*, March 20, 1994, A6.

Basuchoudhary, Atin, Paul Pecorino, and William F. Shughart II. "Reversal of Fortune: the Politics and Economics of the Superconducting Supercollider." *Public Choice* 100 (1999), 189.

Bazell, Robert. "Quark Barrel Politics." *New Republic*, June 22, 1987, 9–10.

Bednorz, J. G., and K. A. Muller. *Zeitschrift fur Physik B64* (1986), 189.

Begley, Sharon. "Sparing Those Sacred Cows." *Newsweek*, March 1, 1993, 17.

Beil, Laura. "Proposal Outlines Use for Collider in Radiation Therapy for Cancer." *Dallas Morning News*, June 4, 1994, A33.

"Between an Atom and a Hard Place." Editorial. *Newsweek*, December 2, 1991.

Bevill, Tom. "Big Science." Editorial, *Dallas Morning News*, May 29, 1991.

———. *Congressional Record—House*, June 28, 1989, H-3308.

Bjorken, James D. "Physics Issues and the VBA." Unpublished. May, 1976. Fermilab Archives.

Boehlert, Sherwood. "Boehlert Says Briefing on SSC Strengthens His Resolve to Require International Funding of the Project." Press release. February 2, 1988. SSC files of Sherwood Boehlert and David Goldston.

———. "Collider Project Doomed Itself by Excess." *New York Times*, Letter to the editor, November 4, 1993.

———. *Congressional Record*, May 2, 1990, H-1958.

———. Conversation with Michael Riordan and Steven Weiss. October 30, 1997. Riordan notes.

———. "Floor Statement on Energy and Water Appropriations." May 29, 1991. SSC files of Boehlert and Goldston.

———. Letter to Allan Gottlieb, Canadian Ambassador to United States. October 29, 1987. SSC files of Boehlert and Goldston.

"Boehlert Is 'Outraged' about Conference Action on SSC." Press release, office of Congressman Sherwood Boehlert. October 14, 1993, SSC files of Boehlert and Goldston.

Briggs, Michael. "Record Shows Atom Smasher Price Tag May Skyrocket." *Chicago Sun-Times*, November 4, 1988.

Broad, William J. "Atom-Smashing Now and in the Future: A New Era Begins." *New York Times*, February 3, 1987, C1.

———. "Giant Atom Smasher Encounters Vexing Technical Obstacles." *New York Times*, Science Times, December 13, 1988, 23, 28.

———. "Limping Accelerator May Fall to Budget Ax." *Science*, N.S. 213:4510 (1981), 846–47, 850.

———. "Magnet Failures Imperil New Accelerator." *Science*, N.S. 210:4472 (1980), 875–78.

———. "Physicists Compete for the Biggest Project of All." *New York Times*, September 20, 1983, C1.

———. "Supermagnet Design Chosen for a 60-mile Atom Smasher." *New York Times*, September 19, 1985, 1.

———. "Vast Sums for New Discoveries Pose a Threat to Basic Science." *New York Times*, May 27, 1990. 1.

Broder, David S. "Sen. Robert Byrd's Vanished Ethic." *Washington Post*, July 1, 2010, A15.

Bromley, D. Allan. *The President's Scientists: Reminiscences of a White House Advisor*. New Haven and London: Yale University Press, 1994.

Brown, George Jr. *Congressional Record—House*, May 29, 1991, H-3670.

Brown, Laurie M., Robert Brout, Tian Yu Cao, Peter Higgs, and Yoichiro Nambu. "Panel Session: Spontaneous Breaking of Symmetry." In *The Rise of the Standard Model: Particle Physics in the 1960s and 1970s*. Edited by Lillian Hoddeson, Laurie Brown, Michael Riordan, and Max Dresden, 478–522. New York: Cambridge University Press, 1997.

Brown, Laurie M., Michael Riordan, Max Dresden and Lillian Hoddeson, "The Rise of the Standard Model, 1964–1979." In *The Rise of the Standard Model*, 3–35.

Browne, Malcolm W. "Building a Behemoth against Great Odds." *New York Times*, March 23, 1993, C1.

Brüning, Oliver, and Paul Collier. "Building a Behemoth." *Nature* 448, July 19, 2007, 285–89.

Brumfiel, Geoff. "The Machine Maker." *Nature* 456:7224 (2008), 862–68.

Bumpers, Dale. *Congressional Record—Senate*, August 3, 1992, S-11144–S-11146.

Burton, Bill. Memorandum to Mack McLarty, "President's Discussion with Secretary O'Leary re SSC." November 8, 1993. FOIA Request 2010-450-F, Clinton Library Digital Archives.

———. Memorandum to President Clinton, "Energy Secretary's Memo re SSC." November 6, 1993. FOIA Request 2010-450-F, Clinton Library Digital Archives.

Bush, George H. W. Letter to Toshiki Kaifu. October 11, 1991. SSC files of John Metzler. In file of Riordan.

Bush, George, and Brent Scowcroft. *A World Transformed*. New York: Alfred A. Knopf, 1998.

Bush, Thomas. "Inflated Japanese Cost Estimates." Memorandum to Roy Schwitters. January 31, 1992. SSC Collection, Fermilab Archives.

———. *Status of the Superconducting Super Collider Program*. Testimony. *Hearing Before the Investigations and Oversight Committee of the House Science, Space and Technology Committee*. May 9, 1991.

Bush, Vannevar. *Science, the Endless Frontier: A Report to the President on a Program for Postwar Scientific Research*. Washington: National Science Foundation, 1945.

Butler, Declan. "Japan Consoled with Contracts as France Snares Fusion Project." *Nature* 435:7046 (2005), 1142–43.

Byerly, Radford, Jr., and Robert A. Pielke Jr. "The Changing Ecology of United States Science." *Science*, N.S. 269:5230 (1995), 1531–32.

Campbell, Philip, and Peter Rodgers. "Europe Beckons US High-Energy Physicists as SSC is Cancelled." *Physics World* (November 1993), 5–6.

Cartwright, Jon. "Home Run Complete, LHC Set to Repeat It Backwards." *Physics World*, September 10, 2010. http://physicsworld.com/cws/article/news/2008/sep/10/home-run-complete-lhc-set-to-repeat-it-backwards.

———. "LHC Milestone Day Gets Off to a Fast Start." *Physics World*, September 10, 2010. http://physicsworld.com/cws/article/news/2008/sep/10/lhc-milestone-day-gets-off-to-fast-start.

———. "LHC Report Confirms Electrical Fault." *Physics World*, October 17, 2008. http://physicsworld.com/cws/article/news/2008/oct/17/lhc-report-confirms-electrical-fault.

———. "Mission Complete for LHC Team." *Physics World*, September 10, 2010. http://physicsworld.com/cws/article/news/2008/sep/10/mission-complete-for-lhc-team.

CERN. "Large Hadron Collider in the LEP Tunnel." In *Proceedings of the ECFA-CERN Workshop on Hadron Colliders in the LEP Tunnel*. Lausanne and Geneva, March 21–27, 1984, vol. 1.

"CERN Council Gives Go-ahead for Large Hadron Collider." CERN press release, December 16, 1994.

Chalmers, Matthew. "Hunt for the Higgs Enters Endgame." *Physics World online*, August 22, 2011. http://physicsworld.com/cws/article/news/2011/aug/22/hunt-for-the-higgs-enters-endgame.

Chanowitz, Michael S. "Electroweak Symmetry Breaking: Unitarity, Dynamics and Experimental Prospects." *Annual Reviews of Nuclear and Particle Science* 38 (1988) 327–28.

Chanowitz, Michael S., and Mary K. Gaillard. "The TeV Physics of Strongly Interacting W's and Z's." *Nuclear Physics* B261 (1985): 379–431.

Chatrchyan, S., et al. (CMS Collaboration). "Study of the Mass and Spin-Parity of the Higgs Boson Candidate via its Decays to Z Boson Pairs." *Physical Review Letters* 110 (2013): 081803.

Chao, Alex, and Maury Tigner. "Requirements for Dipole Field Uniformity and Beam Tube Correction Windings." Report No. SSC-N-183. May 27, 1986. SSC Collection, Fermilab Archives.

Chen, Edwin. "Big Science Faces Big Troubles." *Los Angeles Times*, June 5, 1991, A1.

Cho, Adrian. "Fermilab to End Its Quest for Higgs Particle This Year." *Science* 331 (6014): January 14, 2011, 131.

———."Higgs Boson Makes Its Debut after Decades-Long Search." *Science* N.S. 337:6091 (2012), 141–43.

———. "Higgs or Bust? Fermilab Weighs Adding 3 Years to Tevatron Run." *Science* N.S. 329:5997 (2010), 1266–67.

———. "Last Hurrah: Final Tevatron Data Show Hints of Higgs Boson." *Science* N.S. 335:6073 (2012), 1159.

———."More Bad Connections May Limit LHC Energy or Delay Restart." *Science*, N.S. 325:5940 (2009), 522–23.

———. "Search for Higgs Boson Yields a Definite Maybe." *Science*, N.S. 333 (2011), 1482–83.

Cipriano, Joseph R. Memorandum for the Secretary of Energy. September 14, 1990. SSC files of Sherwood Boehlert and David Goldston.

———. Memorandum to Hazel O'Leary. Undated. Box J3a6, Lederman Collection, Fermilab Archives.

———. Memorandum to the Secretary of Energy. January 12, 1991. SSC files of Boehlert and Goldston.

———. Memorandum to James D. Watkins. March 8, 1991. SSC files of Boehlert and Goldston.

Clery, Daniel, Dennis Normile, Gong Yidong, and Andrey Allakhverdov. "ITER Finds a Home: With a Whopping Mortgage." *Science*, N.S. 309:5731 (2005), 28–29.

Clinton, William J. "Remarks by the President and Prime Minister Miyazawa in Joint Press Availability." April 16, 1993. Clinton Digital Library Archives, http://clinton6.nara.gov/1993/04/1993-04-16-clinton-p-m-miyazawa-joint-press-availability.html.

Clinton, President William J., and Vice President Albert J. Gore Jr. "Technology for America's Growth: A New Direction to Build America's Strength." February 22, 1993. Clinton Library Digital Archives.

Close, Frank. *The Infinity Puzzle: Quantum Field Theory and the Hunt for an Orderly Universe*. New York: Basic Books, 2011.

Cole, F. T., and R. Donaldson, eds. *Proceedings of the 12th International Conference on High-Energy Accelerators*. August 11–16, 1983. Batavia, IL, Fermi National Accelerator Laboratory.

Commissariat, Tushna. "Things That Go Bump in the Night." *Physics World*, April 26, 2011. http://blog.physicsworld.com/2011/04/26/things-that-go-bump-in-the-nig/.

Comptroller General Report to Congress. US General Accounting Office. *Increasing Costs, Competition May Hinder U.S. Leadership Position in High-Energy Physics*. GAO Report No. EMD-SO-58. September 16, 1990.

Congressional Budget Office. *Risks and Benefits of Building the Superconducting Super Collider*. Washington, DC, October, 1988.

"Convention May Boost Martin's Campaign." Staff and wire reports. *Greensboro News & Record*, August 19, 1988.

Conway, John. "The End of the Tevatron." Cosmic Variance blog, *Discover* magazine, http://blogs.discovermagazine.com/cosmicvariance/2011/01/10/the-end-of-the-tevatron/.

Coonan, Jean. "DOE Boosts Budget for Physics, Especially Materials." *Physics Today* 36:4 (1983), 49–52.

———. "Latest Budget Cuts Arouse Concern and Recommendations." *Physics Today* 34:12 (1981), 47–49.

Cowan, Edward. "16 States Forced to get U.S. Loans to Pay the Jobless." *New York Times*, July 19, 1982, 1.

Cranford, John R. "Defeat of Budget Amendment Fans Anti-Deficit Flames." *Congressional Quarterly*, June 13, 1992. 1683–89.

———. "A Winner at Last." *Congressional Quarterly*, June 13, 1992, 1688.

Crawford, Mark. "Accelerator Labs Face Austere Year." *Science*, N.S. 234:4781 (1986), 1195.

———. "Budget Crunch Stalls Super Collider." *Science*, N.S. 240:4848 (1988), 17–18.

———. "California Gears Up to Bid for the SSC." *Science*, N.S. 228:4704 (1985), 1181.

———. "CBO Cautions Congress on SSC." *Science*, N.S. 242:4876 (1988), 186.

———. "CBO Lists Options for Cutting R&D." *Science*, N.S. 243:4894 (1989), 1001.

———. "House Committee Questions SSC." *Science*, N.S. 228:4697 (1985) 309.

———. "Lab Report Puts SSC Magnets in Limbo." *Science*, N.S. 245:4920 (1989), 809–10.

———. "Reagan Okays the Supercollider." *Science*, N.S. 235:4789 (1987), 625.

———. "SSC Cost Estimates Climb." *Science*, N.S. 249:4968 (1990), 473.

———. "SSC Report Attacked." *Science*, N.S. 242:4882 (1988), 1243.

———. "SSC's Forlorn Quest for Foreign Partners." *Science* 252:5002 (1991), 25.

———. "SSC Sites: Then There Were Eight." *Science*, N.S. 239:4835 (1988), 133–34.

———. "The SSC's Price Tag Troubles Congress." *Science*, N.S. 235:4791 (1987), 837–38.

———. "States Want More Time to Prepare SSC Bids." *Science*, N.S. 236:4807 (1987), 1422.

———. "Super Collider Advocates Tangle with Cost Cutters." *Science*, N.S. 247:4939 (1990), 152–53.

———. "Supercollider Faces Budget Barrier." *Science*, N.S. 236:4799 (1987), 246–48.

———. "Texas Lands the SSC." *Science*, N.S. 242:4880 (1988), 1004.

———. "Watkins Named Energy Secretary." *Science*, N.S. 243:4889 (1989), 309.

———. "Watkins Takes Helm at DOE." *Science*, N.S. 243:4895 (1989), 1136.

———. "Will Magnet Problems Delay the SSC?" *Science*, N.S. 243:4897 (1989), 1425–26.

Crease, Robert P. "Choosing Detectors for the SSC." *Science*, N.S. 250:4988 (1990), 1648–50.

———. *Making Physics: A Biography of Brookhaven National Laboratory, 1946–1972*. Chicago: University of Chicago Press, 1999.

———. "Quenched! The ISABELLE Saga, Part I." *Physics in Perspective* 7:3 (2005), 330–76.

———. "Quenched! The ISABELLE Saga, Part II." *Physics in Perspective* 7:4 (2005), 404–52.

Crease, Robert P., and Charles C. Mann. *The Second Creation: Makers of the Revolution in Twentieth-Century Physics*. New York: Macmillan and Co., 1986.

Crenson, Matt. "Tunnel Vision; Geologists Lobby for Underground Lab at SSC Site." *Dallas Morning News*, December 26, 1994, D6.

Cronin, James. "The Case for the Supercollider." *Bulletin of the Atomic Scientists* 42:5 (1986), 8–11.

———. "Dear Colleague" Letter. August 22, 1988. Cronin SSC correspondence file, Fermilab Archives.

———. Letter to John H. Marburger III. August 19, 1988. Cronin SSC correspondence file, Fermilab Archives.

Crowley-Milling, Michael C. *John Bertram Adams: Engineer Extraordinary*. Amsterdam: Gordon and Breach, 1993.

Crown, Judith. "Illinois' Fermi Pushing Hard for Super Collider." *Houston Chronicle*, October 30, 1988.

Culliton, Barbara J. "Alvin Trivelpiece of DOE is Named New Executive Officer of AAAS." *Science*, N.S. 235:4791 (1987), 840.

———. "Frank Press Calls Budget Summit." *Science*, N.S. 214:4521 (1981), 634–35.

———. "Keyworth Gives First Policy Speech." *Science*, N.S. 213:4504 (1981), 183–84.

Davis, Mike. *City of Quartz: Excavating the Future in Los Angeles*. New York: Vintage Books, 1992.

Dawson, Sally, Estia Eichten, and Chris Quigg. "Search for Supersymmetric Particles in Hadron-Hadron Collisions." *Physical Review D* 31 (1985), 1581–1640.

Dean, Cornelia. "Scientific Savvy? In U.S., Not Much." *New York Times*, August 30, 2005.

Deckard, Barbara. "State Party Delegations in the U.S. House of Representatives—a Comparative Study of Group Cohesion." *Journal of Politics* (February 1972), 202–22.

———. "State Party Delegations in the U.S. House of Representatives—an Analysis of Group Action." *Polity* (Spring), 312–34.

Decker, James F. Letter to F. E. Low, December 21, 1989. In *Report of the 1990 HEPAP Subpanel on SSC Physics*, Appendix B. DOE Report No. DOE/ER-434. Washington, DC. January, 1990.

———. "Remarks to High Energy Physics Advisory Panel." January 12, 1990. Irwin Goodwin Collection, AIP Niels Bohr Library & Archives.

———. Statement before House Subcommittee on International Scientific Cooperation. March 16, 1988. SSC files of Sherwood Boehlert and David Goldston.

Democratic Study Group Legislative Report. "Schedule for the Week of April 23, 1990," 19. SSC files of Sherwood Boehlert and David Goldston.

"Description of Reorganization Plan for URA." URA files, General Correspondence 1986–1989. December 1987. Fermilab Archives.

Design Study of the Large Hadron Collider (LHC). LHC Study Group. CERN Report No. 91-03. Geneva: CERN, May 2, 1991.

Deutch, John M. Memorandum to Members of the Board of Overseers of the SSC. April 19, 1984. "Superconducting Supercollider." Copy in files of George A. Keyworth, file OA94676-1 (Box 1), Ronald Reagan Library.

Dickson, David. *The New Politics of Science*. Chicago: University of Chicago Press, 1988.

———. "A Political Push for Scientific Cooperation." *Science*, N.S. 224:4655 (1984), 1317–19.

Diebold, Robert. "The Desertron: Colliding Beams at 20 TeV." *Science*, N.S. 222:4619 (1983), 13–19.

———. "Meeting of Sciulli HEPAP Subpanel at SSCL." Memorandum. January 23, 1990.

———. "SSC Scientific Policy Committee Meeting." Memorandum. January 5, 1990.

Diebold, Robert, et al. "'Conventional' 20-TeV, 10-Tesla, $p^{\pm}p$ Colliders." In *Proceedings of the 1982 DPF Summer Study on Elementary Particle Physics and Future Facilities*. Snowmass, CO. June 28–July 16, 1982. Edited by Rene Donaldson, Richard Gustafson, and Frank Paige, 307–14. Batavia, IL: Fermi National Accelerator Laboratory, 1983.

Dietrich, William S. *In the Shadow of the Rising Sun: The Political Roots of American Economic Decline.* University Park, PA: Pennsylvania State University Press, 1991.

"Editorial: Disappointing End for the SSC." *Nature* 365:6449 (1993), 771.

Divine, Robert A. *Eisenhower and the Cold War*. New York: Oxford University Press, 1981, 143–52.

"Documentation on SSC, [Machine Advisory Committee] Meeting #1 (December 4–5, 1989)." Unpublished. SSC files of Lyndon Evans.

"DOE Group Studies High-Energy Accelerators." *Physics Today* 33:2 (1980), 92.

Domestic Policy Council. "Memorandum for the President. Superconducting Super Collider." January 14, 1987. Domestic Policy Council file 459161, folder 2, Ronald Reagan Library.

Domestic Policy Council. Memorandum for the President, "Superconducting Super Collider." January 27, 1987; plus meeting notice and participant list, January 29, 1987. Documents courtesy of Alvin W. Trivelpiece, who received copies from the Ronald Reagan Library.

Donaldson, Rene, ed. "Supercollider R&D, The First Two Years." Berkeley, CA: SSC Central Design Group, 1985. CDG Collection, Fermilab Archives.

———. "The Superconducting Super Collider." Berkeley, CA: SSC Central Design Group, 1985. CDG Collection, Fermilab Archives.

———. "To the Heart of Matter." Berkeley, CA: SSC Central Design Group, 1984. CDG Collection, Fermilab Archives.

Donaldson, Rene, and J. G. Morfin, eds. *Proceedings of the 1984 Summer Study on the Design and Utilization of the Superconducting Super Collider*. Snowmass, CO. June 23–July 13, 1984. Batavia, IL: Fermi National Accelerator Laboratory.

Donaldson, Rene, Richard Gustafson, and Frank Paige, eds. *Proceedings of the 1982 DPF Summer Study on Elementary Particles and Future Facilities*. Batavia, IL: Fermi National Accelerator Laboratory, 1983.

Dowd, Maureen. "Kaifu Visits Bush and Brings Word of War Payment." *New York Times*, July 12, 1991, A1.

Drell, Sidney. "Some Thoughts on the SSC and the Management of Science." *Physics Today* 46:7 (1993), 73–75.

Dugan, Gerald F., and James R. Sanford, eds. *Superconducting Super Collider: A Retrospective Summary, 1989–1993*. SSC Laboratory Report No. SSCL-SR-1235. April 1994.

Dyson, Freeman. "Alternatives to the Superconducting Super Collider." *Physics Today* 41:2 (1988), 77.

Easterbrook, Gregg. "Big Science on Easy Street." *New York Times*, November 29, 1991.

———. "Radio Free Watkins and the Crisis at Energy." *Washington Post Magazine*, February 18, 1990, 16–40.

Eckart, Dennis. *Congressional Record—House*, May 29, 1991, H-3682.

———. *Congressional Record—House*, June 17, 1992, H-4810–4811.

Editorial. *Los Angeles Times*, February 3, 1987, B4.

Edwards, Helen. "SSC Site Specific Design." Copies of transparencies of workshop presentation. October 3, 1989. Irwin Goodwin Collection, AIP Niels Bohr Library & Archives.

———. "The Tevatron Energy Doubler: A Superconducting Accelerator." *Annual Reviews of Nuclear and Particle Science* 35: (1985), 605–60.

Eichten, E., I. Hinchliffe, K. Lane, and C. Quigg. "Supercollider Physics." *Reviews of Modern Physics* 56: (1984), 597–707.

Eig, Jonathan. "Legislators Reject Plan for SSC Site; Some Criticize Decision to Scrap Cancer Center, Sell Land, Assets." *Dallas Morning News*, May 16, 1995, A17.

Elioff, Tom. "A Chronicle of Costs." SSCL-SR-1242, April 1994.

Ellis, J., G. Ridolfi, and F. Zwirner. "Radiative Corrections to the Masses of Supersymmetric Higgs Bosons." *Physics Letters* 257 (1990): 83–88.

Ellis, J., M. K. Gaillard, and D. V. Nanopoulos. "A Historical Profile of the Higgs Boson," http://arXiv.org/1201.6045v1.pdf.

Ellis, J., M. K. Gaillard, and D. V. Nanopoulos. "A Phenomenological Profile of the Higgs Boson." *Nuclear Physics* B106 (1976): 292–340.

Ellis, John, and Tevong You. "Global Analysis of the Higgs Candidate with mass ~125 GeV." *Journal of High Energy Physics* (September, 2012), 123.

Englert, F., and R. Brout. "Broken Symmetry and the Mass of Gauge Vector Mesons." *Physical Review Letters* 13 (1964): 321–23.

"The Environmental Impact Statement on The Superconducting Super Collider." Department of Energy Facts. September, 1988.

"The Epilogue: Super Collider's Assets Will Not Be Reused." Editorial. *Dallas Morning News*, May 17, 1995, A20.

"Equipment That Might Be Provided by Japan." Undated list. SSC files of John Metzler, copy in Riordan files.

"Europe 3, US Not Even Z-Zero." Editorial. *New York Times*, June 6, 1983, A16.

Evans, Lyndon, ed. *The Large Hadron Collider: A Marvel of Technology*. Lausanne: EPFL Press, 2009.

Evans, Rowland, and Robert Novak. *The Reagan Revolution*. New York: E. P. Dutton, 1981.

Fang, Y., et al. "Observation of a $\gamma\gamma$ Resonance at a Mass in the Vicinity of 115 GeV/c^2 at ATLAS and its Higgs Interpretation." CERN: ATLAS Internal Note No. ATL-COM-PHYS-2011-415.

Feder, Toni. "CERN Grapples with LHC Cost Hike." *Physics Today* 54:12 (2001), 21–22.

Fehner, Terrence R., and Jack M. Holl. "Department of Energy, 1977–1994." Report No. DOE/HR-0098. (November, 1994), 17–31. http://www.energy.gov/media/Summary_History.pdf.

Flam, Faye. "Is There Life After the SSC?" *Science* N.S. 262:5134 (1993), 644–47.

———. "The Message from CERN: Help Wanted." *Science*, N.S. 263:5148 (1994), 749.

———. "Panel Presents Vision for Physics after the Supercollider." *Science*, N.S. 264:5164 (1994), 1397.

———. "Physicists Struggle for Consensus about the Future." *Science*, N.S. 263:5148 (1994), 749.

———. "The SSC: Radical Therapy for Physics." *Science*, N.S. 254:5029 (1991), 194–96.

———. "Taking a Gamble on the Top Quark." *Science*, N.S. 264:5159 (1994), 658–69.

Flawn, Peter T. "Night-and-Day Work Pays Off for Collider Commission." *Austin American Statesman*, November 11, 1988, A17+.

———. *The Story of the Texas National Research Laboratory Commission: How Texas Won . . . and Lost.* Austin, TX: Bureau of Economic Geology, 2003.

"For the Selection of a Management and Operating Contractor for the Establishment, Management, and Initial Operation of the Superconducting Super Collider Laboratory." Request for Proposals No. DE-RP02-88ER40486. SSC Collection, Fermilab Archives.

Franklin, Ben A. "Reagan to Press for $6 Billion Atom Smasher." *New York Times*, February 2, 1987, 1.

Freedman, Allan. "Second Thoughts on the Super Collider." *D Magazine* (July, 1991).

Frenkel, Karen A. "Nobelist Steven Weinberg Calls for Bigger Science, More Taxes." *Science Now*, June 6, 2011. http://news.sciencemag.org/2011/06/nobelist-steven-weinberg-calls-bigger-science-more-taxes.

Fukushima, Glen S. "Bush-Miyazawa Meeting Is 'Staged' for Success." *Los Angeles Times*, January 2, 1992.

"Fund-Raiser for Particle Accelerator Resigns." *New York Times*, July 9, 1991, A14.

"Future Accelerators Seminar in Japan." *CERN Courier* (October, 1984), 319–22.

Gaillard, Mary K. *A Singularly Unfeminine Profession*. Singapore: World Scientific Publishing Co., 2015.

Galison, Peter. *Image and Logic: A Material Culture of Microphysics*. Chicago: University of Chicago Press, 1997.

Galison, Peter, and Bruce Hevly, eds. *Big Science: The Growth of Large-Scale Research*. Stanford, CA: Stanford University Press, 1992.

Gates, Max. "Politics a Key to Super Collider Site Selection." *Ann Arbor News*, February 21, 1988.

———. "State Must Try Harder to Win SSC, Expert Says." *Ann Arbor News*, March 13, 1988.

Gibbons, John H. White House press briefing, February 23, 1993. Clinton Digital Library Archives, http://clinton6.nara.gov/1993/02/1993-02-23-press-briefing-by-john-gibbons-bowman-cutter.html.

———. Memorandum for John D. Podesta, "O'Leary Draft Letter from President Clinton to Kiichi Miyazawa." February 24, 1993. FOIA Request 2010–450–F, Clinton Library Digital Archives.

Gibson, David V., and Everett M. Rogers. *R&D Collaboration on Trial: The Microelectronics and Computer Technology Corporation*. Boston: Harvard Business Review Press, 1994.

Gillette, Robert. "Supercollider's Cost Could Rise by 50%." *Los Angeles Times*, November 8, 1988, 4.

Gilchriese, M. G. C., and K. Metropolis, eds. *Site Specific Design of the Superconducting Super Collider*. Report No. SSCL-SR-1055. July, 1990.

Glashow, Sheldon, and Leon M. Lederman. "The SSC: A Machine for the Nineties." *Physics Today* 38:3 (March 1985), 28–37.

Glauthier, T. J., Keith Mason, and Sylvia Mathews. Memorandum for the President and Mack McLarty, "Summary of Richards/O'Leary Meeting." February 27, 1993. FOIA Request 2010-450-F, Clinton Library Digital Archives.

Golden, Gayle. "Elections Could Boost Odds for Texas Super Collider Bid." *Dallas Morning News*, July 31, 1988.

———. "Image Woes Plague North Carolina Collider Bid." *Dallas Morning News*, July 14, 1988.

Goldston, David. Comments at APS meeting, Washington, DC. April 29, 2001. Riordan notes.

Goldstone, J. "Field Theories with 'Superconductor' Solutions." *Nuovo Cimento* 19 (1961): 154–64.

Goodwin, Irwin. "After a Wait, Hunter Joins DOE, Keyworth, Bernthal to New Jobs." *Physics Today* 41:11 (1988), 52.

———. "After Agonizing Death in the Family, Particle Physics Faces Grim Future." *Physics Today* 47:2 (1994), 89–94.

———. "Amazing Race: the SSC Contest Generates Disorder and Discord." *Physics Today* 41:5 (1988), 69–74.

———. "As SSC Project Accelerates, Its Cost Exceeds $8.2 Billion." *Physics Today* 44:3 (1991), 51–53.

———. "As the SSC Faces 'Live or Die' Vote, DOE's O'Leary Shakes up Management." *Physics Today* 46:9 (1993), 52–53.

———. "At Last, Congress Agrees to Build SSC, after Texas-Type Wheeling and Dealing." *Physics Today* 42:10 (1989), 51–52.

———. "At Long Last, Graham Becomes Reagan's Science Advisor." *Physics Today* 39:11 (1986), 57–58.

———. "Bush, His Faith in R&D, Raises Hope for Higher 1992 Budget in Hard Times." *Physics Today* 44:4 (1991), 79–86.

———. "CERN Leaders Troubled by Mixed Reactions in Washington on Visit to Raise Funds for LHC." *Physics Today* 49:4 (1996), 47–48.

———. "Cheers for Bush's 1993 R&D Budget Cut Short by Problems and Pessimism." *Physics Today* 45:6 (1992), 55–61.

———. "Clinton's Budget Boosts Technology, Making Research Scientists Jittery." *Physics Today* 46:6 (1993), 83–89.

———. "Clinton's Hands-on Economic Plan: Technology Gains, Big Science Loses." *Physics Today* 46:4 (1993), 43–46.

———. "Collaborators Await European Approval of LHC." *Physics Today* 47:11 (1994), 80–81.

———. "Collision Course: Besieged by Congress, SSC Awaits Key Decisions on Its Fate." *Physics Today* 46:8 (1993), 43–45.

———. "Congress Cancels SSC and Allocates High Budgets for Technology in 1994." *Physics Today* 46:11 (1993), 77–80.

———. "Congressmen Review SSC with Budget Deficits on Their Minds." *Physics Today* 38:12 (1985), 55–58.

———. "Conversation with D. Allan Bromley on Major Issues in Scientific Research." *Physics Today* 43:7 (1990), 49–55.

———. "Despite Retirements and Defeats, Congress Retains Friends of Science." *Physics Today* 46:2 (1993), 71–72.

———. "DOE Answers to Congress as it Officially Kills Brookhaven CBA." *Physics Today* 36:12 (1983), 41–43.

———. "DOE Offers SSC Site Document, but Sidesteps Its Endorsement." *Physics Today* 38:9 (1985), 53–55.

———. "DOE Picks Texas for 'Gippertron' Amid Political and Managerial Collisions." *Physics Today* 42:2 (1989), 95–98.

———. "DOE Submits 36 SSC Site Bids While House Seeks to Micro-manage Project." *Physics Today* 40:11 (1987), 45–46.

———. "Four Years after SSC's Demise, US Reaches Agreement on 'Unprecedented' Collaboration in CERN's LHC." *Physics Today* 51:1 (1998), 43–44.

———. "Future Shocks: Bush's 1991 Budget Boosts R&D but Deficit Threatens It." *Physics Today* 43:6 (1990), 51–58.

———. "Happer Leaves DOE under Ozone Cloud for Violating Political Correctness." *Physics Today* 46:6 (1993), 89–91.

———. "HEPAP and Its Subpanel Approve Redesign and Higher Cost of SSC." *Physics Today* 43:2 (1990), 67–68.

———. "Hunter and Nelson Named to DOE Posts." *Physics Today* 40:8 (1987), 53.

———. "Hunter Departs DOE after Riling Key Lawmakers and Top Texans." *Physics Today* 43:1 (1990), 49.

———. "LHC 'Back On Track' as DOE Proposes to Ante Up $400 million to $500 million." *Physics Today* 49:5 (1996), 61.

———. "Making News by Calling It Quits, Bucy Leaves a Message for the SSC." *Physics Today* 44:8 (1991), 52–53.

———. "Numbers Game: Bush's 1990 R&D Budget Uses Reagan's Figures in Making Deals." *Physics Today* 42:5 (1989), 43–49.

———. "President Clinton Picks John Gibbons as Science Adviser to Reinvent Policy." *Physics Today* 46:3 (1993), 73–74.

———. "R&D Budget for Fiscal 1987: Life at the Threshold of Pain." *Physics Today* 39:5 (1986), 55–60.

———. "Race for the Ring: DOE Reacts to Congress's Anxieties on SSC." *Physics Today* 40:8 (1987), 47–50.

———. "Reagan Endorses the SSC, a Colossus Among Colliders." *Physics Today* 40:3 (1987), 47–49.

———. "Reagan's R&D Budget Looks Great, But Congress Has Some Other Ideas." *Physics Today* 41:4 (1988), 55–61.

———. "Senate Rescues SSC, But Final Act Awaits Conference." *Physics Today* 46:10 (1993), 110–11.

———. "Senator Bennett Johnston Talks About Physicists in Politics." *Physics Today* 49:1 (1996), 51–55.

———. "SSC Cost and Size Perplex Congress." *Physics Today* 37:5 (1984): 64.

———. "SSC: Progress on Magnets, Uncertainty on Foreign Collaboration." *Physics Today* 38:3 (1985), 63–66.

———. "Talk With Allan Bromley: On Life in the White House Science Fast Lane." *Physics Today* 44:10 (1991), 93–99.

———. "Tigner Named to Direct R&D Program for SSC." *Physics Today* 37:8 (1984), 69.

———. "Toll Heads URA; Moore Leaves NSF and Sanchez Joins; Foster at DOD." *Physics Today* 43:2 (1990), 72.

———. "Trying Times: Cost of Remodeling SSC Causes Texans to Circle Their Wagons." *Physics Today* 43:1 (1990), 45–47.

———. "2100 Physicists Use a Democratic Process for the SSC." *Physics Today* 45:8 (1992), 59.

———. "Victim of House Budget Balancing War, SSC Now Faces Uncertain Fate in Senate." *Physics Today* 45:7 (1992), 53 54.

———. "What's Gone Wrong with the SSC? It's Political, not Technological." *Physics Today* 45:8 (1992), 58–60.

———. "White House Forum on Science Produces only a Reassuring Hug." *Physics Today* 47:3 (1994), 41–42.

———. "Will High-T_c Superconductivity Affect the SSC's Design?" *Physics Today* 40:8 (1987), 50–52.

Goodwin, Irwin, with Bertram Schwarzschild. "Good News for the SSC as Senate Approves Funds and Magnets Work." *Physics Today* 45:9 (1992), 54–56.

Graham, William R. "Superconducting Super Collider (SSC)." Memorandum to Chief of Staff Donald T. Regan. November 19, 1986. William R. Graham file A04261-1 (CFoA 990), Ronald Reagan Library.

Gray, William H., III. "The Superconducting Super Collider Authorization Act of 1990 (H.R. 4380)." Whip Advisory sheet. April 20, 1990. SSC files of Sherwood Boehlert and David Goldston.

Greenberg, Daniel S. "Are We Really Shortchanging Science?" *Washington Post*, February 12, 1991.

———. *Science, Money, and Politics: Political Triumph and Ethical Error*. Chicago: University of Chicago Press, 2001.

Griffin, Sean. "Teamwork Aids Arizona Supercollider Effort." *Phoenix Gazette*, June 20, 1988, A1.

Grunder, Hermann, Thomas Fields, Boyce McDaniel, Rich Orr, Paul Reardon, Richard Taylor, and Maury Tigner. Letter to Lab Directors. Reprinted in *Report of the DOE Review Committee on the Reference Designs Study*, Appendix A. November 17, 1983.

Gunion, John F., Howard Haber, Gordon Kane, and Sally Dawson. *The Higgs Hunter's Guide*. Menlo Park, CA: Addison-Wesley Publishing Co., 1990, 233–40.

Guralnik, G. S., C. R. Hagen, and T. W. B. Kibble. "Global Conservation Laws and Massless Particles." *Physical Review Letters* 13 (1964): 585–87.

Gwertzman, Bernard. "McFarlane Took Cake and Bible to Tehran, ex-C.I.A. Man Says." *New York Times*, January 11, 1987.

Haber, H., and R. Hempfling. "Can the Mass of the Lightest Higgs Boson of the Minimal Supersymmetric Model Be Larger than m_z?" *Physical Review Letters* 66 (1991): 1815–18.

Hackerman, Norman. Letter to H. Guyford Stever. July 3, 1983. URA files, Lederman Collection, Fermilab Archives.

Hager, George. "Is the Deficit Now Too Big for Congress to Tame?" *Congressional Quarterly*, May 2, 1992, 1140–47.

———. "Rejection of Walls Bill Spells Spending Squeeze at Home." *Congressional Quarterly*, April 4, 1992, 866–67.

———. "Surprise Vote Sounds Alarm for 1993 Spending Bills." *Congressional Quarterly*, June 20, 1992, 1773–77.

Hailey, Charles J., Gordon R. Freeman, Pedro M. Echenique, and Sheldon L. Glashow. "Superconducting Super Collider." *Physics Today* 39:12 (1986), 11–15.

Hamilton, David P. "Ad Hoc Team Revives SSC Competition." *Science*, N.S. 252:5013 (1991), 252.

———. "Allocating the Pain in Energy Science." *Science*, N.S. 253:5027 (1991), 1482.

———. "CERN's Horserace with the SSC." *Science*, N.S. 254:5029 (1991), 194.

———. "DOE Research Funding Freeze." *Science* N.S. 254:5063 (1992), 1507.

———. "Lightning Strikes the SSC." *Science*, N.S. 256:5065 (1992), 1752–53.

———. "Senate Subcommittee Stalls SSC Vote." *Science* N.S. 257:5066 (1992), 19.

———. "The SSC Gets Its (Official) Price Tag: $8.3 Billion." *Science*, N.S. 251:4995 (1991), 741.

———. "The SSC Takes On a Life of Its Own." *Science*, N.S. 249:4970 (1990), 731–32.

Happer, William. "Diversity Needed in Federal Support of Basic Science after the SSC." *APS News* (March, 1994), 12.

Harré, R. *The Philosophies of Science*. New York: Oxford University Press, 1985.

Harris, David. "The End of the HERA Era." *Symmetry* 4:7 (2007). www.symmetrymagazine.org/cms/?pid=1000532.

Hart, William. "Evaluation Leaves State Super Sore." *Dallas Morning News*, February 14, 1988.

"Hats Off to Waxahachie." *FermiNews*, November 18, 1988, 1–7.

Havens, Harry. "Gramm-Rudman-Hollings: Origins and Implementation." *Public Budgeting and Finance* (Autumn 1986), 4–24.

Hayner, Don. "Fermilab Collider Boosted." *Chicago Sun-Times*, February 16, 1988, 5.

Heilbron, J. L., and Robert W. Seidel. *Lawrence and His Laboratory: A History of the Lawrence Berkeley Laboratory*. Berkeley, CA: University of California Press, 1989.

Heitowit, Ezra. Letter to David J. Norton. September 8, 1988.

———. "Minutes of the Joint Informational Meeting of the URA Board of Trustees and the SSC Board of Trustees (*sic*)." December 19, 1989. Panofsky Collection, SLAC Archives.

Herrington, John S. "Superconducting Super Collider (SSC) Project." Memorandum to the Domestic Policy Council. December 15, 1986. William R. Graham file A04261-1 (CFoA), Ronald Reagan Library.

Herman, R. "High-energy Physics Splits the Scientists." *New Scientist*, June 20, 1985, 10.

Hermann, Armin, John Krige, Ulrike Mersits, and Dominique Pestre. *History of CERN*. Amsterdam: North Holland Publishing Co., 1987.

Higgs, Peter W. "Broken Symmetries and the Mass of Gauge Vector Bosons." *Physical Review Letters* 13 (1964): 508–09.

———. "Broken Symmetries, Massless Particles and Gauge Fields." *Physics Letters* 12 (1964): 132–33.

———. "Spontaneous Breaking of Symmetry and Gauge Theories." In *The Rise of the Standard Model: Particle Physics in the 1960s and 1970s*. Edited by Lillian Hoddeson, Laurie Brown, Michael Riordan, and Max Dresden, 506–10. New York: Cambridge University Press, 1997.

———. "Spontaneous Symmetry Breakdown without Massless Bosons." *Physical Review* 145 (1966): 1156–63.

Hirano, Yukihiro. "Public and Private Support of Basic Research in Japan." *Science*, N.S. 258:5082 (1992), 582–83.

Hobsbawm, Eric. *The Age of Extremes: A History of the World, 1914–1991*. New York: Vintage Books, 1994.

Hoddeson, Lillian. "The First Large-Scale Application of Superconductivity: The Fermilab Energy Doubler, 1972–1983." *Historical Studies in the Physical Sciences* 18:1 (1987), 25–54.

Hoddeson, Lillian, and Adrienne W. Kolb. "The Superconducting Super Collider's Frontier Outpost, 1983–1988." *Minerva* 38:3 (2000), 271–310.

Hoddeson, Lillian, Adrienne W. Kolb, and Catherine Westfall. *Fermilab: Physics, the Frontier, and Megascience*. Chicago: University of Chicago Press, 2008.

Hoddeson, Lillian, Laurie Brown, Michael Riordan, and Max Dresden, eds. *The Rise of the Standard Model: Particle Physics in the 1960s and 1970s*. New York: Cambridge University Press, 1997.

Hoddeson, Lillian, Paul W. Henriksen, Roger A. Meade, and Catherine Westfall. *Critical Assembly: A Technical History of Los Alamos during the Oppenheimer Years, 1943–1945*. New York: Cambridge University Press, 1993.

Hodel, Donald. Letter to Alvin Trivelpiece. August 16, 1984. SSC Collection, Fermilab Archives.

Holden, Constance. "Former South Carolina Governor to Head DOE." *Science*, N.S. 211:4482 (1981), 555.

House Committee on Science, Space and Technology. US House of Representatives. *Status of the Superconducting Super Collider Program: Hearing Before the Subcommittee on Investigations and Oversight*, May 9, 1991. Washington, DC: US Government Printing Office, 1991.

"House Vote Revives Fermilab Injector." *Science* (June 7, 1991), 1373.

Hudgens, Gayle. "Muons and Megabucks: Super Collider—Super Bust?" *The Nation*, March 19, 1990. 365ff.

Hughes, Thomas P. *Rescuing Prometheus: The Story of the Mammoth Projects—SAGE, ICBM, ARPANET/Internet, and Boston's Central Artery/Tunnel—That Created New Styles of Management, New Forms of Organization, and a New Vision of Technology*. New York: Pantheon Books, 1998.

Huson, F. Russell, and Peter M. McIntyre. "Widening the Field of SSC Magnet Competitors." Letter to the editor. *Physics Today* 41:8 (1988), 14.

Ianniello, Louis C. "Trip to Japan and Korea." Memorandum. June 7, 1990. SSC files of John Metzler; copy in Riordan files.

"IASSC Member Companies." February, 1991. Irwin Goodwin Collection, AIP Niels Bohr Library & Archives.

Idelson, Holly. "Conferees Save Supercollider But Can't Resolve Test Ban." *Congressional Quarterly*, September 19, 1992, 2806–08.

———. "House Denies Atom Smasher Its 1993 Expense Account." *Congressional Quarterly*, June 20, 1992, 1773–85.

———. "Less Energy Spending Spreads the Pain Around." *Congressional Quarterly*, June 13, 1992, 1692–93.

———. "Nuclear Weapons Complex Braces for Overhaul." *Congressional Quarterly*, April 25, 1992, 1066–73.

Illinois Congressional Delegation. Letter to James D. Watkins. October 17, 1991. SSC files of Sherwood Boehlert and David Goldston.

Independent Cost Estimating Staff. *Independent Cost Estimate for the Superconducting Super Collider*. Washington, DC. September, 1990. Irwin Goodwin Collection, AIP Niels Bohr Library & Archives.

"International Collaboration and the SSC." Undated position paper attached to memorandum dated February 27, 1990. Panofsky Collection, SLAC Archives.

"It's Here!" *Waxahachie Daily Light*, November 10, 1988, 1.

Jackson, J. David. CDG Notebook B4. CDG Collection, Fermilab Archives.

———. CDG Notebook B6, 82–89, 104–108. CDG Collection, Fermilab Archives.

———. CDG Notebook B9, 92–93. CDG Collection, Fermilab Archives.

Jackson, J. David, Maury Tigner, and Stanley Wojcicki. "The Superconducting Supercollider." *Scientific American* 254:3 (1986), 66–77.

Jackson, Judy. "Down to the Wire." *SLAC Beam Line* (Spring 1993), 14–21.

Jackson, Mike. "Remains of Collider Project up for Sale." *Dallas Morning News*, October 5, 1997, A9.

———. "State Reaches Deals on 2 Super Collider Buildings." *Dallas Morning News*, February 12, 1998, A35.

Jacobson, Sherry. "Superconductor Staff Reunites 10 Years Later." *Dallas Morning News*, July 23, 2005; DallasNews.com.

Jacoby, Mary. "Superconductor Already Colliding with Some Texans." *Washington Post*, July 6, 1990, A8.

"The Japanese asked a number of questions concerning the nature of the 'partnership' being proposed by the United States." Unclassified Department of Energy cable. June 5, 1990. SSC files of John Metzler; copy in Riordan files.

Jaroff, Leon. "Crisis in the Labs." *Time*, August 26, 1991, 45–51.

Jaynes, Gregory. "Super Deal." *Life* 12:2 (February, 1989), 100–109.

Johnsen, Kjell. "The CERN Intersecting Storage Rings: The Leap into the Hadron Collider Era." In *The Rise of the Standard Model: Particle Physics in the 1960s and 1970s*. Edited by Lillian Hoddeson, Laurie Brown, Michael Riordan, and Max Dresden, 285–98. New York: Cambridge University Press, 1997.

Johnston, J. Bennett. *Congressional Record*, August 3, 1992, S-11150.

———. Opening remarks, *Joint Hearing Before the Committee on Energy and Natural Resources and the Subcommittee on Energy and Water Development on Appropriations, United States Senate*. August 4, 1993. Senate Hearing 103–185, US Government Printing Office, 3.

Jones, Lawrence W. Letter to Adrienne Kolb. September 16, 2009. Fermilab Archives.

Jones, Richard. "Looking Backward: Why the SSC Was Terminated." FYI No. 142, October 27, 1993. Washington, DC: American Institute of Physics Public Information Division, 1993.

Karpenko, Victor. Memorandum to Maury Tigner. December 7, 1987. CDG Collection, Fermilab Archives.

Kelley, Chris. "Officials Settle SSC Claims with U.S. Government." *Dallas Morning News*, August 27, 1996. A18.

Kenworthy, Tom, and Curt Suplee. "House Panels Vote To Kill Space Station, Trim Super Collider." *Washington Post*, May 16, 1991.

Kevles, Daniel J. "Big Science and Big Politics in the United States: Reflections on the Death of the SSC and the Life of the Human Genome Project." *Historical Studies in the Physical and Biological Sciences* 27:2 (1997), 269–97.

———. "The Death of the Superconducting Super Collider in the Life of American Physics." Preface to second edition of *The Physicists: The History of a Scientific Community in Modern America*. Cambridge, MA: Harvard University Press, 1995.

———. "Good-bye to the SSC: On the Life and Death of the Superconducting Super Collider." *Caltech Engineering & Science* 58:2 (Winter 1995): 16–25.

Keyworth, George A. "Japan 1984." Memorandum in George A. Keyworth file OA94720. Ronald Reagan Library.

———. Letter to Dr. John M. Deutch. June 11, 1984. George A. Keyworth file OA94676-1 (Box 1), Ronald Reagan Library.

———. "The Role of Science in a New Era of Competition." *Science*, N.S. 217:4560 (1982), 606–09.

"Keyworth Decries Scientists' Negative Reaction to ABM." *Physics Today* 36:6 (1983), 45.

Kilday, Anne Marie, and Beth Silver. "House Panel Gives a $620 Million Vote of Approval." *Dallas Morning News*, June 18, 1993, A4.

King, Seth S. "Joblessness Remains at 9.5%; Rate for Adult Men Increases." *New York Times*, July 3, 1982, 1.

Kissinger, Henry. *On China*. New York: Penguin Press, 2011.

Knapp, Edward A. Letter to James R. Bieschke. December 29, 1988. Douglas Pewitt files, SSC Collection, Fermilab Archives.

"Knapp Succeeds Stever in URA Presidency." *Ferminews*, August 8, 1985, 1–3.

Knott, Stephen, Russell Riley, and James S. Young. "Interview with James Miller." Miller Center of Public Affairs, University of Virginia, November 4, 2001.

Kolb, Adrienne W. "A Chronology: VBA (ICFA) → SSC (US-DOE)." Fermilab Report No. FN-415. January 1985. Revised Fermilab Report No. FN-415. April 9, 1989.

Kolb, Adrienne W., and Lillian Hoddeson. "The Mirage of the 'World Accelerator for World Peace' and the Origins of the SSC, 1953–1983." *Historical Studies in the Physical Sciences* 24:1 (1993), 101–24.

Kondo, Jiro. Private conversation with Riordan. October 31, 2000. Riordan notes.

"Korean Participation in S.S.C. Project (To be discussed on June 5, 1990)." List. SSC files of John Metzler; copy in Riordan files.

Koszczuk, Jackie. "Collider's Backers Start Spinoff Talk." *Fort Worth Star Telegram*, December 10, 1991, A18.

Kotulak, Ronald. "Fermi's the Favorite in Race for Mammoth Atom Smasher." *Chicago Tribune*, October 23, 1983, E1.

———. "Frontier of Physics." *Chicago Tribune*, December 2, 1984, E1.

Kramer, David. "Budget Ax Hangs over Major Projects." *Inside Energy*, September 30, 1991, 1.

———. "New Bid to Cut Fermilab Draws Protest from Illinois Delegation." *Inside Energy*, October 28, 1991, 9.

———. "OMB Said to Challenge SSC Cost Projections." *Inside Energy*, October 8, 1990, 1.

———. "Parsons Brinkerhoff-Morrison Knudsen Win $1 Billion SSC Contract." *Inside Energy*, February 26, 1990, 3–4.

———. "SLAC's Richter, in SSC Postmortem, Urges Shorter Construction Times." *Inside Energy*, March 7, 1994, 7–8.

———. "Study Puts SSC Cost at Nearly $12 Billion." *Inside Energy*, August 6, 1990, 1.

———. "Watkins Sees More Global R&D Partners." *Inside Energy*, September 23, 1991, 1–2.

———. "Watkins to Ask for More Funds for Research." *Inside Energy*, December 2, 1991, 1–2.

———. "Watkins to Weigh Changes in Cost of SSC." *Inside Energy*, December 18, 1989, 1.

Krauss, Clifford. "Knocked Out by the Freshmen." *New York Times*, October 26, 1993, C3.

Krieger, Joel. *Reagan, Thatcher, and the Politics of Decline*. Cambridge: Polity Press, 1986.

Krige, John. *American Hegemony and the Postwar Reconstruction of Science in Europe*. Cambridge, MA: MIT Press, 2006.

———. "CERN from the Mid-1960s to the Late 1970s." In *History of CERN, III*. Edited by John Krige, 3–38. Amsterdam: Elsevier/North Holland Publishing Co., 1996.

———. "Distrust and Discovery: The Case of the Heavy Bosons at CERN." *Isis*, 92:3 (2001), 517–40.

———."The ppbar Project. I. The Collider." In *History of CERN, IIII*. Edited by John Krige, 207–50. Amsterdam: Elsevier/North Holland Publishing Co., 1996.

———. "La science et la sécurité civile de l'Occident." In *Les sciences pour la guerre, 1940–1960*. Edited by A. Dahan and Dominique Pestre, 369–97. Paris: EHESS, 2004.

Krige, John, and Dominique Pestre. "The How and the Why of the Birth of CERN." In *History of CERN, I*. Edited by Armin Hermann, John Krige, Ulrike Mersits, and Dominique Pestre, 523–44. Amsterdam: North Holland Publishing Co., 1987.

Krumhansl, James M. Letter to John S. Herrington. February 19, 1987. Irwin Goodwin Collection, AIP Niels Bohr Library & Archives.

LaFraniere, Sharon. "Energy Dept. Official Urges Firing Super Collider Chief." *Washington Post*, August 2, 1993, A3.

———. "Super Collider Quest Mired in Murky Cost Equations." *Washington Post*, September 6, 1993, A1.

Larbalestier, David. Presentation at American Physical Society meeting, Anaheim, California, May 1, 2011. As described by George Zimmerman in "History of Superconductivity Sessions." *History of Physics Newsletter* 11:4 (Spring 2011): 4–7.

Leary, Warren E. "Supercollider Outlook: Higher Cost, Less Income." *New York Times*, May 10, 1991, A20.

Lederman, Leon M. "Fermilab and the Future of HEP." In *Proceedings of the 1982 DPF Summer Study on Elementary Particle Physics and Future Facilities*. Snowmass, CO. June 28–July 16, 1982. Edited by Rene Donaldson, Richard Gustafson, and Frank Paige, 125–27. Batavia, IL: Fermi National Accelerator Laboratory, 1983.

———. Letter to Edward Knapp. October 1, 1985. Lederman Collection, Fermilab Archives.

———. Letter to the editor, *New York Times*. Unpublished. March 10, 1986. Lederman Collection, Fermilab Archives.

———. "New Orleans—A Proposal." Unpublished. ICFA files, 1975. Lederman Collection, Fermilab Archives.

———. "An Open Letter to Colleagues Who Publicly Opposed the SSC." *Physics Today* 47:3 (1994), 9–11.

———. "Scientific Retreat: Demise of the SSC Is a National Loss." *Chicago Tribune*, November 18, 1993, 31.

———. "Scientific Retreat: Demise of the SSC Is a National Loss." *The Scientist*, November 29, 1993, 12.

———. "Status and Outlook for International Collaboration on Future Accelerators." In *Proceedings of the 12th IEEE Particle Accelerator Conference*. Edited by E. R. Lindstrom and L. Taylor. Piscataway, NJ: IEEE, 1988.

———. "Submicroscopic Nature Needs Megascience." *History of Physics Newsletter* 9:5 (Fall 2005), 15.

———. "VBA." *IEEE Transactions on Nuclear Science* NS-24:3 (1977): 1903–08.

Lederman, Leon M., and Chris Quigg. *Appraising the Ring: Statements in Support of the Superconducting Super Collider*. Washington: Universities Research Association, 1988.

Lederman, Leon M., with Dick Teresi. *The God Particle: If the Universe Is the Answer, What Is the Question?* New York: Houghton Mifflin, 1993.

Lee, Benjamin W., C. Quigg, and H. B. Thacker. "Strength of Weak Interactions at Very High Energies and the Higgs Boson Mass." *Physical Review Letters* 38 (1977): 883–85.

Lee, Gary. "Clinton Offers Package to 'Halt Global Warming.'" *Washington Post*, October 20, 1993, A4.

Lehman, Daniel. "Lessons from the SSC: An ER-65 Critique." PowerPoint presentation. February, 1994. Fermilab Archives.

Leiss, James. Letter to H. Guyford Stever. March 1, 1984. URA file, Lederman Collection, Fermilab Archives.

Lemonick, Michael D. "The $2 Billion Hole." *Time*, November 1, 1993, 69.

———. "The Ultimate Quest." *Time*, April 16, 1990, 50–56.

Lepkowski, Wil. "Recent Discoveries Stir Debate over Superconducting Super Collider." *Chemical and Engineering News* 65:19 (1987), 14.

Levi, Barbara G. "A Look at the Future of Particle Physics." *Physics Today* 36:1 (1983), 19–21.

Levi, Barbara Goss, and Bertram Schwarzschild. "Super Collider Magnet Program Pushes Toward Prototype." *Physics Today* 41:4 (1988), 17–21.

"LHC Cost Review to Completion." October 16, 2001. http://user.web.cern.ch/user/LHCCost/2001-10-16/LHCCostReviewToCompletion (downloaded October 19, 2009).

"LHC: The Guide." Brochure published by CERN, 2008.

Lineberry, Danny. "No-funds Collider May Enrich Plan." *Durham Morning Herald*, April 6, 1988.

Lippman, Thomas W. "Energy's 'Mountain Building Up.'" *Washington Post*, February 12, 1991.

———. "Super Collider's Cost, Technology Criticized." *Washington Post*, May 10, 1991.

"Lisbon Conference." *CERN Courier* (September, 1981), 283–88.

Llewellyn-Smith, Chris. "How the LHC Came to Be." *Nature* 448:7151 (2007), 281–84.

Lloyd, Therese. "SSC Faces Uncertain Future." *The Scientist*, February 23, 1987, 1.

Lock, W. O. "Origins and Early Years of ICFA." Unpublished draft manuscript. December, 1982. ICFA files, Lederman Collection, Fermilab Archives.

Logsdon, John. *Together in Orbit: The Origins of International Participation in the Space Station*. NASA Monographs in Aerospace History no. 11. Washington, DC: National Aeronautics and Space Administration, 1988.

Loveless, Bill, and David Kramer. "Moore Finds Korea Bullish over SSC; Japan's Reaction is Restrained." *Inside Energy*, July 11, 1990.

Low, F. E. Letter to James Decker. *Report of the 1990 HEPAP Subpanel on SSC Physics*, Appendix B. DOE Report No. DOE/ER-434. Washington, DC. January 12, 1990.

Lowi, Theodore J., Benjamin Ginsberg, and E. J. Feldman. *Poliscide: Big Government, Big Science, Lilliputian Politics*. New York: Macmillan, 1976.

Lubkin, Gloria B. "Accelerator Superconducting Magnets Give Headaches." *Physics Today* 34:4 (1981), 17–20.

———. "DOE Boosts Particle-Physics Funds." *Physics Today* 35:4 (1982), 20–21.

———. "Panel Says Go for a Multi-TeV Collider and Stop Isabelle." *Physics Today* 36:9 (1983), 17–20.

———. "R&D Funding for the Super Collider." *Physics Today* 37:10 (1984), 21.

———. "SSC Design Goes to DOE; ICFA Discusses CERN Hadron Collider." *Physics Today* 37:6 (1984), 17–19.

———. "Tristan e^+e^- Collider in Japan Yields 50 GeV Center of Mass." *Physics Today* 40:1 (1987), 21–23.

———. "UA1 at CERN Says It Has Candidates for Sixth Quark, Top." *Physics Today* 37:8 (1984), 17–18.

Luce, Thomas W. Letter to J. Bennett Johnston. May 30, 1989. SSC files of Riordan.

Lynch, H. L. "Detector R&D." In *Superconducting Super Collider: A Retrospective Summary, 1989–1993*. Report No. SSCL-SR-1235 (1994), 227–29.

MacLean, John N. "Early Decision Urged on Super Collider." *Chicago Tribune*, April 10, 1987, 14.

———. "Illinois Gains in Bid for Supercollider." *Chicago Tribune*, December 30, 1987, 1.

———. "New York Withdraws Collider Bid." *Chicago Tribune*, January 15, 1988, 1.

Magner, Mike. "Collider Finalists Trying to Boost Support for Project." *Kalamazoo Gazette*, March 24, 1988.

Magnet Aperture Workshop. Report No. SSC-TR- 2001 (November, 1984).

"Management of the R&D and Conceptual Design Phase of the Superconducting Super Collider." Undated. URA file, Box J3a3, Lederman Collection, Fermilab Archives.

Mantsch, Paul. "Spirit of Snowmass Spreads Across Land." *Ferminews*, June 14, 1984, 2–3.

"Many Show Interest in SSC Contract, But Few Are Expected to Bid." *Inside Energy*, September 12, 1988.

Marbach, William D. "When Protons—and Politics—Collide." *Newsweek*, July 6, 1987, 44.

Marburger, John H., III. *Science Policy Up Close*. Edited by Robert P. Crease. Cambridge, MA: Harvard University Press, 2015.

———. "The Superconducting Super Collider and US Science Policy." *Physics in Perspective* 16 (2014), 218–49.

Mark, Hans. "Site Selection Criteria for Site Selection Process." *Proc. Philosophical Society of Texas* 56:37 (1993).

———. *The Space Station: A Personal Journal*. Durham, NC: Duke University Press, 1987.

Marshall, Eliot. "Big Versus Little Science in the Federal Budget." *Science*, N.S. 236:4799 (1987), 249.

———. "'Black Book' Threatens Synfuels Projects." *Science*, N.S. 211:4485 (1981), 903 04, 906.

———. "Clinton Backs SSC, Space Station." *Science*, N.S. 260:5116 (1993), 1873.

———. "An Early Test of Reagan's Economics." *Science*, N.S. 211:4477 (1981), 29–31.

———. "Space Station Science: Up in the Air." *Science*, N.S. 246:4934 (1989), 1110–12.

Marshall, Eliot, and Christopher Anderson. "Clinton's Mixed Broth for R&D." *Science*, N.S. 260:5106 (1993), 284–85.

Martin, Joseph D. "Fundamental Disputations: The Philosophical Debates that Governed American Physics, 1939–1993." *Historical Studies in the Natural Sciences*, vol. 45:5 (2015).

McAuley, James. "The City with a Death Wish in Its Eye." *New York Times*, November 16, 2013. http://www.nytimes.com/2013/11/17/opinion/sunday/dallas-role-in-kennedys-murder.html.

McDaniel, Boyce. Memo to Cornell Laboratory of Nuclear Studies personnel. June 22, 1984. SSC Collection, CDG 1984–1988 file, Fermilab Archives.

———. "SSC Project Cost Estimates." Attached to letter to John Toll. December 11, 1989. Panofsky Collection, SLAC Archives.

———. "Trip Report, Visit to Dallas 6/21–22." July 26, 1989. Panofsky files, SLAC Archives.

McDonald, Kim A. "Reagan Backs Giant $4.4 Billion Particle Accelerator; Scientists Face Major Hurdles in Promoting the Device." *Chronicle of Higher Education*, February 11, 1987, 7–9.

———. "States Spend Millions in Stiff Competition to Provide Site for Proposed Supercollider." *Chronicle of Higher Education*, April 8, 1987, 4–6.

———. "Supercollider Scientists Are Embroiled in Dispute; Outcome Could Raise Project's $8-Billion Cost." *Chronicle of Higher Education*, April 10, 1991, A19.

———. "3 Projects Pushed by White House, Reversing Energy Dept.'s Position." *Chronicle of Higher Education*, May 5, 1993, A31–34.

———. "Who Killed the SSC?" *Chronicle of Higher Education*, March 2, 1994, A10.

McIlwain, Colin. "Barrage of Mud Fails to Stick to Super Collider." *Nature* 364:6433 (1993), 92.

———. "SSC Decision Ends Post-war Era of Science-Government Partnership." *Nature* 365:6449 (1993), 773–74.

———. "SSC Falls Victim to Congressional Austerity." *Nature* 364:632 (1993), 6.

———. "Stanford Accelerator Takes Lead in Race to Quantify CP Violation." *Nature* 403:6770 (2000), 586–87.

Meese, Edwin. "Memorandum for the Domestic Policy Council on the Superconducting Super Collider (SSC)." January 30, 1987. Domestic Policy Council file 459161 (series FG010-03), folder #2, Ronald Reagan Library.

Memorandum from Laboratory Directors to Hermann Grunder et al. December 14, 1983. Lederman Collection, Fermilab Archives.

Mervis, Jeffrey. "Bromley Memoirs Take Off the Gloves." *Science*, N.S. 265:5177 (1994), 1357.

———. "The Endgame." *Science*, N.S. 262:5134 (1993), 646.

———. "Scientists Are Long Gone, But Bitter Memories Remain." *Science* N.S. 302:5642 (2003), 40–41.

———. "A Strategic Message from Mikulski." *Science*, N.S. 263:5147 (1994), 604.

———. "Supercollider Suffering Birth Pangs." *The Scientist*, October 2, 1989, 1.

Mervis, Jeffrey, and Charles Seife. "Lots of Reasons, But Few Lessons." *Science* 302, October 3, 2003, 38–40.

Metzler, John. Draft of letter from President George H. W. Bush to Toshiki Kaifu. May 25, 1990. SSC files of John Metzler; in Riordan files.

Michel, Robert, and J. Dennis Hastert. Letter to James D. Watkins. October 22, 1991. SSC files of Riordan.

Mikulski, Barbara A. "Science in the National Interest." *Science*, N.S. 264:5156 (1994), 221–22.

Mills, Mike. "Collider 'Problems' Lead O'Leary to Replace Its Main Contractor." *Congressional Quarterly*, August 7, 1993, 2156.

———. "Negotiators Preserve Collider, Setting Up House Showdown." *Congressional Quarterly*, October 16, 1993, 2800–01.

———. "Oversight Questions Nag Efforts to Save Super Collider Again." *Congressional Quarterly*, July 31, 1993, 2031–34.

———. "Super Collider, Advanced Reactor Withstand Assaults in Senate." *Congressional Quarterly*, October 2, 1993, 2628–30.

Meeting Minutes, City Council, Waxahachie, TX. February 6, 1989.

Meeting Minutes, Fermilab Board of Overseers. November 10, 1993. URA Collection, Fermilab Archives.

Minutes of the Meeting of the SSC Board of Overseers. Chicago, Illinois. August 19–20, 1988. Panofsky Collection, SLAC Archives.

Minutes of the Meeting of the SSC Board of Overseers. October 4–5, 1989. Panofsky Collection, SLAC Archives.

Minutes of the Sixth ICFA Meeting. Unpublished draft. Serpukhov, Russia. November 5, 1981. Lederman Collection, Fermilab Archives.

Molella, Arthur. *Report on Places of Invention: The First Lemelson Institute*. Incline Village, Nevada. August 16–18. Washington, DC: Lemelson Center for the Study of Invention and Innovation.

Morgenthaler, George W., and Uriel Nauenberg, eds. "SSC Status Report to the Nation: a National Symposium on the Superconducting Super Collider." University of Colorado, October 24, 1988.

Morris, Charles R. *The Trillion-Dollar Meltdown: Easy Money, High Rollers, and the Great Credit Crash*. New York: Public Affairs, 2008.

Munson, Richard. *The Cardinals of Capitol Hill: The Men and Women Who Control Government Spending*. New York: Grove Press, 1993.

Myers, Christopher. "Congressional Actions Suggest Lawmakers Are Skeptical About Continued Support for Big-Science Projects." *Chronicle of Higher Education*, May 29, 1991. A15ff.

Myers, Frederick Shaw. "SSC: The Japan That Can Say No." *Science*, N.S. 254:5038 (1991), 1579.

Myers, Steven. "The Contribution of John Adams to the Development of LEP." John Adams Memorial Lecture. November 26, 1990. http://sl-div.web.cern.ch/sl-div/history.lep-doc.html.

Myers, S., and W. Schnell. "Preliminary Performance Estimates for a LEP Proton Collider." CERN/LEP Note 440. April 11, 1983.

Nagorka, Jennifer. "More than 100 Scout Out SSC Equipment; Agency Awaits Ideas on Use of Project Assets." *Dallas Morning News*, March 23, 1994, A29.

———. "SSC Site May Be Reborn as Medical Research Center." *Dallas Morning News*, November 27, 1994, A33.

———. "State Approves Super Collider Settlement; Texas Would Get $210 Million in Cash, $510 Million Worth of Land, Equipment." *Dallas Morning News*, August 2, 1994, A17.

———. "Super Collider Dying a Quiet Death." *Dallas Morning News*, January 6, 1994, A1.

———. "Technical Experts Discuss Uses for Collider Assets." *Dallas Morning News*, January 6, 1994, A25.

———. "Texas Comes out Ahead in SSC Deals, Officials Say; Congress, Clinton Still Must Approve Settlement." *Dallas Morning News*, July 23, 1994, A1.

———. "3 Options Chosen for SSC Assets; State Panel Focusing on Technological Uses." *Dallas Morning News*, February 22, 1994, A21.

Nambu, Y., and G. Jona-Lasinio. "Dynamical Model of Elementary Particles Based on an Analogy with Superconductivity." *Physical Review* 122 (1961): 345–58.

"News Conference of President Bush and Prime Minister Toshiki Kaifu of Japan." Kennebunkport, Maine, July 11, 1991. The American Presidency Project, UC Santa Barbara. http://www.presidency.ucsb.edu/ws/?pid=19775.

Norman, Colin. "Commerce to Inherit Energy Research." *Science*, N.S. 215:4529 (1982), 147–48.

———. "The Making of a Science Advisor." *Science*, N.S. 218:4573 (1982), 658–60.

———. "Reagan Administration Prepares Budget Cuts." *Science*, N.S. 211:4485 (1981), 901, 903.

———. "Science Advisor Post Has Nominee in View." *Science*, N.S. 212:4497 (1981), 903–04.

———. "Science Budget: Growth Amid Red Ink." *Science*, N.S. 251:4994 (1991), 616–18.

———. "Science Increases Will Test New Regime." *Science*, N.S. 251:4994 (1991), 617.

Norman, Colin, John Walsh, Marjorie Sun, Eliot Marshall, M. Mitchell Waldrop, and Constance Holden. "Science Budget: Coping with Austerity." *Science*, N.S. 215:4535 (1982), 944–45.

Norman, Colin, Marjorie Sun, John Walsh, R. Jeffery Smith, Eliot Marshall, and M. Mitchell Waldrop. "Reagan's Budget Boosts Basic Research." *Science*, N.S. 219:4585 (1983): 747–51.

Northern Illinois University Center for Governmental Studies Public Opinion Laboratory. "An Overview of Citizen Reactions to the Proposed Superconducting Super Collider." April 1987. Box N4b7, Lederman Collection, Fermilab Archives.

Novell, James R. "The Railroad Commission of Texas: Its Origin and History." *Southwestern Historical Quarterly* 68 (April 1965): 465–80.

Office of Management and Budget. "A Vision of Change for America." February 17, 1993.

O'Leary, Hazel R. Letter to John H. Dingell. September 1, 1993. SSC files of Riordan.

———. Letter to John H. Marburger III. June 24, 1993. URA Collection, Box J3a6, Fermilab Archives.

———. Letter to Stanley Wojcicki. November 4, 1993. Report No. DOE/ER-0614P, Appendix B.

———. Memorandum for the President, "Status of the Superconducting Super Collider." October 21, 1993. FOIA Request 2010-450-F. Clinton Library Digital Archives.

———. Memorandum for the President, "Successful Closeout of the Superconducting Super Collider." November 5, 1993. FOIA Request 2010-450-F. Clinton Library Digital Archives.

———. Memorandum to Mack McLarty, "Weekly Report." August 5, 1993, Clinton Library Digital Archives.

———. Memorandum to Mack McLarty, "Weekly Report." November 5, 1993. Clinton Library Digital Archives.

———. Memorandum to Mack McLarty, "Weekly Report." November 18, 1993. Clinton Library Digital Archives.

———. Memorandum to Mack McLarty, "Weekly Report." November 12, 1993. Clinton Library Digital Archives.

———. Testimony in *Joint Hearing Before the Committee on Energy and Natural Resources and the Subcommittee on Energy and Water Development on Appropriations, United States Senate*. August 4, 1993. Senate Hearing 103-185, US Government Printing Office, 90.

O'Leary, Hazel R., and George E. Brown Jr. "Resuming the Pursuit of Knowledge." *Los Angeles Times*, November 21, 1993.

O'Malley, Kathy, and Hanke Gratteau. "Inc." *Chicago Tribune*, May 23, 1988, 24.

Overbye, Dennis. "Chasing the Higgs Boson." *New York Times*, March 5, 2013, D1ff.

———. "Collider Sets Record, and Europe Takes U.S.'s Lead." *New York Times*, December 10, 2009, D1.

———. "European Collider Begins Its Subatomic Exploration." *New York Times*, March 30, 2010, D1.

Panetta, Leon. Phone conversation with Michael Riordan. June 25, 2002. Riordan notes.

Panetta, Leon, with Jim Newton. *Worthy Fights: A Memoir of Leadership in War and Peace*. New York: Penguin Press, 2014.

Panofsky, Wolfgang K. H. "Changes in Top Management Structure of the SSC." Memorandum. October 8, 1990. Panofsky Collection, SLAC Archives.

———. "Directorial Nominee for URA Proposal." Memorandum to Board of Overseers. September 6, 1988. Panofsky Collection, SLAC Archives.

———. "Discussions with Knapp and McDaniel." Memo to Board of Overseers. August 1, 1988. Panofsky Collection, SLAC Archives.

———. "Information from Lloyd Sides." Letter to Board of Overseers. August 26, 1988. Panofsky Collection, SLAC Archives.

———. "International Collaboration for the SSC." Memorandum to the URA Board of Overseers. February 13, 1985. Panofsky Collection, SLAC Archives.

———. Letter to William Wallenmeyer. January 22, 1987. Panofsky Collection, SLAC Archives.

———. Memo, Board of Overseers Files. December 20, 1988. Panofsky Collection, SLAC Archives.

———. "Meeting of the SSC Board of Overseers Executive Committee." November 21, 1989. Panofsky Collection, SLAC Archives.

———. "A Personal View on SSC Project Cost Growth and Its Control." December 13, 1989. Panofsky Collection, SLAC Archives.

———. "Remarks on International Participation in SSC." Memorandum. October 24, 1988. Panofsky Collection, SLAC Archives.

———. "Search for Directorial Nominee by URA." Memorandum to Board of Overseers. August 26, 1988. Panofsky Collection, SLAC Archives.

———. "The SSC's End: What Happened? And What Now?" Letter to the editor. *Physics Today* 47:3 (1994), 13–15, 88–92.

———. "Talking Paper: Discussions with Trilling and McDaniel." August 8, 1988. Panofsky Collection, SLAC Archives.

Park, Robert L. "What's New." APS Office of Public Affairs Newsletter, April 19, 1991.

Pells, D. L. "Meeting Minutes: PMS Workshop Group Meeting of April 11, 1989." Pewitt files, SSC Collection, Fermilab Archives.

Peoples, John. "Introduction." In *Proceedings of the 1984 Summer Study on the Design and Utiliza-*

tion of the Superconducting Super Collider. Snowmass, CO. June 23–July 13, 1984. Edited by R. Donaldson and J. G. Morfin.
———. "Snowmass '84," APS/DPF Newsletter, July 30, 1984.
———. "Status of the SSC Superconducting Magnet Program." Report No. SSC-185, submitted to Applied Superconductivity Conference, San Francisco, CA. September 21–26, 1988. SSC Collection, Fermilab Archives.
Perkins, Donald. 1997. "Gargamelle and the Discovery of Weak Neutral Currents." In *The Rise of the Standard Model: Particle Physics in the 1960s and 1970s*. Edited by Lillian Hoddeson, Laurie Brown, Michael Riordan, and Max Dresden, 428–46. New York: Cambridge University Press, 1997.
Perricone, Mike. "A Banner Day for the Main Injector." *Fermi News* June 18, 1999, 2–5.
Pestre, Dominique. "The Difficult Decision, Taken in the 1960s, to Construct a 3–400 GeV Proton-Synchrotron in Europe." In *History of CERN, III*. Edited by John Krige. Amsterdam: Elsevier/North Holland Publishing Co., 1996, 65–96.
———. "The Second Generation of Accelerators for CERN, 1956–1965: The Decision-Making Process." In *History of CERN, II*. Edited by A. Hermann, L. Weiss, D. Pestre, U. Mersits, and J. Krige. Amsterdam: Elsevier Publishing Co., 1990, 679–80.
Pestre, Dominique, and John Krige. "Some Thoughts on the Early History of CERN." In *Big Science: The Growth of Large-Scale Research*. Edited by Peter Galison and Bruce Hevly, 78–99. Stanford, CA: Stanford University Press, 1992.
———. "Some Thoughts on the Early History of CERN." In *History of European Scientific and Technological Cooperation*. Edited by John Krige and Luca Gazetti, 37–60. Luxembourg: Office for Official Publications of European Community, 1997.
Pewitt, N. Douglas. Letter to the editor. *Physics Today* 47:3 (1994), 88–89.
———. "Red Team Briefing Document." September 30, 1988. SSC Collection, Fermilab Archives.
———. "SSC O&M Proposal." Memorandum to Edward Knapp. August 24, 1988. SSC Collection, Fermilab Archives.
———. "Sverdrup Corporation Performance." Memorandum. October 11, 1989. SSC files of Riordan.
Pewitt, N. Douglas, and Roy Schwitters. "Rationale for Teaming." October 7, 1988. SSC Collection, Fermilab Archives.
Pianin, Eric. "House Deals a Big Defeat to Atom Smasher in Texas." *Washington Post*, October 20, 1993, A1.
Pianin, Eric, and Sharon LaFraniere. "Proponents of Collider Give Up." *Washington Post*, October 21, 1993, A1, A12.
Pianin, Eric, and Tom Kenworthy. "Super Collider Takes Last Breath." *Washington Post*, October 21, 1993, A11.
Pickering, Andrew. *Constructing Quarks: A Sociological History of Particle Physics*. Edinburgh: Edinburgh University Press, 1984; republished by University of Chicago Press, 1999.
Pitrelli, Nico. "No Reserved Communication Lanes for High Energy." Editorial, *Journal of Science Communication* 5:2 (June 2006).
"Politics Takes Toll on SSC." *Science*, N.S. 261:5121 (1993), 539.
Pool, Robert. "SSC Gets New Detector." *Nature* 351:6329 (1991), 681.
The President's News Conference with Prime Minister Kiichi Miyazawa of Japan in Tokyo. January 9, 1992. http://www.presidency.ucsb.edu/ws/?pid=20437.
Proceedings of the Fifth International Workshop on Next-Generation Linear Colliders. Stanford Linear Accelerator Center. October 13–21, 1993. Report No. SLAC-436.
Proceedings of the 1984 ICFA Seminar on Future Perspectives in High-Energy Physics. ICFA Report No. 84-14, KEK, May 14–20, 1984. Fermilab Archives.
"Proposal for a Dedicated Collider at the Fermi National Accelerator Laboratory." Batavia, IL: May 1983. Fermilab Archives.

"Proton-Proton Collider Upgrade (Main Injector, New Tevatron)." Fermilab Project No. 90-CH-400, Technical Components and Civil Construction. May, 1988. Fermilab Archives.

PSSC: Physics at the Superconducting Super Collider Summary Report. Batavia, IL: Fermilab, 1984. Bruce Winstein Collection, Fermilab Archives.

PSSC Records. Bruce Winstein Collection, Fermilab Archives.

"Quark Barrel Politics." *The Economist,* March 26, 1988, 24.

Quigg, Chris. "Elementary Particles and Forces." *Scientific American* 252:4 (1985), 84–95.

Quigg, Chris, and Roy Schwitters. "Elementary Particle Physics and the SSC." *Science,* N.S. 231:4745 (1986), 1522–27.

Racine, John. "Texas Commission to Price $250 Million of Lease Bonds for Super Collider Project." *Bond Buyer,* December 10, 1991, E114.

Ramo, Simon. *The Business of Science: Winning and Losing in the High-Tech Age.* New York: Hill & Wang, 1988.

———. *America's Technology Slip.* New York: John Wiley & Sons, 1980.

"Reagan Calls Super Collider 'Doorway to New World.'" *Los Angeles Times,* March 31, 1988, A4.

Reardon, Paul. Letter to Joseph R. Cipriano. October 15, 1991. Reardon files, SSC Collection, Fermilab Archives.

Regalado, Antonio. "With Quark Discovery, Truth Comes out on Top—Twice." *Science,* N.S. 267:5203 (1995), 1423.

Reifenberg, Anne. "SSC's Fall Linked to Its Complexity." *Dallas Morning News,* October 24, 1993, A1–A15.

Reference Designs Study for US Department of Energy: Superconducting Super Collider. Draft II (May 8, 1984). Fermilab Archives.

"Remarks by the President in Meeting with Supporters of the Superconducting Super Collider Program." Text of speech, White House press release. March 30, 1988. SSC Collection, Fermilab Archives.

"Report of the Ad Hoc Committee on SSC Physics." December 11, 1989. SSCL Report No. SSC-250.

Report of the DOE Review Committee on the Conceptual Design of the Superconducting Super Collider. Report No. DOE/ER-0267 (May 1986).

Report of the HEPAP Subpanel to Review Recent Information on Superferric Magnets. Report No. DOE/ER-0272 (May 1986).

"Report of the SSC Machine Advisory Committee, December 4–5, 1989." Draft. SSC files of Riordan.

"Report of the SSC Machine Advisory Committee, November 9–10, 1991." SSC files of Riordan.

Richter, Burton. "Japan and the SSC." Memorandum to John Gibbons. January 14, 1993. Richter Collection, SLAC Archives.

———. Memorandum to Ralph DeVries and Wallace Kornak. September 1, 1983. Lederman Collection, Fermilab Archives.

———. Personal notebook. Richter Collection, SLAC Archives.

Riordan, Michael. "The Demise of the Superconducting Super Collider." *Physics in Perspective* 2:4 (December, 2000), 411–25.

———. *The Hunting of the Quark: A True Story of Modern Physics.* New York: Simon & Schuster, 1987.

———. "No Monopoly on Innovation." *Harvard Business Review* 83:12 (2005), 18–20.

———. Notes of IASSC meeting, February 5, 1991. Transcribed and recorded February 10, 1991. Riordan notes.

———. "A Tale of Two Cultures: Building the Superconducting Super Collider, 1988–1993." *Historical Studies in the Physical and Biological Sciences* 32:1 (2001), 125–44.

———. "A Vision Unfulfilled: The Hopeful Birth and Painful Death of the Superconducting Super Collider." In *Challenges and Goals for Accelerators in the 21st Century.* Edited by Oliver Brüning and Steven Meyers. Singapore: World Scientific Publishing Co., 2015.

———. "Why Are We Building the SSC?" *SLAC Beam Line* (Summer 1991): 1–8.

Riordan, Michael, and David N. Schramm. *The Shadows of Creation: Dark Matter and the Structure of the Universe*. New York: W. H. Freeman, 1991.

Riordan, Michael, and Kasuke Takahashi. "Cooperation in High Energy Physics between the United States and Japan." *SLAC Beam Line* (Spring 1992): 1–9.

Riordan, Michael, and Lillian Hoddeson. *Crystal Fire: The Birth of the Information Age*. New York: W. W. Norton, 1997.

———. "Transistor's Father Knew How to Tie Basic Industrial Research to Development." *Research and Technology Management* 41:1 (1998), 9–11.

Riordan, Michael, Guido Tonelli, and Sau Lan Wu. "The Higgs at Last?" *Scientific American* (October, 2012), 66–73.

Riordan, Michael, P. C. Rowson, and Sau Lan Wu. "The Search for the Higgs Boson." *Science*, N.S. 291:5502 (2001), 259–60.

Ritson, David. "Demise of the Texas Supercollider." *Nature* 366:6456 (1993), 607–10.

Ritter, Don. "Quark Barrel Politics (Should We Spend $5 Billion for the Higgs Boson?)." *Policy Review* (Spring 1998): 70–72.

Robinson, Arthur L. "CERN Sets Intermediate Vector Boson Hunt." *Science*, N.S. 213:4504 (1981), 191–94.

———. "Physicists Give ISABELLE a 'Yes, But . . .'" *Science*, N.S. 214:4522 (1981), 769–70.

Robinson, Clay. "Super Collider Decision Is Big Hit in Texas." *Houston Chronicle*, November 11, 1988, A25.

Rodgers, Peter. "How the Super Collider Ground to a Halt." *Physics World*, November 1993, 6–7.

Rogers, David. "Deficit Law Turns the Budget Debate Into Process of Cutting and Spending." *Wall Street Journal*, February 7, 1991, A6.

———. "House Panels Vote to Kill 1992 Funds for Space Station, Trim Collider Request." *New York Times*, May 16, 1991.

"Rooting for Fermilab." Editorial. *Chicago Tribune*, October 22, 1983, 8.

Rosenbaum, David E. "Southwest to Get Economic Benefits in Savings Bailout." *New York Times*, June 25, 1990. D5.

Ross, Ian M. "R&D in the United States: Its Strengths and Challenges." *Science*, N.S. 217:4555 (1982), 130–31.

Ross (*sic*), G., and T. Kirk. "Report of the SSC Collider Dipole Review Panel." SSCL Report No. SSC-SR-1040. ("Ross" should have been spelled "Voss.")

Rossi, Lucio. "The Longest Journey: The LHC Dipoles Arrive on Time." *CERN Courier*, October 5, 2006.

Rossi, Lucio, and Ezio Todesco. "Superconducting Magnets." In *The Large Hadron Collider: A Marvel of Technology*. Edited by Lyndon Evans, 68–85. Lausanne: EPFL Press, 2009.

Rubbia, Carlo, "The Physics Frontier of Elementary Particles and Future Accelerators." *IEEE Transactions on Nuclear Sciences* NS-28 (1981): 3541–48.

Rubin, Alissa J. "Supercollider Surges Ahead Despite Concerns over Cost." *Congressional Quarterly*, (1991), 856–57.

Russo, Arturo. "The Intersecting Storage Rings: The Construction and Operation of CERN's Second Large Machine and a Survey of Its Experimental Program." In *History of CERN, III*. Edited by John Krige, 97–170. Amsterdam: Elsevier/North Holland Publishing Co., 1996.

Saegert, Laura K. *Records Appraisal Report: Texas National Research Laboratory Commission*. Texas State Library and Archives Commission, August 4, 1998. http://www.tsl.state.tx.us/arc/appraisal/tnrlc.html.

Salam, A. "Weak and Electromagnetic Interactions." In *Elementary Particle Theory*. Edited by N. Svartholm, 367–77. New York: John Wiley & Sons, 1968.

Samios, Nicholas P. "High Energy Physics at Brookhaven National Laboratory." In *Proceedings of the 1982 DPF Summer Study on Elementary Particle Physics and Future Facilities*. Snowmass, CO.

June 28–July 16, 1982. Edited by Rene Donaldson, Richard Gustafson, and Frank Paige, 140–45. Batavia, IL: Fermi National Accelerator Laboratory, 1983.

Sample, Ian. *Massive: The Missing Particle That Sparked the Greatest Hunt in Science*. New York: Basic Books, 2010.

Sanford, James R. "Isabelle, a Proton-Proton Colliding Beam Facility at Brookhaven." *IEEE Transactions on Nuclear Science* NS-24:3 (1977), 1845–48.

Sanford, James R., and D. M. Mathews, eds. *Site-Specific Conceptual Design of the Superconducting Super Collider*. Report No. SSCL-SR-1056. July, 1990.

Sanger, David E. "Japan Wary as U.S. Science Comes Begging." *New York Times*, October 27, 1991, E13.

Sarpalius, William. *Congressional Record—House*, May 29, 1991. H-3673.

Sasser, Jim. "The Budget Follies Have to Stop." *New York Times*, July 29, 1991.

Schlesinger, Jacob M. "U.S. Asks Tokyo to Fund Basic Science." *Wall Street Journal*, May 31, 1990.

Schmemann, Serge. "Clamor in the East: East Germany Opens Frontier to the West for Migration or Travel; Thousands Cross." *New York Times*, November 10, 1989.

Schopper, Herwig. "LEP and Future Options." In *Proceedings of the 12th International Conference on High-Energy Accelerators*. Edited by F. T. Cole and Rene Donaldson, 658–63. Batavia, IL: Fermi National Accelerator Laboratory, August 11–16, 1983.

———. *LEP—The Lord of the Collider Rings at CERN, 1980–2000: The Making, Operation and Legacy of the World's Largest Scientific Instrument*. Dordrecht, The Netherlands: Springer Verlag, 2009.

———. Lecture at American Physical Society meeting. Washington, DC. April 29, 2001.

Schultz, Todd. Memorandum to Republican Members of the ISC Subcommittee. March 16, 1988. SSC files of Boehlert/Goldston.

Schwarzschild, Bertram. "Panel Reaffirms High-Field Magnet Choice for Supercollider." *Physics Today* 39:7 (1986), 21–23.

———. "SSC Design Revisions Call for Thinner Beams and Fatter Magnets." *Physics Today* 43:1 (1990), 47–49.

———. "Tunnel Boring Begins at Superconducting Super Collider." *Physics Today* 46:3 (1993), 19.

Schweber, Silvan S. "Physics, Community and the Crisis in Physical Theory." *Physics Today* 46:11 (1993), 34–40.

Schwitters, Roy F. "Decision Memorandum—on Aspects of the Initial Scientific Program for the SSC." January 4, 1991. SSC files of Riordan.

———. Letter to John H. Marburger III. October 28, 1993. URA Files, 1993 Correspondence, Fermilab Archives.

———. "Project Management Changes." SSCL Memorandum. March 12, 1990. SSC files of Riordan.

———. "Report to the PAC." May 3, 1991. SSC files of Riordan.

———. "Super Collider." *American Politics* (July, 1986): 5–7.

———. Telegram to W. Panofsky. January 19, 1989. SSC files of Riordan.

"Secretary James Watkins, 1991 Trip to Japan, December 3–6." In "Briefing Book: Global Partnership for the Superconducting Super Collider and Energy Cooperation." November 29, 1991. SSC files of John Metzler; in Riordan files.

Seidel, Robert W. "Accelerating Science: The Postwar Transformation of the Lawrence Radiation Laboratory." *Historical Studies of the Physical Sciences* 13 (1983), 375–400.

———. "Accelerators and National Security: The Evolution of Science Policy for High-Energy Physics, 1947–1967." *History and Technology* 11 (1994), 361–91.

Seife, Charles. "Physics Tries to Leave the Tunnel." *Science* 302 (October 3, 2003), 36–38.

Seigenthaler, Kathy. "Collider's Foes Get Sympathy." *Chicago Tribune*, February 17, 1988, B6.

———. "Opposing Forces Clash Again at Hearings on Collider." *Chicago Tribune*, October 8, 1988.

Senior Administration Official (unidentified). Background Briefing on Clinton-Miyazawa summit meeting. White House Office of the Press Secretary. April 16, 1993. Clinton Library Digital Archives, http://Clinton.nara.gov/1993/04/1993-04-16-background-briefing-on-japan.webarchive.

Shepard, Irving. *Jack London's Tales of Adventure*. New York: Doubleday, 1956.

Siegfried, Tom. "Super Collider's Death Is Sign of Bad Times for Basic Science." *Dallas Morning News*, November 1, 1993, D9.

Siemann, Robert, et al. "Analysis of the Fermilab Tevatron as an Injector for the SSC." RTK—A Joint Venture. October 1987 (unpublished). Fermilab Archives.

Site Proposal for the Superconducting Super Collider in Illinois. SSC Collection, Fermilab Archives.

Slakey, Francis. "Painful Lessons from the SSC." *Physics World* (October, 1993), 19–20.

Slansky, Richard. *SSC Site Atlas*, 2nd edition. Los Alamos Report No. LA-UR-84-3893, 1984. SSC Collection, Fermilab Archives.

Slattery, Jim. *Congressional Record*, June 17, 1992, H–4819.

Smith, R. Jeffery. "Supercollider Could Face Cutbacks." *Washington Post*, November 19, 1989, A1.

Smith, Robert W. *The Space Telescope*. New York: Oxford University Press, 1989.

"SSC at Fermilab Means $3.28 Billion Savings." *SSC for Fermilab Newsletter* (Spring 1988), 1, 9.

SSC Central Design Group. *Conceptual Design of the Superconducting Super Collider*. Report No. SSC-SR-2020. March, 1986. Fermilab Archives.

———. *Superconducting Super Collider: Siting Parameters Document*. Report No. SSC-SR-2040. February, 1987. Fermilab Archives.

SSC Laboratory. *SSCL Termination Project Monthly Report*. December 1993. Fermilab Archives.

SSC Monthly Report. August 1985. CDG Collection, Fermilab Archives.

SSC Monthly Report. September 1985. CDG Collection, Fermilab Archives.

SSC News, August 1992. www.hep.net/ssc/new/history/sscnews.

SSC News, December 1992. www.hep.net/ssc/new/history/sscnews.

SSC News, February 1993. www.hep.net/ssc/new/history/sscnews.

SSC News, May 1993. www.hep.net/ssc/new/history/sscnews.

SSC News, November 1992. www.hep.net/ssc/new/history/sscnews.

SSC News, October 1992. www.hep.net/ssc/new/history/sscnews.

SSC Project Monthly Progress Report. January 1992. SSC Collection, Fermilab Archives.

SSC Project Monthly Progress Report. September 1993. SSC Collection, Fermilab Archives.

SSC Reference Design Charter. *SSC Newsletter*, February 15, 1984. Published by American Physical Society/Division of Particles & Fields. Fermilab Archives.

SSC Reference Design Charter. *SSC Newsletter*, March 15, 1984. Published by American Physical Society/Division of Particles & Fields. Fermilab Archives.

SSC Reference Designs Study. March 1984. Tigner files, CDG Collection, Fermilab Archives.

SSCL Technology Transfer Office. *Not for Scientists Only: Technology Spin-offs from High Energy Physics and the Super Collider*. December, 1991. SSCL Report No. SSCL-Pub-0001.

"SSC Status Report to the Nation: A National Symposium on the Superconducting Super Collider." Draft of proceedings. Edited by George W. Morgenthaler and Uriel Nauenberg. University of Colorado, October 24, 1988.

Staley, Kent W. *The Evidence for the Top Quark: Objectivity and Bias in Collaborative Experimentation*. Cambridge: Cambridge University Press, 2004.

State of Illinois. "Review and Comment on the Site Selection Evaluations by the US Department of Energy for the Superconducting Super Collider." March 1989. SSC Collection, Fermilab Archives.

Steining, Rae. "Some Thoughts about a 20 TeV Proton Synchrotron." In *Possibilities and Limitations of Accelerators and Detectors*. Edited by Lee Teng. Fermilab, October 15–21, 1978.

Stern, Todd, and John D. Podesta. Memorandum for Mack McLarty, "Hazel O'Leary/Superconducting Super Collider." March 18, 1993. FOIA Request 2010-450-F, Clinton Library Digital Archives.

Stork, Donald H. "SSC Design." *Science*, N.S. 231:4734 (1986), 103.

Subcommittee on Oversight and Investigation, Committee on Energy and Commerce, US House of Representatives. *Out of Control: Lessons Learned from the Superconducting Super Collider*. Washington: US Government Printing Office, December, 1994.

Sullivan, Walter. "Physicists Hope to Build a 30-Mile Atom Device to Explore Matter." *New York Times*, October 10, 1976, 30.

———. "Team Reports Breakthrough in Conductivity of Electricity." *New York Times*, February 16, 1987, 1.

"Summary Conclusions of the Versailles Summit Follow-on Meeting for High Energy Physics." Washington, DC, October 3–4, 1983. Unpublished draft dated October 27, 1983. Richter Collection, SLAC Archives.

"Summary of the Meeting and Recommendations of the Superconducting Super Collider Program Advisory Committee." SSC Laboratory internal memorandum. Snowmass, CO. July, 1990.

Sumner, Jane. "Hollywood Pirates on the Prairie; Abandoned Super Collider Site Finds New Life as Film Set." *Dallas Morning News*, February 13, 1999, C1.

Sun, Marjorie. "No Boost in Sight for Science Budgets." *Science*, N.S. 214:4519 (1981), 420–21.

"The Supercollider: It's Crunch Time." *Business Week* (May 6, 1991), 131.

"Supercollider: Magnet Decision." *CERN Courier* (November, 1985), 383–84.

"Super Collider Plans." *Washington Post*, March 17, 1990, A18.

"Super Collider Review Trims 7, Leaving 36 Sites in Running." *Science and Government Report*, October 1, 1987, 1–4.

"Super Collider: Settlement Greatly Improves Site's Options." Editorial. *Dallas Morning News*, July 23, 1994, A28.

Super Collider Site Evaluation Committee (NAS and NAE). *Siting the Superconducting Super Collider*. Washington: National Academy Press, 1988.

"Superconducting Super Collider: A New Frontier for Science and Technology." Americans for the SSC. Undated. Irwin Goodwin Collection, AIP Niels Bohr Library & Archives.

"Superconducting Super Collider: An International Partnership in Basic Research." Undated. SSC files of John Metzler; in Riordan files.

"Superconducting Super Collider: Framework for International Participation." Unclassified draft. April, 1990. SSC files of Boehlert/Goldston.

Superconducting Super Collider. Hearings Before the House Committee on Science, Space, and Technology. April 7–9, 1987, 288–93.

"Superconducting Super Collider Status Report." October, 1990. SSC files of John Metzler; in Riordan files.

"Super Hasty on the Supercollider." Editorial. *New York Times*, April 28, 1988.

"Superheroes." Editorial. *Dallas Morning News*, November 14, 1988.

Susskind, Charles M. "Alvin W. Trivelpiece: AAAS Executive Officer." *Science*, N.S. 236:4800 (1987), 377.

Svartholm, N., ed. *Elementary Particle Theory: Relativistic Groups and Analyticity*. New York: John Wiley & Sons, 1968.

Swanson, Stevenson. "Collider Foes Speak Out, and Governor Answers." *Chicago Tribune*, February 19, 1988, 1.

Swanson, Stevenson, and Andrew Bagnato. "State Shows Where Collider Would Go." *Chicago Tribune*, January, 13, 1988, 1.

Swanson, Stevenson, and Katherine Siegenthaler. "U.S. Scientist Team to Visit Collider Site." *Chicago Tribune*, May 17, 1988, 1–2.

Sweet, William. "Abragam and Rubbia Reports Chart Future for CERN." *Physics Today* 40:9 (1987), 71–75.

———. "1994 Expected to Be Year of Decision for European Super Collider." *Physics Today* 47:2 (1994), 93–96.

Taubes, Gary. "Collision over the Supercollider." *Discover* (July, 1985), 60–69.

———. "Fight Heats Up Over SSC's Remains." *Science*, N.S. 262:5139 (1993), 1506–07.

———."Onward to the Desertron." *Physics Today* 36:12 (1983), 9.

———. "SSC Detectors Desperately Seeking Donors." *Science*, N.S. 259:5096 (1993), 757.

———. "The Supercollider: How Big Science Lost Favor and Fell." *New York Times*, October 26, 1993, C1–C3.

———. "Young Physicists Hear Wall Street Calling." *Science*, N.S. 264:5155 (1994), 22.

Taylor, George Rogers, ed. *The Turner Thesis: Concerning the Role of the Frontier in American History*. Boston: D. C. Heath and Co., 1956.

Taylor, Robert E. "President Will Request Funds to Build World's Largest Particle Accelerator." *Wall Street Journal*, February 2, 1987, 1.

Telegdi, V. L. "Conclusions." In *ICFA Seminar on Future Perspectives in High-Energy Physics*. KEK, Tsukuba, Japan, May 14–20, 1984, 336–42.

Temple, L. Edward. "Office of Energy Research: Project Performance." Unpublished. February 1986. Fermilab Archives.

"10. Superconducting Super Collider," *Congressional Quarterly*, December 19, 1992, 3867.

Teng, Lee. "A 3 TeV on 3 TeV Proton-Proton Dedicated Collider for Fermilab." Fermilab Report No. TM-1516. March 30, 1988.

Teng, Lee, ed. *Possibilities and Limitations of Accelerators and Detectors*. Batavia, IL: Fermilab, October 15–17, 1978. Fermilab Archives.

TEVNPH Working Group for the CDF and D-zero Collaborations. "Combined CDF and D-zero Upper Limits on Standard Model Higgs Boson Production with up to 8.6 fb-1 of Data." May 13, 2013. Updated September 20, 2011, http://arXiv:1107.5518v21.

t' Hooft, Gerard. "Renormalization of Gauge Theories." In *Rise of the Standard Model*. Edited by Lillian Hoddeson, Laurie Brown, Michael Riordan, and Max Dresden, 179–98. New York: Cambridge University Press, 1997.

Tigner, Maury. *Accelerator Physics Issues for a Superconducting Super Collider*. University of Michigan Report No. UMHE 84-1, 1984.

———. Correspondence with Leon Lederman. 1986. Lederman Collection, Fermilab Archives.

———. Letter to the editor. *Physics Today* 41:8, 1988, 14–15.

———. Memo to Burton Richter. October 15, 1984. Chao Collection, Lawrence Berkeley Laboratory Archives, Berkeley, California.

———. Memo to Nicholas Samios. September 10, 1984. Chao Collection, Lawrence Berkeley Laboratory Archives, Berkeley, California.

———. "Phase 1 Program Milestones." Undated. Copy of transparency. CDG files, Fermilab Archives.

———. "Planning Meeting" notice. September 14, 1988. CDG Collection, Fermilab Archives.

———. *Report of the 20 TeV Hadron Collider Workshop*. Ithaca, NY, 1983. Fermilab Archives.

———. "Research and Development for the Super Collider." Testimony Before the Subcommittee on Energy Development and Applications, Committee on Science and Technology, United States House of Representatives. Report No. SSC-53A. October 29, 1985.

"The Tokyo Declaration on the U.S.-Japan Global Partnership." Official summit document from "Bush Originals—1992." George H. W. Bush Presidential Library, College Station, TX.

Toohig, Timothy. "SSC: The Anatomy of a Failure: A Case of Institutional Amnesia." Unpublished draft manuscript. November 3, 1996.

Transcript of the Department of Energy's debriefing of Illinois. December 7, 1988. SSC files, Library of the Illinois State Geological Survey, Fermilab Archives.

Trines, Dieter. "Constructing HERA: Rising to the Challenge." *CERN Courier*, January 21, 2008.

Traweek, Sharon. *Beamtimes and Lifetimes: The World of High Energy Physicists*. Cambridge: Harvard University Press, 1988.

Trivelpiece, Alvin W. "Prospective Proposers." Letter. April, 1987. SSC Collection, Fermilab Archives.

———. "Some Observations on DOE's Role in 'Megascience.'" *History of Physics Newsletter* 9:5 (Fall 2005), 14.

Trubowitz, Peter. *Defining the National Interest: Conflict and Change in American Foreign Policy*. Chicago: University of Chicago Press, 1998.

Turner, Frederick Jackson. "The Significance of the Frontier in American History." Speech delivered in 1893. Reprinted in *The Turner Thesis: Concerning the Role of the Frontier in American History*. Edited by George Rogers Taylor. Boston: DC Heath and Co., 1956.

"The Unaffordable Atom-Smasher." Editorial. *New York Times*, November 16, 1988.

Universities Research Association. "Proposal for the Selection of a Management and Operating Contractor for the Establishment, Management, and Initial Operation of the Superconducting Super Collider Laboratory." Washington, DC. November 4, 1988. Fermilab Archives.

———. "Proposal to Serve as Contractor for the Construction and Operation of the 'Superconducting Super Collider' Laboratory." Submitted March 2, 1987. URA Collection, Fermilab Archives.

———. "Proposal to Serve as Contractor for the Construction and Operation of the 'Superconducting Super Collider' Laboratory." Revised. Sent to James Decker, February 22, 1988. URA Collection, Fermilab Archives.

Unnervik, Anders. "The Construction of the LHC: Lessons in Big Science Management and Contracting." In *The Large Hadron Collider: A Marvel of Technology*. Edited by Lyndon Evans, 38–55. Lausanne: EPFL Press, 2009.

US Department of Energy. *Best Qualified Sites for the Superconducting Super Collider*. Report of the United States Department of Energy Superconducting Super Collider Site Task Force. January, 1988.

———. "DOE Requests Funding to Build the Superconducting Super Collider." Press release. January 30, 1987. Fermilab Archives.

———. "DOE Selects Super Collider Management and Operating Contractor." Press release. January 19, 1989. Fermilab Archives.

———. *HEPAP Subpanel on Vision for the Future of High-Energy Physics*. Report No. DOE/ER-0614P. May, 1994.

———. *Invitation for Site Proposals for the Superconducting Super Collider*. Report No. DOE/ER-0315. April, 1987.

———. *Report of the 1980 Subpanel on Review and Planning for the U.S. High Energy Physics Program*. Report No. DOE/ER-0066. June, 1980.

———. *Report of the 1983 Subpanel on New Facilities for the U.S. High Energy Physics Program*. Report No. DOE/ER-0169, July, 1983.

———. *Report of the Subpanel on Accelerator Research and Development of the High Energy Physics Advisory Panel*. Report No. DOE/ER-0067. June, 1980.

———. *Report of the Subpanel on Long-Range Planning for the U.S. High Energy Physics Program*. Report No. DOE/ER-0128. January, 1982.

———. *SSC Site Evaluations: A Report by the SSC Task Force*. Report No. DOE/ER-0392. 1988 (also called the "DOE Green Book").

———. "Superconducting Super Collider Management and Implementation Plans." Unpublished. August 30, 1989. SSC Collection, Fermilab Archives.

———. "Superconducting Super Collider Status Report." Unpublished. October, 1990. SSC files of Riordan.

———. "Texas Is Site for Super Collider." Press release. January 18, 1989.

———. "Watkins Names Cipriano Project Manager of the Superconducting Supercollider." Press release. April 19, 1990.

———. *Report of the 1980 Subpanel on Review and Planning for the U.S. High Energy Physics Program*. Report No. DOE/ER-0066. June, 1980.

US Department of Energy, Office of Energy Research. *HEPAP Subpanel Report on Planning for the Future of US High-Energy Physics*. Report No. DOE/ER-0718. February, 1998. http://science.energy.gov/~/media/hep/files/pdfs/hepap_report.pdf.

———. *Report of the HEPAP Subpanel on US High Energy Physics Program for the 1990s*. Report No. DOE/ER-0453P. April, 1990.

———. *Report of the 1990 HEPAP Subpanel on SSC Cost Estimate Oversight*. Report No. DOE/ER-0464P. Washington, DC, July, 1990.

———. *Report on the Superconducting Super Collider Cost and Schedule Baseline*. Report No. DOE/ER-0468P. Washington, DC, January, 1991 (also called the "1981 Green Book").

US Department of Energy, Office of Field Management. *Report of the DOE Review Committee on the Baseline Validation of the Superconducting Super Collider*. Report No. DOE/ER-0594P. August, 1993 (also called the "Scango Report").

US Department of Energy, Office of Inspector General. *Follow-up Audit, Department of Energy's Superconducting Super Collider Program*. Report No. DOE/IG-0305. March 13, 1992.

———. *Summary Audit Report on Lessons Learned from the Superconducting Super Collider Project*. Report No. DOE/IG-0389. April, 1996.

US General Accounting Office. *Federal Research: Determination of the Best Qualified Sites for DOE's Super Collider*. Report No. GAO/RCED-89-18. January, 1989.

———. *Federal Research: Foreign Contributions to the Superconducting Super Collider*. Report No. GAO/RCED-93/75. December, 1992.

———. *Federal Research: Implementation of the Super Collider's Cost and Schedule Control System*. Report No. GAO/RCED-92-242. July 21, 1992.

———. *Federal Research: Information on Site Selection Process for DOE's Super Collider*. Report No. GAO/RCED- 90-33BR. October, 1989.

———. *Federal Research: Super Collider Is Over Budget and Behind Schedule*. Report No. GAO-93-87. February, 1993.

———. *Final Site Selection Process for DOE's Super Collider*. Report No. GAO/RCED-89-129BR. June, 1989.

Van, Jon. "Scientists Hear Plea for New Accelerator." *Chicago Tribune*, May 29, 1985, 5B.

Veltman, Martinus. "The Path to Renormalizability." In *The Rise of the Standard Model: Particle Physics in the 1960s and 1970s*. Edited by Lillian Hoddeson, Laurie Brown, Michael Riordan, and Max Dresden, 145–78. New York: Cambridge University Press, 1997.

von Braun, Wernher. "Crossing the Last Frontier." *Collier's*, March 22, 1952, 24–31.

"Vote Analysis: The Slattery Amendment to Strike SSC Funding from H.R. 2427." URA internal report. SSC files of Riordan.

Waldrop, M. Mitchell. "Congress Questions SSC Cost." *Science*, N.S. 230:4727 (1985), 785.

———. "Gambling on the Supercollider." *Science*, N.S. 221:4615 (1983), 1038–40.

———. "Magnets Chosen for Supercollider." *Science*, N.S. 230:4721 (1985), 50.

———. "The Supercollider, 1 Year Later." *Science*, N.S. 225:4661 (1984), 490–91.

Watkins, James D. Letter to George E. Brown Jr. January 14, 1993. SSC files of Riordan.

———. Letter to James A. Baker III. October 25, 1991. SSC files of John Metzler; in Riordan files.

———. Letter to Tom Bevill. June 27, 1989. SSC files of Riordan.

———. "SSC Management Plan." Memo for Robert O. Hunter Jr. September 7, 1989.

Weakley, Benjamin. "Meeting with Toichi Sakata on SSC." Memorandum to Harold Jaffe. July 13, 1990. SSC files of John Metzler; in Riordan files.

Weinberg, Alvin M. "Criteria for Scientific Choice." *Minerva* (Winter 1963), 159–71.

———. "Impact of Large-Scale Science in the United States." *Science*, N.S. 134:3473 (1961), 161–64.

———. *Reflections on Big Science*. Cambridge, MA: MIT Press, 1967.

Weinberg, Steven. *Dreams of a Final Theory: The Search for the Ultimate Laws of Nature*. New York: Pantheon Books, 1992.

———. "A Model of Leptons." *Physical Review Letters* 19 (1967): 1264–66.

Weingarten, Paul. "Supercollider a Big Win When Texas Needed One Most." *Chicago Tribune*, November 13, 1988, A4.

Weisskopf, Victor. "Keynote Talk." *ICFA Seminar on Future Perspectives in High-Energy Physics*. KEK, Tsukuba, Japan, May 14–20, 1984, 9–16.

Westfall, Catherine. "Panel Session: Science Policy and the Social Structure of Large Laboratories." In *The Rise of the Standard Model: Particle Physics in the 1960s and 1970s*. Edited by Lillian Hoddeson, Laurie Brown, Michael Riordan, and Max Dresden, 364–83. New York: Cambridge University Press, 1997.

———. "Retooling for the Future: Launching the Advanced Light Source at Lawrence's Laboratory, 1980–1986." *Historical Studies in the Natural Sciences* 38:4 (2008), 569–609.

———. "The Site Contest for Fermilab." *Physics Today* 42:1 (1989), 44–52.

Westfall, Catherine, and Lillian Hoddeson. "Thinking Small in Big Science." *Technology and Culture* 37:3 (1996), 457–92.

Westwick, Peter J. *The National Labs: Science in an American System, 1947–1974*. Cambridge, MA: Harvard University, 2003.

Wicker, Tom. "Reagan's Fig Leaf." *New York Times*, July 16, 1982, A27.

Willard, E. Payson. "The Demise of the Superconducting Super Collider: Strong Politics or Weak Management?" In *The Project Management Casebook*. Edited by David I. Cleland, Richard Puerzer, Karen M. Bursic, and A. Yaroslav Vlasak, 493–502. Newton Square, PA: Project Management Institute, 1994.

Williams, John. "Super Collider Funding to Be 'Hardest' Part." *Houston Chronicle*, November 11, 1988, A1ff.

Wilson, Robert R. "Starting Fermilab: Some Personal Viewpoints of a Laboratory Director (1967–1978)." 1987 Fermilab Annual Report.

———. "Superferric Magnets for 20 TeV." In *Proceedings of the 1982 DPF Summer Study on Elementary Particle Physics and Future Facilities*. Snowmass, CO. June 28–July 16, 1982. Edited by Rene Donaldson, Richard Gustafson, and Frank Paige, 330–34. Batavia, IL: Fermi National Accelerator Laboratory, 1983.

———. "The Tevatron." *Physics Today* 30:10 (1977), 23–30.

———. "Toward a World Accelerator Laboratory." Fermilab Report No. TM-811. August 16, 1978.

———."Ultrahigh-Energy Accelerators." *Science*, N.S. 133:3464 (1961), 1602–04, 1607.

———. "A World Laboratory for World Peace." *Physics Today* 28:11 (1975), 120.

———. "A World Organization for the Future of High-Energy Physics." *Physics Today* 37:9 (1984), 9, 112.

Wines, Michael. "Bush in Japan: Bush Reaches Pact with Japan, But Automakers Denounce It." *New York Times*, January 10, 1992, 1.

Wojcicki, Stanley. Memorandum to subpanel members. March 11, 1983. SSC files of Riordan.

———. "The Supercollider: The Pre-Texas Days." *Reviews of Accelerator Science and Technology* 1 (2008): 259–302.

———. "The Supercollider: The Texas Days." *Reviews of Accelerator Science and Technology* 2 (2009): 265–301.

Wolf, Günther. "Physics at HERA." *Annals of the New York Academy of Sciences* 461 (2006): 699–724.

Wolpe, Howard. *Status of the Superconducting Super Collider Program: Hearing Before the Subcommittee on Investigations and Oversight, House Committee on Science, Space and Technology*. May 9, 1991, 1–2.

"Workshop on Master Builders: The Reconstruction of American Science after World War II." Unpublished. April 16–17, 1999. Johns Hopkins University.

"Workshop on Physics and Experiments for the Superconducting Super Collider." Unpublished. Dallas, TX, October 1–4, 1989.

"Workshop on the Feasibility of Hadron Colliders in the LEP Tunnel." Lausanne and CERN. March 21–27, 1984. Draft manuscript. Richter Collection, SLAC Archives.

Wu, Sau Lan. "Hadron Jets and the Discovery of the Gluon." In *The Rise of the Standard Model: Particle Physics in the 1960s and 1970s*. Edited by Lillian Hoddeson, Laurie Brown, Michael Riordan, and Max Dresden, 600–621. New York: Cambridge University Press, 1997.

Yamaguchi, Y. "ICFA: Its History and Current Activities." In *Proceedings of the 1985 International Symposium on Lepton and Photon Interactions at High Energies*, 826–47. Kyoto, 1986.

Yang, C. N., and R. Mills. "Conservation of Isotopic Spin and Isotopic Gauge Invariance." *Physical Review* 96 (1954): 191–95.

———. "Isotopic Spin and a Generalized Gauge Invariance." *Physical Review* 95 (1954): 631.

Zachary, G. Pascal. *Endless Frontier: Vannevar Bush, Engineer of the American Century*. New York: The Free Press, 1999.

Zimmerman, George. "History of Superconductivity Sessions." *History of Physics Newsletter* 11:4 (Spring 2011), 4–7.

Zysman, John, and Laura Tyson, eds. *American Industry in International Competition: Government Policies and Corporate Strategies*. Ithaca, NY: Cornell University Press, 1983.

Index

Throughout this index, the initials SSC stand for the Superconducting Super Collider. Page numbers followed by the letter *f* indicate a figure.